ADVANCES IN CHEMICAL PHYSICS

VOLUME LXXX

EDITORIAL BOARD

Advances in
CHEMICAL PHYSICS

Edited by

I. PRIGOGINE

University of Brussels
Brussels, Belgium
and
University of Texas
Austin, Texas

and

STUART A. RICE

Department of Chemistry
and
The James Franck Institute
The University of Chicago
Chicago, Illinois

VOLUME LXXX

WILEY

A WILEY-INTERSCIENCE PUBLICATION
JOHN WILEY & SONS, INC.
NEW YORK • CHICHESTER • BRISBANE • TORONTO • SINGAPORE

An Interscience® Publication

Library of Congress Cataloging Number: 58-9955

ISBN 0–471–53281–9

Printed in the United States of America
10 9 8 7 6 5 4 3 2 1

CONTRIBUTORS TO VOLUME LXXX

BIMAN BAGCHI, Solid State and Structural Chemistry Unit, Indian Institute of Science, Bangalore, India

AMALENDU CHANDRA, Solid State and Structural Chemistry Unit, Indian Institute of Science, Bangalore, India

C. A. CHATZIDIMITRIOU-DREISMANN, Iwan N. Stranski-Institut für Physikalische und Theoretische Chemie, Technische Universität Berlin, Berlin, FRG

MARKUS EISWIRTH*, Department of Chemistry, Stanford University, Stanford, California

ALBRECHT FREUND*, Department of Chemistry, Stanford University, Stanford, California

Y. FUJIMURA, Department of Chemistry, Faculty of Science, Tohoku University, Sendai, Japan

H. KONO, Department of Basic Technology, Faculty of Engineering, Yamagata University, Yonezawa, Japan

H. MÄTZING, Laboratorium für Aerosolphysik und Filtertechnik I, Kernforschungszentrum Karlsruhe GmbH, Karlsruhe, FRG

Y. NOMURA, Department of Chemistry, Faculty of Science, Tohoku University, Sendai, Japan

JOHN ROSS, Department of Chemistry, Stanford University, Stanford, California

---

*Present address: Fritz-Haber-Institut der MPG, D-1000 Berlin 33, FRG.
**Present address: Institut für Physikalische Chemie der Universität, D-8700 Würzburg, FRG.

v

INTRODUCTION

Few of us can any longer keep up with the flood of scientific literature, even in specialized subfields. Any attempt to do more and be broadly educated with respect to a large domain of science has the appearance of tilting at windmills. Yet the synthesis of ideas drawn from different subjects into new, powerful, general concepts is as valuable as ever, and the desire to remain educated persists in all scientists. This series, *Advances in Chemical Physics*, is devoted to helping the reader obtain general information about a wide variety of topics in chemical physics, which field we interpret very broadly. Our intent is to have experts present comprehensive analyses of subjects of interest and to encourage the expression of individual points of view. We hope that this approach to the presentation of an overview of a subject will both stimulate new research and serve as a personalized learning text for beginners in a field.

I. Prigogine
Stuart A. Rice

CONTENTS

ADVANCES IN CHEMICAL PHYSICS

VOLUME LXXX

COLLECTIVE ORIENTATIONAL RELAXATION IN DENSE DIPOLAR LIQUIDS

BIMAN BAGCHI and AMALENDU CHANDRA

Solid State and Structural Chemistry Unit, Indian Institute of Science, Bangalore India

CONTENTS

Advances in Chemical Physics Volume LXXX, Edited by I. Prigogine and Stuart A. Rice
ISBN 0-471-53281-9 © 1991 John Wiley & Sons, Inc.

I. INTRODUCTION

Orientational relaxation in a dense liquid, where molecules interact with an anisotropic potential, has been a subject of great importance in liquid-phase chemistry. Not only does orientational relaxation play a key role in many important relaxation processes, such as in polarization and dielectric relaxation, but it also profoundly influences the dynamics of many important chemical reactions, like the electron- and the proton-transfer reactions in a dipolar liquid. The orientational relaxation of anisotropic molecules in a liquid clearly depends on the density and the temperature of the liquid and also on the nature of the anisotropy of the interaction potential. It may also be coupled with the translational modes of the liquid and the internal modes of the molecules. Orientational relaxation in a dense liquid can show rich and diverse dynamical behavior.

In this review we shall be concerned primarily with the collective orientational relaxation in a dense dipolar liquid. By collective relaxation we shall mean a process that involves a cooperative motion of several molecules of the liquid. Because of the long-range nature of the dipolar interactions, the orientational motions of different molecules are correlated over a long distance. For example, the bare potential energy of interaction of two parallel point dipoles aligned radially, each of strength 2 Debye and each embedded in a sphere of radius $2\,\text{Å}$ and separated by a distance of $4\,\text{Å}$, is equal to $12 \times 10^{-14}\,\text{erg}$, which is rather large compared to the thermal energy at room temperature (which is about $4 \times 10^{-14}\,\text{erg}$). In fact, dipolar interaction energy per particle in a strongly polar dipolar liquid can be a significant part of the total potential energy, indicating the presence of considerable degree of orientational order in such liquids. The presence of such orientational correlation is also known from theoretical studies and from computer simulations. Thus, the collective nature of orientational relaxation in a dense dipolar liquid is not surprising. In fact, these collective effects can also influence single-particle orientational motion through dielectric friction on a rotating solute in a dipolar liquid. In this sense, the single-particle and the collective motions are not completely disjoint.

Orientational motions are studied directly by many different experimental techniques [1]. Many experiments, such as NMR relaxation, incoherent neutron scattering, and Raman line shapes, are sensitive to the single-particle orientational motion. The other limit of collective orientational motion involving all the molecules of the system is probed by experiments like dielectric relaxation, depolarized light scattering, and coherent neutron scattering. It has not been possible to study directly collective orientation involving a small number of molecules (say, on the order of 10–100). However, indirect information about this intermediate limit can be obtained from such linear spectroscopic techniques as time-dependent fluorescence Stokes shift (TDFSS) measurements of a newly created ion or dipole or such nonlinear optical techniques as Kerr relaxation [2]. In these measurements, some moments of the inverse distances of the solvent molecules from the probe over time- and space-dependent solvent polarization is studied. If the orientational relaxation of the nearest-neighbor molecules is significantly different from those that are far off in the bulk, then these indirect methods can provide reliable information on collective orientational relaxation in the intermediate regime. This is currently an area of active research.

The collective orientational relaxation in the long-wavelength limit involving all the molecules of the system was studied initially by borrowing concepts from the continuum description of electrostatics. In these theories, the dipolar molecule is replaced by a cavity of some simple shape with a point dipole at its center and the liquid is replaced by a frequency-dependent dielectric continuum [3]. The interaction between the rotating dipole and the bulk of the liquid is included through a time-dependent reaction field, which arises from the electric polarization of the solvent by the rotating dipole. The reaction field is obtained by a quasistationary boundary value calculation. This field gives rise to a dielectric friction, which retards the rotation of the dipole and gives rise to a nonexponential decay of the dipolar correlation function.

The advantage of the continuum model theories is that they provide simple expressions for the orientational correlation functions that can be tested against experiments. This is especially true for a complex liquid where a microscopic theory is bound to be complicated. The main drawback of the continuum model theories is that they ignore the intermolecular correlations that are present in a dense dipolar liquid. If these correlations are important in a relaxation process, then the continuum model is obviously inadequate.

A consistent and detailed molecular theory of the orientational relaxation in general, and the dielectric relaxation in particular, has been developed by Madden and Kivelson [4, 5]. These authors have derived several useful theorems on orientational correlation functions, especially on the relationship between the macroscopic and the microscopic correlation functions and they have clarified many aspects of orientational relaxation in a dipolar liquid. They have also discussed the role of translational modes in the decay of the orientational correlation function. We refer to the excellent review of the work of Madden and Kivelson that appeared in this series several years ago [4]. The focus of the present review is different. The emphasis of this work is on understanding the *collective* orientational motion at intermediate-length scales. By intermediate-length processes, we mean a relaxation where a small number of molecules are primarily involved. Such a relaxation can be quite different from either the single-particle orientation or from the fully collective motion involving all the molecules of the system. The molecular treatment developed by Madden and Kivelson [5] has not been extended to the intermediate regime. Another difference of the treatment presented here from that developed by Madden and Kivelson is that the present discussion is based on an extended (and generalized) hydrodynamic description that is reliable at both small and intermediate wavenumbers. In the overdamped limit, it reduces to a diffusion equation of the Smoluchowski–Vlasov type where the effects of the intermolecular interactions are included through a mean-field force term. The mean-field force term is expressed in terms of a convolution integral over the product of the two particle direct correlation function and the density fluctuation. The structure of the mean field force makes it especially convenient to work in the Fourier space with the wavevector $\mathbf{k}$ as the conjugate variable to the position vector $\mathbf{r}$. We find that in a linear (in the density fluctuation) theory, Fourier space provides a simple and natural description of the collective orientational relaxation. In this respect, the present discussion is similar to spatial density relaxation studied extensively in the last two decades [6]. The advantage of working with the direct correlation function representation of intermolecular interactions was discussed by Madden and Kivelson who, however, preferred a radial-distribution-based description because of the intuitive picture that the latter provides. In this respect, the treatment discussed here is complementary with the treatment of Madden and Kivelson [4, 5].

Because of the calculational simplicity of the direct correlation-function-based microscopic approach, it has been possible to explore the different facets of the collective orientational relaxation in detail. For example, one can obtain explicit analytic expressions for the wavevector-dependent relaxation times within a linearized theory. In the overdamped limit, one can also treat the non-Markovian case fairly accurately. It has also been possible to study microscopically the properties of the collective excitations (the dipolarons) in a dense dipolar liquid. A first-principles calculation of the effects of solvent translational modes on the dielectric friction on a rotating dipolar solute has also been carried out.

In addition to the review of Madden and Kivelson [4], there exist several articles on the orientational relaxation. Berne and Pecora [1], Fleming [2], Keyes and Ladanyi [7], Reid and Evans [8], Cole and Evans [9], Steele [10], Coffey [11], and Evans [12] have reviewed different aspects of the orientational relaxation in dense liquids. The major concern of the preceding articles was the single-particle dynamics. The emphasis of the present article is on the collective dynamics at the intermediate-length scale that involves cooperative motion of few (more than 2 but less than 100) molecules.

II. EQUILIBRIUM AND TIME-DEPENDENT ORIENTATIONAL CORRELATION FUNCTIONS

In linear optical spectroscopy, the quantity that is usually measured is a two-point time correlation function of a dynamical quantity. The dynamical quantity that is of interest in this discussion is the collective dipole moment of a dipolar liquid. A dense dipolar liquid is a strongly correlated system, and this correlation is clearly reflected in the equilibrium correlation functions. We briefly discuss these correlation functions here because of the extensive use that will be made of them in this review. The equilibrium correlation functions that are of interest here are the radial distribution functions $g(\mathbf{r}, \mathbf{\Omega}, \mathbf{r}', \mathbf{\Omega}')$, the pair correlation function $h(\mathbf{r}, \mathbf{\Omega}, \mathbf{r}', \mathbf{\Omega}')$, and the two particle direct correlation function $c(\mathbf{r}, \mathbf{\Omega}, \mathbf{r}', \mathbf{\Omega}')$ [6, 13], where $(\mathbf{r}, \mathbf{\Omega})$ denotes the position $(\mathbf{r})$ and the orientation $(\mathbf{\Omega})$ of a molecule, both measured in a laboratory fixed frame. In this discussion we shall assume that the dipolar molecules are rigid so that the orientation of the dipole is given by a unit vector $\hat{a}(\mathbf{\Omega})$ with orientation $\mathbf{\Omega}$. The radial distribution function is defined as

$$g(\mathbf{r}, \mathbf{\Omega}, \mathbf{r}', \mathbf{\Omega}') = \frac{\Omega_0^2}{\rho_0^2} \sum_{i \neq j} \langle \delta(\mathbf{r} - \mathbf{r}_i)\delta(\mathbf{r}' - \mathbf{r}_j)\delta(\mathbf{\Omega} - \mathbf{\Omega}_i)\delta(\mathbf{\Omega}' - \mathbf{\Omega}_j) \rangle \quad (2.1)$$

where $\langle \cdots \rangle$ denotes averaging over an equilibrium ensemble and ρ_0 is the number density of the liquid. Ω_0 is the measure of the angular space. Thus, $\Omega_0 = 4\pi$ for axially symmetric molecules, and it is $8\pi^2$ for nonsymmetric molecules. The pair correlation function is related to the radial distribution function by the usual relation

$$h(\mathbf{r}, \mathbf{\Omega}, \mathbf{r}', \mathbf{\Omega}') = g(\mathbf{r}, \mathbf{\Omega}, \mathbf{r}', \mathbf{\Omega}') - 1 \qquad (2.2)$$

The two distribution function $h(\mathbf{r}, \mathbf{\Omega}, \mathbf{r}', \mathbf{\Omega}')$ and $c(\mathbf{r}, \mathbf{\Omega}, \mathbf{r}', \mathbf{\Omega}')$ are related to each other by the (generalized) Ornstein–Zernike relation [6, 13]

$$h(\mathbf{r}, \mathbf{\Omega}, \mathbf{r}', \mathbf{\Omega}') = c(\mathbf{r}, \mathbf{\Omega}, \mathbf{r}', \mathbf{\Omega}') + \frac{\rho_0}{4\pi} \int d\mathbf{r}'' d\mathbf{\Omega}'' c(\mathbf{r}, \mathbf{\Omega}, \mathbf{r}', \mathbf{\Omega}') h(\mathbf{r}'', \mathbf{r}', \mathbf{\Omega}'', \mathbf{\Omega}') \qquad (2.3)$$

It is usual to expand the two-particle distribution functions in terms of the Wigner matrices [1, 13]:

$$g(\mathbf{r} - \mathbf{r}', \mathbf{\Omega}, \mathbf{\Omega}') = \sum_{l_1 l_2 l} \sum_{m_1 m_2 m} \sum_{n_1 n_2} g(l_1 l_2 l; n_1 n_2; \mathbf{r} - \mathbf{r}')$$
$$C_g(l_1 l_2 l; m_1 m_2 m) D_{l_1 m_1}^{n_1}(\mathbf{\Omega}) D_{l_2 m_2}^{n_2}(\mathbf{\Omega}') Y_{lm}(\mathbf{\Omega}'') \qquad (2.4)$$

where $\mathbf{\Omega}''$ gives the orientation of the vector $\mathbf{r} - \mathbf{r}'$ in the laboratory fixed frame, $C_g(l_1 l_2 l; m_1 m_2 m)$ is the usual Clebsch–Gordan coefficient and $D_{lm}^n(\mathbf{\Omega}(t))$ are the Wigner rotation matrices. The single-particle time correlation functions are defined by

$$S_{l_1 l_2 m_1 m_2}^{n_1 n_2}(t) = \langle D_{l_1 m_1}^{n_1}(\mathbf{\Omega}(0)) D_{l_2 m_2}^{n_2}(\mathbf{\Omega}(t)) \rangle \qquad (2.5)$$

Although only the angular coordinate of the particle in question appears in Eq. (2.5), it depends implicitly on the dynamics of the other variables of the system through the time dependence of $\mathbf{\Omega}(t)$.

The collective orientational correlation function depends on the relative positions of all the particles of the system. It is convenient to work in the Fourier space, and we define the collective dynamical variable [14]

$$D_{lm}^n(\mathbf{k}, t) = \sum_i r^{i\mathbf{k} \cdot \mathbf{r}_i} D_{lm}^n(\mathbf{\Omega}_i(t)) \qquad (2.6)$$

where the sum is over all the molecules of the system. The collective orientational time correlation function is now defined as

$$C_{l_1 l_2 m_1 m_2}^{n_1 n_2}(\mathbf{k}, t) = \langle D_{l_1 m_1}^{n_1}(-\mathbf{k}) D_{l_2 m_2}^{n_2}(\mathbf{k}, t) \rangle \qquad (2.7)$$

In this work, we shall be concerned mostly with homogeneous systems, in the absence of any external field. Therefore,

$$S_{l_1 l_2 m_1 m_2}^{n_1 n_2}(t) = S_{l_1 m_1}^{n_1}(t) \delta_{l_1 l_2} \delta_{m_1 m_2} \delta_{n_1 n_2} \qquad (2.8)$$

and

$$C_{l_1 l_2 m_1 m_2}^{n_1 n_2}(\mathbf{k}, t) = C_{l_1 m_1}^{n_1}(\mathbf{k}, t)\delta_{l_1 l_2}\delta_{m_1 m_2}\delta_{n_1 n_2} \tag{2.9}$$

The correlation functions become rather simple for a spherical diffuser, because the Wigner matrices are now replaced by spherical harmonics:

$$S_{lm}(t) = \langle\, Y_{lm}(\mathbf{\Omega}(0))Y_{lm}(\mathbf{\Omega}(t))\,\rangle \tag{2.10}$$

$$C_{lm}(\mathbf{k}, t) = \langle\, Y_{lm}(-\mathbf{k})Y_{lm}(\mathbf{k}, t)\,\rangle \tag{2.11}$$

We next discuss different experimental techniques that measure $S_{lm}(t)$ and $C_{lm}(\mathbf{k}, t)$.

III.   RELATIONSHIP WITH EXPERIMENTAL OBSERVABLES

As was mentioned in the Introduction, both the single-particle correlation function $S_{lm}(t)$ and the collective correlation function $C_{lm}(\mathbf{k}, t)$ are routinely measured in different experiments. In the following we briefly mention the experiments and the relevant expressions that relate the experimental observables to the orientational correlation functions. For simplicity, the correlation functions will be written in terms of spherical harmonics, so they are applicable to molecules of ellipsoidal and spherical shapes. These can be generalized easily to the case of anisotropic diffusers.

A.   Dielectric Relaxation

Conventional dielectric relaxation experiments measure the collective orientational motion of the dipolar molecules. If we define the dipolar correlation function $\phi(\mathbf{k}, t)$ by

$$\phi(\mathbf{k}, t) = \frac{\sum_{m=-1}^{1} C_{lm}(\mathbf{k}, t)}{\sum_{m=-1}^{1} C_{lm}(\mathbf{k}, 0)} \tag{3.1}$$

then the $k \to 0$ limit of $\phi(\mathbf{k}, t)$ is related to the frequency-dependent dielectric function $\varepsilon(\omega)$ by the following relation [15, 16]

$$\mathscr{L}\left[-\frac{d\phi(t)}{dt}\right] = \frac{[\varepsilon(\omega) - 1][2\varepsilon(\omega) + 1]\varepsilon_0}{[\varepsilon_0 - 1][2\varepsilon_0 + 1]\varepsilon(\omega)} \tag{3.2}$$

where ε_0 is the static dielectric constant of the liquid and $\mathscr{L}[f(t)]$ denotes Laplace transformation of the function $f(t)$. As was shown by Eq. (3.2), all the molecules participate in the $k = 0$ limit. Thus, dielectric relaxation measures the collective relaxation. In view of this, it is rather surprising that the simple noninteracting model introduced by Debye many years ago, which gives for $\varepsilon(\omega)$

$$\varepsilon(\omega) = \varepsilon_\infty + \frac{\varepsilon_0 - \varepsilon_\infty}{1 + i\omega\tau_D} \tag{3.3}$$

where ε_∞ is the infinite-frequency dielectric constant and τ_D is the Debye relaxation time, is rather successful in describing the polar response of many strongly interacting dipolar liquids. We shall discuss this fact subsequently in more detail. Here we note that because all the molecules are involved equally in $\varepsilon(\omega)$, information on the collective relaxation of few molecules are not available from $\varepsilon(\omega)$.

B. Dynamic Light Scattering

A light scattering experiment measures the long-wavelength (the $k \to 0$ limit) response of a liquid and, therefore, it also measures the collective relaxation, like the dielectric relaxation experiments. Like its spatial counterpart, orientational density fluctuations also lead to light scattering. Orientational fluctuations are best studied by depolarized (VH) Rayleigh scattering, because pure transverse fluctuations are observed in this geometry and the spectrum is less complicated. In view of the excellent introduction to the theory of dynamic light scattering and the relevant correlation functions given by Berne and Pecora [1], it is not necessary to go into a detailed discussion of them here; we will just mention the essentials. The intensity of the scattered field is proportional to

$$I_{if}(\mathbf{k}, \omega) = \frac{1}{2\pi} \int_{-\infty}^{\infty} dt\, e^{-i\omega t} I_{if}(\mathbf{k}, t) \tag{3.4}$$

where $I_{if}(\mathbf{k}, t)$ is determined by the fluctuation in the polarizability of the system

$$I_{if}(\mathbf{k}, t) = \langle \delta\alpha_{if}(-\mathbf{k}, 0)\delta\alpha_{if}(\mathbf{k}, t) \rangle \tag{3.5}$$

with

$$\delta\alpha_{if}(\mathbf{k}, t) = \sum_{j=1}^{N} \alpha_{if}^j \exp[i\mathbf{k} \cdot \mathbf{r}_j(t)] \tag{3.6}$$

In the preceding equations (i, f) denote the coordinate axes and α_{if}^j is the (if) component of the polarizability of the jth molecule, measured in a space fixed frame. In the general case, the evaluation of the correlation function [Eq. (3.5)] is complicated. Since light scattering involves only small wavevector processes, the translational contribution to orientational relaxation may be negligible and the error that is involved in assuming statistical independence of molecular rotation and translation may not be significant. If we further assume that interaction between the rotating molecules is negligible, then the correlation function that is measured by the light scattering experiments correspond to only single-particle motion, The relevant correlation functions for several different models have been given in the Berne and Pecora [1].

The assumption of no interaction between the rotating molecules is not correct in a dense dipolar liquid where the molecules possess a dipole in addition to being polarizable. In such a situation, dynamic light scattering experiments measure the collective property. The polarizability can be expanded in terms of spherical harmonics and, for a molecule of ellipsoidal or cylindrical symmetry, the only terms that appear are Y_{2m}, so the dynamical quantity is $Y_{2m}(\mathbf{k}, t)$ and the correlation functions are $C_{2m}(\mathbf{k}, t)$. In the long-wavelength limit, the quantities of interest are $C_{2m}(\mathbf{k} = 0, t)$, which can be significantly different from $S_{2m}(t)$. We shall discuss this point in more detail in Section XIII, where we discuss collective orientation of molecules with ellipsoidal symmetry.

C. Magnetic Resonance Experiments

Nuclear magnetic relaxation due to magnetic dipole–dipole interactions in a liquid can provide direct information on the orientational relaxation time of a rotating molecule. In nuclear magnetic resonance experiments, usually single-particle relaxation is measured. The spin–lattice relaxation time and the spin–spin relaxation time are essentially the time integrals over various components of $S_{2m}(t)$, defined by Eq. (2.10). Line shape experiments can provide information on $S_{2m}(t)$ in the frequency plane. For detailed discussion, we refer to the existing literature [17].

D. Neutron Scattering

The theory and the principles of neutron scattering are similar to those of light scattering. The major difference between these two techniques lie in the range of wavevector k that is sampled. The neutrons have small wavelength, so they can probe large-wavevector processes in contrast to the light scattering experiments, which can probe only the long-wavelength $(k \to 0)$ processes. While light scattering is usually coherent, neutron scattering can be either coherent or incoherent—the former measuring the collective dynamics, whereas the latter measures the single particle. Another important difference is that because of the range of wavevector processes probed by neutron scattering, the translational and the librational modes of the liquid molecule are important here.

E.   Raman and Infrared Line Shapes

Raman line shape measurements usually give information on $S_{20}(t)$, while infrared line shapes give $S_{10}(t)$. However, the analysis is not straightforward, because a decoupling between vibrational and rotational motion is necessary to extract information on rotational motion, and this is not easy for a dipolar liquid especially if the dipole moment has a strong dependence on the normal coordinate being probed. For a nonpolar liquid, the normal vibrational modes of one molecule are uncorrelated with those on another molecule (to a good approximation). In such a situation, Raman and infrared techniques measure the single-particle time correlation function. The situation is more complicated for dipolar liquids. This has been discussed extensively in the literature [18].

F.   Quadrupolar Relaxation

Quadrupolar spin relaxation of ionic nuclei in dipolar and electrolyte solutions can provide important information on the collective dynamics of the liquid. Nuclei with a quadrupole moment interact with the local electric field gradient tensor originating from the neighboring molecules. Quadrupolar relaxation essentially measures the field gradient correlation function

$$g(t) = \langle \mathbf{V}(0) \odot \mathbf{V}(t) \rangle \tag{3.7}$$

where $\mathbf{V}(t)$ is the field gradient tensor at time t and $\odot$ denotes a scalar tensor product. For a solvent of rigid dipoles, the field gradient tensor is given by [19, 20]

$$\mathbf{V}_{\alpha\beta}(t) = \sum_i \left[\mathbf{V}_\alpha \mathbf{V}_\beta \mathbf{V}_\gamma \frac{1}{\mathbf{r}_i} \right] \otimes \boldsymbol{\mu}_i(\boldsymbol{\Omega}) \tag{3.8}$$

where $\mathbf{r}_i$ is the vector connecting the quadrupolar molecule with the solvent molecule and the sum is over all the solvent molecules. $\boldsymbol{\mu}_i(\boldsymbol{\Omega})$ is the dipole moment vector of the ith molecule with orientation $\boldsymbol{\Omega}$ and $\otimes$ is an irreducible tensor product [19].

The expression for $\mathbf{V}(t)$ is unfortunately rather complex. It can be evaluated under certain simplifying approximations; we are not aware of a full microscopic calculation. Nonetheless, Eq. (3.8) clearly shows that the quadrupolar relaxation is dominated essentially by the nearest-neighbor molecules because of the strong dependence [r_i^{-4} in $\mathbf{V}(t)$] on the distance of the solvent molecules from the quadrupolar nuclei. Therefore, the translational and the librational modes of the solvent molecules will play important roles in quadrupolar relaxation and a continuum-model-based theory is not satisfactory.

G.   Solvation Dynamics (Stokes Shift)

For many organic molecules, the dipole moment (both the magnitude and the direction) is significantly different in the excited state from its value in the ground state. Because of their photophysics, some of these molecules can be used as a good probe of polar solvent dynamics. This is conventionally done by exciting the molecule and subsequently monitoring the time-dependent fluorescence Stokes shift of the emission spectrum, which usually shifts to longer wavelengths because of the increased stabilization of the excited state. If the shape of the emission spectra do not change significantly with time, then the experimental data can be analyzed in terms of a solvation time correlation function $C_s(t)$ defined by [21, 22]

$$C_s(t) = \frac{\bar{v}(t) - \bar{v}(\infty)}{\bar{v}(0) - \bar{v}(\infty)} \tag{3.9}$$

where $v(t)$ is the average frequency of the spectrum at time t, defined as the first moment of the spectrum. Thus, Eq. (3.9) gives an average estimate of the time dependence of the progression of solvation. Recently, several detailed experimental and theoretical studies of $C_s(t)$ have been carried out and this is currently an active area of research [23–26].

In order to understand the role of collective orientational relaxation in solvation dynamics, let us remember that a fairly accurate expression for the energy of interaction of the bare electric field, $E_0(r)$, of a polar solute molecule with the polarization, $P(r, t)$, of the dipolar molecule is given by

$$E_{solv}(t) = -\frac{1}{2} \int dr E_0(r) \cdot P(r, t) \tag{3.10}$$

Equation (3.10) can be written in the Fourier space (with k as the Fourier variable conjugate to r)

$$E_{solv}(t) = -\frac{(2\pi)^{-3}}{2} \int dk E_0(k) \cdot P(k, t) \tag{3.11}$$

The time dependence of $P(k, t)$ is controlled by the collective orientational relaxation [27–31]. Under favorable conditions (small dipolar solute in a strongly dipolar liquid of larger solvent molecules), the integration in Eq.

(3.11) may be dominated by the intermediate wavevector processes, so that $E_{solv}(t)$ contains nontrivial information on collective orientational dynamics.

H. Kerr Relaxation

In Kerr relaxation, one studies optically the decay of a transient birefringence created initially by an intense polarized light pulse sent through the medium. The polarized light pulse forces a partial alignment of the anisotropic molecules because of the torque generated by the light field on the anisotropic molecule. The induced orientational anisotropy leads to a difference in the refractive index in the parallel and perpendicular light polarization directions. The relaxation of this anisotropy is now probed by a second light pulse that is polarized at 45° to the pump pulse. The optical Kerr effect essentially measures the nonlinear polarization ($\mathbf{P}^{NL}$) created by the electric field of the laser pulse. This nonlinear polarization depends on the polarizability–polarizability time correlation function defined by

$$C_{ijkl}(t) = \langle \alpha_{ij}(0)\alpha_{kl}(t) \rangle \tag{3.12}$$

Note that it is the same correlation that is measured in depolarized light scattering. The difference is that the Kerr relaxation studies can be carried out directly in the time domain with ultrashort laser pulses, so that the short-time dynamics of the liquid can be studied. Such studies have been carried out recently [32, 33]. It is important to note that Eq. (3.12) is a collective orientational correlation function because $\alpha_{ij}(t)$ is a sum over all the polarizabilities of the system, as in Eq. (3.5).

IV. EXTENDED HYDRODYNAMIC DESCRIPTION OF ORIENTATIONAL MOTION

Although relaxation of collective variables in a dense simple liquid has been a subject of tremendous interest in the last few decades [6, 34], there has been comparatively little work on the relaxation of collective modes in a molecular liquid, where the orientational degrees of freedom are also important. However, one can gain considerable insight from the studies on the spatial part and many of the techniques can be straightforwardly used for orientational relaxation.

The main dynamical quantities involved here are the number density, $\rho(\mathbf{r}, \Omega, t)$, and the spatial and angular momenta densities $g_T(\mathbf{r}, \Omega, t)$ and $g_\Omega(\mathbf{r}, \Omega, t)$ respectively. In a dense liquid, the relaxation of energy fluctuation may not be rate determining and will not be considered here. Since we are interested in the collective orientational relaxation involving not only large but also few (for example, nearest-neighbor) molecules, the traditional

Navier–Stokes hydrodynamic description [6], valid only in the long-wavelength limit, is not adequate. However, this time-honored description can be meaningfully extended to treat dynamics at a molecular length scale. Such an extension has been carried out recently by several groups and is described here.

The main idea in this extended hydrodynamic description is as follows. At molecular length scale, the momentum relaxation depends on the force field experienced by the molecules. This force field comes from intermolecular interaction. Thus, the gradient of the pressure term in Navier–Stokes equation should be modified to include the effects of the intermolecular interactions, which depend on the molecular arrangements. The force field acting on a molecule at position $\mathbf{r}$ with orientation Ω can be obtained from the density functional theory, which gives a general expression for the free energy functional of an nonhomogeneous system. The extended hydrodynamic equations for the number density and the momenta densities are [29, 35]

$$\frac{\partial}{\partial t}\rho(\mathbf{X}, t) + \frac{1}{m}\mathbf{V}_T \cdot \mathbf{g}_T(\mathbf{X}, t) + \frac{1}{I}\mathbf{V}_\Omega \cdot \mathbf{g}_\Omega(\mathbf{X}, t) = 0 \qquad (4.1)$$

$$\frac{\partial}{\partial t}\mathbf{g}_i(\mathbf{X}, t) = -\rho(\mathbf{X}, t)\mathbf{V}_i \frac{\delta F'[\rho(t)]}{\delta\rho(\mathbf{X}, t)}$$

$$-\sum_j \int dt_1 d\mathbf{X}_1 \Gamma_{ij}(\mathbf{X}, \mathbf{X}_1, t - t_1; \rho(t)) \frac{\delta F}{\delta \mathbf{g}_j(\mathbf{X}, t)}$$

$$+ f_i(\mathbf{X}, t) \qquad (4.2)$$

where $\mathbf{X} = (\mathbf{r}, \Omega)$, i stands for T or Ω, m and I are the mass and the moment of inertia, respectively, $\mathbf{V}_T$ and $\mathbf{V}_\Omega$ are the usual spatial and angular gradient operators, $F'[\rho(t)]$ is the interaction part of the free energy for an nonhomogeneous fluid [36]. F and F' are given by the following expressions:

$$F'[\rho(t)] = k_B T\left\{ \int d\mathbf{X}\rho(\mathbf{X}, t)[\ln\rho(\mathbf{X}, t) - 1]\right.$$

$$-\frac{1}{2}\int d\mathbf{X}d\mathbf{X}_1 c_2(\mathbf{X}, \mathbf{X}_1)\delta\rho(\mathbf{X}, t)\delta\rho(\mathbf{X}_1, t)$$

$$-\frac{1}{6}\int d\mathbf{X}d\mathbf{X}_1 d\mathbf{X}_2 c_3(\mathbf{X}, \mathbf{X}_1, \mathbf{X}_2)\delta\rho(\mathbf{X}, t)\delta\rho(\mathbf{X}_1, t)\delta\rho(\mathbf{X}_2, t) + \cdots \qquad (4.3)$$

and

$$F(\rho(t)) = F'[\rho(t)] + \int d\mathbf{X} \left[\frac{g_T^2}{2m\rho} + \frac{g_\Omega^2}{2I\rho} \right] \qquad (4.4)$$

where $\delta\rho(\mathbf{X}, t)$ is the density fluctuations and $c_n(\mathbf{X}, \mathbf{X}_1, \cdots)$ is the n-particle direct correlation function, which is the nth expansion coefficient in the density expansion of free energy. Equation (4.3) is exact. In Eq. (4.2), $f_i(\mathbf{X}, t)$'s are the random force and random torque, which are assumed to be Gaussian and related to the dissipative kernels Γ_{ij} by the second fluctuation–dissipation theorem [37]:

$$\langle f_T(\mathbf{X}, 0) f_T(\mathbf{X}_1, t) \rangle = k_B T \Gamma_{TT}(\mathbf{X}, \mathbf{X}_1, t; \rho(t)) \qquad (4.5)$$

$$\langle f_\Omega(\mathbf{X}, 0) f_\Omega(\mathbf{X}_1, t) \rangle = k_B T \Gamma_{\Omega\Omega}(\mathbf{X}, \mathbf{X}_1, t; \rho(t)) \qquad (4.6)$$

$$\langle f_T(\mathbf{X}, 0) f_\Omega(\mathbf{X}_1, t) \rangle = k_B T \Gamma_{T\Omega}(\mathbf{X}, \mathbf{X}_1, t; \rho(t)) \qquad (4.7)$$

with

$$\Gamma_{T\Omega} = \Gamma_{\Omega T}$$

Equations (4.1)–(4.7) allow a general description of the coupled translational and orientational dynamics in a dense liquid, provided reasonable forms of the dissipative kernels are available. Clearly, similar equations can also be derived from the Mori–Zwanzig projection operator technique with $\{\rho, \mathbf{g}_T, \mathbf{g}_\Omega\}$ as the subset of slow variables [38, 39]. However, we regard the hydrodynamic approach more straightforward and intuitive. This approach is also closely related to the well-known time-dependent Ginzburg–Landau (TDGL) approach to collective dynamics [40].

Because of the complexities of the hydrodynamic equation, a most general calculation is yet to be carried out. However, different limits of these equations have been studied and they reveal a rich dynamical behavior of orientational relaxation. The following limiting situations have been discussed:

(a) The Markovian (or the memoryless) limit where the random forces are delta correlated in time. Also, the random forces are assumed to be totally local and the dissipative kernels are diagonal. That is,

$$\Gamma_{ij}(\mathbf{X}, \mathbf{X}_1, t - t_1; \rho(t)) = \Gamma_{ij}(\bar{\rho})\delta(\mathbf{X} - \mathbf{X}_1)\delta(t - t_1)\delta_{ij} \qquad (4.8)$$

where $\bar{\rho}$ is the average density, equal to $\rho_0/4\pi$ for a spherical diffusor, ρ_0 is the number density of the liquid. Both the overdamped and the inertial limits of this approximation have been studied.

(b) The non-Markovian form for the angular dissipative kernel, but the Markovian form for the spatial kernel. The reason for studying this situation is that the long-range nature of dipolar interactions makes the angular kernel frequency dependent at a lower frequency where the viscoelastic effects are yet to be important, so that the frequency dependence of spatial dissipative kernel can be neglected at such frequencies.

Finally, let us make some comments on the validity of such an extended hydrodynamic approach. The basic idea is that at molecular length scale, the relaxation of density is very slow and it is the most important dynamical quantity [41–43]. The reason is that at this length only the number density is a conserved quantity and both momentum and energy relaxations are much faster than the density relaxation. Thus, a generalized diffusion equation description of the dynamics should be fairly reliable. The validity of an extended hydrodynamic approach for simple monatomic liquids has been demonstrated by de Schepper and Cohen [41] and also by Kirkpatrick and co-worker [42, 43].

V. SIMPLE MARKOVIAN THEORY OF COLLECTIVE ORIENTATIONAL RELAXATION

A self-consistent Markovian theory of the collective orientational relaxation in a dipolar liquid was first presented by Berne [14]. However, Berne's treatment was essentially based on a continuum model, so the details of the molecular properties of the liquid were not included in his theory. Recently, Chandra and Bagchi [29, 44] have presented a microscopic theory for describing the collective orientational relaxation of the dipolar molecules. These authors confirmed at the microscopic level many of Berne's conclusions. There are, however, significant differences between the two theories. We shall discuss the two theories in detail.

Berne's treatment is based on an equation for rotational and translational diffusion with interaction terms that arise from the torques and forces on a molecule resulting from the electric field generated by other charges in the system. The equation of motion used is given by

$$\frac{\partial}{\partial t}\rho(\mathbf{r}, \boldsymbol{\Omega}, t) = D_R \nabla_\Omega^2 \rho(\mathbf{r}, \boldsymbol{\Omega}, t) + D_T \nabla_T^2 \rho(\mathbf{r}, \boldsymbol{\Omega}, t)$$

$$- \beta D_R \mathbf{V}_\Omega \cdot [N(\mathbf{r}, \boldsymbol{\Omega}, t)\rho(\mathbf{r}, \boldsymbol{\Omega}, t)]$$

$$- \beta D_T \mathbf{V}_T \cdot [F(\mathbf{r}, \boldsymbol{\Omega}, t)\rho(\mathbf{r}, \boldsymbol{\Omega}, t)] \qquad (5.1)$$

where D_T and D_R are the translational and rotational diffusion coefficients, respectively. $\mathbf{V}_T$ and $\mathbf{V}_\Omega$ are the spatial and rotational gradient operators. The force and torque are expressed in terms of the molecular charge distribution $Z(s)$ and the electric field E by the following equations:

$$F(\mathbf{r}, \boldsymbol{\Omega}, t) = \int ds Z(s) E(\mathbf{r} + s\boldsymbol{\Omega}, t) \tag{5.2}$$

$$N(\mathbf{r}, \boldsymbol{\Omega}, t) = \int ds Z(s) s\boldsymbol{\Omega} \times E(\mathbf{r} + s\boldsymbol{\Omega}, t) \tag{5.3}$$

From Eqs. (5.1)–(5.3), Berne [14] found that in the limit of zero wavevector $(k \to 0)$ and for a liquid of nonpolarizable molecules the correlation function $\phi(t)$ [defined by Eq. (3.1)] is biexponential. $\phi(t)$ is given by

$$\phi(t) = \mu^2 [\tfrac{2}{3} e^{-2D_R t} + \tfrac{1}{3} e^{-2D_R(1 + 3Y)t}] \tag{5.4}$$

and

$$Y = (4\pi/9)\beta\mu^2\rho_0 \tag{5.5}$$

This result indicates that the dipole–dipole interactions give rise to a marked difference between the decay of the longitudinal and the transverse fluctuations. If the dipole–dipole interactions are neglected, as was originally done by Debye, then $\phi(t)$ is single exponential and is given by

$$\phi(t) = \exp[-2D_R t] \tag{5.6}$$

Building on the effort of Berne, Chandra and Bagchi [29, 44] developed a microscopic theory of orientational relaxation. In their microscopic calculations, Chandra and Bagchi have shown that the relaxation of $\phi(\mathbf{k}, t)$ is modified by polar forces in several ways. Their study has confirmed several of Berne's conclusions. They have also found two types of relaxation times for $\phi(\mathbf{k}, t)$, except that their relaxation times are wavevector dependent and are functions of the appropriate direct correlation functions, which will be discussed subsequently. They have shown that some of the assumptions made by Berne are not justified in the microscopic theory. The Markovian theory of Chandra and Bagchi [29, 44] is based on a generalized Smoluchowski equation (GSE) to describe the dynamics of the position- and orientation-dependent density of the dipolar liquid molecules. The derivation of the kinetic equation is based on the fact that in a dense polar liquid the relaxations of both the spatial momentum and the angular momentum are very fast and only the number conservation is important on the time scale of interest. So, the starting equation is the continuity equation given by Eq. (4.1). One now neglects the non-Markovian effects and the random fluctuations in the momentum relaxation. That means the equation of motion of g_T is now given by

$$\frac{\partial}{\partial t} g_T(\mathbf{r}, \mathbf{\Omega}, t) = -\rho(\mathbf{r}, \mathbf{\Omega}, t) \nabla_T \frac{\delta F'[\rho(t)]}{\delta \rho(\mathbf{r}, \mathbf{\Omega}, t)}$$

$$- \int d\mathbf{r}' d\mathbf{\Omega}' \Gamma_T(\mathbf{r}, \mathbf{r}', \mathbf{\Omega}, \mathbf{\Omega}'; \rho(t)) \frac{\delta F[\rho(t)]}{\delta g_T} \qquad (5.7)$$

and a similar equation for $\mathbf{g}_\Omega$. $F'[\rho(t)]$ is the free energy of an inhomogeneous equilibrium fluid minus its kinetic energy contribution. The functional derivative on the right-hand side of Eq. (5.7) is given by

$$\frac{\delta F'[\rho(t)]}{\delta \rho(\mathbf{r}, \mathbf{\Omega}, t)}$$

$$= k_B T \left\{ \ln \rho(\mathbf{r}, \mathbf{\Omega}, t) - \int d\mathbf{r}' d\mathbf{\Omega}' c(\mathbf{r}, \mathbf{\Omega}, \mathbf{r}', \mathbf{\Omega}') \, \delta \rho(\mathbf{r}', \mathbf{\Omega}', t) \right.$$

$$\left. - \frac{1}{2} \int d\mathbf{r}' d\mathbf{\Omega}' d\mathbf{r}'' d\mathbf{\Omega}'' c_3(\mathbf{r}, \mathbf{r}', \mathbf{r}'', \mathbf{\Omega}, \mathbf{\Omega}', \mathbf{\Omega}'') \, \delta \rho(\mathbf{r}', \mathbf{\Omega}', t) \, \delta \rho(\mathbf{r}'', \mathbf{\Omega}'', t) + \cdots \right\}$$

$$(5.8)$$

where $c(\mathbf{r}, \mathbf{r}', \mathbf{\Omega}, \mathbf{\Omega}')$ and $c_3(\mathbf{r}, \mathbf{r}', \mathbf{r}'', \mathbf{\Omega}, \mathbf{\Omega}', \mathbf{\Omega}'')$ are, respectively, the two-particle and the three-particle direct correlation functions of the liquid. To proceed further, one needs the bare nonlinear dissipative kernel Γ_T and Γ_Ω. It is assumed [29] that

$$\Gamma_T^{ij} = \delta_{ij} \delta(\mathbf{r} - \mathbf{r}') \delta(\mathbf{\Omega} - \mathbf{\Omega}') k_B T / D_T$$

$$\Gamma_\Omega^{ij} = \delta_{ij} \delta(\mathbf{r} - \mathbf{r}') \delta(\mathbf{\Omega} - \mathbf{\Omega}') k_B T / D_R \qquad (5.9)$$

where i, j denote the cartesian components. Next, one neglects the momenta relaxation, in the spirit of time-dependent Ginzburg–Landau theories, and uses the resultant expressions for $\mathbf{g}_T$ and $\mathbf{g}_\Omega$ in the continuity equation to obtain the following nonlinear Smoluchowski equation [29, 44, 45]:

$$\frac{\partial}{\partial t} \delta \rho(\mathbf{r}, \mathbf{\Omega}, t) = D_R \nabla_\Omega^2 \delta \rho(\mathbf{r}, \mathbf{\Omega}, t) + D_T \nabla_T^2 \delta \rho(\mathbf{r}, \mathbf{\Omega}, t)$$

$$- D_R \nabla_\Omega \cdot \rho(\mathbf{r}, \mathbf{\Omega}, t) \nabla_\Omega \int d\mathbf{r}' d\mathbf{\Omega}' c(\mathbf{r}, \mathbf{r}', \mathbf{\Omega}, \mathbf{\Omega}') \, \delta \rho(\mathbf{r}', \mathbf{\Omega}', t)$$

$$- D_T \nabla_T \cdot \rho(\mathbf{r}, \mathbf{\Omega}, t) \nabla_T \int d\mathbf{r}' d\mathbf{\Omega}' c(\mathbf{r}, \mathbf{r}', \mathbf{\Omega}, \mathbf{\Omega}') \, \delta \rho(\mathbf{r}', \mathbf{\Omega}', t) \qquad (5.10)$$

where, for pure liquid, $\delta \rho(\mathbf{r}, \mathbf{\Omega}, t) = \rho(\mathbf{r}, \mathbf{\Omega}, t) - \rho_0/4\pi$.

In writing Eq. (5.10), the terms containing the triplet and higher-order direct correlation functions are neglected. It should be made clear here that Eq. (5.10) describes decay of fluctuations off a homogeneous state. Thus, we are studying orientational relaxation in a homogeneous liquid. It is of interest to compare Eq. (5.10) with the equation of motion of Berne [14]. By comparing Eqs. (5.1) and (5.10), one can easily obtain molecular expressions (in terms of the direct correlation function) for the force and the torque terms. The important point here is that Eq. (5.10) is microscopic as it contains the microstructural information of the liquid. As will be discussed, this is especially important at intermediate wavevectors. In fact, it is improper to consider intermediate to large wavevector relaxation processes without including an appropriate description of the liquid structure. Such neglect leads to an inaccurate description of relaxation dynamics. Note that previous investigations [46–48] have shown that a Smoluchowski equation with a mean-field force term due to intermolecular interactions can describe large wave number processes rather accurately in a dense liquid. In the long-wavelength limit, the translational contribution to density relaxation becomes insignificant compared to rotational relaxation. This limit is correctly described by Eq. (5.10). It is straightforward to generalize the preceding treatment to multicomponent systems, which will be discussed in Section XV.

The nonlinear Smoluchowski equation (5.10) is next linearized to obtain

$$\frac{\partial}{\partial t} \delta\rho(\mathbf{r}, \mathbf{\Omega}, t) = D_R \nabla_\Omega^2 \delta\rho(\mathbf{r}, \mathbf{\Omega}, t) + D_T \nabla_T^2 \delta\rho(\mathbf{r}, \mathbf{\Omega}, t)$$

$$- D_R(\rho_0/4\pi)\nabla_\Omega^2 \int d\mathbf{r}' d\mathbf{\Omega}' c(\mathbf{r} - \mathbf{r}', \mathbf{\Omega}, \mathbf{\Omega}')\delta\rho(\mathbf{r}', \mathbf{\Omega}', t)$$

$$- D_T(\rho_0/4\pi)\nabla_T^2 \int d\mathbf{r}' d\mathbf{\Omega}' c(\mathbf{r} - \mathbf{r}', \mathbf{\Omega}, \mathbf{\Omega}')\delta\rho(\mathbf{r}', \mathbf{\Omega}', t) \quad (5.11)$$

Since it is convenient to work in wavevector space, Eq. (5.11) is Fourier transformed to obtain

$$\frac{\partial}{\partial t} \delta\rho(\mathbf{k}, \mathbf{\Omega}, t) = D_R \nabla_\Omega^2 \delta\rho(\mathbf{k}, \mathbf{\Omega}, t) - D_T k^2 \delta\rho(\mathbf{k}, \mathbf{\Omega}, t)$$

$$- D_R(\rho_0/4\pi)\nabla_\Omega^2 \int d\mathbf{\Omega}' c(\mathbf{k}, \mathbf{\Omega}, \mathbf{\Omega}') \delta\rho(\mathbf{k}, \mathbf{\Omega}', t)$$

$$+ D_T(\rho_0/4\pi)k^2 \int d\mathbf{\Omega}' c(\mathbf{k}, \mathbf{\Omega}, \mathbf{\Omega}') \delta\rho(\mathbf{k}, \mathbf{\Omega}', t) \quad (5.12)$$

Now, one expands $\delta\rho(\mathbf{k}, \mathbf{\Omega}, t)$ in terms of the spherical harmonics

$$\delta\rho(\mathbf{k}, \mathbf{\Omega}, t) = \sum_{lm} a_{lm}(\mathbf{k}, t) Y_{lm}(\mathbf{\Omega}) \tag{5.13}$$

where $a_{lm}(\mathbf{k}, t) = \int d\mathbf{\Omega} \, Y_{lm}^*(\mathbf{\Omega}) \, \delta\rho(\mathbf{k}, \mathbf{\Omega}, t)$. It is straightforward to show that the correlation function $C_{lm}(\mathbf{k}, t)$ is given by

$$C_{lm}(\mathbf{k}, t) = \langle a_{lm}(\mathbf{k}, t) a_{lm}(-\mathbf{k}, 0) \rangle \tag{5.14}$$

where $\langle \cdots \rangle$ means average over an equilibrium ensemble of homogeneous states of the liquid. The direct correlation function $c(\mathbf{k}, \mathbf{\Omega}, \mathbf{\Omega}')$ can also be expanded in terms of the spherical harmonics:

$$c(\mathbf{k}, \mathbf{\Omega}, \mathbf{\Omega}') = \sum_{l_1 l_2 m} c(l_1 l_2 m; \mathbf{k}) Y_{l_1 m}(\mathbf{\Omega}) Y_{l_2 m}(\mathbf{\Omega}') \tag{5.15}$$

where $\mathbf{k}$ is taken parallel to the z axis. If one substitutes Eq. (5.15) and (5.13) into Eq. (5.12) and takes the scalar product of the resulting equation with $Y_{lm}(\mathbf{\Omega})$, one obtains

$$\frac{\partial}{\partial t} a_{lm}(\mathbf{k}, t) = -[D_R l(l+1) + D_T k^2] a_{lm}(\mathbf{k}, t)$$

$$+ \frac{\rho_0}{4\pi} (-1)^m [D_R l(l+1) + D_T k^2] \sum_{l_2} c(l l_2 m; \mathbf{k}) a_{l_2 m}(\mathbf{k}, t) \tag{5.16}$$

This is the primary equation of Chandra and Bagchi, which gives the orientational relaxation in a pure liquid [29, 44]. It is clear that each $a_{lm}(\mathbf{k}, t)$ is coupled to several other coefficients. The equation of motion for the correlation function relevant for dielectric relaxation is obtained by setting $l = 1$ in Eq. (5.16).

A considerable simplification in Eq. (5.16) results if one assumes that the two-particle direct correlation function of the dipolar liquid is given by the linearized equilibrium theory. Examples of such linear theories are the mean spherical approximation model (MSA) or the linearized hypernetted chain (LHNC) model [6, 13]. These models predict that the only nonvanishing $c(l_1 l_2 m; k)$'s are $c(000; k)$, $c(110; k)$ and $c(111; k)$. Within this approximation, Eq. (5.16) reduces to a simpler equation of the form [29, 44]

$$\frac{\partial}{\partial t} a_{lm}(\mathbf{k}, t) = -[D_R l(l+1) + D_T k^2] a_{lm}(\mathbf{k}, t)$$

$$+ \frac{\rho_0}{4\pi} (-1)^m [D_R l(l+1) + D_T k^2] c(llm; k) a_{lm}(k) [\delta_{l1} + \delta_{l0}] \tag{5.17}$$

So the orientational correlation function, $C_{lm}(k, t)$ is given by

$$C_{lm}(\mathbf{k}, t) = \langle |a_{lm}(\mathbf{k})|^2 \rangle \exp[-\{D_R l(l+1) + D_T k^2\}$$
$$\times \{1 - (\rho_0/4\pi)(-1)^m c(llm; k)(\delta_{l0} + \delta_{l1})\}] \qquad (5.18)$$

Equation (5.18) shows that the only correlation functions which are affected by the dipolar forces are $C_{1m}(k, t)$; the higher-order correlation functions are totally unaffected. This conclusion, of course, depends on the assumption that $c(l_1 l_2 m; k)$ can be given by a linearized theory. We shall return to the validity of this assumption. Here we note that the prediction of Eq. (5.18) is expected to be valid when the dipolar interactions are weak; hence, it would be valid when the liquid is weakly polar. The dipolar correlation function $\phi(\mathbf{k}, t)$, defined by Eq. (3.1), can be calculated from Eq. (5.18) and is given by [29]

$$\phi(\mathbf{k}, t) = A_1 \exp[-(2D_R + D_T k^2)(1 - (\rho_0/4\pi)c(110; k))t]$$
$$+ A_2 \exp[-(2D_R + D_T k^2)(1 + (\rho_0/4\pi)c(111; k))t] \qquad (5.19)$$

where

$$A_1 = \langle |a_{10}(k)|^2 \rangle / (\langle |a_{10}(k)|^2 \rangle + 2\langle |a_{11}(k)|^2 \rangle) \qquad (5.20)$$

$$A_2 = 2\langle |a_{11}(k)|^2 \rangle / (\langle |a_{10}(k)|^2 \rangle + 2\langle |a_{11}(k)|^2 \rangle) \qquad (5.21)$$

Equation (5.18) predicts that the only multipole moment that contributes to the dipolar orientational relaxation is the dipole moment. Note that there are two relaxation times for the relaxation of the dipolar correlation function. The longitudinal dipolar correlations are given by $C_{10}(\mathbf{k}, t)$ and the transverse correlations by $C_{11}(\mathbf{k}, t)$. The two relaxation times are [29]

$$\tau_{10}^{-1}(k) = 2D_R\{(1 + p'(k\sigma)^2)(1 - (\rho_0/4\pi)c(110; k))\} \qquad (5.22)$$

and

$$\tau_{11}^{-1}(k) = 2D_R\{(1 + p'(k\sigma)^2)(1 + (\rho_0/4\pi)c(111; k))\} \qquad (5.23)$$

where σ is the molecular diameter and $p' = D_T/2D_R\sigma^2$. Thus, p' is a measure of the relative importance of the translational modes in the orientational relaxation. Equations (5.22) and (5.23) give the decay of the transverse and the longitudinal fluctuations, respectively. This result is similar to that obtained by Berne [14], but in his calculation τ_{11} was simply $(2D_R)^{-1}$—there was no wavevector dependence of the rate constants. Chandra and Bagchi have found that both τ_{10} and τ_{11} are modified by polar interactions, the former is affected more strongly than the latter. For the complete determination of $\phi(\mathbf{k},t)$ one needs the expressions for the equilibrium correlation functions $\langle|a_{10}|^2\rangle$ and $\langle|a_{11}|^2\rangle$. These are determined by the microscopic structure of the liquid and are given by [29]

$$\langle|a_{10}(k)|^2\rangle = (N/4\pi)\{1 + (\rho_0/4\pi)h(110;k)\} \tag{5.24}$$

$$\langle|a_{11}(k)|^2\rangle = (N/4\pi)\{1 - (\rho_0/4\pi)h(111;k)\} \tag{5.25}$$

where N is the total number of molecules in the system and $h(110;k)$ and $h(111;k)$ are the different components in the spherical harmonic expansion of the total pair correlation function. $h(llm;k)$'s are related to $c(llm;k)$'s by the following Ornstein–Zernike relation:

$$h(llm;k) = c(llm;k) + \frac{\rho_0}{4\pi}(-1)^m \sum_{l_1} c(ll_1m;k)h(l_1lm;k) \tag{5.26}$$

Berne assumed that $\langle|a_{10}|^2\rangle = \langle|a_{11}|^2\rangle$. Numerical calculations show that this equality is not exact.

As was emphasized previously, the theory discussed is valid when a linearized equilibrium theory is used for the two-particle direct correlation function [29, 44]. For detailed numerical calculations MSA was used because simple analytical expressions are available for $c(110;k)$ and $c(111;k)$ [49]. Because of the use of MSA, the numerical results are strictly valid only for weakly polar liquids. The formalism, however, holds where the equilibrium correlation functions are given by a linearized theory. It is believed that the qualitative aspects will not change significantly if a different theory is used to compute the direct correlation function coefficients.

In Figures 1 and 2, the wavevector dependence of the relaxation times, $\tau_{10}(k)$ and $\tau_{11}(k)$, are shown for several different values of the parameter p'. For comparison, the case $p' = 0$ (when translational contribution is totally absent) is also included. It is seen clearly that the translational diffusion plays an important role in the intermediate values of the wavevector. The relaxation times decrease with increase in the translational contribution. In Figure 3, the calculated $\phi(\mathbf{k},t)$ is plotted against time for several different values of the wavevector k, including $k = 0$. The value of p' is kept fixed. In the limit $k \to 0$, translational effects are not present. As k is increased, relaxation becomes faster, which shows that translational diffusion contributes significantly in the intermediate values of the wavevector. In Figure 4 $\phi(\mathbf{k},t)$ is plotted against time for $k\sigma = 6.2$ and for several different values of the parameter p'. The case $p' = 0$ is also included. It is seen that at this intermediate

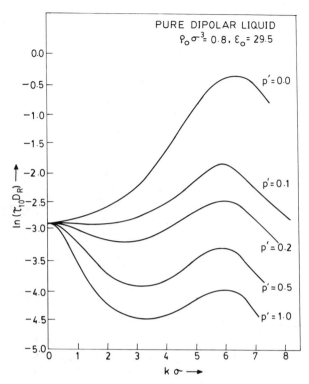

FIGURE 1. The dependence of the longitudinal relaxation time on wavevector and on translational diffusion. The longitudinal relaxation time, $\tau_{10}(k)$, is plotted against wavevector k for several different values of the translational parameter, p'. The values of the static dielectric constant, ε_0, and the reduced density, $\rho_0\sigma^3$, are 29.5 and 0.8, respectively. (From Ref. 29.)

value of the wavevector, k, translational effects influence $\phi(\mathbf{k}, t)$ to a great extent.

It is interesting to note that when translational motions are neglected, then $C_{10}(\mathbf{k}, t)$ has a pronounced slowing down at the intermediate value of the wavevector k. This slowing down of relaxation at intermediate wavevectors is similar to the well-known de Gennes' narrowing [50] of the dynamic structure factor of a dense liquid at intermediate wavevectors. In the latter case, the slowing down occurs near the wavevector where the static structure factor is sharply peaked and is because of the short-range order that is present in a dense liquid. In the case of orientational relaxation, the slowing down of relaxation also occurs near the first peak of the static structure factor and this may also be attributed to the short-range *orientational correlations*

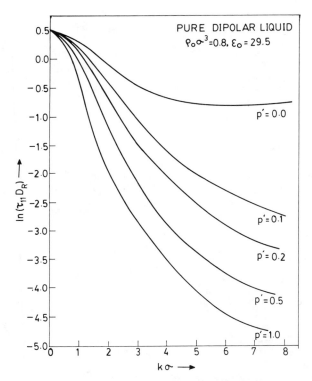

FIGURE 2.   The dependence of the transverse relaxation time on the wavevector and on the translational diffusion. The calculated values of $\tau_{11}(k)$ are plotted against wavevector, k, for several different values of the translational parameter, p'. The values of the static dielectric constant and the density are the same as in Fig. 1. (From Ref. 29.)

present in a dense polar liquid. This slowing down of relaxation becomes progressively weaker when the translational modes are important in the density relaxation.

In their classic study on dielectric relaxation, Titulaer and Deutch [51] pointed out that a linearized Smoluchowski equation treatment of orientational relaxation may not be reliable if the torque–torque correlation function decays on the same time scale as the orientational correlation function $\phi(\mathbf{k}, t)$. This calls for two additional studies. First, a non-Markovian theory may be needed in some cases. Such a theory will be discussed in Section IX. Second, it is thus important to obtain an estimate of the contributions of the nonlinear terms. We are not aware of any complete study of the latter problem. A preliminary analysis of the nonlinear terms

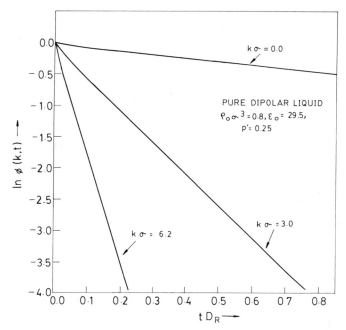

FIGURE 3. The wavevector and time dependence of the correlation function $\phi(\mathbf{k}, t)$ of a pure liquid. The calculated values of $\phi(\mathbf{k}, t)$ are plotted against time for several different values of the wavevector, k. The value of p' is 0.25. The values of the static dielectric constant and the reduced density are the same as in Fig. 1. (From Ref. 29.)

has been carried out recently [29]. This analysis shows that this is a nontrivial problem. The nonlinear terms lead to complicated equations (even when the triplet contribution is neglected), which involve many terms. Some of the important aspects of this analysis are given in the Appendix C of Ref. 29. A detailed perturbative analysis to obtain a quantitative measure of the nonlinear effects is yet to be carried out. In the following, some conclusions of general nature are discussed.

The modified equation of motion for $a_{10}(\mathbf{k}, t)$ and $a_{11}(\mathbf{k}, t)$, with all the nonlinear contributions [29], shows the absence of any coupling between coefficients $a_{1m}(\mathbf{k}, t)$, even in the nonlinear terms. Hence, it is reasonable to conclude that, at least in the dipolar orientational relaxation, the dynamic nonlinear effects may not be very important. It is also clear that for $a_{1m}(\mathbf{k}, t)$ the important nonlinear term involves $a_{00}(\mathbf{k}, t)$, which is the number density fluctuation and is expected to vary slowly in time, more so near the value of the wavevector where the static structure factor is peaked. It is rather

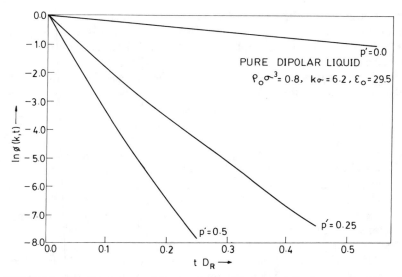

FIGURE 4. The dependence of the correlation function $\phi(\mathbf{k}, t)$ of a pure liquid on time and on translational diffusion. The calculated values of $\phi(\mathbf{k}, t)$ are plotted against time for several different values of the translational parameter, p'. The value of the wavevector, $k\sigma$, is 6.2. The values of the static dielectric constant and the reduced density are the same as in Fig. 1. (From Ref. 29.)

curious that such a coupling between spatial density fluctuation and orientational fluctuation is present at the nonlinear level. It will be of considerable interest to obtain a quantitative measure of the influence of this term on orientational relaxation.

Finally, we summarize the main features of the Markovian theories of collective orientational relaxation in a dipolar liquid. The dipolar correlation function, $\phi(\mathbf{k}, t)$, is biexponential with wavevector-dependent relaxation times. At $k = 0$, one time constant is identical with the transverse relaxation time of the liquid. The second relaxation time is the longitudinal relaxation time. For dipolar hard spheres, the contribution of the longitudinal component is small. The wavevector-dependent relaxation times show rich behavior. The longitudinal relaxation undergoes a pronounced slowing down at intermediate values of the wavevector k, if the translational contribution is not large. This slowing down of orientational relaxation at intermediate wavevectors is similar to de Gennes' narrowing of the dynamic structure factor of a dense, simple liquid [50]. In the latter case the narrowing occurs because of the slowing down in decay of density fluctuations at intermediate wavevectors (where the static structure factor is sharply peaked). Both in the present case and in the case of de Gennes' narrowing, the slowing down of relaxation at intermediate wavevectors is because of short-range correlations that are present in a dense dipolar liquid. The translational motion of dipolar molecules is found to be an important mode of orientational relaxation at intermediate wavevectors. The dipolar forces do not affect higher-order orientational correlation functions $(l > 1)$ when the following restrictive conditions are assumed. First, the molecules are assumed to be spherical and second, a linearized description of the direct correlation function is used. In general, the polar forces are expected to affect the higher-order correlations functions as well. The preliminary analysis of the nonlinear terms reveals an interesting coupling between the spatial density fluctuations and the orientational fluctuations. It is found that the a_{1m} terms are not directly coupled to themselves in the nonlinear equations. The most important coupling appears to be that between a_{1m} and a_{00}. In studying the nonlinear equations, the contributions of the triplet term have been neglected. For a system of hard spheres (no dipolar interactions, so no orientation dependence), the triplet term is known to change the density relaxation rate by only about 10% [52] if Kirkwood superposition approximation is used to evaluate c_3. The present case is, of course, more complicated because of the orientation dependence of the direct correlation function. It is straightforward to extend the present theory to treat the situation where molecules are represented by dipolar ellipsoids. We shall discuss this in Section XIII.

VI. DIELECTRIC RELAXATION:
THE MARKOVIAN THEORY

The time-dependent response of a dense dipolar liquid to an external electric field is a subject of great interest in liquid-phase chemistry. In the limits of long wavelength and of long time, the response is governed primarily by the orientational motions of the dipolar molecules. In this limit, a continuum-model-based description of the liquid polarization relaxation is often adequate [53, 54]. However, in many chemical processes of great interest, such as the solvation of a newly created charge or of an instantaneously changed dipole, the opposite limit of short time and intermediate wavevector response of the polar liquid is required [23, 27–31, 55, 56]. The continuum-model theories cannot be used to describe the collective polarization relaxation at intermediate wavevectors because the local (such as the nearest-neighbor) correlations are important in this regime. Moreover, at intermediate wavevectors, the translational modes of solvent become important [27–31]. Thus, we need to understand both the wavevector and the frequency dependence of the dielectric function. Therefore, the microscopic structure of the liquid must be included to obtain a consistent theoretical description of $\varepsilon(k, \omega)$.

Earlier, Hubbard, Kayser, and Stiles (HKS) [57, 58] and van der Zwan and Hynes (vdZH) calculated the wavevector and frequency-dependent dielectric function of a polar liquid by using a continuum theory. These authors [57–60] derived an expression for the frequency- and wavevector-dependent dielectric function, which is given by

$$\varepsilon(k, \omega) = \frac{\varepsilon_0 + \varepsilon_0 D_L \tau_D k^2 - i\omega\tau_D}{1 + D'_T \tau_D k^2 - i\omega\tau_D} \tag{6.1}$$

where ε_0 is the static dielectric constant and τ_D is the Debye relaxation time of the polar liquid. The two diffusion coefficients, D_L and D'_T, are related to the diffusion of the polarization charge and of the polarization velocity, respectively [57–60]. While in general D_L and D'_T need not be equal, both Hubbard et al. and van der Zwan and Hynes argued that they cannot differ very much. Recent studies have shown that the expression (6.1) of $\varepsilon(k, \omega)$ may be, in fact, a representation of the transverse part of the dielectric function. When the dielectric function is wavevector dependent, then we must make the distinction between the transverse and the longitudinal components [61–65] and caution is needed in the use of appropriate expressions. Pollock and Alder [61, 62] carried out interesting computer simulation studies of $\varepsilon(k, \omega)$ of a Stockmayer fluid. Neumann [64] carried out computer simulation of $\varepsilon(k)$ of dipolar hard spheres. Loring and Mukamel [65] studied the wavevector- and frequency-dependent dielectric function of the Zwanzig lattice model [66] where the material is composed of interacting point dipoles located at the sites of a cubic lattice. In addition to experiencing the electric field by all other dipoles, each dipole undergoes rotational Brownian motion as a result of interactions with a bath. The advantage of studying such an idealized model is that one can carry out detailed analysis without taking recourse to approximations. Even in this case, Loring and Mukamel could extract only limited information on $\varepsilon(k, \omega)$. Recently a microscopic theory of the dielectric relaxation in a dense dipolar liquid was presented [67, 68, 44]. A Markovian theory similar to the one described in Section V was used and the closed-form expression for the frequency- and wavevector-dependent dielectric function was obtained. It was shown that Eq. (6.1) is not appropriate at intermediate values of k because it is based on a continuum model that ignores the details of interparticle interactions that are important in a dense polar liquid. Another important result of this study is that it includes the effects of the translational modes [67]. The study of Chandra and Bagchi [67, 68] have revealed many new features of the wavevector- and frequency-dependent dielectric function. An especially interesting result is the negative value of the static longitudinal dielectric constant at intermediate values of the wavevector. This result, however, does not violate the requirements of system stability [69]; Kramers–Kronig relation only forbids a value of $\varepsilon(k)$ between zero and unity, which will be discussed.

Formal expressions for the longitudinal dielectric function, $\varepsilon_L(k, \omega)$, and the transverse dielectric function, $\varepsilon_T(k, \omega)$, can be obtained by using the linear response theory [61, 62]. We consider a single-component system and we define the collective dipole moment of the dipolar liquid by

$$\mathbf{M}(\mathbf{k}, t) = \sum_j \boldsymbol{\mu}_j(t) \exp\left[i\mathbf{k} \cdot \mathbf{r}_j(t)\right] \tag{6.2}$$

where the sum is over all the molecules of the system. $\boldsymbol{\mu}_j$ is the dipole moment vector and $\mathbf{r}_j$ is the position vector of the jth molecule. It can be easily shown that the collective dipole moment is equal to the collective polarization fluctuation

$$\mathbf{M}(\mathbf{k}, t) = \delta \mathbf{P}(\mathbf{k}, t) \tag{6.3}$$

$\delta \mathbf{P}(\mathbf{k}, t)$ is defined by

$$\delta \mathbf{P}(\mathbf{k}, t) = \mu \int d\boldsymbol{\Omega} \dot{a}(\boldsymbol{\Omega}) \delta\rho(\mathbf{k}, \boldsymbol{\Omega}, t) \tag{6.4}$$

where $\dot{a}(\boldsymbol{\Omega})$ is a unit vector defined by angle $\boldsymbol{\Omega}$ and $\delta\rho(\mathbf{k}, \boldsymbol{\Omega}, t)$ is the fluctuation in density. Next, a correlation function (a tensor) for the collective dipole moment $\mathbf{M}(\mathbf{k}, t)$ is defined by the following relation

$$\mathbf{C}_M(\mathbf{k}, t) = \langle \mathbf{M}(-\mathbf{k})\mathbf{M}(\mathbf{k}, t) \rangle$$
$$= \langle \delta \mathbf{P}(-\mathbf{k})\delta \mathbf{P}(\mathbf{k}, t) \rangle \tag{6.5}$$

where $\mathbf{C}_M(\mathbf{k}, t)$ is a tensor containing both transverse and longitudinal components; $\langle \cdots \rangle$ denotes averaging over an equilibrium ensemble. Now, the linear response theory provides the following expression for the polarizability tensor $\boldsymbol{\alpha}(\mathbf{k}, \omega)$ [61–65]:

$$\boldsymbol{\alpha}(\mathbf{k}, \omega) = \frac{\beta}{V} \int d\omega e^{-i\omega t} \left[-\frac{d\mathbf{C}_M(\mathbf{k}, t)}{dt} \right] \tag{6.6}$$

where V is the total volume of the system. The longitudinal and the transverse components of the dielectric function are related to the longitudinal and transverse components of the polarizability tensor by the following relations:

$$\varepsilon_L(\mathbf{k}, \omega) = [1 - 4\pi\alpha_L(\mathbf{k}, \omega)]^{-1} \tag{6.7a}$$

$$\varepsilon_T(\mathbf{k}, \omega) = 1 + 4\pi\alpha_T(\mathbf{k}, \omega) \tag{6.7b}$$

where

$$\alpha_L(\mathbf{k}, \omega) = \frac{\beta}{V}[C_{ML}(\mathbf{k}, 0) - i\omega\tilde{C}_{ML}(\mathbf{k}, \omega)] \tag{6.8a}$$

$$\alpha_T(\mathbf{k}, \omega) = \frac{\beta}{V}[C_{MT}(\mathbf{k}, 0) - i\omega\tilde{C}_{MT}(\mathbf{k}, \omega)] \tag{6.8b}$$

with

$$\tilde{C}_{ML}(\mathbf{k}, \omega) = \hat{k} \cdot \tilde{\mathbf{C}}_M(\mathbf{k}, \omega) \cdot \hat{k} \tag{6.9a}$$

$$\tilde{C}_{MT}(\mathbf{k}, \omega) = \hat{u} \cdot \tilde{\mathbf{C}}_M(\mathbf{k}, \omega) \cdot \hat{u}, \tag{6.9b}$$

where $\hat{k}$ is a unit vector parallel to $\mathbf{k}$ and $\hat{u}$ is a unit vector orthogonal to $\mathbf{k}$ and $\tilde{\mathbf{C}}_M(\mathbf{k}, \omega)$ is the Laplace transform of $\mathbf{C}_M(\mathbf{k}, t)$.

Equation (6.4) relates the polarization fluctuation in a dipolar liquid to the fluctuations in the position- and orientation-dependent number density, $\delta\rho(\mathbf{k}, \boldsymbol{\omega}, t)$. A reliable description of the dynamics of density fluctuations is given by the modified Smoluchowski equation discussed in Section 5 [Eq. (5.12)].

In this section we shall be concerned mostly with molecules of spherical shape; the formal expressions are equally applicable to linear rigid molecules. We do not consider less symmetric molecules here, although the formalism discussed here can be easily extended to this case as well. By using Eq. (6.4) in Eq. (5.12), one obtains the following equation of motion for the polarization fluctuation:

$$\frac{\partial}{\partial t}\delta\mathbf{P}(\mathbf{k}, t) = -(2D_R + D_Tk^2)\delta\mathbf{P}(\mathbf{k}, t) + \tfrac{1}{3}\rho_0(2D_R + D_Tk^2)\mathbf{C}(k) \cdot \delta\mathbf{P}(\mathbf{k}, t), \tag{6.10}$$

where $\mathbf{C}(k)$ is the Fourier transform of the correlation tensor $C(\mathbf{r}, \mathbf{r}')$, which is defined by [6, 13]

$$C(\mathbf{r} - \mathbf{r}', \boldsymbol{\Omega}, \boldsymbol{\Omega}') = C_{\text{iso}}(\mathbf{r} - \mathbf{r}') + \hat{\alpha}(\boldsymbol{\Omega}) \cdot \mathbf{C}(\mathbf{r} - \mathbf{r}') \cdot \hat{\alpha}(\boldsymbol{\Omega}') + \cdots, \tag{6.11}$$

The longitudinal and the transverse polarization correlation functions are now given by the following expressions:

$$\langle P_L(-\mathbf{k})P_L(\mathbf{k},t)\rangle = \langle P_L(-\mathbf{k})P_L(\mathbf{k})\rangle e^{-t/\tau_L(k)} \qquad (6.12)$$

$$\langle P_T(-\mathbf{k})P_T(\mathbf{k},t)\rangle = \langle P_T(-\mathbf{k})P_T(\mathbf{k})\rangle e^{-t/\tau_T(k)} \qquad (6.13)$$

The longitudinal and the transverse relaxation times are given by Eqs. (5.22) and (5.23) with

$$\tau_L(k) = \tau_{10}(k)$$
$$\tau_T(k) = \tau_{11}(k) \qquad (6.14)$$

Now one needs the equilibrium correlation functions $\langle P_L(-\mathbf{k})P_L(\mathbf{k})\rangle$ and $\langle P_T(-\mathbf{k})P_T(\mathbf{k})\rangle$ for complete determination of the polarization correlation function. The equilibrium polarization fluctuation correlation functions are given by

$$\langle P_L(-\mathbf{k})P_L(\mathbf{k})\rangle = \tfrac{1}{3}N\{1 + \tfrac{1}{3}\rho_0(h_\Delta(k) + 2h_D(k))\} \qquad (6.15)$$

$$\langle P_T(-\mathbf{k})P_T(\mathbf{k})\rangle = \tfrac{1}{3}N\{1 + \tfrac{1}{3}\rho_0(h_\Delta(k) - h_D(k))\} \qquad (6.16)$$

where h_Δ abd h_D are the anisotropic components of the pair correlation function $h(\mathbf{k}, \mathbf{\Omega}, \mathbf{\Omega}')$, which are related to the spherical harmonic components by the following relations:

$$h_\Delta(k) + 2h_D(k) = (3/4\pi)h(110; k)$$
$$h_\Delta(k) - h_D(k) = -(3/4\pi)h(111; k) \qquad (6.16a)$$

The analytic expressions for the longitudinal and the transverse components of the dielectric function are obtained by combining the preceding equations. The expressions for $\alpha_L(\mathbf{k}, \omega)$ and $\alpha_T(\mathbf{k}, \omega)$ are given by

$$\alpha_L(\mathbf{k}, \omega) = (\beta/V)\langle P_L(-\mathbf{k})P_L(\mathbf{k})](1 + i\omega\tau_L(k))^{-1} \qquad (6.17)$$

$$\alpha_T(\mathbf{k}, \omega) = (\beta/V)\langle P_T(-\mathbf{k})P_T(\mathbf{k})\rangle(1 + i\omega\tau_T(k))^{-1} \qquad (6.18)$$

If Eqs. (6.17) and (6.18) are substituted into Eqs. (6.7), then one obtains the following two expressions for $\varepsilon_L(k, \omega)$ and $\varepsilon_T(k, \omega)$:

$$\varepsilon_L(k, \omega) = \frac{(1-a) + \omega^2\tau_L^2(k)}{(1-a)^2 + \omega^2\tau_L^2(k)} - i\frac{\omega\tau_L(k) - (1-a)\omega\tau_L(k)}{(1-a)^2 + \omega^2\tau_L^2(k)} \qquad (6.19a)$$

and

$$\varepsilon_T(k, \omega) = \frac{(1 + b) + \omega^2 \tau_T^2(k)}{1 + \omega^2 \tau_T^2(k)} - i\frac{(1 + b)\omega\tau_T(k) - \omega\tau_T(k)}{1 + \omega^2 \tau_T^2(k)} \qquad (6.19b)$$

where

$$a = \frac{4\pi\beta\mu^2}{V}\langle P_L(-\mathbf{k})P_L(\mathbf{k})\rangle \qquad (6.20)$$

$$= 3Y[1 + \tfrac{1}{3}\rho_0(h_\Delta(k) + 2h_D(k))] \qquad (6.21)$$

and

$$b = 3Y[1 + \tfrac{1}{3}\rho_0(h_\Delta(k) - h_D(k))] \qquad (6.22)$$

where $3Y = 4\pi\beta\mu^2\rho_0/3$.

We now divide $\varepsilon_L(k, \omega)$ and $\varepsilon_T(k, \omega)$ into real and imaginary parts in the following way:

$$\varepsilon_L(k, \omega) = \varepsilon_L'(k, \omega) - i\varepsilon_L''(k, \omega)$$

and

$$\varepsilon_T(k, \omega) = \varepsilon_T'(k, \omega) - i\varepsilon_T''(k, \omega)$$

Then, Eqs. (6.19a) and (6.19b) give the analytic expressions for the real and imaginary parts of $\varepsilon_L(k, \omega)$ and $\varepsilon_T(k, \omega)$, respectively. Equations (6.19) can also be written in the following form:

$$\varepsilon_L(k, \omega) = 1 + \frac{\varepsilon_L(k) - 1}{1 + i\omega\tau_L(k)\varepsilon_L(k)} \qquad (6.23a)$$

$$\varepsilon_T(k, \omega) = 1 + \frac{\varepsilon_T(k) - 1}{1 + i\omega\tau_T(k)} \qquad (6.23b)$$

Thus we see that in this Markovian theory the dispersion of both the transverse and the longitudinal dielectric functions are of simple Debye form. Now we discuss different limiting behaviors of $\varepsilon_L(k)$ and $\varepsilon_T(k)$. We will consider three limiting cases: (a) $k \to 0$, $\omega \to 0$; (b) $k \to 0$, ω is fixed; and (c) k is fixed, $\omega \to 0$.

A. $k \to 0$, $\omega \to 0$

In this case, we start with Kirkwood's exact expression of the static dielectric constant ε_0 in terms of the pair correlation function $h(\mathbf{r}, \Omega, \Omega')$. This expression is given by [70]

$$\frac{(\varepsilon_0 - 1)(2\varepsilon_0 + 1)}{9\varepsilon_0 Y} = 1 + \rho_0 \int d\mathbf{r} \langle h(\mathbf{r}, \mathbf{\Omega}, \mathbf{\Omega}') \cos \gamma \rangle_{\Omega\Omega'} \tag{6.24}$$

where γ is the angle between two particles separated by a vector $\mathbf{r}$. Relation (6.24) can be expressed in terms of the k-space anisotropy function $h_\Delta(k)$ [13], for $k = 0$,

$$\frac{(\varepsilon_0 - 1)(2\varepsilon_0 + 1)}{9\varepsilon_0 Y} = 1 + \tfrac{1}{3}\rho_0 h_\Delta(k = 0). \tag{6.25}$$

Another route to ε_0 from the correlation function $g(\mathbf{r}, \mathbf{\Omega}, \mathbf{\Omega}')$ is based on the formally exact asymptotic relation for large $\mathbf{r}$,

$$h(\mathbf{r}, \mathbf{\Omega}, \mathbf{\Omega}') = \frac{(\varepsilon_0 - 1)^2}{3Y} [-\beta U_a(\mathbf{r}, \mathbf{\Omega}, \mathbf{\Omega}')]/\varepsilon_0 \tag{6.26}$$

Equation (6.26) follows from the formally exact relation

$$c(\mathbf{r}, \mathbf{\Omega}, \mathbf{\Omega}') = -\beta U_a(\mathbf{r}, \mathbf{\Omega}, \mathbf{\Omega}') \tag{6.27}$$

where $U_a(\mathbf{r}, \mathbf{\Omega}, \mathbf{\Omega}')$ is the asymptotic r^{-3} dipole–dipole potential. Equation (6.26) can be reexpressed in the following from [16]:

$$\frac{(\varepsilon_0 - 1)^2}{3\varepsilon_0 Y} = -\rho_0 h_D(k = 0) \tag{6.28}$$

Both Eqs. (6.25) and (6.26) independently lead to the following MSA expression for ε_0:

$$\varepsilon_0 = q_+/q_- \tag{6.29}$$

where $q_+ = q_{PY}(2\xi), q_- = q_{PY}(-\xi)$ and $q_{PY}(\xi) = (1 + 2\xi)^2/(1 - \xi)^4$, where ξ is determined by the solution of the following equation:

$$\frac{(1 + 4\xi)^2}{(1 - 2\xi)^4} - \frac{(1 - 2\xi)^2}{(1 + \xi)^4} = 3Y \tag{6.29a}$$

If we combine Eqs. (6.25) and (6.28), then we obtain the following relations:

$$3Y\{1 + \tfrac{1}{3}\rho_0[h_\Delta(0) + 2h_D0)]\} = 1 - 1/\varepsilon_0 \tag{6.30}$$

$$3Y\{1 + \tfrac{1}{3}\rho_0[h_\Delta(0) - h_D(0)]\} = \varepsilon_0 - 1, \tag{6.31}$$

Therefore, from Eqs. (6.19) and (6.20), we finally find the exact relation

$$\varepsilon_L(k=0, \omega=0) = \varepsilon_T(k=0, \omega=0) = \varepsilon_0 = q_+/q_- \qquad (6.32)$$

where the second part is the MSA expression of the static dielectric constant. Thus, we see that $\varepsilon_L(k, \omega)$ and $\varepsilon_T(k, \omega)$ have the correct behavior in the limit $k \to 0$ and $\omega \to 0$.

B. $k \to 0$, ω Fixed

In this limit, we should recover the exact relation between the collective dipole moment correlation function, $C_M(t)$ and the frequency-dependent dielectric function, $\varepsilon(\omega)$. Fatuzzo and Mason [15] showed that the auto-correlation function of the net dipole moment of a sphere embedded in a medium of the same dielectric constant is related to the frequency-dependent dielectric function by the Fattuzo–Mason relation [Eq. (3.2) of Section III].

In the present case, it follows from Eqs. (6.8a) and (6.8b) that

$$\mathcal{L}\left[-\frac{d\phi(t)}{dt} \right] = \frac{V}{\beta\phi(0)} [\alpha_L(\omega) + 2\alpha_T(\omega)] \qquad (6.33)$$

Now, from Kirkwood's formula [70] relating the mean square dipole moment to the static dielectric constant ε_0, we have the following expression:

$$\phi(0) = \langle |M(k=0)|^2 \rangle = \frac{(\varepsilon_0 - 1)(2\varepsilon_0 + 1)}{\varepsilon_0} \frac{V}{4\pi_0}. \qquad (6.34)$$

Again, from Eqs. (6.7) and (6.8), the following relation is obtained:

$$\alpha_L(\omega) + 2\alpha_T(\omega) = \frac{1}{4\pi} \frac{[\varepsilon(\omega) - 1][2\varepsilon(\omega) + 1]}{\varepsilon(\omega)} \qquad (6.35)$$

The Fattuzo–Masson relation is now obtained by combining the preceding equations. Thus, we see that the frequency- and wavevector-dependent dielectric functions, defined by Eqs. (6.7) and (6.8), correctly go over to the expression of Fatuzzo and Mason [15] as the wavevector k approaches zero at fixed ω.

C. $\omega \to 0$, k Fixed

This limit gains an additional interest in view of the result that $\varepsilon_L(k, \omega=0)$ is negative for $k\sigma \gtrsim 0.8$. This unusual result immediately raises the question: Does this contradict the requirements of system stability? The answer is a clear no. We next proceed to discuss this point of admissible sign of the static, longitudinal dielectric function. It has been pointed out recently by Kirzhnitz and co-workers [69] that for an arbitrary momentum $k \neq 0$, the dielectric function does not obey the Kramers–Kronig relation. It is $1/\varepsilon_L(k, 0)$ that obeys the Kramers–Kronig relation, which is now given by [69]

$$\frac{1}{\varepsilon_L(k, 0)} = 1 + \frac{2}{\pi} \int_0^\infty \frac{d\omega}{\omega} \operatorname{Im} \frac{1}{\varepsilon_L(k, \omega)} \tag{6.36}$$

If we take into account of the exact inequalities

$$\operatorname{Im} \varepsilon_L(k, \omega) \geqslant 0, \qquad \operatorname{Im} \frac{1}{\varepsilon_L(k, \omega)} \leqslant 0 \tag{6.37}$$

following from a direct relation of the quantity $1/\varepsilon_L(k, \omega)$ with the structure factor of the system, then we obtain inequalities of the following form:

$$\varepsilon_L(k, 0) \geqslant 1, \qquad \varepsilon_L(k, 0) \leqslant 0 \tag{6.38}$$

Thus, the causality conditions *do not preclude negative values of the static dielectric function of the system.* Only the values between 0 and 1 turn out to be forbidden. An analysis of the dynamic stability of the system with respect to the polarization fluctuations has been carried out [71]. It leads to the inequality (6.38) for the longitudinal case. For the transverse fluctuations, the requirements of system stability gives the condition $\varepsilon_T(k) \geqslant 1$. This analysis is complementary to the analysis based on Kramers–Kronig relation.

In fact, negative values of $\varepsilon_L(k)$ were encountered earlier by Hansen and McDonald [72] in their simulation of the hydrogen plasma. Dolgov et al. [69] showed that negative sign of the static, longitudinal dielectric function is valid for a rather wide class of condensed media (simple metals, nonideal plasma).

The Kramers–Kronig relation for the dielectric function itself can be written only for $k = 0$ and then we obtain the known inequality

$$\varepsilon(k = 0, \omega = 0) \geqslant 1 \tag{6.39}$$

The numerical results always obey the inequality (6.38). This lends further credence to this microscopic calculation of $\varepsilon(k, \omega)$.

The analytic expressions for dielectric functions are given by Eqs. (6.19)–(6.23). The Only unknown quantities in these equations are the h and c functions. For dipolar hard spheres, these functions are known analytically under the MSA theory. It is thus straightforward to obtain explicit numerical results.

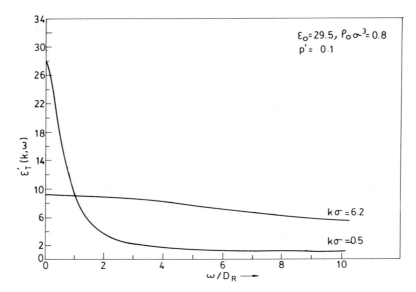

FIGURE 5. The dependence of the real part of the transverse dielectric constant, $\varepsilon_{T'}(k, \omega)$, on wavevector k and frequency ω. The calculated values of $\varepsilon'_T(k, \omega)$ (for $p' = 0.1$) are plotted against frequency, ω, for two different values of the wavevector k. The values of the static dielectric constant ε_0 and the reduced density $\rho_0 \sigma^3$ are 29.5 and 0.8, respectively. (From Ref. 67.)

In Fig. 5, the real part of the transverse dielectric function, $\varepsilon'_T(k, \omega)$ is plotted against frequency ω for two different values of the wavevector k. In Fig. 6, the imaginary part of the transverse dielectric function is plotted against frequency ω. It is seen that the relaxatons are widely different at intermediate wavevectors as compared to the relaxations at small wavevectors.

In Figs. 7 and 8, the real and the imaginary parts of the longitudinal dielectric constant are plotted. Figure 8 shows that $\varepsilon'_L(k, \omega)$ is a positive decreasing function of frequency for small values of k, the behavior is almost Lorentzian. As the values of k are increased to about unity, there an abrupt change appears in $\varepsilon'_L(k, \omega)$—it becomes negative. Note that at large values of k $\varepsilon_L(k, \omega)$ approaches unity, as frequency is increased, in a manner which is entirely different from that at small values of k. The behavior of the imaginary component, $\varepsilon''_L(k, \omega)$, however, shows nothing unusual, as is evident from Fig. 8. As the value of the wavevector increases, the maximum in $\varepsilon''_L(k, \omega)$ versus ω curve shifts to larger values of ω, indicating faster relaxation, which is clearly the effect of the translational modes that are important at the large wavevectors. Note that a similar shift in the maximum also occurred in the case of $\varepsilon''_T(k, \omega)$ (Fig. 6).

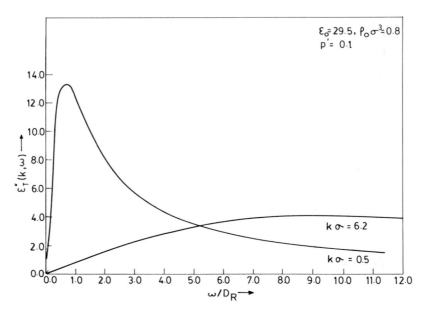

FIGURE 6. The dependence of the imaginary part of the transverse dielectric function $\varepsilon_T''(k, \omega)$, on wavevector k and frequency, ω. The calculated values of $\varepsilon_T''(k, \omega)$ (for $p' = 0.1$) are plotted against frequency, ω, for two different values of the wavevector k. The values of ε_0 and $\rho_0\sigma^3$ are the same as in Fig. 5. (From Ref. 67.)

In Fig. 9 we have shown the Cole–Cole plot for the wavevector-dependent longitudinal dielectric function $\varepsilon_L(k, \omega)$. The imaginary part of the dielectric function is plotted against the real part calculated at different frequencies. The plots are perfect semicircles, as we expect for the Debye dispersion. In Fig. 10 the Cole–Cole plot for the transverse dielectric function is shown at different wavevectors.

Figure 11 depicts the dependence of the static dielectric constants on the wavevector k. The behavior of the longitudinal component, $\varepsilon_L(k)$, is quite interesting. It increases initially from ε_0 as k is increased from zero, then it abruptly becomes negative in sign and large in magnitude. The transition occurs around $k\sigma \sim 0.8$. It is also clear from Fig. 11 that there are two branches in $\varepsilon_L(k)$. The negative value of $\varepsilon_L(k)$ was encountered earlier by Hansen and McDonald [6] in their computer simulation of the dynamics of hydrogen plasma. These authors [6] attributed the negative value to the intermolecular interactions. The negative value of $\varepsilon_L(k)$ was studied in great detail by Dolgov et al. [69], who concluded that there is nothing unusual about the negative value of $\varepsilon_L(k)$. At very large wavevectors the dielectric

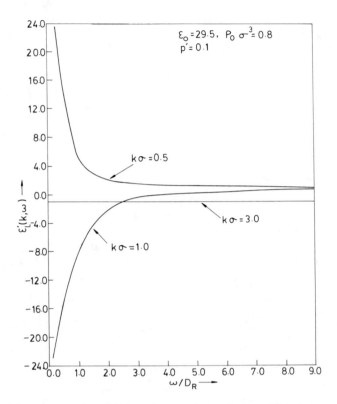

FIGURE 7. The dependence of the real part of the longitudinal dielectric function $\varepsilon'_L(k, \omega)$, on wavevector, k and frequency ω. The calculated values of $\varepsilon'_L(k, \omega)$ (for $p' = 0.1$) are plotted against frequency ω for three different values of the wavevector k. The values of ε_0 and $\rho_0\sigma^3$ are the same as in Fig. 6. (From Ref. 67.)

function should approach unity because the system cannot screen an electric field that is oscillating with a very small wavelength. Numerically, this correct asymptotic limit has not been possible to achieve because of the point dipole approximation made in the definition of the polarization vector. The correct behavior (obtained for several other systems, such as a one-component plasma) is shown in Fig. 11 by the solid line.

We now summarize the main results of this section. We have discussed the microscopic expressions for the wavevector- and frequency-dependent longitudinal and transverse dielectric constants of a pure dipolar liquid. The starting point is the linear response theory expression that relates the frequency- and wavevector-dependent dielectric function to the polarization

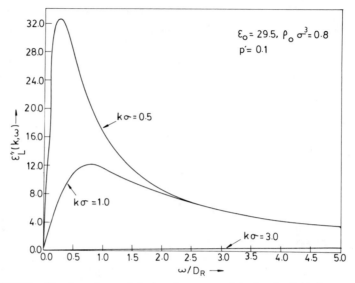

FIGURE 8. The dependence of the imaginary part of the longitudinal dielectric function $\varepsilon_L''(k, \omega)$, on wavevector k and the frequency ω. The calculated values of $\varepsilon_L''(k, \omega)$ are plotted against frequency for three different values of the wavevector k. The value of p' is 0.1. The values of the other parameters are the same as in Fig. 5. (From Ref. 67.)

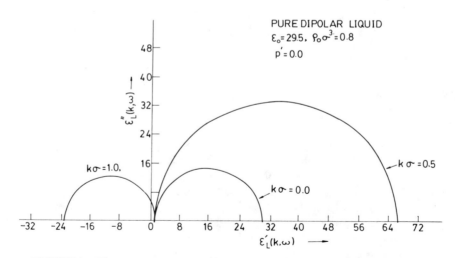

FIGURE 9. The wavevector-dependent Cole–Cole plots of $\varepsilon_L(k, \omega)$. The imaginary part of the longitudinal dielectric function is plotted against the real part for different values of the wavevector k, which are indicated on the figure. The values of the other parameters are the same as in Fig. 5. (From Ref. 28.)

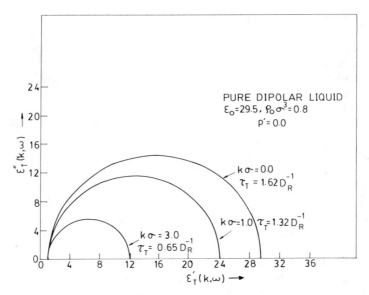

FIGURE 10. The wavevector-dependent Cole–Cole plot of the transverse dielectric function $\varepsilon_T(k, \omega)$. The imaginary part of the transverse dielectric function is plotted against the real part for different values of the wavevector, k, which are indicated on the figure. The values of the other parameters are the same as in Fig. 5. (From Ref. 28.)

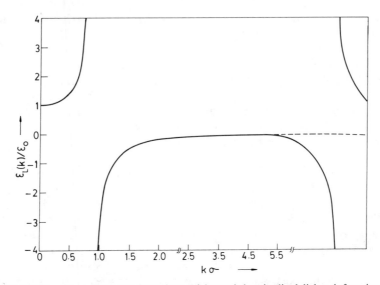

FIGURE 11. The wavevector dependence of the static longitudinal dielectric function, $\varepsilon_L(k)$. The calculated values of $\varepsilon_L(k)$ are plotted against wavevector k. The values of ε_0 and $\rho_0\sigma^3$ are 18.0 and 0.8, respectively. Note that $\varepsilon_L(k)$ is negative at intermediate wavevectors. The dashed line at large wavevectors shows the behavior obtained from the microscopic calculation based on the point dipole approximation, which breaks down at large k. (See Ref. 68 for a detailed discussion.) The correct behavior of $\varepsilon_L(k)$ at large k is shown by the solid line.

time correlation function. The Smoluchowski equation was then used to derive an equation of motion for the polarization fluctuations. This equation was solved to obtain the time dependence of the polarization correlation function, which was found to decay with wavevector-dependent time constants. In the Markovian theory discussed here, the frequency-dependent dielectric function is given by a simple Debye form. The wavevector-dependent static longitudinal dielectric function becomes negative at intermediate wavevectors. The direct experimental verification of this prediction is yet to come. The computer simulations of Hansen and McDonald and of Neumann show that the wavevector-dependent longitudinal dielectric function, at zero frequency, is negative at intermediate values. This wavevector dependence of the dielectric function has far-reaching consequences in the solvation and charge-transfer processes in dipolar liquids.

VII. INERTIAL EFFECTS IN POLARIZATION AND DIELECTRIC RELAXATION: COLLECTIVE EXCITATIONS IN DIPOLAR LIQUIDS

The theory discussed in Section VI is valid in the overdamped limit where the inertial motions of the liquid molecules are not important. However, in some cases the inertial motion of the solvent molecules can modify the short time relaxation significantly. In this section we discuss the inertial effects on the solvent polarization relaxation and on the dielectric dispersion in a dipolar liquid.

An interesting dynamic feature of a dense dipolar liquid is the possible existence of collective excitations in the polarization fluctuations, commonly known as dipolarons in analogy with the plasmons, which are well-known excitations in Coulomb systems. The possibility of such collective dipolaronic modes was pointed out first by Lobo et al. [73] who used a continuum model to study the high-frequency behavior of long-wavelength, frequency-dependent, dielectric function $\varepsilon(\omega)$. These authors extended the calculations of Nee and Zwanzig [16] to include the effects of inertial motion of the liquid molecules on $\varepsilon(\omega)$ and found well-defined dipolaronic modes arising from the competition between the reaction field of a dipolar molecule and the inertial motion. Lobo et al. proposed that the dipolarons can be of importance in the high-frequency response of dipolar liquids. Subsequently, Ascarelli [74] claimed to have observed dipolaronic behavior in the reflection spectra of a solution of nitromethane in carbon tetrachloride. Pollock and Alder [61, 62] demonstrated the existence of such dipolaronic modes in the longitudinal component of polarization relaxation by carrying out a molecular dynamics simulation of a system of dipolar particles interacting via the Stockmayer potential. Pollock and Alder found that, in contrast to the longitudinal mode, the transverse mode was strongly damped. As was pointed out by Lobo et al. [73], dipolaronic modes may play a role in the observed infrared resonances in several common dipolar liquids.

Recently, a microscopic theoretical study of the collective polarization fluctuation in dipolar liquids, with the emphasis on the intermediate wavelength behavior, was presented [75]. This work was also based on an extended hydrodynamic description of the density relaxation discussed in earlier chapters. In partial agreement with Pollock and Alder [61, 62] and with Lobo et al. [73], it was found that the dipolaronic modes are important for certain values of the relevant parameters. In this study, the scheme of Zwanzig for defining and identifying the collective dipolaronic modes (or elementary excitations) was used. This scheme is based on the idea that a collective mode should be nearly periodic in time. So, if $A(t)$ is the dynamical variable of interest, then the condition for $A(t)$ to be a good collective mode is given approximately by

$$\frac{d}{dt} A(t) \simeq i\omega A(t) \tag{7.1}$$

where ω is some appropriate frequency. Equation (7.1) actually implies that the real (or the damping) part of the time derivative of A is much smaller than ω. As pointed out by Zwanzig [76, 77], a good collective mode ought to be an approximate eigenfunction of the Liouville operator (L), and its frequency ought to be the corresponding eigenvalue:

$$LA(t) \simeq \omega A(t) \tag{7.2}$$

In their interesting computer simulation study on collective excitation in a Stockmayer liquid, Pollock and Alder [62] pointed out that a maximum in $\mathrm{Im}\,[1/\varepsilon_L(k, \omega)]$ can indicate the existence of a dipolaron, where $\varepsilon_L(k, \omega)$ is the frequency- and wavevector-dependent longitudinal dielectric function of the liquid. A similar conclusion holds for the transverse component also. A well-defined dipolaron peak in $\mathrm{Im}\,[1/\varepsilon_L(k, \omega)]$ was indeed observed by Pollock and Alder [62] in their computer simulation study of Stockmayer fluid. Linear response theory provides the relationship between the dielectric functions $\varepsilon_L(k, \omega)$ and $\varepsilon_T(k, \omega)$ and the polarization correlation functions [61], as discussed in Section VI. The polarization relaxation is, in turn, related to the density relaxation by Eq. (6.4). Because of Eq. (6.4), both the polarization correlation functions and the dielectric functions are determined essentially by the density–density time correlation function. Naturally, the rotational and the translational motions of the liquid molecules play important roles in the time dependence of these correlation functions.

Both Lobo et al. [73] and Pollock and Alder [61, 62] considered dipolarons as the natural counterpart in dipolar liquids of the plasmons, which are the elementary excitations in Coulomb systems. In a one component system, the plasma frequency is given by

$$\omega_p^2 \approx 4\pi\rho q^2/m \qquad (7.3)$$

where ρ is the density and q and m are, respectively, the charge and the mass of the plasma particles. The frequency of the collective oscillation, the so-called plasmon mode, is given by [6]

$$\omega(k) = \omega_p \left(1 + \tfrac{1}{2} \frac{C_s^2 k^2}{\omega_p^2} \right) \qquad (7.4)$$

where C_s is the adiabatic sound velocity. Equation (7.4) shows that the frequency of the collective oscillation does not vanish in the $k \to 0$ limit, but tends to the value ω_p, in marked constrast to the hydrodynamic modes. This situation is a hallmark of the plasmon mode and is a consequence of the long-range nature of the Coulomb potential. Since the dipolar interaction potential is of shorter range than the Coulomb potential, the existence of a plasmonlike mode is not obvious *a priori*. However, by using a non-Markovian theory, Lobo et al. found the existence of dipolaronic modes in the dipolar liquid. The dipolaronic frequency is approximately given by

$$\omega_d^2 \approx 4\pi\rho_0\mu^2/(I^*\varepsilon_\infty) \qquad (7.5)$$

where ρ_0 is the density of dipolar molecules and I^* is the relevant moment of inertia. This result is to be compared to the result for the plasmon [Eq. (7.4)]. Madden and Kivelson [4] have discussed the importance of the dipolarons and have concluded that these modes may not be important in the polar liquid dynamics, in contradiction to the prediction of Lobo et al. Recently, a microscopic study on collective excitations was presented [75] that is in fair agreement with the conclusions of Madden and Kivelson. We next proceed to discuss the microscopic treatment developed recently.

The microscopic study is based on the observation that the polarization vector is related to the density by Eq. (6.4). Thus, the starting equation is the continuity equation for the density [Eq. (4.1)]. The relaxation of spatial (T) and angular (Ω) momenta are assumed to be given by the following two equations:

$$\frac{\partial}{\partial t} g_T(\mathbf{r}, \mathbf{\Omega}, t) = -\rho(\mathbf{r}, \mathbf{\Omega}, t)\nabla_T \frac{\delta F'[\rho(t)]}{\delta \rho(\mathbf{r}, \mathbf{\Omega}, t)} - \int dt' \Gamma_T(t - t') \frac{\delta F}{\delta g_T} \qquad (7.6)$$

$$\frac{\partial}{\partial t} g_{\Omega}(\mathbf{r}, \mathbf{\Omega}, t) = - \rho(\mathbf{r}, \mathbf{\Omega}, t) \mathbf{V}_{\Omega} \frac{\delta F'[\rho(t)]}{\delta \rho(\mathbf{r}, \mathbf{\Omega}, t)} - \int dt' \Gamma_{\Omega}(t - t') \frac{\delta F}{\delta g_{\Omega}} \qquad (7.7)$$

The different quantities in the preceding equations have been defined in Section V. Next, the non-Markovian effects are neglected, that is, $\Gamma_i(t)$ is replaced by $\Gamma_i(t) = \Gamma_i \delta(t)$. The rationale for this (Markovian) approximation is the following. The collective excitations are expected at quite large frequencies ($\omega = 10^{12} - 10^{13} \, \text{s}^{-1}$). At such large frequencies, the contribution of the dielectric friction (which is the main contributor to the non-Markovian part in this frequency range) will be rather small compared to the Stokes friction arising from the short-range part of the interaction potential [78]. Also, the frequency is still not large enough for the viscoelastic effects to become important at the viscosities and for the liquids considered, so that one can safely assume that the Stokes friction to be independent of frequency. The microscopic treatment is different from that of Lobo et al. [73], who assumed a non-Markovian dielectric friction but neglected the mean-field force term. At the frequency range where collective excitations were obtained in Ref. 73, it is more appropriate to consider the mean-field force term and neglect the frequency dependence of the dielectric friction than doing the opposite. Additional support for the microscopic treatment comes from the observation that even a moderately sizable translational component in polarization relaxation will significantly reduce the contribution of the dielectric friction to the total rotational friction [78].

Now one uses expressions (5.8) and (7.6)–(7.7) in the continuity equation [Eq. (4.1)] to obtain the following equation of motion for the longitudinal (L) and transverse (T) parts of the polarization vector:

$$\frac{\partial}{\partial t} \mathbf{P}_i(\mathbf{k}, t) = - ABk^2 f_i(k) \int_0^t dt' P_i(\mathbf{k}, t') e^{-AB(t - t')/2p'}$$

$$- 2Bf_i(k) \int_0^t dt' P_i(\mathbf{k}, t') e^{-B(t - t')} \qquad (7.8)$$

where $i = L$ for the longitudinal and $i = T$ for the transverse component. The functions $f_L(k), f_T(k)$, and other parameters are given by

$$f_L(k) = 1 - (\rho_0/4\pi) c(110; k) \qquad (7.9a)$$

$$f_T(k) = 1 + (\rho_0/4\pi) c(111; k) \qquad (7.9b)$$

$$A = I/m\sigma^2 \qquad (7.9c)$$

$$B = k_B T / I D_R^2 \qquad (7.9d)$$

and

$$p' = D_T/2D_R\sigma^2 \qquad (7.9e)$$

As before, $c(110;k)$ and $c(111;k)$ are the (110) and (111) components of the spherical harmonic expansion of the two-particle direct correlation function and are the solvent molecular diameters. In Eq. (7.9), time (t) is expressed in unit of D_R^{-1}. It is straightforward to obtain an analytic expression of $P_i(\mathbf{k}, t)$ from Eq. (7.9), which is now Laplace transformed (with z as the complex frequency variable) to obtain the following expression for the wavevector- and frequency-dependent polarization fluctuation:

$$\tilde{P}_i(\mathbf{k}, z) = P_i(\mathbf{k}, t = 0)p(z)/q_i(z) \qquad (7.10)$$

where

$$p(z) = (z + AB/2p')(z + B) \qquad (7.10a)$$

$$q_i(z) = z^3 + z^2\{AB/2p' + B\} + z\{AB^2/2p' + ABk^2f_i(k) + 2Bf_i(k)\}$$
$$+ AB^2k^2f_i(k) + AB^2f_i(k)/p' \qquad (7.10b)$$

$\tilde{P}_i(\mathbf{k}, z)$ is the Laplace transform of $P_i(\mathbf{k}, t)$. Now, to find the three roots of the denominator, $q_i(z)$, one writes $p(z)/q_i(z)$ in terms of partial fractions and carry out the Laplace inversion to obtain the following analytic expression for $P_i(\mathbf{k}, t)$:

$$P_i(\mathbf{k}, t) = P_i(\mathbf{k}, t = 0)g_i(\mathbf{k}, t) \qquad (7.11)$$

The explicit form of $g_i(\mathbf{k}, t)$ depends on the nature of the three roots (z_{i1}, z_{i2}, and z_{i3}) of $q_i(z)$ and three different scenarios are possible. When all the three roots are distinct, $g_i(\mathbf{k}, t)$ is given by the following expression [75]:

$$g_i(\mathbf{k}, t) = \sum_j A_{ij}e^{z_{ij}t}, \qquad j = 1, 3 \qquad (7.12)$$

where

$$A_{ij} = \frac{p(z_{ij})}{[q_i(z)/(z - z_{ij})]_{z=z_{ij}}} \qquad (7.12a)$$

This happens for some values of the parameters A, B, p', and the wavevector, k. In this case, there is a smooth decay of polarization fluctuations. In the second situation, only one root may be real and the two other roots are complex (and conjugate to each other). In such a case, there will be oscillations in the decay of the polarization fluctuations. In this case, there is a possibility of dipolaronic mode provided the lifetime of the mode is large compared to the period of oscillation. The third possibility is that the roots may be degenerate. Then the form of $g_i(\mathbf{k}, t)$ will be slightly different and can be obtained by expressing $p(z)/q_i(z)$ in terms of the appropriate partial fractions. For doubly degenerate roots $(z_{i2} = z_{i3})$, the expression for $g_i(\mathbf{k}, t)$ is given by

$$g_i(\mathbf{k}, t) = A_{i1} e^{z_{i1}t} + B_{i1} e^{z_{i2}t} + B_{i2} t e^{z_{i2}t} \qquad (7.13)$$

where A_{i1} is given by Eq. (7.12a). B_{i1} and B_{i2} are given by the following equations:

$$B_{i1} = \frac{d}{dz}\left[\frac{p(z)}{q_i(z)/(z - z_{i2})^2}\right]_{z = z_{i2}} \qquad (7.13a)$$

$$B_{i2} = p(z_{i2})/[q_i(z)/(z - z_{i2})^2]_{z = z_{i2}} \qquad (7.13b)$$

Thus, one can obtain the time dependence of the polarization correlation functions and substitute them into Eqs. (6.7–6.8) to obtain the following expressions for the longitudinal and transverse dielectric functions:

$$1 - \varepsilon_L^{-1}(k, \omega) = \frac{9Y}{\rho_0\mu^2}\left\{\frac{A_{L1}}{1 + i\omega\tau_{L1}} + \frac{A_{L2}}{1 + i\omega\tau_{L2}} + \frac{A_{L3}}{1 + i\omega\tau_{L3}}\right\}\langle P_L(-\mathbf{k})P_L(\mathbf{k})\rangle$$

$$(7.14)$$

$$\varepsilon_T(k, \omega) - 1 = \frac{9Y}{\rho_0\mu^2}\left\{\frac{A_{T1}}{1 + i\omega\tau_{T1}} + \frac{A_{T2}}{1 + i\omega\tau_{T2}} + \frac{A_{T3}}{1 + i\omega\tau_{T3}}\right\}\langle P_T(-\mathbf{k})P_T(\mathbf{k})\rangle$$

$$(7.15)$$

where $\tau_{ij} = -z_{ij}^{-1}(i = L, T$ and $j = 1, 3)$ and $Y = \frac{4}{9}\pi\rho_0\beta\mu^2$. The equilibrium correlation functions are given by Eqs. (6.14) and (6.15). Thus, $\varepsilon_L(k, \omega)$ and $\varepsilon_T(k, \omega)$ are now completely determined. In the long-wavelength limit (the $k = 0$ limit), the longitudinal and the transverse components become identical with each other and the appropriate formula for $\varepsilon(\omega)$ in that limit may be obtained by putting $k = 0$ into Eq. (7.15).

Therefore, the predicted dielectric dispersion is of multiple Debye form, with three relaxation times, τ_{T1}, τ_{T2}, and τ_{T3}. For certain values of the different parameters, two roots among z_{Tj} may be complex. In that case there can be a macroscopic resonance in the dielectric dispersion. This resonance signals the appearance of collective dipolaronic oscillations in the system as the system parameters are changed. If the nature of the complex root is such that the imaginary part is much larger than the real (i.e., the decaying part) part, then we may have a true dipolaronic behavior in the system. Next we discuss some numerical results.

Numerical results on the collective excitations and on dielectric relaxation have been obtained by using the mean spherical approximation for the direct correlation function. Experimental data on moment of inertia and rotational diffusion coefficients show that for common dipolar liquids the value of the parameter B [defined by Eq. (7.16c)] is usually between 10^2 and 10^4. For that reason different values of B were used in the numerical calculations. Use of the theoretical expression for the moment of inertia of a sphere gives a value 0.1 for the parameter A. However, use of experimental results for moment of inertia and for the molecular diameter (σ) give a much smaller value for A. (This is because the real polyatomic molecules are not spherical and the mass distribution is not uniform.) So, the value of A has also been varied between 0.025 and 0.1. The former value was taken by Pollock and Alder for the parameter A [62].

In Fig. 12 the calculated decays of the longitudinal polarization fluctuations for two different values of the parameter B are shown. The value of

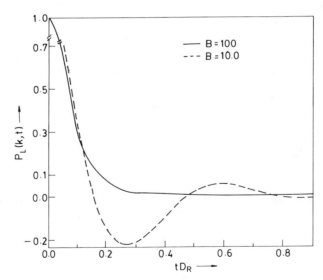

FIGURE 12. The time dependence of the normalized longitudinal polarization fluctuation, $P_L(k, t)$, is shown for different values of the inertial parameter, B. The different values of B are indicated on the figure. The value of the wavevector, k, is 1.0 and those of A and p' are 0.025 and 0.0, respectively. The values of the static dielectric constant (ε_0) and the reduced density ($\rho_0\sigma^3$) are 17.8 and 0.8, respectively. (From Ref. 75.)

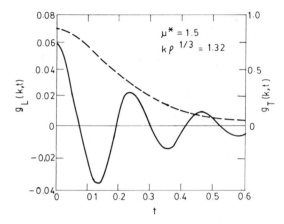

FIGURE 13.   The time dependence of the longitudinal $[g_L(k, t)]$ and the transverse $[g_T(k, t)]$ polarization fluctuations in a Stockmayer fluid at a finite wavevector where $\mathbf{g}(k, t) = (N\mu^2)^{-1}$ $\langle \mathbf{P}(\mathbf{k}, t)\mathbf{P}(-\mathbf{k}, 0)\rangle$. $g_L(k, t)$ is shown by the solid line and $g_T(k, t)$ is shown by the dashed line. The values of the reduced density and the parameter A are 0.8 and 0.025, respectively. The values of the temperature, T^* ($= 1/\beta \varepsilon_{LJ}$), and the dipole moment, μ^* [$= \mu/\varepsilon_{LJ}\sigma^3)^{1/2}$], is 1.35 and 1.5, respectively, where ε_{LJ} is the usual Lennard–Jones parameter. For these set of parameter values, the value of $3Y$ is 5.58. (From Ref. 62.)

the wavevector, $k\sigma$, is 1.0 and the value of A is 0.025. it is seen that, when B is large, the decay is smooth and the decay oscillates for small values of B. Such oscillations have been found to exist for $B < 30$ for k equal to 1.0. However, for most common dipolar liquids the value of B is in the range of 10^2–10^4.

In their computer simulation studies, Pollock and Alder found the presence of oscillations in the longitudinal polarization and a smooth decay in the transverse polarization at finite wavevectors ($k\sigma \sim 1$). Their result is shown in Fig. 13. The value of B in their work is 16.8 and the value of A is 0.025. The value of the dielectric parameter is $3Y = 5.58$. Both the longitudinal and the transverse polarizations for nearly these parameter values were calculated by using the microscopic theory [75] and the results are shown in Fig. 14. The value of p' is 0.5 and $k\sigma = 1.0$. It is clearly seen that there are oscillations in the longitudinal part, whereas the transverse polarization decays smoothly. These results are in good agreement with the computer simulation results of Pollock and Alder. However, for larger B values, which correspond to real liquids, both the longitudinal and the transverse polarization decay smoothly. Therefore, we are forced to conclude that the dipolaronic behavior observed by Pollock and Alder in their computer simulation may not be easily observed in real experiments.

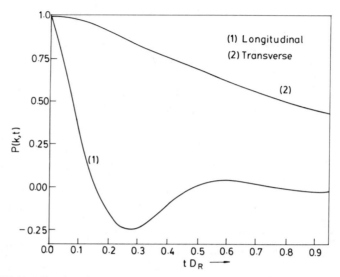

FIGURE 14. The time dependence of the normalized longitudinal (1) and transverse (2) polarization fluctuations. The values of the wavevector, k, and the parameters A and B are 1.0, 0.025, and 10.0, respectively. The value of the translational parameter, p', is 0.5. The value of the dielectric constant, ε_0, is 17.8. This value of the dielectric constant corresponds to $3Y = 6.198$, in the MSA theory. The value of the reduced density is the same as in Fig. 12. (From Ref. 75.)

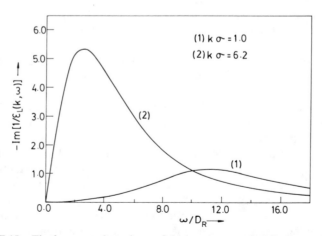

FIGURE 15. The frequency dependence of the imaginary part of the inverse longitudinal dielectric function, $\text{Im}[1/\varepsilon_L(k, \omega)]$, is shown at different wavevectors, which are indicated on the figure. The values of the parameters, p', A, and B are 0.5, 0.025, and 10.0, respectively. The values of the static dielectric constant and the reduced density are the same as in Fig. 1. (From Ref. 75.)

Figure 15 shows the frequency dependence of the imaginary part of $1/\varepsilon_L(k, \omega)$ against frequency ω for different values of the wavevector k. The value of B is 10.0. As was pointed out by Pollock and Alder, the presence of a sharp peak in the behavior of imaginary part of $1/\varepsilon_L(k, \omega)$ corresponds to a dipolaronic mode in the underlying decay of polarization fluctuations. This figure clearly shows the presence of such peaks (that is, the presence of dipolarnoic modes in the dipolar liquid). This result is in fair agreement with the computer simulation results of Pollock and Alder. Figure 16 shows the effect of polarity on the existence of dipolaronic modes. Here the imaginary part of $1/\varepsilon_L(k, \omega)$ is plotted against ω for different values of the static dielectric constant ε_0. It is seen that the sharpness of the peak is increased with an increase in the dielectric constant of the medium. This means that the dipolaronic modes are more likely to be present in strongly polar liquids. Pollock and Alder also found similar results in their computer simulation studies.

In Fig. 17, the real part of the long-wavelength dielectric function $[\varepsilon'(\omega)]$ is plotted against the imaginary part $[\varepsilon''(\omega)]$ calculated at different frequencies. The values of the parameters A and B are 0.025 and 10^3, respectively. The figure shows that the dispersion is nearly Debye with a single relaxation time. In this case the two relaxation times in Eq. (7.15) are

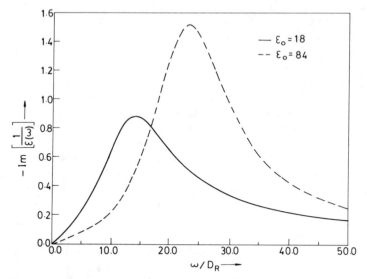

FIGURE 16.   The frequency dependence of the imaginary part of the inverse longitudinal dielectric function is shown for different values of the static dielectric constant (ε_0) in the long-wavelength limit (that is, $k = 0$). The different values of ε_0 are indicated on the figure. The values of the parameters, p', A, and B are 0.0, 0.025, and 15.0, respectively. The value of the reduced density is the same as in Fig. 12. (From Ref. 75.)

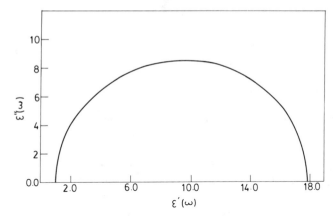

FIGURE 17.   The Cole–Cole plot of the long-wavelength frequency-dependent dielectric function. The imaginary part of the dielectric function [$\varepsilon''(\omega)$] is plotted against the real part [$\varepsilon'(\omega)$] calculated at different frequencies. The value of the wavevector, k, is zero. The values of the parameters, p', A, and B are 0.0, 0.025, and 10^3, respectively. The values of the static dielectric constant and the reduced density are the same as in Fig. 12. (From Ref. 75.)

extremely small and thus the dielectric dispersion is dominated by the slowest relaxation time and we obtain a simple Debye behavior. Figure 18 shows the Cole–Cole plot for a smaller value of the parameter B ($B = 10$). Here the dispersion is markedly different from the simple Debye form. However, such a small value of B is rather unlikely for any real liquid. No complex z_T has been found even for $B = 5$ (in the $k = 0$ limit). At about $B = 2$, Chandra and Bagchi [75] have found the existence of complex roots (and a correspondingly macroscopic resonance in the dielectric dispersion) in the transverse component.

In view of the preceding results, especially the good agreement between the microscopic theory and the simulation results of Pollock and Alder, the observation of Ascarelli [74] of a dipolaronic mode in reflection spectroscopic study of nitromethane at infrared frequencies poses a dilemma. There appear to be only two possible explanations for Ascarelli's result. First, nitromethane may be characterized by a very low value of the parameter B (that is, $B < 10$). Unfortunately, we do not have accurate values of D_T and D_R for this system to check this explanation. The second explanation is that the oscillatory modes observed are not the true collective modes. This view is further substantiated by the large width of the peaks observed in this experiment, which implies short lifetimes for these modes.

In order to check that the non-Markovian effects are indeed not important in the frequency range where dipolaronic behavior is expected, a calculation of the frequency dependence of the dielectric friction was carried out by using

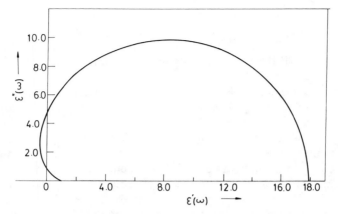

FIGURE 18. The Cole–Cole plot of the long-wavelength frequency-dependent dielectric function calculated at a small value of the inertial parameter B. The imaginary part of the dielectric function $[\varepsilon''(\omega)]$ is plotted against the real part $[\varepsilon''(\omega)]$ calculated at different frequencies. The value of the parameter B is 5.0. The values of the wavevector, k, and the parameters p' and A are 0.0, 0.0, and 0.025, respectively. The values of the static dielectric constant and the reduced density are the same as in Fig. 12. (From Ref. 75.)

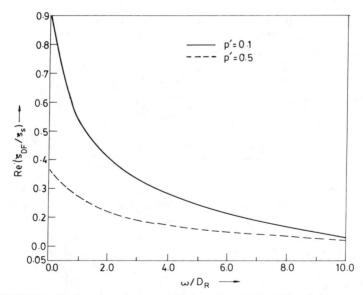

FIGURE 19. The frequency dependence of the real part of the dielectric friction calculated for two different values of the translational parameter, p'. ζ_{DF} is the frequency-dependent dielectric friction and ζ_s is the Stokes friction. The different values of p' are indicated on the figure. The values of the static dielectric constant and the reduced density are the same as in Fig. 12. (From Ref. 75.)

the method discussed in Section IX. The results are shown in Fig. 19 for two values of the translational parameter p'. The friction is seen to decrease at a faster rate at small values of frequency, but varies slowly at larger frequency. Also, the value of the dielectric friction is rather small at large values of frequency ($\omega > 10D_R$). These results indicate that the frequency dependence of friction may not play an important role in the dipolaronic behavior, which is best understood as a manifestation of inertial motion in an effective many body potential (like the density functional potential used here).

Finally, let us summarize the main points of this section. It is found that the dipolaronic modes can exist only in a dipolar liquid that is characterized by a large ε_0 and a small value of the parameter B. A sizable translational component ($p' > 0.5$) can give rise to intermediate wavenumber dipolarons at larger B values. Thus, dipolaronic behavior has been observed in the longitudinal polarization relaxation at $k\sigma = 6$ at about 80 for $p' = 1.0$. Such oscillations in the transverse mode have also been observed but at significantly smaller values of B. however, study of dipolarons at such small B values may be of academic interest only; most common dipolar liquids have a much larger B value. The difficulty in observing collective oscillations for real liquids seem to arise from two sources. First, the strength of the intermolecular potential is not sufficient to sustain the collective oscillations for a long duration. Second, the motion is largely in the overdamped limit. In fact, these two reasons are related. We suggest that in order to detect dipolarons, experiments should be carried out in liquids with large ε_0 and small B. The latter condition implies that we must work with liquids with large rotational diffusion constant and large moment of inertia. Such conditions may not be easy to satisfy, but unequivocal experimental detection of dipolarons is clearly an interesting problem.

VIII. SINGLE-PARTICLE ORIENTATIONAL DYNAMICS AND THE DIELECTRIC FRICTION

In a dense dipolar liquid, the orientation of a single dipole is coupled to the motion of the neighboring molecules. Therefore, strictly speaking, single-particle correlation also contains contributions from collective motions. However, it is still meaningful to consider the correlation functions $S_{lm}(t)$ introduced in Section II and defined by Eq. (2.10), since they can be measured experimentally in some cases. The structure of $S_{lm}(t)$ can be considerably different from $C_{lm}(\mathbf{k}, t)$. In this regard we would like to point out the assertion of Madden and Kivelson [4] that $S_{lm}(t)$ may be of same functional form as $C_{lm}(\mathbf{k}, t)$. This point will be discussed in Section X.

The study of single-particle orientational dynamics is intimately connected with the study of the rotational friction on the solute molecule. This friction should be considered frequency (or time) dependent because of the long-range nature of dipolar interactions. The frequency (or time) dependence of this friction plays an important role in determining the form of the orientational correlation function, $S_{lm}(t)$. To understand the nature of the frictional resistance to the rotational motion of a dipolar solute molecule in a dense dipolar liquid, one must consider that the total friction may have a significant contribution solely from the dipole–dipole interactions between the solute and the solvent molecules. This contribution to the total friction is called the dielectric friction. Sometimes, the magnitude of dielectric friction may even be comparable to the part arising from short-range interactions of the solute with the solvent molecules. In recent years, several theoretical [15, 16, 79, 80] and experimental [81, 82] studies were devoted to understand the effects of dielectric friction on molecular orientation.

In one of the early studies, Nee and Zwanzig [16] presented a clear exposition of the concept of dielectric friction. These authors developed a continuum-model-based theory for a slowly rotating dipole where the source of the dielectric friction is the lag of the reaction field in following the rotation of the tagged dipole. Hubbard and Wolynes [79] generalized the theory of Nee and Zwanzig in several directions and obtained interesting predictions about the dielectric friction on a dipolar solute. If the total friction can be decomposed as a sum of friction from the short-range interactions (ζ_s) and from the dipolar interactions (ζ_{DF})

$$\zeta = \zeta_s + \zeta_{DF} \tag{8.1}$$

then Hubbard and Wolynes showed that in the limit of slow solute reorientation, the dipolar contribution is given by the following expression:

$$\frac{\zeta_{DF}}{\zeta_s} = \frac{6D_R^s \mu^2(\varepsilon_0 - 1)}{k_B T a^3 (2\varepsilon_0 + 1)^2} \tau_D \tag{8.2}$$

where D_R^s, μ, and a are, respectively, the rotational diffusion coefficient, the dipole moment, and the radius of the solute molecule, and ε_0 and τ_D are, respectively, the static dielectric constant and Debye relaxation time of the solvent. Equation (8.2) is valid for a nonpolarizable liquid. Hubbard and Wolynes [79] also obtained a result for the general orientational correlation function $S_{lm}(t) = \langle Y_{lm}(\mathbf{\Omega}(0)) Y_{lm}(\mathbf{\Omega}(t)) \rangle$ and showed that these correlation functions are, in general, nonexponential and that dielectric friction has a decreased effect for higher l correlation functions. Subsequently, Madden and Kivelson [80] presented a molecular theory of dielectric friction starting from the Mori continued fraction [83]. This theory included the effects of molecular translations and anisotropic relaxation of the polarization induced by the rotating dipole in the surrounding medium. The theory of Madden and Kivelson [80] constitutes an important improvement over the previous continuum-model-based theories. On the experimental side, Kenney-Wallace and co-workers [81, 82] have used picosecond and femtosecond laser spectroscopy to probe the role of dielectric friction on reorientation dynamics of polar dye molecules in pure and binary dipolar liquids.

A consistent microscopic theory for the orientational motion of a single-particle embedded in a dipolar liquid has been presented recently [78]. This theory is similar in spirit to that of Hubbard and Wolynes [79] in that both are based on a Smoluchowski equation; they also differ in several aspects. The most important difference is that the effects of the solvent translational modes are included in the latter treatment [78]. These translational modes can drastically reduce the magnitude of the dielectric friction.

In a dense dipolar liquid the equation of motion for solute orientational density distribution is coupled to the equation of motion for position- and orientation-dependent solvent density distribution. Such coupled equations of motion can be derived by using methods of time-dependent density functional theory [13]. The two coupled equations for the solute density $[\rho_s(\mathbf{\Omega}, t)]$ and the solvent density $[\rho_0(\mathbf{r}, \mathbf{\Omega}, t)]$ are given by [78]

$$
\frac{\partial}{\partial t} \rho_s(\mathbf{\Omega}, t) = D_R^s \nabla_{\mathbf{\Omega}}^2 \rho_s(\mathbf{\Omega}, t) - D_R^s \nabla_{\mathbf{\Omega}} \left[\rho_s(\mathbf{\Omega}, t) \nabla_{\mathbf{\Omega}} \int d\mathbf{r}' d\mathbf{\Omega}' \right.
$$

$$
\left. \times C_{so}(\mathbf{r}', \mathbf{\Omega}, \mathbf{\Omega}') \delta \rho_0(\mathbf{r}', \mathbf{\Omega}', t) \right] \tag{8.3}
$$

$$
\frac{\partial}{\partial t} \rho_0(\mathbf{r}, \mathbf{\Omega}, t) = D_R^0 \nabla_{\mathbf{\Omega}}^2 \rho_0(\mathbf{r}, \mathbf{\Omega}, t) + D_T^0 \nabla_T^2 \rho_0(\mathbf{r}, \mathbf{\Omega}, t)
$$

$$
- D_R^0 \nabla_{\mathbf{\Omega}} \cdot \left(\rho_0(\mathbf{r}, \mathbf{\Omega}, t) \nabla_{\mathbf{\Omega}} \int d\mathbf{r}' d\mathbf{\Omega}' [c(\mathbf{r}, \mathbf{\Omega}, \mathbf{r}', \mathbf{\Omega}') \delta \rho_0(\mathbf{r}', \mathbf{\Omega}', t) \right.
$$

$$
\left. + C_{so}(\mathbf{r}', \mathbf{\Omega}, \mathbf{\Omega}') \delta \rho_s(\mathbf{\Omega}', t)] \right)
$$

$$
- D_T^0 \nabla_T \cdot \left(\rho_0(\mathbf{r}, \mathbf{\Omega}, t) \nabla_T \int d\mathbf{r}' \, d\mathbf{\Omega}' [c(\mathbf{r}, \mathbf{\Omega}, \mathbf{r}', \mathbf{\Omega}') \delta \rho_0(\mathbf{r}', \mathbf{\Omega}', t) \right.
$$

$$
\left. + C_{so}(\mathbf{r}', \mathbf{\Omega}, \mathbf{\Omega}') \delta \rho_s(\mathbf{\Omega}', t)] \right) \tag{8.4}
$$

where D_R and D_T are the rotational and translational diffusion coefficients of the solvent molecules, $C_{so}(\mathbf{r}, \mathbf{\Omega}, \mathbf{\Omega}')$ is the two particle direct correlation function between the solute and solvent molecules, and $c(\mathbf{r}, \mathbf{\Omega}, \mathbf{r}', \mathbf{\Omega}')$ is the two-particle direct correlation function of the solvent. Equations (8.3) and (8.4) have the same structure as the rotational diffusion equations used by Hubbard and Wolynes [79], who used a continuum model to evaluate the torque term. Equation (8.3) gives the following molecular expression of the total torque:

$$\tau(\mathbf{\Omega}, t) = \frac{1}{\beta} \nabla_\Omega \int d\mathbf{r}' d\mathbf{\Omega}' c_{SO}(\mathbf{r}', \mathbf{\Omega}, \mathbf{\Omega}') \delta\rho_0(\mathbf{r}', \mathbf{\Omega}', t) \qquad (8.5)$$

The dielectric friction, ζ_{DF}, is given by a time integral of the correlation function of the normal component of the torque vector. The correlation function involved is $\langle(\hat{a}(\mathbf{\Omega}(0)) \times \tau(\mathbf{\Omega}, 0)) \cdot (\hat{a}(\mathbf{\Omega}(t)) \times \tau(\mathbf{\Omega}, t))\rangle$, where $\hat{a}(\mathbf{\Omega}(t))$ is a unit vector in the direction of the dipole whose orientation is given by $\mathbf{\Omega}(t)$. Therefor, the frequency-dependent dielectric friction is given by

$$\zeta_{DF}(\omega) = \beta \int_0^\infty dt \, e^{i\omega t} \langle(\hat{a}(\mathbf{\Omega}(0) \times \tau(\mathbf{\Omega}, 0)) \cdot (\hat{a}(\mathbf{\Omega}(t) \times \tau(\mathbf{\Omega}, t))\rangle \qquad (8.6a)$$

In the limit of slowly rotating solute (that is, the collective solvent dynamic is much faster than the solute reorientation), the dielectric friction is given by the following simplified expression:

$$\zeta_{DF}(\omega) = \frac{\beta}{2} \int_0^\infty dt \, e^{i\omega t} \langle \tau(\mathbf{\Omega}, 0) \cdot \tau(\mathbf{\Omega}, t)\rangle \qquad (8.6b)$$

Equation (8.6b) is obtained from Eq. (8.6a) by using the rule of vector product and assuming that for slow solute reorientation, $\hat{a}(\mathbf{\Omega}(0)) \cdot \hat{a}(\mathbf{\Omega}(t)) \simeq 1$. Substitution of Eq. (8.5) into Eq. (8.6) gives the following final expression of the dielectric friction:

$$\zeta_{DF}(\omega) = \frac{1}{2(2\pi)^4 \beta} \int dk \{c_{SO}^2(110; k) \langle |a_{10}(k)|^2 \rangle \tau_{10}(k)/(1 + i\omega\tau_{10}(k))$$
$$+ 2c_{SO}^2(111; k) \langle |a_{11}(k)|^2 \rangle \tau_{11}(k)/(1 + i\omega\tau_{11}(k)) \qquad (8.7)$$

where the solvent relaxation times τ_{10} and τ_{11} are given by Eqs. (5.22) and (5.23). $\langle |a_{1m}|^2 \rangle$ is the equilibrium correlation function of the anisotropic component of solvent density in its spherical harmonic expansion. In deriving Eq. (8.7) it has been assumed that the direct correlation functions are given by a linearized equilibrium theory. Thus, the expression (8.7) may not be valid for a strongly polar liquid. It is important to note that Eq. (8.7) incorporates contributions from all the wavevectors, and we know that solvent relaxation dynamics depend strongly on the wavevector k. This is contrast to the continuum-model-based theories, which includes solvent dynamics only at $k = 0$. Also, the dielectric friction includes contributions from both (110) and (111) components of solvent polarization relaxation.

Figure 20 illustrates the dependence of the dielectric friction on the translational diffusion of the medium. ζ_{DF} is plotted against p' ($= D_T/2D_R\sigma^2$) for two different values of the static dielectric constant ε_0. The numerical calculations are performed using the MSA solution of the direct correlation functions. It is assumed that the dipole moments and the sizes of the solute and the solvent molecules are the same. As can be seen from the figure, translational diffusion has a drastic effect on the magnitude of the dielectric friction. The effect of the solvent translational modes will decrease if the size of the solute is made larger than the solvent molecules, whereas it will increase if the change is made in the reverse direction. The frequency dependence of the dielectric function is given in Section VII.

Now we discuss the dynamics of the orientational correlation function of the solute dipole. The normalized orientational correlation function is

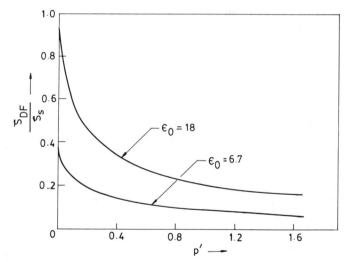

FIGURE 20. The effects of the translational diffusion on the dielectric friction on a rotating dipolar hard sphere. ζ_s is the part of friction coefficient arising from short-range interactions and is related to D_R^0 by the Stokes–Einstein relation. The dielectric friction is plotted against p' for two different values of the dielectric constant ε_0. The sizes and the dipole moments of the solute and the solvent molecules have been assumed to be the same. The reduced density of the solvent, $\rho_0\sigma^3$, is 0.8. (From Ref. 78.)

defined as

$$\phi(t) = \sum_{m=-1}^{1} S_{1m}^s(t) \bigg/ \sum_{m=-1}^{1} S_{1m}^s(t=0)$$

where

$$S_{1m}^s(t) = \langle A_{1m}(0)A_{1m}(t) \rangle$$

and $A_{1m}(t)$ is the coefficient of the solute density in its spherical harmonic expansion. The time dependence of $\phi(t)$ can be obtained from the coupled equations (8.3) and (8.4). The final expressions of the solute orientational correlation functions are as follows [78]:

$$S_{10}^s(t) = e^{-2D_R^s t} \left\{ S_{10}^s(t=0) + \frac{D_R^s}{(2\pi)^4} \int' dk \, \frac{c_{so}(110;k)c_{10}^{SO}(k)}{2D_R^s - 1/\tau_{10}(k)} \right.$$

$$\left. \times e^{-[2D_R^s - 1/\tau_{10}(k)]t} - \frac{D_R^s}{(2\pi)^4} \int' dk \, \frac{c_{so}(110;k)c_{10}^{SO}(k)}{2D_R^s - 1/\tau_{10}(k)} \right\} \qquad (8.8)$$

$$S_{11}^s(t) = e^{-2D_R^s t} \left\{ S_{11}^s(t=0) - \frac{D_R^s}{(2\pi)^4} \int' dk \, \frac{c_{so}(111;k)c_{11}^{SO}(k)}{2D_R^s - 1/\tau_{11}(k)} \right.$$

$$\left. \times e^{-[2D_R^s - 1/\tau_{11}(k)]t} + \frac{D_R^s}{(2\pi)^4} \int' dk \, \frac{c_{so}(111;k)c_{11}^{SO}(k)}{2D_R^s - 1/\tau_{11}(k)} \right\} \qquad (8.9)$$

with

$$c_{1m}^{SO}(k) = \frac{(-1)^m h^{SO}(11m;k)}{(4\pi)^2} \qquad (8.10)$$

$$S_{1m}^s(t=0) = 1/4\pi \qquad (8.11)$$

The prime in the integration of k means that $k = 0$ contribution to the integral is neglected. Finally, the required expression for $\phi(t)$ is obtained from the above equations.

Figure 21 shows the time dependence of the orientational correlation function (for $l = 1$). The decay is markedly non exponential, especially because there is a slow long-time component in the decay that is primarily because of short-range correlations between the solute and the solvent molecules. As the translational diffusion increases, the relaxation becomes faster, which is expected.

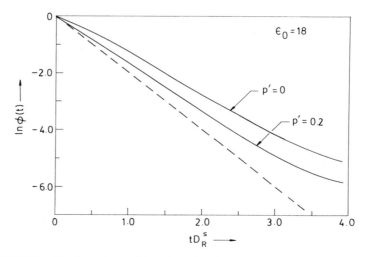

FIGURE 21. The dependence of the orientational correlation function $\phi(t)$ of the dipolar solute on time. The calculated values of $\phi(t)$ are plotted against time for two different values of the parameter p'. The value of the dielectric constant is 18.0. The results of pure rotational diffusion limit [where $\ln \phi(t) = -2D_R^s t$] is shown by the dashed line. The density of the liquid is the same as in Fig. 20. (From Ref. 78.)

Now we summarize the main points of this section. We have discussed a microscopic theory of the effect of dipolar interactions on solute reorientation dynamics in a dense dipolar liquid. The theory is limited in scope to the extent that it is based on a generalized diffusion equation. The advantage of this approach is that the effects of short-range (and long-range) dipolar correlations present in a dense liquid can be included rather accurately and self-consistently. The translational modes of solvent relaxation can drastically reduce the effect of dipolar interactions on the total friction. The dynamics of the orientational correlation function is nonexponential, in general. The translational modes significantly enhance the relaxation of this correlation function.

IX. MEMORY EFFECTS IN COLLECTIVE ORIENTATIONAL RELAXATION

Because of the long-range nature of dipolar interactions, the orientational motion of a dipolar molecule at a given time can depend on its orientation at an earlier time. In the language of the continuum model, the reaction field because of the solvent polarization lags behind the orientation of the rotating molecule. This memory of the past orientation can have a profound effect on the collective orientational relaxation of a dense dipolar liquid. The effect of this memory has been studied by several groups in the context of dielectric relaxation, some of which has been discussed in the Section VIII. In this section we shall discuss a microscopic treatment that incorporates and extends the earlier theoretical studies.

Madden and Kivelson [4] pointed out earlier that the major hurdle in understanding the problem of dielectric relaxation is a quantitative understanding of the torque–torque correlation function (TTCF). In their well-known three-variable theory, the relaxation of TTCF was included phenomenologically through a time constant. The other two time constants of the three-variable theory described the inertial motion and the orientational relaxation. The merit of the microscopic study discussed here is that it includes a theoretical calculation of the torque–torque correlation function. This calculation is a generalization of the TTCF presented in Section VIII to include non-Markovian effects.

We discussed in Section VI that a Markovian treatment of the orientational relaxation gives rise to a simple exponential decay of the orientational correlation function and a Debye form for the dielectric function $\varepsilon(\omega)$. A dense, highly polar, dipolar liquid is a system where molecules interact strongly and a Markovian theory for the polarization relaxation may not be reliable. In a non-Markovian theory an exponential relaxation of the correlation function and a Debye form for $\varepsilon(\omega)$ is not expected *a priori*. Much theoretical effort has gone into explaining non-Debye behavior in polymeric and glassy systems where non-Debye dielectric behavior is well known. The situation is rather different for dipolar liquids. Except for the detailed study of Madden and Kivelson [4, 5], no microscopic study has been carried out. Experimental results on many dense dipolar liquids also pose dilemmas. For many dipolar liquids, $\varepsilon(\omega)$ can be described rather well by a simple Debye form (or a sum of two Debye forms). For example, $\varepsilon(\omega)$ of liquid water at room temperature seems to be dominated by a relaxation time of 6.8 ps [84, 85]. This is true for many other systems [86]. While accurate data are still needed to understand the high-frequency behavior, the apparent validity of the Debye form over a large frequency range is indeed suprising and deserves further study.

As discussed in Section VIII, because of the presence of long-range orientational correlations in a dipolar liquid, each molecule experiences a dielectric friction (ζ_{DF}), in addition to the usual Stokes friction (ζ_s), which arises from the short-range interactions. Because of the long-range nature of the dipolar interaction, this dielectric friction is frequency-dependent. This, in turn, can lead to a nonexponential decay of the correlation function $C_{lm}(\mathbf{k}, t)$

and a strong non-Debye form of $\varepsilon(\omega)$, as discussed elegantly by Nee and Zwanzig [16] and also by Hubbard and Wolynes [79]. However, these treatments were based on a continuum model and neglected the effects of translational modes on $\zeta_{DF}(\omega)$. A recent non-Markovian theory [35,87] has shown that the translational modes of the liquid molecules can significantly reduce the nonexponential nature of the collective orientational relaxation and thus can give rise to a simple Debye form of the dielectric dispersion. The basic idea is that these translational modes can drastically reduce the importance of $\zeta_{DF}(\omega)$ relative to ζ_s, by giving rise to significantly faster decay of the orientational correlation. Therefore, $\varepsilon(\omega)$, which would be non-Debye in the absence of translational modes (such as in the lattice model of Zwanzig [66]), may actually show a Debye dispersion when translational modes are important.

The theory is based on the extended hydrodynamic equations with the memory terms discussed in Section IV [see Eqs. (4.1)–(4.2)]. In the treatment that we discuss here, the dissipative kernels $\Gamma_T(\mathbf{X}, \mathbf{X}'; t - t'; \rho[t])$ and $\Gamma_\Omega(\mathbf{X}, \mathbf{X}', t - t_1'; \rho[t])$ are approximated by their values at equilibrium number density, ρ_0, that is, the nonlinearities in Γ's are neglected. The non-Markovian effect is included only in the angular kernel, but neglected in the spatial part. The rational for this approximation is that one is mostly interested in the frequency range where the viscoelastic effects are still not important in the liquids usually considered (e.g., $10^{12}\,\text{s}^{-1}$ in liquid water). In the overdamped limit, it is straightforward to solve the non-Markovian hydronamic equations in the frequency domain, and the final expression of the wavevector- and frequency-dependent longitudinal and transverse orientational correlation functions, $C_{10}(\mathbf{k}, \omega)$ and $C_{11}(\mathbf{k}, \omega)$, are given by

$$C_{10}(\mathbf{k}, \omega) = C_{10}(\mathbf{k}, t = 0)\left[i\omega + (2k_B T/\zeta(\omega) + D_T k^2)(1 - (\rho_0/4\pi)c(110; k))\right]^{-1}$$

$$(9.1)$$

$$C_{11}(\mathbf{k}, \omega) = C_{11}(\mathbf{k}, t = 0)\left[i\omega + (2k_B T/\zeta(\omega) + D_T k^2)(1 + (\rho_0/4\pi)c(111; k))\right]^{-1}$$

$$(9.2)$$

The equilibrium correlation functions $C_{10}(\mathbf{k}, 0)$ and $C_{11}(\mathbf{k}, 0)$ are given by Eqs. (5.24) and (5.25). $\zeta(\omega)$ is the total rotational friction (Stokes + dielectric). If $\zeta_{DF}(\omega)$ is replaced by $\zeta_{DF}(\omega = 0)$, we recover the orientational relaxation at the Markovian limit and a Debye form for $\varepsilon(\omega)$. An expression for $\zeta_{DF}(\omega)$ can be obtained from the torque–torque correlation function. Its calculation simplifies if we consider the limit where the single-particle orientation is slower than the torque–torque correlation function. In this limit $\zeta_{DF}(\omega)$ is given by [78]

$$\zeta_{DF}(\omega) = (32\pi^4\beta)^{-1} \int dk \sum_{m=-1}^{1} \langle |a_{1m}(k)|^2 \rangle c^2(11m; k)\tau_{1m}(k, \omega)$$

$$\times [1 + i\omega\tau_{1m}(k, \omega)]^{-1} \tag{9.3}$$

where $a_{1m}(k)$ and $c(11m; k)$ have been defined previously, the transverse (τ_{11}) and longitudinal (τ_{10}) relaxation times are given by

$$\tau_{1m}^{-1}(k, \omega) = 2D_{RO}\left[\frac{k_B T}{\zeta(\omega)D_{RO}} + p'(k\sigma)^2\right]\left[1 - \frac{\rho_0}{4\pi}(-1)^m c(11m; k)\right] \tag{9.4}$$

where $p' = D_T/(2D_{RO}\sigma^2)$, D_T is the translational diffusion coefficient $D_{RO} = k_B T/\zeta_s$, and σ is the molecular diameter. p' is a relative measure of the importance of translational diffusion in the polarization relaxation of the dipolar liquid. Equations (9.3) and (9.4) are to be solved self-consistently for $\zeta_{DF}(\omega)$ and $\tau_{1m}(k, \omega)$. This was done numerically by using a nonlinear rootsearch technique (finite difference Levenberg–Marquardt algorithm) to obtain

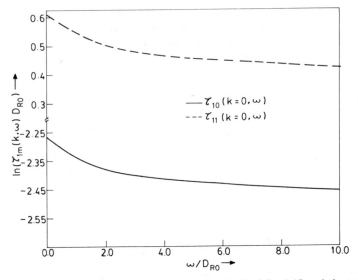

FIGURE 22. The frequency dependence of the longitudinal $[\tau_{10}(\omega)]$ and the transverse $[\tau_{11}(\omega)]$ relaxation times. The real parts of $\tau_{10}(k, \omega)$ and $\tau_{11}(k, \omega)$ are plotted against frequency at the zero wavevector limit. The value of the translational parameter p' is 0.05. The values of the static dielectric constant and the reduced density are 18.0 and 0.8, respectively.

frequency- and wavevector-dependent relaxation times. Numerical results of the frequency dependent relaxation times. are shown in Fig. 22. In this figure the real part of the longitudinal and transverse relaxation times are plotted against frequency at the long-wavelength limit. It is seen that both the relaxation times decrease with increase in the frequency. Thus, the short-time dynamics of the orientational relaxation may be quite different from that at long times.

The dielectric function can be calculated from the $k = 0$ part of the polarization correlation function. The results of the numerical calculations of $\zeta(\omega)$ is shown in Fig. 23. As usual, $\varepsilon'(\omega)$ and $\varepsilon''(\omega)$ are the real and imaginary parts of $\varepsilon(\omega)$, respectively. In the absence of any translational contribution ($p' = 0$), the graph is markedly non-Debye. It can be fitted well to a Cole–Cole equation [88] with the exponent $(1 - \alpha)$ equal to 0.37 and τ equal to $1.35 D_R^{-1}$. As the translational contribution is increased, the Cole–Cole plot approaches Debye behavior and for $p' > 0.5$, the plot is virtually indistinguishable from the simple Debye form.

The predicted strong non-Debye behavior of a dipolar liquid in the absence of any translational contribution ($p' = 0$) is somewhat surprising. It can be reconciled with the lattice calculation (with no translation) of Zwanzig [66] who showed the emergence of non-Debye behavior because of dipolar interactions. We discussed here that the translational modes, always present in a liquid but neglected in earlier calculations, can remove (or at least weaken) the non-Debye behavior predicted for $p' = 0$ [16, 66]. This may be a plausible mechanism for Debye behavior observed (over the most of the

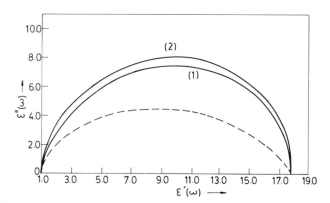

FIGURE 23. The Cole–Cole plot of the frequency-dependent dielectric function, $\varepsilon(\omega)$, where the imaginary part, $\varepsilon''(\omega)$, is ploted against the real part, $\varepsilon'(\omega)$, for several different values of the translational parameter, p'. The dashed line is for p' equal to zero. Curves 1 and 2 are for p' equal to 0.05 and 0.25, respectively. The values of the static dielectric constant and the reduced density are the same as in Fig. 22. (From Ref. 35.)

frequency domain) for many strong dipolar liquids. The values of p' for common dipolar liquids can be sufficiently large to ensure Debye dispersion. For example, approximate values of p' for water, methanol, and acetonitrile are 0.3, 0.7, and 0.1, respectively. Thus, the translational modes play an extremely important hidden role in the long-wavelength dielectric relaxation in a dipolar liquid.

X. RELATIONSHIP BETWEEN MICROSCOPIC AND MACROSCOPIC ORIENTATIONAL RELAXATIONS

The relationship between the single-particle orientational relaxation time τ_m and the many-body orientational relaxation time, τ_M, of dipolar molecules in a dense liquid has been a subject of much discussion in the recent past [1, 4, 5, 89]. Another related question is the relationship between τ_m and the dielectric relaxation time τ_D of a pure dipolar liquid [79, 90–93]. The dielectric relaxation time contains some amount of many-body effects, although the precise relations between τ_m, τ_M, and τ_D are still not clearly understood. Madden and Kivelson [4, 5] suggested that the functional forms of these two correlation functions may not be very different although the relaxation times can differ substantially. The first macro–micro relation was perhaps proposed by Debye [3, 90] who used his continuum theory of dielectric relaxation to propose the following relation:

$$\tau_m = \frac{n^2 + 2}{\varepsilon_0 + 2} \tau_D \tag{10.1}$$

where ε_0 is the static dielectric constant and n is the refractive index. The Debye form is inadequate for strongly polar solvents. Glarum [91] modified Eq. (10.1) for strongly polar liquids and his expression is given by

$$\tau_m = \frac{2\varepsilon_0 + \varepsilon_\infty}{3\varepsilon_0} \tau_D \tag{10.2a}$$

where ε_∞ is the infinite frequency dielectric constant. Equation (10.2) is based on Onsager's model of static dielectric constant, and we shall refer to this as the Onsager–Glarum expression. Equation (10.2) predicts that for large ε_0, $\tau_m = 0.67\,\tau_D$. In fact, Onsager–Glarum expression is a limiting form of a more general relation derived by Powles [92] several years earlier. This form is given by

$$\tau_m = \frac{2\varepsilon_0 + \varepsilon_\infty}{3\varepsilon_0} \frac{\tau_D}{g} \tag{10.2b}$$

where g is the well-known Kirkwood's g factor [4], which is a measure of short-range correlations in the dense dipolar liquid; these correlations are neglected in a Onsager-type theory. Thus, Powles' modification takes into account the effects of the short-range correlations. Equation (10.2a) reduces to Eq. (10.2) when $g = 1$. An analytic expression for g can be given in terms of an integration over the anisotropic part of the radial distribution function of the dipolar liquid [4, 13]. In their study on macro–micro relations, Madden and Kivelson [4] made several interesting observations. To understand their work, we need to define the macroscopic and the microscopic correlation functions $C_M(t)$ and $C_m(t)$ by

$$C_m(t) = \langle \boldsymbol{\mu}_i(0) \cdot \boldsymbol{\mu}_i(t) \rangle \tag{10.3}$$

$$C_M(t) = \left\langle \sum_i \boldsymbol{\mu}_i(0) \cdot \sum_j \boldsymbol{\mu}_j(t) \right\rangle \tag{10.4}$$

where $\boldsymbol{\mu}_i$ is the dipole moment of the ith molecule and the sum is over all the N molecules of the system and the average is over an equilibrium ensemble. The final result of the mirco–macro theorem of Kivelson and Madden is that if $C_m(t)$ can be expanded as a sum of exponentials, then $C_M(t)$ takes the same form with scaled parameters. Although there are certain approximations that restrict the validity of this theorem, it is clearly an important theorem that deserves further study. Madden and Kivelson [4, 5] have shown that τ_D and τ_m are related by the following equation:

$$\tau_m = \frac{\beta \mu^2 \rho_0}{3\varepsilon_0(\varepsilon_0 - 1)} g' \tau_D \tag{10.5}$$

where $\beta = (k_B T)^{-1}$, k_B is Boltzmann's constant and T is the temperature. μ is the magnitude of the dipole moment of liquid molecules and ρ_0 is the equilibrium density of the liquid. g' is the dynamic coupling parameter [4]. Recently it has been shown that the macro–micro theorem of Madden and Kivelson holds in a Markovian theory based on a generalized Smoluchowski equation [44]. The validity of such a kinetic equation has been discussed in Section V. This theory shows that neither Eq. (10.1) nor Eq. (10.2) provides a correct description at large ε_0, and it also shows that, although the dipolar correlation function is biexponential, the frequency-dependent dielectric function has a simple Debye form with τ_D equal to the transverse polarization relaxation time of the dipolar liquid.

The frequency-dependent dielectric function $\varepsilon(\omega)$ may be calculated by using either Fatuzzo–Mason relation [15] or by using more general linear response relations recently given by Pollock and Alder [61], discussed in

Section VI. It is also discussed in Section V that even for a system of dipolar hard spheres, $C_M(t)$ is biexponential with one component that relaxes with transverse polarization time constant (τ_{11}), which is of the order of $(2D_R)^{-1}$. The second component, however, relaxes much faster, with a time constant equal to the longitudinal relaxation time (τ_{10}). The two collective relaxation times are given by Eqs. (5.22) and (5.23) with $k = 0$. The single-particle orientational relaxation time is obtained by taking the $k \to \infty$ limit of Eq. (5.23) (with $D_T = 0$). The frequency-dependent dielectric function, however, is found to have a simple Debye form with a relaxation time equal to the transverse polarization relaxation time. Thus, at least in this model, one can define a unique Debye relaxation time that can be related to other relaxation times of the polar liquid. The dielectric relaxation time, τ_D, is equal to the transverse polarization relaxation time, τ_T, given by Eq. (6.14) with $k = 0$. Thus Eqs. (5.23) and (6.14) give the following relationship between the microscopic (τ_m) and the macroscopic (τ_D) relaxation times:

$$\tau_m = (2D_R)^{-1}$$

$$\tau_D = \{1 + (\rho_0/4\pi)c(111;0)\}^{-1}\tau_m \qquad (10.6)$$

Thus, Eq. (10.6) predicts a slowing down of the collective orientational relaxation compared to the single-particle motion with increase in liquid polarity. This feature has already been observed in computer simulations [62]. An interesting consequence of the preceding analysis is that, if one compares Eq. 10.2b with Eq. (10.6), then an analytic expression for the dynamic coupling parameter g' can be obtained in terms of the direct correlation functions. This is given by

$$g' = \frac{3}{\beta\mu^2\rho_0}\varepsilon_0(\varepsilon_0 - 1)\left\{1 + \frac{\rho_0}{4\pi}c(111;0)\right\} \qquad (10.7)$$

Since the microscopic expressions of the orientational relaxation times are now available, it is instructive to check the reliability of the continuum theoretic expressions of Debye [Eq. (10.1)] and of Glarum [Eq. (10.2)]. Figure 24 shows the dependence of Debye relaxation time on the static dielectric constant of the dipolar liquid for all the three expressions. The MSA solution is used to evaluate the direct correlation function components. At low values of the static dielectric constant, all three expressions agree with each other. As the value of ε_0 is increased, the three predictions deviate from each other. The Debye equation is known to be inaccurate at large ε_0 because it neglects the reaction field effects. The continuum-model prediction [Eq. (10.2), referred to as Onsager–Glarum expression] correctly predicts a saturation in the dependence of τ_D on ε_0, but the saturation is predicted much too early. In fact, the microscopic theory discussed here predicts that the real behavior falls in between the two continuum predictions.

In the Markovian theory discussed previously, both $C_m(t)$ and $C_M(t)$ are single exponential with different time constants. Thus, the macro–micro theorem of Madden and Kivelson is satisfied in the Markovian theory discussed previously. However, the derivation of Eq. (10.6) was based on several approximations that are rather difficult to remove satisfactorily. Thus, it will be of interest to explore the macro–micro relations in an exactly solvable system. Such a study was carried out recently for the kinetic Ising model with Glauber [94] dynamics. Although this model may not be ideal for dipolar liquid dynamics, similar models have been used previously to explain dielectric relaxation in glassy polymers [95, 96]. Also, using this kinetic model Skinner [96] could provide a nice quantitative model for the earlier work of Glarum [97] and Bordewijk [98] on defect diffusion models of dielectric relaxation. In this model, the spin–spin correlation function is highly nonexponential. The relaxation of the *collective* dipole moment is,

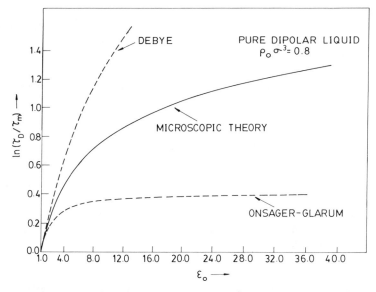

FIGURE 24. The dependence of the dielectric relaxation time, τ_D, on the static dielectric constant ε_0 of the dipolar liquid. The solid line gives the result from the microscopic theory. Predictions of Debye [Eq. (10.1)] and Onsager–Glarum [Eq. (10.2)] are also plotted for comparison. (From Ref. 44.)

however, exponential and $\varepsilon(k, \omega)$ is of simple Debye form. Thus, in this model system, the macroscopic and microscopic correlation functions are rather different. In this context, it is of interest to recollect an earlier analysis of Oxtoby [99], who also considered relations between the macroscopic and the microscopic correlation functions and showed that because of cross-correlations, a macroscopic correlation function may decay faster than a microscopic correlation function.

Glauber's kinetic Ising model consists of a one-dimensional array of spins, each having two possible orientations identified by the spin variable $\sigma = \pm 1$. The energy of the system is given by the Ising Hamiltonian

$$H = -J\sum_{ij}{}' \sigma_i \sigma_j \tag{10.8}$$

where the prime denotes a sum over the nearest neighbors. The coupling constant J is taken to be positive, indicating that the lowest energy state is the one of perfect alignment. Note that the relevant dimensionless parameter is $K = J/k_B T$. As pointed out by Budimir and Skinner [96], the Hamiltonian in Eq. (10.8) is to be regarded as a free energy when applying it to real systems, because all other degrees of freedom have been integrated out. The parameter K is a measure of the static correlation length through $\xi = 1/\ln[\coth(K)]$.

The probability of a given spin $(\sigma_1, \sigma_2, \sigma_3, \ldots, \sigma_N)$ at time t is denoted by $p(\sigma_1, \sigma_2, \sigma_3, \ldots, \sigma_N)$, where N is the number of spins in the system. The time rate of change of these probabilities is assumed to be given by the Glauber "master equation" [94].

$$\frac{d}{dt}p(\sigma_1, \sigma_2, \ldots, \sigma_N, t) = -\sum_j W_j(\sigma_j)p(\sigma_1, \sigma_2, \ldots, \sigma_N, t)$$
$$+ \sum_j W_j(-\sigma_j)p(\sigma_1, \sigma_2, \ldots, -\sigma_j, \ldots, \sigma_N, t) \tag{10.9}$$

where the sums go over all the spins. The transition probability $W_i(\sigma_i)$ is the probability per unit time of flipping the ith spin; W_i obviously depends on the spin state of the chain. The most general form of a nearest-neighbor-dependent transition probability for such a model is given by [94]

$$W_i(\sigma_i) = \tfrac{1}{2}\alpha[1 + \delta\sigma_{i-1}\sigma_{i+1} - \tfrac{1}{2}\gamma(1 + \delta)\sigma_i\{\sigma_{i-1} + \sigma_{i+1}\}] \tag{10.10}$$

where $\gamma = \tanh(2K)$. As discussed by Budimir and Skinner [96], this transition probability allows for the following processes:

$$\uparrow\downarrow\uparrow \underset{k_3}{\overset{k_1}{\rightleftarrows}} \uparrow\uparrow\uparrow$$

$$\uparrow\uparrow\downarrow \underset{k_2}{\overset{k_2}{\rightleftarrows}} \uparrow\downarrow\downarrow \qquad (10.11)$$

where $k_3 = k_1 \exp(-4K)$ to satisfy detailed balance. Note that rates k_1, k_2, and k_3 are functions of the variable α, γ, and δ. k_1 and k_2 are not determined *a priori*. Glauber [94] chose $k_1/k_2 = 1 + \tanh(2K)$. It appears that it is the only choice that leads to an analytic solution of the problem. Skinner [96] pointed out that if one makes the choice $k_1 = 0$ (which corresponds to $\delta = 0$), then the relaxation can be interpreted as a random walk of domain walls and one recovers an interactive Glarum model [97] of defect diffusion. Both Glauber and Skinner considered the dynamics at constant temperature. However, α and γ are obviously functions of temperature. We are interested in the single-particle spin–spin correlation function $[C_m(t)]$ and the collective spin–spin correlation function $[C_M(t)]$, which are defined by

$$C_m = \langle \sigma_i(0)\sigma_i(t) \rangle$$

$$C_M(k,t) = \sum_{ij} e^{ika(i-j)} \langle \sigma_i(0)\sigma_j(t) \rangle \qquad (10.12)$$

where a is the spacing of the lattice.

It was shown by Glauber [94] that for the particular choice $k_1/k_2 = 1 + \tanh(2k)$ of the transition probabilities, the correlation function $C_m(t)$ is given by

$$C_m(t) = e^{-\alpha t} \sum_{j=-\infty}^{\infty} \eta^{|j|} I_j^{(\alpha\gamma t)} \qquad (10.13)$$

where $I_k(Z)$ is the modified Bessel function of order k and argument Z. η is the short-range-order parameter, equal to $\tanh(K)$. The collective spin–spin correlation function [or the dynamic structure factor, $C_M(k,t)$] can be calculated and is given by [100]

$$C_M(t) = C_M(0) \exp[-\Gamma(k=0)t] \qquad (10.14)$$

with

$$\Gamma(k) = \alpha(1 - \gamma \cos ka) \qquad (10.15)$$

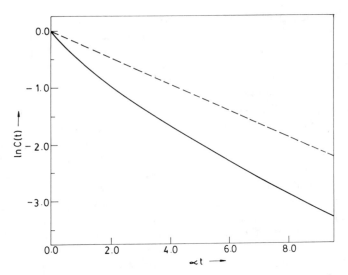

FIGURE 25.   The time dependence of the microscopic and the macroscopic correlation functions for the kinetic Ising model. The microscopic correlation function, $C_m(t)$, is shown by the solid line and the macroscopic correlation function, $C_M(t)$, is shown by the dashed line. The value of the coupling parameter, K, is 0.5. (From Ref. 100.)

Thus, the macroscopic, collective correlation function is a single exponential, whereas the microscopic, single-particle correlation function is nonexponential. These two correlation functions are compared in Fig. 25. It can be seen that the two functions are vastly different at all times. The microscopic correlation function is highly nonexponential and it decays faster than the macroscopic correlation function.

Skinner and Budimir [96] have pointed out that the correlation function $C_m(t)$ can be fitted to the well-known stretched exponential form of Kohlrausch–Williams–Watts (KWW), with exponents significantly less than unity for nonzero values of the interaction parameter K. For example, $C_m(t)$ can be fitted well to KWW form with $\beta = 0.675$ for $K = 1$. On the other hand, the macroscopic correlation function $C_M(t)$ is a single exponential for all values of K. As expected, in the limit of $K \to 0$ (no interaction between spins, the Debye limit), $C_m(t)$ and $C_M(t)$ become equal to each other. The preceding results clearly show that in the present problem, the macro–micro theorem of Madden and Kivelson [4] is not applicable.

It is now natural to ask: What happens in higher dimensions that have not been possible to solve exactly? The computer simulation studies suggest that in the two-dimensional Ising model with Glauber dynamics, $S(k = 0, t)$

is single exponential at temperatures above the critical temperature, T_c. Below T_c, there is considerable nonexponentiality in the relaxation that is partly due to the difference in the dynamical behavior between the spins in large clusters and in small and isolated clusters. Thus, the dielectric relaxation is exponential above T_c, but nonexponential below T_c.

The result obtained in the present study differs from previous results in the exact exponential behavior of $C_M(t)$. Several comments on this difference are in order. First, the difference may be a consequence of Glauber dynamics. Second, the solutions of $C_m(t)$ and $C_M(t)$ is exact. Third, the long-range nature of dipolar interactions are absent here. As discussed by Nee and Zwanzig [16], this long-range interaction gives rise to a time-varying reaction field that is responsible for the nonexponential decay of $C_M(t)$. However, the translational modes of the liquid, if important, can reduce the impact of the reaction field and return a single exponential decay.

We now summarize the main points of this section. A microscopic analysis of solvent orientational relaxation reveals an interesting relation between the microscopic and the macroscopic orientational relaxation times, in agreement with the analysis of Madden and Kivelson [4] and with the computer simulation results of Pollock and Alder [62]. However, in systems characterized by short-range interactions, the situation can be rather different, as revealed by the study of the kinetic Ising model discussed above.

XI. ROLE OF COLLECTIVE ORIENTATIONAL RELAXATION IN SOLVATION DYNAMICS AND ELECTRON TRANSFER REACTIONS

We discussed in Section III that one of the experimental probes to study the collective orientational relaxation is the study of the dynamics of solvation of a newly created charge or of an instantaneously changed dipole. This is because the process of solvation is primarily the orientational rearrangement of the solvent molecules around the polar solute particle. In many chemical reactions, the dynamics of polar solvation play a direct role. For example, the rate of a liquid-phase charge transfer reaction may be controlled by the dynamics of solvation of the changing charge distribution in the reactant and the product molecules. Thus the study of solvation dynamics is important to understand not the only solvent orientational dynamics but also many liquid-phase chemical reactions, especially the charge transfer reactions.

As we will discuss subsequently, within the linear response of the dipolar solvent, dynamic solvent effects on polar solvation and outer sphere electron transfer reaction are controlled by the collective orientational relaxation of the liquid. However, if the solvent is distorted significantly by the polar solute, then the connection with the orientational dynamics of the pure liquid is not so clear. Even in the latter situation, the collective solvent dynamics will continue to play an important role.

In this section we discuss briefly the role of collective (k-dependent) orientational relaxation in the dynamics of solvation and in the charge transfer reactions. For detailed discussions on these topics we refer to the recent reviews [23–26].

A.   Solvation Dynamics

Recent theoretical and experimental studies have led to a considerably improved understanding of the process of solvation. The early theoretical studies on solvation dynamics used a continuum model for the solvent. These continuum theories predict that if the solvent dielectric relaxation is given by a simple Debye form with the relaxation time τ_D, then the solvation time of an ion is given by [21–23]

$$\tau_{\text{ion}} = \frac{\varepsilon_\infty}{\varepsilon_0} \tau_D \tag{11.1}$$

and that of a dipole is given by

$$\tau_{\text{dipole}} = \frac{2\varepsilon_\infty + \varepsilon_c}{2\varepsilon_0 + \varepsilon_c} \tau_D \tag{11.2}$$

where ε_c is the dielectric constant of the spherical cavity that contains the solute molecule. Subsequently, these simple continuum theories were generalized in several directions to include the effects of non-Debye relaxation of the solvent, the nonspherical shape of the solute cavity, and the dielectric inhomogeneity of the solvent around the solute. These theories are phenomenological, based on macroscopic concepts and have no microscopic basis. Moreover, these continuum-model theories were found to be inadequate in explaining the recent experimental observations on the dynamics of polar solvation. One important reason for this failure is that the electric field of the solute probe solvent dynamics at all wavevectors; the continuum model includes only the long-wavelength, that is, the zero-wavevector contribution. The first microscopic theory of the dynamics of solvation was given by Wolynes [101]. He generalized the equilibrium MSA theory of ion solvation to the time domain and obtained an approximate expression of the time-dependent solvation energy. Later, Nichols and Calef [56] and Rips, Clafter and Jortner [102, 103] derived exact analytic expressions for the dynamics of dipolar and ionic solvation within the nonequilibrium MSA model of Wolynes. This nonequilibrium MSA theory of solvation dynamics is a significant improvement over the continuum-model theories. However, this theory is limited in the sense that it considers only the orientational motion of the solvent molecules and the other solvent modes (like translational, librational, etc.) of polarization relaxation are not included in this theory. Because the nonequilibrium MSA theory is not based on a kinetic equation, the results of this theory are not easily interpretable. Recently, a microscopic theory of solvation dynamics based on a kinetic equation was presented [27–31, 104–108]. This theory also includes the effects of translational modes in the dynamics of solvation. This theory reveals the intimate connection of solvation dynamics with the collective orientational relaxation.

The time dependence of the solvation energy is given by Eq. (3.10). We discussed in Section VI that the orientational correlations and translational modes play important roles in the collective polarization relaxation. Thus, it is clear that the dynamics of solvation is strongly correlated with the orientational and translational modes of the solvent. In a linear theory of the solvent polarization relaxation, the explicit expression for the time-dependent solvation energy for solvation of an ion of charge Ze is given by [31]

$$E_{solv}^{ion}(t) = \frac{(Ze)^2}{\pi\sigma} \int_0^\infty dq \left(1 - \frac{1}{\varepsilon(q)}\right) e^{-t/\tau_L(q)} \left[\frac{\sin(q(R+1)/2)}{q(R+1)/2}\right]^2 \tag{11.3}$$

and that of a dipole is given by [105]

$$E_{solv}^{dipole}(t) = \frac{2\mu^2}{3\pi\sigma^3} \left\{ \frac{8}{(R+1)^2} \int_0^\infty dq \left(1 - \frac{1}{\varepsilon(q)}\right) j_1^2\left(\frac{q}{2}(R+1)\right) e^{-t/\tau_L(q)} \right.$$
$$+ \int_0^\infty dq \, q^2 \left(1 - \frac{1}{\varepsilon(q)}\right) e^{-t/\tau_L(q)} \left[\frac{\sin q(R+1)/2}{q(R+1)/2}\right.$$
$$\left.\left. + \int_{q(R+1)/2}^\infty dx \frac{\cos x}{x}\right]^2 \right\} \tag{11.4}$$

where R is the solute–solvent size ratio equal to $2a/\sigma$ where a is the radius of the solute, $q = k\sigma$, and $j_1(z)$ is the spherical Bessel function of order 1. The other terms in Eqs. (11.3) and (11.4) have been defined previously. Note that the dynamics of solvation depends critically on the longitudinal relaxation time $\tau_L(k)$. This is the connection with orientational dynamics. It is seen from these expressions that the dynamics of solvation depend on many factors, namely, the dielectric constant of the medium, the solute–solvent size ratio, the rotational and translational diffusion coefficients, and also the microscopic structure of the liquid. This theory has been found to be quite useful in explaining many experimental results, which have been discussed in detail

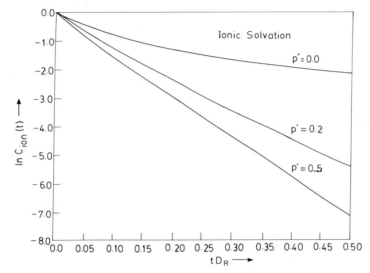

FIGURE 26.   The effects of translational modes on the time dependence of the normalized solvation energy of an ionic solute. $C_{ion}(t)$ is plotted against time for several different values of the parameter p'. The value of the solute–solvent size ratio, R, is 1.0. The values of the static dielectric constant and the reduced density are 18.0 and 0.8, respectively. (From Ref. 106.)

in a recent review [23]. Figure 26 and 27 show the role of translational diffusion on ionic and dipolar solvation dynamics, respectively. It is seen that the translational diffusion accelerates the process of solvation in both cases. However, the effect of translation is more pronounced for the dipolar case than for the ionic case. The reason is that the intermediate wavevector contributions are more important in the dipolar solvation because of the short-range nature of the dipolar field, as compared to the ionic field. In fact, Eq. (11.4) shows that the zero wavevector itself does not make any contribution to the dipolar solvation though it contributes considerably to the ionic case. A pictorial description of the role of translational diffusion mechanism in the solvation dynamics is shown in Fig. 28.

The effects of the molecular size in solvation dynamics have also been discussed in detail in a recent paper [107]. As the size of the solute molecule (as compared to that of the solvent molecules) is progressively increased, the solvation dynamics probe more and more of the small-wavevector processes. In the absence of translational diffusion these small-wavevector processes relax at a faster rate than the intermediate-wavevector processes. Thus, for zero translational diffusion, the dynamics of solvation become faster as the size of the solute is increased. This conclusion is in agreement with the

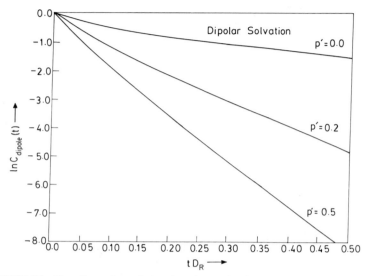

FIGURE 27.   The effects of translational modes on the time dependence of the normalized solvation energy of a dipolar solute. C_{dipole} is plotted against time for several different values of the parameter p'. The value of the solute–solvent size ratio, R, is 1.0. The values of the static dielectric constant and the reduced density are 18.0 and 0.8, respectively. (From Ref. 106.)

predictions of the nonequilibrium MSA model, which neglects the translational contribution. However, when the translational contribution is significant ($p' > 0.5$), then the small-wavevector processes relax at a slower rate than the intermediate-wavevector processes and thus the dynamics of solvation become slower as the solute size is increased. This reversal of the solute size dependence has been observed in a recent experimental study on solvation dynamics [109].

We now summarize the main points of the ionic and dipolar solvation dynamics. The dynamics is dependent on many solvent and solute parameters. It depends on the dielectric constant, the solute–solvent size ratio, the rotational and translational diffusion coefficients and also the microscopic structure of the liquid. The translational modes of the solvent molecules can enhance the rate of solvation. This translational effect also depends on the size of the solute. For a very large solute (compared to the solvent molecules), the dynamics is mainly governed by the zero-wavevector processes and thus in such situations the translational modes do not play any important role in the dynamics of solvation. The dynamics is generaly nonexponential for both ionic and dipolar solvation. However, the details of the two dynamics are widely different mainly because ionic and dipolar electric fields probe

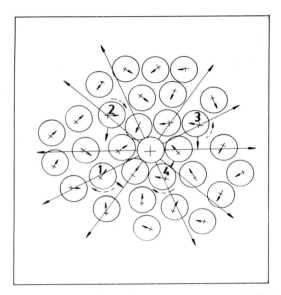

FIGURE 28. A pictorial description of the role of translational diffusion mechanism in the solvation dynamics of a positive ion, which is at the center of the figure. The solvent molecules, numbered 1, 2, 3, 4, can relax either by rotating (dashed arrows) at fixed position or by translating (solid arrows) a small amount in the required direction. The molecules close to the ion experience mostly the large-wavevector field where orientational relaxation is slow, so the tranlational relaxation mechanism is important. (From Ref. 23.)

different regimes of the wavevector-dependent solvent relaxation. Because of this, the translational modes are more important for dipolar solvation than for the ionic case. Finally, the dynamics is controlled by the solvent orientational relaxation.

B.   The Electron Transfer Reactions

The possible role of polar solvent dynamics in influencing the rate of an electron transfer reaction has been a subject of several theoretical studies in recent years [110–126]. While the majority of the theoretical predictions are yet to be confirmed, recent experimental results seem to indicate a definite role of polar solvent dynamics in determining the rate of an outer-sphere electron transfer reaction [24–26, 127–132].

The electron transfer may involve a passage over a high activation barrier or it may also be an activationless [23]. By activationless reactions we mean the cases where the barrier height is much smaller than the thermal energy. For clarity, the relevant potential energy surfaces are shown in Fig. 29. In

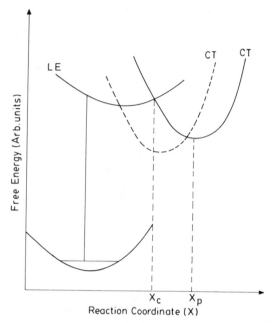

FIGURE 29. A schematic representation of the one-dimensional free energy surfaces of high barrier and barrierless reactions. The diabatic surface corresponding to the locally excited (LE) and the charge transfer state (CT) cross at a point. If the free energy of the CT state is given by the solid curve, then crossing occurs at X_c and the reaction involves a high barrier. The reaction is barrierless when the free energy of the CT state is given by the dashed curve.

both cases the dynamics of electron transfer are strongly coupled with the solvent dynamics because the transfer of electrons occurs from one site to the another through the solvent.

Most of the earlier theoretical studies on the electron transfer reactions were based on a continuum model of the solvent and thus the molecular aspects of the solvent dynamics are missing in those theories. In this section we discuss a recent molecular theory of dynamic solvent effects on both an outer-sphere high-barrier adiabatic and a barrierless (or low-barrier) electron transfer reaction, which extends the earlier theoretical studies [1–4] in several directions. This theory includes the microscopic structure and the dynamics of the solvent self-consistently. This theory also includes the effects of the translational modes of the solvent in the dynamics of electron transfer, which were neglected in the previous studies. These translational modes can significantly enhance the rate over that given by the rotational modes alone.

The appropriate reaction coordinate for an outer-sphere charge transfer reaction is given by [112]

$$X(t) = \int d\mathbf{r}[\mathbf{E}_0^P(\mathbf{r}) - \mathbf{E}_0^R(\mathbf{r})] \cdot \delta\mathbf{P}(\mathbf{r}, t) \qquad (11.5)$$

where $\mathbf{E}_0^R(\mathbf{r})$ and $\mathbf{E}_0^P(\mathbf{r})$ are the bare electric fields of the reactant and the product, respectively. So, $X(t)$ is the difference, for a given polarization, of the Coulomb potential energy of interaction of the reactant and the product solute pairs with the dipolar solvent. Equation (11.5) quantifies the importance of the collective solvent orientational dynamics in the charge transfer reaction, since the reaction coordinate itself is explicitly coupled with the solvent polarization relaxation.

For any theoretical study of a charge transfer reaction, we first have to model the reacting system. First we discuss the high-barrier adiabatic case and after that we shall discuss the barrierless (or low-barrier) case.

1. Adiabatic High-Barrier Reactions

In this case, two different models have been used to describe outer-sphere charge transfer reactions. As expected, the predictions of the two different models are qualitatively the same. In the first model, the reaction involves an isolated redox couple

$$Ox + e \rightleftharpoons Red \qquad (11.6a)$$

This model has been studied recently by McManis and Weaver [133] who used the dynamic mean spherical approximation (DMSA) of Wolynes [101] to find the dynamic solvent effects on the adiabatic barrier crossing frequency. As was pointed out by McManis and Weaver [133], this reaction can be thought of as an electrochemical exchange process, with the reactant located sufficiently far from the metal surface. The net free energy driving force for this reaction is zero, so that the free energy of activation, $\Delta G^{\neq}$, is the "intrinsic" outershell (i.e., the solvent) part of the barrier energy.

In the second model, we consider the model electron transfer reaction

$$A^{-1/2}B^{1/2} \rightleftharpoons A^{1/2}B^{-1/2} \qquad (11.6b)$$

for a solute pair AB, with $A = B$. This model has been studied recently by Zichi et al. [120], who performed molecular dynamics simulations to study the departure of electron transfer rate from the Marcus theory [134, 135] predictions. Significant deviation was found in the adiabatic limit because of dynamic solvent effects.

A simple and elegant theoretical formulation to include dynamic solvent effects on a high-barrier adiabatic electron transfer reaction has been presented recently by Hynes [113]. The merit of Hynes' formulation is that the two different limiting situations, namely, the weakly and the strongly adiabatic reactions, can be treated within the same framework. In this theory, the rate of a symmetric adiabatic electron transfer reaction is given by

$$k_{et}^{-1} = k_b^{-1} + 2k_W^{-1} \qquad (11.7)$$

where k_b is the rate of the crossing the activation barrier and k_W is the rate of solvent polarization relaxation within the potential wells. Both k_b and k_W can be expressed in terms of the reaction coordinate time correlation function $\Delta_X(t)$ defined by

$$\Delta_X(t) = \langle X(0)X(t) \rangle_i / \langle X(0)X(0) \rangle_i \qquad (11.8)$$

where the average is over the solvent degrees of freedom at equilibrium in the presence of the reaction system in the state i. The rate constant k_b is given by the well-known Grote–Hynes [136, 137] formula for a parabolic barrier top:

$$k_b = k_{GH} = k^{TST}(\lambda_X/\omega_b) \qquad (11.9)$$

where ω_b is the barrier frequency and λ_X is given by

$$\lambda_X = \frac{\omega_b^2}{\lambda_X + \zeta_X(\lambda_X)} \qquad (11.10)$$

The transition state rate, k^{TST}, is given by

$$k^{TST} = \frac{\omega_X}{2\pi} \exp[-\beta \Delta G^{\neq}] \qquad (11.11)$$

where ω_X is the frequency of the reactant harmonic well. The frequency (s) dependent friction, $\zeta_X(s)$, is related to the Laplace transform of the reaction time correlation function via generalized Langevin equation. In the Laplace plane, this relation given by

$$\Delta_X(s) = \frac{s + \zeta_X(s)}{s^2 + \omega_X^2 + s\zeta_X(s)} \qquad (11.12)$$

The reactant well frequency ω_X is given, for harmonic well, by [113, 120]

$$\omega_X = \langle \dot{X}(0)\dot{X}(0) \rangle / \langle X(0)X(0) \rangle \tag{11.13}$$

When the barrier region is cusped, then the barrier frequency ω_b may be too large for solvent dynamic effects to be important and k_b approaches k^{TST}. In the overdamped limit and for a large barrier, $k_b \ll k_W$, so k_{et} will be given by k_b alone, that is, $k_{el} \sim k^{TST}$. In this limit, Marcus' theory [134, 135] is applicable. In the opposite, strongly adiabatic limit, ω_b may be small $(\leqslant \omega_X)$ and solvent dynamic effects will be important in electron transfer.

The expression for the well relaxation rate, k_W, has been given by Hynes [113] in terms of the reaction coordinate time correlation function, $\Delta_X(t)$:

$$
\begin{aligned}
k_W^{-1} = \exp\left(\beta \Delta G^{\neq}\right) \int_0^{\infty} dt \{ [1 - \Delta_X^2(t)]^{-1/2} \\
\times \exp\left(-\beta \Delta G^{\neq}[1 - \Delta_X(t)]^2/[1 - \Delta_X^2(t)]\right) \\
- \exp\left(-\beta \Delta G^{\neq}\right) \}
\end{aligned}
\tag{11.14}
$$

Thus the calculation of the rate constant now reduces to the calculation of the reaction coordinate time correlation function. Under certain approximations (to be discussed), the reaction coordinate time coorelation function is given by

$$
\begin{aligned}
\Delta_X(t) = \int_0^{\infty} d\mathbf{k}\, E_0^2(\mathbf{k}) \langle \delta P_L(-\mathbf{k}) \delta P_L(\mathbf{k}, t) \rangle \\
\times \left[\int_0^{\infty} d\mathbf{k}\, E_0^2(\mathbf{k}) \langle \delta P_L(-\mathbf{k}) \delta P_L(\mathbf{k}) \rangle \right]^{-1}
\end{aligned}
\tag{11.15}
$$

where $E_0(\mathbf{k})$ is the Fourier transform of the bare electric field of the reacting system. In writing Eq. (11.15), two approximations have been made. The anisotropy of the liquid created because of the local distortion of the solvent by the electric field has been neglected and it has been assumed that the solvent dipoles are rigid. The general conclusions will not be significantly affected by these approximations.

Now, in a linear theory, the decay of the polarization time correlation function is exponential:

$$\langle \delta P_L(-\mathbf{k}) \delta P_L(\mathbf{k}, t) \rangle = \langle \delta P_L(-\mathbf{k}) \delta P_L(\mathbf{k}) \rangle e^{-t/\tau_L(k)} \tag{11.16}$$

where the wavevector-dependent relaxation times are given by Eq. (5.22).

The electric field in the reaction coordinate time coorelation function is that of an ion for model I and is that of a dipole for model II. Because of the short-range nature of the dipolar potential as compared to the ionic potential, the solvent translational modes will be more important in charge transfer reactions described by model II than that in model I. For complete determination of the rate constant of the electron transfer reaction, one needs the value of the barrier height, $\Delta G^{\#}$. The free energy of activation, $\Delta G^{\#}$, for model I is given by the following expression [138]:

$$\Delta G^{\#} = \frac{Ze^2}{4\pi\sigma} \int_0^\infty dq \left(1 - \frac{1}{\varepsilon_L(q)}\right) \left[\frac{\sin(q(R+1)/2)}{q(R+1)/2}\right]^2 \qquad (11.17)$$

with $q = k\sigma$. The preceding expression correctly reduces to the corresponding dielectric continuum formula for the intrinsic barrier when $\varepsilon(k)$ is approximated by ε_0 and when R is very large. For model II, the expression of the barrier height is also given by the expression for the solvation energy of a dipole [105]. The frequency factor, v, of a chemical reaction is defined by

$$k_{et} = v \exp[-\beta \Delta G^{\neq}] \qquad (11.18)$$

Now we discuss numerical results of v for both weakly adiabatic and strongly adiabatic cases [138]. In the former case, the assumed value of ω_b is, $\omega_b = 4\omega_X$ and for the latter case $\omega_b = \omega_X$; ω_X is given by Eq. (11.12). The frequency-dependent friction, $\zeta_X(s)$, has been calculated by using Eq. (11.12) in the overdamped limit, thus ignoring the acceleration term. In all the numerical calculations, the mean spherical approximation (MSA) has been used for the two-particle direct correlation function [138].

Figure 30 and 31 show the results of the calculations [138] of v for model I. In Fig. 30, the dependence of v on the translational parameter, p' in the *weakly* adiabatic limit ($\omega_b = 4\omega_X$) is shown. It is seen that the rate of electron transfer increases with increase in the value of p'. The increase in the electron transfer rate is higher for smaller solute size, since when the size of the reacting system is small, the reaction probes more of the intermediate-wavevector processes than when the size is large. The relaxation of the solvent polarization at the intermediate wavevectors becomes increasingly faster as the value of the translational parameter p' is increased. In Fig. 30, the continuum-model (which includes only the $k = 0$ mode contribution toward the solvent polarization relaxation) prediction of the rate, given by $v = (2\pi\tau_L)^{-1}$, is also shown by a dashed line. The calculated rate is less than the continuum-model prediction for $p' = 0$, but becomes greater as p' is increased.

Figure 31 shows the dependence of v on translation, but in the *strongly* adiabatic limit ($\omega_b = \omega_X$). In this limit the electron transfer occurs at a rate lower than that in the weakly adiabatic case and also the dependence on p'

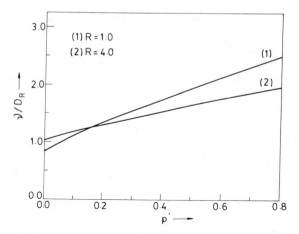

FIGURE 30.   The dependence of the barrier crossing frequency, v, of a weakly adiabatic electron transfer reaction (described by model I) on the translational diffusion of the solvent molecules. The calculated values of v are plotted against p' for two different values of the solute–solvent size ratio, R. The barrier frequency, ω_b, is equal to $4\omega_x$, where ω_x is the well frequency. The value of the quantity $\beta e^2/\sigma$ is 140. The values of the static dielectric constant and the reduced density are the same as in Fig. 20. (From Ref. 138.)

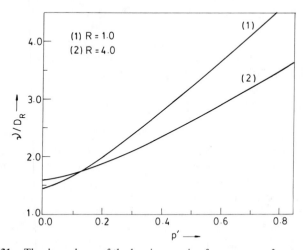

FIGURE 31.   The dependence of the barrier crossing frequency, v, of a strongly adiabatic electron transfer reaction (described by model I) on the translational diffusion of the solvent molecules. The calculated values of v are plotted against p' for two different values of the solute–solvent size ratio, R. The barrier frequency, ω_b, is equal to the well frequency ω_x. The values of the other parameters are the same as in Fig. 30. (From Ref. 138.)

is somewhat weaker. The reason is that, in the weakly adiabatic limit, the reaction probes the high-frequency components of the solvent response. At large p', translational modes populate these high-frequency components, so the dependence on p' is stronger for a weakly adiabatic reaction than that for a strongly adiabatic reaction [138]. Also, the reactive friction is less for the former than for the latter. The results of the calculations based on model II are qualitatively the same as that obtained from model I.

Now we summarize the main points of this section. We have discussed the effects of the collective solvent dynamics on the high-barrier charge transfer reactions. The collective solvent dynamics enters again through longitudinal relaxation time $\tau_L(k)$. We have discussed two somewhat different models of outer-sphere adiabatic electron transfer reactions. For both the models, the translational modes of the dipolar solvent can significantly enhance the rate of the reaction [138]. Moreover, these dynamic effects are significantly dependent on the adiabaticity of the reaction.

2. The Barrierless Electron Transfer Reactions

Here we consider an outer-sphere barrierless electron transfer reaction and discuss the role of collective orientational dynamics. Examples of barrierless electron transfer reactions are the photoinduced electron transfer reactions in many conjugated aromatic molecules. These reactions are very fast, with lifetimes of the order of few picoseconds. Experiments indicate that these reactions may also be coupled strongly to the polarization relaxation of the solvent. Initial experiments of Kossower and Huppert [131] suggested that the time constants of such barrierless reactions may be directly proportional to the longitudinal relaxation time τ_L of the solvent. More detailed experiments performed by Barbara and co-workers [139, 140] and by Su and Simon [129, 130] have shown that the dependence of the rate of electron transfer on τ_L may not be so simple. However, in all these later experiments a correlation was found between the rate of electron transfer and the rate of solvation.

The theoretical description of barrierless electron transfer reactions has been developed in several different ways. Sumi, Nadler, and Marcus [114, 115] used a modified Smoluchowski equation to describe the motion of the reactant (electron + solvent in the initial state) on a harmonic surface with a position-dependent sink, which accounts for the decay resulting from the electron transfer to the product state. This model is similar to the one used by Bagchi, Fleming, and Oxtoby [141, 143] to describe barrierless photoisomerization reactions. Alternatively, one can use the stochastic Liouville equation, as used by Rips and Jortner [117]. In this section we discuss a recently developed microscopic theory, which is based on the former approach [144].

For a microscopic calculation of the dynamic solvent effects on barrierless reactions, we consider the following simple model of an intramolecular electron transfer reaction:

$$A - B \xrightarrow{hv} [A^{\delta +} - B^{\delta -}]* \xrightarrow{ET} A^+ - B^-$$

where A and B are the two segments of the molecule and electron transfer takes place from one segment to the other. The intermediate step is the formation of a locally excited state. This locally excited state may have a small dipole moment, although the ground state may be totally neutral. The motion in the reactant well (here the locally excited state) is described by the following Smoluchowski equation for the time evolution of the probability distribution function, $P_{LE}(X, t)$, of the reaction coordinate [141, 143]:

$$\frac{\partial}{\partial t} P_{LE}(X, t) = \frac{k_B T}{\zeta_X} \frac{\partial}{\partial X} \left[\frac{\partial}{\partial X} + \frac{\partial \beta F}{\partial X} \right] P_{LE}(X, t) \tag{11.19}$$

where ζ_X is the effective friction on motion along the reaction coordinate, $\beta = (k_B T)^{-1}$, and F is the free energy of the locally excited state. It is assumed that the reactant surface is harmonic and that the free energy is given by $F = \frac{1}{2}\omega_X X^2$, ω_X being the frequency of the harmonic surface. Now the reaction near $X = X_c$ is described by a sink term. So, the modified Smoluchowski equation is given by

$$\frac{\partial}{\partial t} P_{LE}(X, t) = \frac{k_B T}{\zeta_X} \frac{\partial^2 P_{LE}(X, t)}{\partial X^2} + \frac{\omega_X^2}{\zeta_X} \frac{\partial}{\partial X} X P_{LE}(X, t) - k_0 S(X) P_{LE}(X, t) \tag{11.20}$$

where k_0 is the rate of electron transfer from the origin (where the rate is maximum for a symmetric sink) and $S(X)$ is the sink function that gives the position dependence of the rate. Sumi and Marcus [114] pointed out that $k_0 S(X)$ is determined by the nature of the coupling between the two diabatic electronic surfaces. If the coupling is strong (the adiabatic limit), then $k_0 S(X)$ may be approximated by an absorbing barrier at $X = X_c$. On the other hand, for a weak coupling (the nonadiabatic limit), $k_0 S(X)$ is a finite sink [141–143] located at $X = X_c$.

The methods of solution of Eq. (11.20) are available [141–143], and the results for several different forms of the sink function have been discussed in detail in Ref. 143. The important step now is to calculate the friction parameter ζ_X and the frequency ω_X from first principles. They can be calculated from a knowledge of the time dependence of the reaction coordinate correlation function as we have discussed. Now we discuss both adiabatic and nonadiabatic cases, separately.

a.   Adiabatic Limit. It was pointed out by Sumi and Marcus [114] that in barrierless adiabatic electron transfer, the two surfaces interact strongly to create an "avoided crossing" near $X = X_c$, so that the electron transfer occurs with unit probability when the critical value $X = X_c$ is reached. Thus, this situation corresponds to the presence of an absorbing barrier at $X = X_c$. If $X_c = 0$ (that is, zero barrier limit), then one recovers the symmetric pinhole sink model of Bagchi et al. [141–143]. For this case Eq. (11.4) can be solved exactly and the total population remaining in the reactant well after time t, $P_{LE}(t)$, is given by the following expression:

$$P_{LE}(t) = \int_0^\infty dX' [P_0(X') + P_0(-X')] \operatorname{erf} F(X', t) \qquad (11.21)$$

where

$$F(X, t) = \omega_X X [2k_B T \{1 - \exp(-2\omega_X^2 t/\zeta_X)\}]^{-1/2} \exp(-\omega_X^2 t/\zeta_X) \qquad (11.22)$$

The error function, erf a, is defined by

$$\operatorname{erf} a = \frac{2}{\sqrt{\pi}} \int_0^a dy e^{-y^2} \qquad (11.23)$$

$P_0(X)$ is the initial probability distribution in the reactant well. If $P_0(X)$ is given by the equilibrium, Boltzmann, probability distribution on the *LE* surface, then $P_{LE}(t)$ can be given by the following, simpler, expression [143]:

$$P_{LE}(t) = \frac{2}{\pi} \sin^{-1}(e^{-\omega_X^2 t/\zeta_X}) \qquad (11.24)$$

Expressions (11.21) and (11.24) show that the decay of reactant population is, in general, nonexponential. However, it can be easily seen that the long-time decay is single exponential. Therefore, one can define a long-time rate, k_{et}^L, of electron transfer that is given by $k_{et}^L = \omega_X^2/\zeta_X$. It has been shown that k_{et} is exactly the *average* rate of solvation of the newly formed charge transfer state if the solvent distortion by the reaction system is neglected. We shall return to this point subsequently. The effects of the short-range correlations and of the translational modes of the solvent are contained in both ω_X and ζ_X. The rate of electron transfer becomes faster as the translational diffusion of the solvent molecules is increased. The rate also depends on the relative sizes of the reacting and solvent molecules. This is clearly shown in Fig. 32, where the rate of electron transfer is plotted against the translational parameter p' for several different values of the solute–solvent

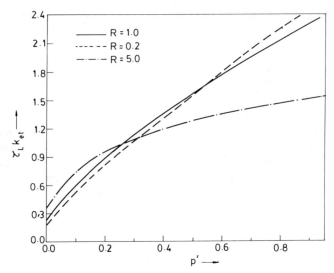

FIGURE 32. The dependence of the long-time rate of a barrierless adiabatic reaction on translational diffusion. The product of the long-time-rate of electron transfer with the longitudinal relaxation time is plotted against p' for three different values of the reactant–solvent size ratio, R. The value of the static dielectric constant is 18.0. (From Ref. 144.)

size ratio R. It is seen that the effects of translational modes become more important as the size of the reacting molecule is decreased. This is because the intermediate-wavevector processes become more important as the size of the reacting molecule is decreased.

If the avoided crossing occurs at a position such that X_c is greater than zero, so that a very small activation barrier exists for electron transfer, then it is not possible to obtain an analytical solution for the reactant well population. One can, however, solve for the average rate. Numerical calculations have shown that in this case also the collective solvent dynamics play an important role in the electron transfer, especially, the strong dependence of the translational diffusion remains unaltered.

b. Nonadiabatic Limit. In this case, the zero-order surfaces interact weakly so that both the surfaces participate in the reaction, and there is only a finite probability of electron transfer as the system arrives at the critical configuration. In this case, the form of the sink function, $S(X)$, and the magnitude of the intrinsic decay rate, k_0, play important roles in the electron transfer. One can envisage two limiting forms of $S(X)$: a narrow sink at X_c or a broad sink centered at X_c. As was pointed out by Sumi and Marcus, the form of $S(X)$ will largely by determined by the nonreacting modes of the solute, especially the vibrational modes that will usually lead to a broadening of the sink. These authors suggested that a Gaussian form can be a sensible approximation for $S(X)$. This form can be represented by

$$S(X) = \exp\left[-(X - X_c)^2/a_s^2\right] \tag{11.25}$$

Note that if one takes the simultaneous limits $k_0 \to \infty$ and (the width parameter) $a_s \to 0$, then one recovers the pinhole sink discussed previously. This suggests the possibility of a continuous transition from a nonadiabatic to an adiabatic electron transfer, mediated by the system parameters k_0 and a_s.

The analytic solution of Eq. (11.20) with the sink function given by Eq. (11.25) has not been possible. Fortunately, an extensive numerical solution of this problem has already been carried out, with $X_c = 0$ [114, 141] and with a small X_c. A scaling analysis of Eq. (11.4) suggests that the dynamics in this case will be controlled largely by the dimensionless parameter $\tilde{k}_0$, defined by

$$\tilde{k}_0 = k_0 \zeta_X/\omega_X^2 \tag{11.26}$$

In the following paragraphs, we summarize the main results for the present situation.

If the sink is narrow (what Sumi and Marcus termed "narrow window"), then an exponential decay is always obtained at sufficiently long times. The long-time rate is proportional to ω_X^2/ζ_X for large $\tilde{k}_0$ ($\tilde{k}_0 > 5$). The rate is independent of ω_X^2/ζ_X at small $\tilde{k}_0$, so that there is a region of nonlinear dependence of the electron transfer rate on ω_X^2/ζ_X. We refer to a recent review [143] for a detailed discussion of this point. The nonlinear dependence of the rate of electron transfer on ω_X^2/ζ_X also implies a nonlinear and a nontrivial dependence of rate on the long-wavelength longitudinal relaxation time τ_L.

It is clear that both the dynamics of solvation and the dynamics of electron transfer are critically dependent on the collective orientational relaxation of the dipolar solvent. Now we discuss how much the two dynamics differ. We define the average solvation time by the following expression:

$$\langle \tau_s \rangle = \int_0^\infty dt\, C_s(t) \tag{11.27}$$

where $C_s(t)$ is the normalized time-dependent solvation energy. It may be shown [144] that in a linear theory the following relationship holds between the average solvation time and the long-time rate of electron transfer:

$$\langle \tau_s \rangle^{-1} = \frac{\omega_X^2}{\zeta_X} = k_{et}^L \tag{11.28}$$

This is an important relation. It shows that the long-time rate of electron transfer is equal to the average rate of solvation. This theoretical prediction supports the recent experimental observations of Barbara et al. and of Su and Simon. However, the details of the dynamics of these two processes can still be very different from each other [144].

We now summarize the main points of this section. We have discussed a molecular theory of the effects of collective solvent dynamics on barrierless (and very-low-barrier) electron transfer reactions in dipolar solvents. The rate of electron transfer is critically dependent on solvent motions. Especially, the translational modes of the solvent can significantly enhance the rate of electron transfer. For a barrierless reaction, the long-time rate of electron transfer is exactly the *average* rate of solvation of the newly formed charge transfer state within the one-dimensional linear theory described here.

XII.   KERR RELAXATION

It was mentioned in Section III that Kerr relaxation can be used as a probe for short-time dynamics of the collective orientational relaxation in a dense liquid. The quantity that is measured in Kerr relaxation is the time dependence of the total polarizability–polarizability correlation function

$$C_{ijkl}(t) = \langle \alpha_{ij}(0)\alpha_{kl}(t) \rangle \tag{12.1}$$

where $\alpha_{ij}(t)$ is sum over the polarizability of all the molecules of the system. The important point is that the polarizability of a molecule contains the interaction-induced effects. The total polarizability can be decomposed in the following way:

$$\boldsymbol{\alpha}(t) = \sum_i \boldsymbol{\alpha}^{(i)}(t) + \sum_{i,j}' \boldsymbol{\alpha}^{(ij)}(t) + \cdots \tag{12.2}$$

where, to the lowest order in the interaction between the molecules, $\boldsymbol{\alpha}^{(ij)}(t)$ is given by

$$\boldsymbol{\alpha}^{(ij)}(t) = \boldsymbol{\alpha}^{(i)}(t)\mathbf{T}^{(ij)}(t)\boldsymbol{\alpha}^{(j)}(t) \tag{12.3}$$

where $\mathbf{T}^{(ij)}$ is the usual dipole–dipole tensor

$$
\begin{aligned}
\mathbf{T}^{(ij)}(t) &= (4\pi\varepsilon_0)^{-1}\boldsymbol{\nabla}\boldsymbol{\nabla}\mathbf{r}_{ij}^{-1}(t) \\
&= \frac{1}{4\pi\varepsilon_0}\left\{ \frac{1}{r_{ij}^3(t)}\left(\mathbf{1} - \frac{3\mathbf{r}_{ij}(t)\mathbf{r}_{ij}(t)}{r_{ij}^2(t)} \right) \right\}.
\end{aligned} \tag{12.4}
$$

The important point to note here is that the interaction-induced polarizability in a given molecule (i) by any other molecule (j) of the liquid depends strongly on the distance between the ith and the jth molecules. As a result, the major contribution to the interaction-induced polarizability comes from the nearest neighbors. This implies that Kerr relaxation will be dominated by the inter-mediate-wavevector processes and the translational modes of the liquid can be important.

A detailed study of the interaction-induced effects on light scattering has been carried out by Madden and co-workers [145–147]. They have pointed out that in the interaction-induced effects, reorientation and relative transla-tion of the participating molecules can be important. The interaction-induced part can have a field-theoretic representation:

$$\sum_{(ij)}'\boldsymbol{\alpha}^{(ij)}(t) = \rho^2 \int d\mathbf{r}d\Omega d\mathbf{r}'d\Omega'g(\mathbf{r},\mathbf{r}',\Omega,\Omega')\boldsymbol{\alpha}(\Omega)\cdot\mathbf{T}(\mathbf{r}-\mathbf{r}',t)\cdot\boldsymbol{\alpha}(\Omega')$$

$$= \rho^2(2\pi)^{-3}v\int d\Omega d\Omega'd\mathbf{k}g(\mathbf{k},\Omega,\Omega')\boldsymbol{\alpha}(\Omega)\cdot\mathbf{T}(\mathbf{k},t)\cdot\boldsymbol{\alpha}(\Omega') \quad (12.5)$$

where we have assumed that the important time dependences are in $\boldsymbol{\alpha}$'s and in $\mathbf{T}(k)$. Now, $g(\mathbf{k},\Omega,\Omega')$ $\mathbf{T}(k)$ will derive its maximum contribution from the nearest neighbors, that is, where the structure factor is sharply peaked. As discussed previously, in this range the relaxation may be dominated by the translational modes of the liquid molecules.

Recently, Kenney-Wallace and co-workers [32, 33] have reported femto-second optical Kerr studies in simple pure liquids like CS_2 and chloroform and also in binary liquid solutions with CS_2 as the probe. These studies seem to reveal an intermediate-time response with a 400–600 fs time constant, which was attributed to the relaxation of the "solvent cage" primarily via relative translational motion. In the language adopted here, it means that this intermediate time response is dominated by the dynamics of the $\mathbf{T}(\mathbf{k},t)$ term. However, some caution is warranted in such interpretation because the separation of time scales between different effects observed in experiments is not clear. And also, a quantitative theory of the interaction-induced effects in dense liquids is needed for a reliable interpretation of the results, as emphasized repeatedly by Madden et al. [145–149].

XIII.　ORIENTATIONAL RELAXATION IN A LIQUID OF ELLIPSOIDAL MOLECULES

In this section we discuss the collective orientational relaxation in a dense liquid of ellipsoidal molecules. Our reason for studying this system follows.

Many molecules can be approximately represented as ellipsoids [150]. The orientational correlation induced by the nonspherical shape of the molecules in a dense liquid is expected to be rather important. The model of hard ellipsoids is expected to mimic some aspects of this correlation. This is especially true of orientational dynamics at the molecular length scale, which is of primary interest here.

The orientational motion of ellipsoidal molecules has been studied by several workers. Hu and Zwanzig [151] presented the hydrodynamic calculation of frictional drag on a rotating ellipsoid. In this classic work, it was shown that hydrodynamics can be used even at the molecular level to explain various experimental results. This calculation, however, is for single-particle rotation only. Berne and Pecora [1] have extensively used the ellipsoidal model to explain experiments on depolarized light scattering. However, the collective orientational effects were not included in their discussion. Warchol and Vaughan [152a] have extended the work of Berne to dipolar ellipsoidal molecules. They have shown that the orientational correlation functions of different spherical harmonic components are coupled because of the dipolar potential and the geometry of the ellipsoidal molecules; However, their treatment was phenomenological and only the long-wavelength limit was considered in their study.

Cole and Evans [9] presented a detailed microscopic study of several aspects of collective orientational relaxation of ellipsoids and spherocylinders. These authors used Enskog kinetic theory to arrive at several interesting predictions on the collective orientational relaxation [9]. Allen and Frenkel [152b] carried out a molecular dynamics simulation of orientational relaxation in a system of prolate hard ellipsoids near the isotropic-nematic liquid-crystal phase transitions. They observed a pronounced slowing down of the collective orientational dynamics compared to single-particle motion [152b].

Recent advances in femtosecond laser spectroscopy have opened the possibility of studying the collective orientational motion directly in the time domain. Nonlinear optical techniques, such as the measurements of optical Kerr relaxation [32, 33] in anisotropic molecules, have revealed fast (400–600 fs) response, which may be attributed to the collective relaxation of nearest-neighbor molecules. Simple rotational diffusion models may not be useful in explaining this collective relaxation. An extended hydrodynamic approach valid at molecular length scale is needed here.

The extended hydrodynamic approach discussed in Sections IV and V has been used recently to study the orientational relaxation in a liquid of ellipsoidal molecules [153]. In this approach, the interaction between the different molecules is included through a force field derived from a generalized free energy functional, which is given by the density functional theory.

Another important ingredient of this approach is that it includes the contribution of the translational modes in the collective orientational relaxation. These translational modes have already been found to be important in the femtosecond optical Kerr effect studies. This study [153] reveals a rich structure in $C_{lm}(\mathbf{k}, t)$. First, different $C_{lm}(\mathbf{k}, t)$'s are affected differently by intermolecular correlations. Second, the relaxation of $\phi_l(\mathbf{k}, t)[= \sum_m C_{lm}(\mathbf{k}, t)/ \sum_m C_{lm}(k)]$ is, in general, nonexponential. Third, the translational modes play an important role in the orientational relaxation at intermediate wavevectors. The decay of intermediate-wavevector orientational fluctuations is considerably accelerated by the translational modes. Fourth, the collective orientational relaxation is slower than the single-particle orientational relaxation. This slowing down of the collective orientational relaxation increases with increase in the anisotropy of the molecular shape and also with increase in the density of the liquid.

We now discuss the details of the calculations of the orientational time correlation functions, $C_{lm}(\mathbf{k}, t)$, defined by Eq. (2.11). The analysis of the orientational relaxation is based on a Smoluchowski equation, which is valid in the overdamped limit. As we have discussed previously, in the overdamped limit, the spatial and angular fluxes are given by [153]

$$\frac{1}{m} g_T(\mathbf{r}, \boldsymbol{\Omega}, t) = - \mathbf{D}_T \rho(\mathbf{r}, \boldsymbol{\Omega}, t) \mathbf{V}_T \frac{\delta F'[\rho(t)]}{\delta \rho(\mathbf{r}, \boldsymbol{\Omega}, t)} \tag{13.1}$$

$$\frac{1}{I} g_\Omega(\mathbf{r}, \boldsymbol{\Omega}, t) = - \mathbf{D}_R \rho(\mathbf{r}, \boldsymbol{\Omega}, t) \mathbf{V}_\Omega \frac{\delta F'[\rho(t)]}{\delta \rho(\mathbf{r}, \boldsymbol{\Omega}, t)} \tag{13.2}$$

where $\mathbf{D}_R$ and $\mathbf{D}_T$ are the rotational and translational diffusion tensors, respectively. $F'[\rho(\mathbf{r}, \boldsymbol{\Omega}, t)]$ is the free energy functional of the inhomogeneous fluid. Thus, an equation of motion of the density field can be obtained by evaluating the functional derivatives in Eqs. (13.1) and (13.2) and substituting them into the continuity equation for the density. It is assumed that the rotational and translational diffusion tensors are diagonal and are given by

$$\mathbf{D}_R = \begin{pmatrix} D_R^\perp & 0 & 0 \\ 0 & D_R^\perp & 0 \\ 0 & 0 & D_R^\| \end{pmatrix} \quad \text{and} \quad \mathbf{D}_T = \begin{pmatrix} D_T^\perp & 0 & 0 \\ 0 & D_T^\perp & 0 \\ 0 & 0 & D_T^\| \end{pmatrix} \tag{13.2a}$$

where $D_R^\perp$ and $D_R^\|$ are the diffusion coefficients for rotation about the minor and major axes, respectively. They are independent of the angles $\boldsymbol{\Omega}$. $D_T^\perp$ and $D_T^\|$ are similar terms. if Eqs. (13.1) and (13.2) are substituted into the continuity equation (4.1), and the resulting equation of motion is linearized in the density fluctuation, then one obtains the following equation for the density fluctuation in wavevector space [153]:

$$\frac{\partial}{\partial t}\delta\rho(\mathbf{k},\mathbf{\Omega},t) = [D_\perp \nabla_\Omega^2 + (D_\parallel - D_\perp)L_z^2 - D_T k^2]\delta\rho(\mathbf{k},\mathbf{\Omega},t)$$

$$- \frac{\rho_0}{4\pi}[D_\perp \nabla_\Omega^2 + (D_\parallel - D_\perp)L_z^2 - D_T k^2]$$

$$\times \int d\mathbf{\Omega}' c(\mathbf{k},\mathbf{\Omega},\mathbf{\Omega}')\delta\rho(\mathbf{k},\mathbf{\Omega}',t) \tag{13.3}$$

where L_z is the z component of the angular momentum operator. In Eq. (13.3), $D_\perp$, $D_\parallel$, and D_T stand for $D_R^\perp$, $D_R^\parallel$, and $D_T^\parallel$, respectively, of Eq. (13.2a). The vector $\mathbf{k}$ is taken along the z axis. The density $\delta\rho(\mathbf{k},\mathbf{\Omega},t)$ and the direct correlation function $c(\mathbf{k},\mathbf{\Omega},\mathbf{\Omega}')$ are expanded in terms of the spherical harmonics as usual. The final equation of motion of the coefficients $a_{lm}(\mathbf{k},t)$ is given by

$$\frac{\partial}{\partial t}a_{lm}(\mathbf{k},t) = -[D_\perp l(l+1) + (D_\parallel - D_\perp)m^2 + D_T k^2]a_{lm}(\mathbf{k},t)$$

$$+ \frac{\rho_0}{4\pi}(-1)^m[D_\perp l(l+1) + (D_\parallel - D_\perp)m^2 + D_T k^2]$$

$$\times \sum_{l_2} c(ll_2 m; k)a_{l_2 m}(\mathbf{k},t) \tag{13.4}$$

From Eq. (13.4) one can calculate the time dependence of the correlation function $C_{lm}(\mathbf{k},t)$ and $\phi_l(\mathbf{k},t)$. Because of the up–down symmetry of the ellipsoidal molecules, the only nonzero $c(l_1 l_2 m; k)$'s are those for which both l_1 are even numbers. Thus, Eq. (13.4) shows that the relaxation of $\phi_1(\mathbf{k},t)$ is not affected by the orientational correlations of the liquid molecules. However, the relaxation of $\phi_2(\mathbf{k},t)$ is modified by the orientational correlations present among the ellipsoidal molecules. Numerical calculations show that for ellipsoidal molecules $c(200;k)$ is very small at all wavevectors. Thus the cross correlations in Eq. (13.4) can be neglected and the solution for $C_{2m}(\mathbf{k},t)$ is given by

$$C_{2m}(\mathbf{k},t) = C_{2m}(\mathbf{k},0)\exp[-t/\tau_{2m}(k)] \tag{13.5}$$

where the relaxation time $\tau_{2m}(k)$ is given by

$$\tau_{2m}^{-1} = [6D_\perp + (D_\parallel - D_\perp)m^2 + D_T k^2][1 - (\rho_0/4\pi)(-1)^m c(22m;k)] \tag{13.6}$$

For complete evaluation of $C_{lm}(\mathbf{k},t)$, one needs the value of the equilibrium correlation function $C_{lm}(\mathbf{k})$ and also the values of the different components in the spherical harmonic expansion of the direct correlation function. The analytic expression of the equilibrium correlation function $\langle a_{l_1 m}(\mathbf{k}) a_{l_2 m}(-\mathbf{k}) \rangle$ is given by

$$\langle a_{l_1 m}(\mathbf{k}) a_{l_2 m}(-\mathbf{k}) \rangle = \frac{N}{4\pi} \left\{ \delta_{l_1 l_2} + \frac{\rho_0}{4\pi} (-1)^m h(l_1 l_2 m; k) \right\} \qquad (13.7)$$

An approximate calculation of the direct correlation function can be performed in the following way. The scheme is based on the idea of factorizing the translational and angular variables [154]. The direst correlation function is approximated as

$$C(\mathbf{r} - \mathbf{r}', \Omega, \Omega') = C_0(\mathbf{r} - \mathbf{r}') f(\Omega, \Omega') \qquad (13.8)$$

where, for $C_0(\mathbf{r} - \mathbf{r}')$ one uses the DCF given by the Percus–Yerick approximation, evaluated at the same density. The angle-dependent part $f(\Omega, \Omega')$ results from the intrinsically anisotropic hard ellipsoidal interactions. If one uses the expression of Berne and Pechukas [150] for the angle-dependent contact distance of two ellipsoids, then the expression of $f(\Omega, \Omega')$ is given by [154]

$$f(\Omega, \Omega') = \left\{ \frac{1 - \chi^2 (\hat{e} \cdot \hat{e}')^2}{1 - \chi^2} \right\}^{1/2} \qquad (13.9)$$

where, $\chi = (l^2 - d^2)/(l^2 + d^2)$, where l and d are, respectively, the lengths of the major and minor axis of the ellipsoidal molecules. $\hat{e}$ and $\hat{e}'$ are the unit vectors along the principal axis of the two ellipsoidal molecules. The coefficients can now be calculated by numerically carrying out the angular integrations over Ω and Ω'. Next, we discuss some numerical results on $C_{2m}(\mathbf{k}, t)$ and $\phi_2(\mathbf{k}, t)$ that have been obtained by following the preceding procedure.

The time dependence of the calculated $C_{20}(t)$ [$= C_{20}(k = 0, t)$] is shown in Fig. 33 for different values of the density of the liquid. $C_{20}(t)$ describes the collective orientational relaxation in the long-wavelength limit. The value of the aspect ratio, l/d, is 2.0. It is seen that the relaxation of $C_{20}(t)$ becomes slower as the density of the liquid is increased, which is a manifestation of the increasing orientational order. Figure 34 shows the effects of the translational modes on the relaxation of $\phi_2(k, t)$ at the intermediate wavevectors. Here $\phi_2(k, t)$ is plotted against time for three different values of the translational parameter p ($= D_T/D_\perp d^2$). The values of the wavevector kd is 3.6. It can be seen from the figure that the relaxation becomes considerably faster

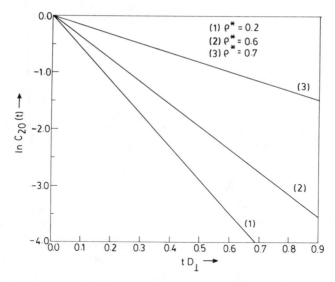

FIGURE 33.   The time dependence of the collective orientational correlation function $C_{20}(t)$ $[= C_{20}(k = 0, t)]$. The normalized $C_{20}(t)$ is plotted against time for three different values of the reduced density, ρ^* ($= \rho_0 l d^2$), which are indicated on the figure. the value of the aspect ratio, l/d, is 2.0.

as the value of the parameter p is increased. Thus, the translational diffusion plays an important role in the orientational relaxation at intermediate wavevectors.

The theoretical results discussed here have important consequences in experimental investigations. It has already been noted by Kenny-Wallace and co-workers [32, 33] that Kerr relaxation may contain contributions from translational motion of the neighboring molecules. Since the interaction-induced effects that have been studied in these experiments depend strongly on the inverse distance of the probe from the solvent molecules, the intermediate-wavevector processes naturally make a major contribution. In fact, one can easily obtain an *approximate* estimate of the time constant of the relaxation of the intermediate-wavevector processes. We consider only the translational contribution here because it is expected to dominate the relaxation at intermediate wavevectors. Let us take the typical values, $kd = 2\pi$, $d = 2$ Å, $D_T = 2 \times 10^{-5}$ cm^2/s, then the time constant, given approximately by $(D_T k^2)^{-1}$, is equal to 500 fs, which is in the correct range.

So far we have considered only the nonpolar ellipsoids. An interesting question here is the effect of molecular shape on the dielectric relaxation. This is an important problem in view of the large amount of experimental

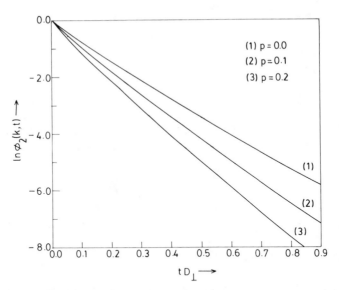

FIGURE 34. The effects of the translational diffusion on the time dependence of the correlation function, $\phi_2(k, t)$. $\phi_2(k, t)$ is plotted against time for three different values of the translational parameter, p, which are indicated on the figure. The value of the wavevector, kd, is 3.6. The values of the reduced density and the aspect ratio are 0.6 and 1.4, respectively. The value of the ratio $D_{\parallel}/D_{\perp}$ is 1.3.

work that has been and is still being done on the liquid crystals [155]. In this case we need to consider ellipsoids or hard rods with a permanent dipole moment. The dielectric relaxation of such systems can show rich and diverse behavior. Although we are not aware of any proper microscopic study of the collective orientational relaxation in such systems, there are some conclusions that can be drawn on the basis of the work discussed here. First, the $c(llm; k)$ coefficients will be nonzero now because the up–down symmetry is lost and the relaxation at $l = 1$ and $l = 2$ [where l is the rank of the spherical harmonics) will be coupled. So, the collective orientational relaxation will be nonexponential even in the linearized Markovian theory and $\varepsilon(\omega)$ will be non-Debye. In this case, the translational modes can play an even more important role in $\varepsilon(\omega)$ as the orientational relaxation can be very slow. This interesting point was perhaps first discussed by Hubbard et al. [57].

Another important problem where the collective orientation of anisotropic molecules is important is the behavior of depolarized (VH) Raleigh scattering at low temperature near the glass transition. This field has been reviewed recently by Patterson and Munoz-Rojas [156]. In general, the role and the behavior of orientational relaxation near a glass transition and in the supercooled regime are interesting, unexplored problems. The orientational relaxation of a system of ellipsoids at high density can serve as a good model for these problems.

We now summarize the main conclusions of this section. We have discussed a microscopic theory for the collective orientatioal relaxation in a dense liquid of ellipsoidal molecules where the orientational correlation functions $\phi_1(k, t)$ have been calculated, At the linear level, the relaxation of $\phi_1(k, t)$ is not affected by the orientational correlations present in the liquid. However, the relaxation of $\phi_2(k, t)$ is affected significantly by the orientational correlations. The orientational correlations slow down the collective orientational relaxation as compared to the single-particle motion. This slowing down of the collective orientational relaxation has important consequences in the pretransitional dynamics near the isotropic-nematic liquid-crystal transition. The translational modes play an important role in the orientational relaxation at intermediate wavevectors. The relaxation at the intermediate wavevectors become faster as the contribution of the translational modes increases. As discussed, the orientational relaxation of a system of ellipsoids can be of importance in many problems of interest.

XIV. ORIENTATIONAL RELAXATION IN A SUPERCOOLED DIPOLAR LIQUID

As the temperature of a liquid is progressively lowered toward its glass transition temperature, the molecular motions (both rotation and translation) become increasingly sluggish. In a dense liquid of nonspherical molecules (and also of molecules that interact with anisotropic potential like the dipole–dipole interaction), the orientational relaxation becomes increasingly dependent on the motion of the nearest molecules. That is, the relaxation becomes collective. This collectivity of the relaxation is manifested dramatically in the dielectric relaxation, which shows a bifurcation in the relaxation time near the glass transition temperature. It has been observed experimentally that below a certain temperature T_B (which is above the glass transition temperature T_g), the dielectric loss spectrum rather suddenly develops a two-peak structure. Above T_B, there is only one peak in the dielectric loss. Of the two peaks, one peak is a continuation of the peak above T_B, but the second new peak appears at much lower frequency and obviously corresponds to slow relaxation times. As the liquid is further supercooled, the new peak again disappears at (or very near) the glass transition. These observations have lead to the conclusion that the peak above T_B corresponds to single-particle-like dynamics and is termed β relaxation. The second peak at lower frequency that exists for temperatures $T_B > T > T_g$ arises from collective dynamics, which is also responsible for the glass transition. This relaxation is termed α or primary relaxation, since it carries most of the relaxation strength. The other branch is also called secondary relaxation. A two-peak dielectric loss spectrum is shown in Fig. 35. An important point is that the (αβ) relaxation and the bifurcation are observed not only for polymeric liquids, but also for molecular liquids composed of small, rigid molecules like pyridine.

The bifurcation of the rotational relaxation at T_B for small rigid molecules was first observed by Johari and Goldstein [157–159] and has been discussed elegantly by Johari in a well-known review [160]. Despite its importance, very little quantitative study was done on this until recently when Kivelson and Kivelson [161] presented a statistical mechanical analysis of the (αβ) relaxation. These authors proposed a model in which the β relaxation is attributed to rapid angular diffusion within a long-lived torsional potential well and the α relaxation to random restructuring of the torsional potential

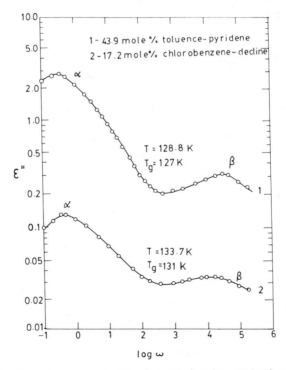

FIGURE 35. The imaginary part of the dielectric function, $\varepsilon''(\omega)$, of two glass-forming liquids is plotted against frequency on a logarithmic scale at a temperature several degrees above the glass transition temperature. (From Ref. 160.)

itself. In this model, above the bifurcation temperature T_B, only the β relaxation occurs and $\tau_1/\tau_2 \approx 3$ in accordance with rotational diffusion model, where τ_1 is the relaxation time for the lth order spherical harmonic. The α relaxation appears below T_B and signals the emergence of longer length scales.

Although the analysis of Kivelson and Kivelson [161] seems reasonable, no detailed microscopic calculation leading the experimentally observed behavior was presented. Recently, Bagchi, Chandra, and Rice (BCR) [162] presented a microscopic theory that seems to explain some aspects of the $(\alpha\beta)$ relaxation phenomena.

The analysis of BCR is based on the same extended hydrodynamic approach discussed in the preceding sections and was based on a model that treats the solute–solvent interactions explicitly. The advantage of this model is that the contributions of the short-range local motions and of the long-range collective motions separates out nicely and one can clearly identify the α and the β relaxation processes. Here we briefly discuss the calculational procedure. An equation of motion is written for the solute orientation, which contains the force from the interaction of the solute with the solvent molecules. Thus, the dynamics of solute orientational motion is coupled with the rotational and the translational motion of the solvent molecules. The coupled equations can be solved analytically in the influence of the solute on the solvent dynamics is neglected. The final expression for the solute orientational function is given by [162]

$$
\begin{aligned}
S_{lm}^s(t) = {} & S_{lm}^s(t=0)\exp[-l(l+1)D_R^s t] \\
& + \frac{(-1)^m}{2}(2\pi)^{-4}\int d\mathbf{k}\, \frac{c_{so}(llm;k)c_{lm}^{so}(k,t=0)\exp[-t/\tau_{lm}(k)]}{1-[l(l+1)D_R^s\tau_{lm}(k)]^{-1}} \\
& - \tfrac{1}{2}(2\pi)^{-4}(-1)^m\int d\mathbf{k}\, \frac{c_{so}(llm;k)C_{lm}^{so}(k,t=0)}{1-[l(l+1)D_R^s\tau_{lm}(k)]^{-1}} \\
& \times \exp[-l(l+1)D_R^s t]
\end{aligned}
\tag{14.1}
$$

$C_{lm}^{so}(k,t=0)$ is the equilibrium cross-correlation between the solute and the solvent molecules and is given by

$$
C_{lm}^{so}(k,t=0) = (-1)^m \frac{h^{so}(llm;k)}{(4\pi)^2}
\tag{14.2}
$$

where $h^{so}(llm;k)$ is the Fourier transform of the spherical harmonic coefficient of the pair correlation function. D_R^s is the solute rotational diffusion coefficient.

$\tau_{lm}(k)$ is the solvent orientational relaxation time constants given by

$$\tau_{lm}^{-1}(k) = \{l(l+1)D_R + D_T k^2\}\left\{1 - (-1)^m \frac{\rho_0}{4\pi} c(llm; k)\right\} \qquad (14.3)$$

where D_R and D_T are the solvent rotational and translational diffusion coefficients. $S_{lm}^s(t=0) = 1/4\pi$ for spherical diffusors. Note that the preceding procedure for separating the dynamics into a single-particle part and a collective part can also be done for a pure liquid. In fact, this is a reliable way to treat the dynamics by extending the set of slow variables but working in the Markovian limit.

Equation (14.1) has an interesting structure. It shows that because of intermolecular interactions, the orientational correlation function consists of two parts. One part decays exponentially with the time constant $[1(1+1)D_R^s]^{-1}$ as expected from a rotational diffusion model. The second part decays nonexponentially and depends on the solvent diffusion constants. Clearly, the second term contains the effects of the collective solvent dynamics on the solute orientational dynamics. The first part is because of the single-particle motion and is to be identified with the β relaxation. As observed by Kivelson and Kivelson, the ratio of $\tau_1/\tau_2 \approx 3$ for this relaxation. The second, nonexponential, part with solvent effects is to be identified as the α relaxation. In the latter case, simple relaxation times like τ_1 and τ_2 are not meaningful.

The theory of BCR [162] can be used to make some qualitative comments on the temperature dependence of α and β relaxation processes. Since D_R^s is assumed to contain only short-range effects, the nature of the temperature dependence of β relaxation is not predicted to change significantly, even in the vicinity of the glass transition temperature. The situation with α relaxation is completely different. Since collective effects enter through $\tau_{lm}(k)$'s, which are strong functions of temperature, the α relaxation is predicted to have a strong temperature dependence. If the temperature dependence of $\tau_{lm}(k)$'s are given by something like a Vogel–Fulcher expression, then α relaxation is also predicted to slow down in a similar fashion. These conclusions are in good agreement with experiments.

In order to carry out a quantitative comparison with experiment, we need accurate values of the direct correlation functions, which, unfortunately, are available only for dipolar hard spheres. Use of MSA solution for the direct correlation functions in Eq. (14.1) lead to the expected bimodal structure in $\varepsilon''(\omega)$ as the ratio D_R/D_R^s is made much smaller than unity. Note that the theory of BCR is not a theory of glass transition; it uses the dependence of the liquid transport properties near T_g to explain the $(\alpha\beta)$ relaxation phenomena. It certainly will be worthwhile to connect this theory with the bulk liquid dynamics near the glass transition.

XV.   ORIENTATIONAL RELAXATION IN DIPOLAR MIXTURES

Many natural chemical and biological processes occur in multicomponent polar liquids. Thus, a study of the dynamics of multicomponent polar liquids is of vital importance in understanding liquid-phase chemistry. Berne [14] earlier discussed orientational relaxation in a multicomponent system. His discussion was based on a continuum model. In this section, we discuss a recently developed microscopic theory of collective orientational relaxation in a multicomponent dipolar liquid [29, 163]. We shall explicitly work out the binary case. However, the formalism discussed here is also applicable to higher-order multicomponent liquids. In the first part of this section we discuss the collective orientational relaxation and in the second part we discuss the dielectric relaxation in a dense binary dipolar liquid.

A.   Collective Orientational Relaxation

Let us consider a liquid mixture consisting of n components some (or all) of which are dipolar. We now consider the general equation of motion of the coefficients $a_{lm}^{(i)}(k, t)$ of the ith species. If $\rho^{(i)}(\mathbf{r}, \mathbf{\Omega}, t)$ denotes the density distribution function for the location and orientation of molecules of the ith kind and $\delta\rho^{(i)}(\mathbf{r}, \mathbf{\Omega}, t)$ is the fluctuation in density, defined by

$$\delta\rho^{(i)}(\mathbf{r}, \mathbf{\Omega}, t) = \rho^{(i)}(\mathbf{r}, \mathbf{\Omega}, t) - \rho_0^{(i)}/4\pi \qquad (15.1)$$

then the linearized equation of motion for $\delta\rho^{(i)}(\mathbf{r}, \mathbf{\Omega}, t)$ can be written as

$$\frac{\partial}{\partial t} \delta\rho^{(i)}(\mathbf{r}, \mathbf{\Omega}, t) = D_{Ri}\nabla_\Omega^2 \delta\rho^{(i)}(\mathbf{r}, \mathbf{\Omega}, t) + D_{Ti}\nabla_T^2 \delta\rho^{(i)}(\mathbf{r}, \mathbf{\Omega}, t)$$

$$- D_{Ri}(\rho_0^{(i)}/4\pi)\nabla_\Omega^2 \sum_j \int d\mathbf{r}' d\mathbf{\Omega}' c^{(ij)}(\mathbf{r} - \mathbf{r}', \mathbf{\Omega}, \mathbf{\Omega}')\delta\rho^{(j)}(\mathbf{r}', \mathbf{\Omega}', t)$$

$$- D_{Ti}(\rho_0^{(i)}/4\pi)\nabla_T^2 \sum_j \int d\mathbf{r}' d\mathbf{\Omega}' c^{(ij)}(\mathbf{r} - \mathbf{r}', \mathbf{\Omega}, \mathbf{\Omega}')\delta\rho^{(j)}(\mathbf{r}', \mathbf{\Omega}', t)$$

$$(15.2)$$

where D_{Ri} and D_{Ti} are the rotational and translational diffusion coefficients of component i. $c^{(ij)}(\mathbf{r} - \mathbf{r}', \mathbf{\Omega}, \mathbf{\Omega}')$ is the two-particle direct correlation function between two molecules of the components i and j. By following essentially the same steps as outlined in Section V, one obtains the following general equation for the orientational relaxation in a mixture [29]:

$$\frac{\partial}{\partial t} a_{lm}^{(i)}(\mathbf{k}, t) = -[D_{Ri}l(l+1) + D_T k^2] a_{lm}^{(i)}(\mathbf{k}, t)$$

$$+ (-1)^m \frac{\rho_0}{4\pi} [D_{Ri}l(l+1) + D_{Ti}k^2] \sum_j \sum_{l_2} c^{(ij)}(ll_2 m; k) a_{l_2 m}^{(j)}(\mathbf{k}, t)$$

$$(15.3)$$

In the derivation of Eq. (15.3), $\mathbf{k}$ is taken parallel to the z axis. It is seen from Eq. (15.3) that the orientational relaxation of one component is coupled with that of other components in the mixture. Next we consider the orientational relaxation in a fluid mixture composed only of two kinds of molecules, denoted 1 and 2, with different dipole moments μ_1 and μ_2 and different molecular diameters σ_1 and σ_2. In a linearized theory (like MSA) for the direct correlation function $c^{(ij)}(\mathbf{k}, \Omega, \Omega')$ for a binary dipolar mixture, the only nonvanishing spherical harmonic components are $c^{ij}(000; k)$, $c^{(ij)}(110; k)$, and $c^{(ij)}(111; k)$. So, in a linearized theory for $c^{(ij)}(\mathbf{k}, \Omega, \Omega')$ only the time evolution of $a_{lm}^{(i)}(k, t)$ is affected by the dipolar forces and the higher correlations ($l > 2$) are not at all affected by dipolar interactions. However, the coefficients $a_{10}^{(1)}(\mathbf{k}, t)$, $a_{10}^{(2)}(\mathbf{k}, t)$, $a_{1\pm 1}^{(1)}(\mathbf{k}, t)$, and $a_{1\pm 1}^{(2)}(\mathbf{k}, t)$ are affected profoundly. The two coupled equations for the coefficients $a_{10}^{(1)}(\mathbf{k}, t)$ and $a_{10}^{(2)}(\mathbf{k}, t)$, can be written in the following form:

$$\frac{\partial}{\partial t} \mathbf{a}_{10}(\mathbf{k}, t) = \mathbf{M} \mathbf{a}_{10}(\mathbf{k}, t) \tag{15.4}$$

where $\mathbf{a}_{10}(\mathbf{k}, t)$ is the column matrix, $(a_{10}^{(i)}(k, t)$, $i = 1, 2$, and $\mathbf{M}$ is the square matrix [29]:

$$\mathbf{M} = \begin{pmatrix} (1 + \lambda_{11})\gamma_1 & \lambda_{12} \\ \lambda_{21} & (1 + \lambda_{22})\gamma_2 \end{pmatrix} \tag{15.4a}$$

where

$$\lambda_{11} = -(\rho_0^{(1)}/4\pi)c^{(11)}(110; k), \qquad \lambda_{12} = [2 + p(k\sigma_1)^2]D_{R1}(\rho_0^{(1)}/4\pi)c^{(12)}(110; k)$$

$$\lambda_{22} = -(\rho_0^{(2)}/4\pi)c^{(22)}(110; k), \qquad \lambda_{21} = [2q + r(k\sigma_1)^2]D_{R1}(\rho_0^{(2)}/4\pi)c^{(21)}(110; k)$$

$$\gamma_1 = -[2 + p(k\sigma_1)^2]D_{R1}, \qquad \gamma_2 = -[2q + r(k\sigma_1)^2]D_{R1}$$

$$(15.5)$$

p, q, and r are dimensionless quantities defined as

$$p = D_{T1}/(D_{R1}\sigma_1^2), \quad q = D_{R2}/D_{R1}, \quad \text{and} \quad r = D_{T2}/(D_{R1}\sigma_1^2) \qquad (15.6)$$

where σ_1 is the molecular diameter of the first type of molecule, which is, in general, different from that of the second type. The information of the disparate molecular radius, of course, is contained in the direct correlation functions. Equation (15.4) can be solved to obtain the following analytic expression for $(\mathbf{a}_{10}(\mathbf{k}, t))$ [29]:

$$\begin{pmatrix} a_{10}^{(1)}(\mathbf{k}, t) \\ a_{10}^{(2)}(\mathbf{k}, t) \end{pmatrix} = \frac{1}{s_+ - s_-} \begin{pmatrix} [(1 + \lambda_{11})\gamma_1 - s_-]a_{10}^{(1)}(\mathbf{k}) + \lambda_{12}a_{10}^{(2)}(\mathbf{k}) \\ \lambda_{21}a_{10}^{(1)}(\mathbf{k}) + [(1 + \lambda_{22})\gamma_2 - s_-]a_{10}^{(2)}(\mathbf{k}) \end{pmatrix} e^{s+t}$$

$$- \frac{1}{s_+ - s_-} \begin{pmatrix} [(1 + \lambda_{11})\gamma_1 - s_+]a_{10}^{(1)}(\mathbf{k}) + \lambda_{12}a_{10}^{(2)}(\mathbf{k}) \\ \lambda_{21}a_{10}^{(1)}(\mathbf{k}) + [(1 + \lambda_{22})\gamma_2 - s_+]a_{10}^{(2)}(\mathbf{k}) \end{pmatrix} e^{s-t}$$

$$(15.7)$$

where s_+ and s_- are the two eigenvalues of matrix $\mathbf{M}$, which are given by

$$s_\pm = \tfrac{1}{2}[(1 + \lambda_{11})\gamma_1 + (1 + \lambda_{22})\gamma_2]$$
$$\pm \tfrac{1}{2}\{[(1 + \lambda_{11})\gamma_1 - (1 + \lambda_{22})\gamma_2]^2 + 4\lambda_{12}\lambda_{21}\}^{1/2} \qquad (15.8)$$

By following the same procedure, we obtain the following expression for $(\mathbf{a}_{1\pm 1}(\mathbf{k}, t))_,$:

$$\begin{pmatrix} a_{1\pm 1}^{(1)}(\mathbf{k}, t) \\ a_{1\pm 1}^{(2)}(\mathbf{k}, t) \end{pmatrix} = \frac{1}{u_+ - u_-} \begin{pmatrix} [(1 + \lambda'_{11})\gamma_1 - u_-]a_{1\pm 1}^{(1)}(\mathbf{k}) + \lambda'_{12}a_{1\pm 1}^{(2)}(\mathbf{k}) \\ \lambda'_{21}a_{1\pm 1}^{(1)}(\mathbf{k}) + [(1 + \lambda'_{22})\gamma_2 - u_-]a_{1\pm 1}^{(2)}(\mathbf{k}) \end{pmatrix} e^{u+t}$$

$$- \frac{1}{u_+ - u_-} \begin{pmatrix} [(1 + \lambda'_{11})\gamma_1 - u_+]a_{1\pm 1}^{(1)}(\mathbf{k}) + \lambda'_{12}a_{1\pm 1}^{(2)}(\mathbf{k}) \\ \lambda'_{21}a_{1\pm 1}^{(1)}(\mathbf{k}) + [(1 + \lambda'_{22})\gamma_2 - u_+]a_{1\pm 1}^{(2)}(\mathbf{k}) \end{pmatrix} e^{u-t}$$

$$(15.9)$$

where

$$\lambda'_{11} = (\rho_0^{(1)}/4\pi)c^{(11)}(111; k), \qquad \lambda'_{12} = -D_{R1}(\rho_0^{(1)}/4\pi)[2 + p(k\sigma_1)^2]c^{(12)}(111; k)$$
$$\lambda'_{22} = (\rho_0^{(2)}/4\pi)c^{(22)}(111; k), \qquad \lambda'_{12} = -D_{R1}(\rho_0^{(2)}/4\pi)[2q + r(k\sigma_1)^2]c^{(21)}(111; k)$$

u_+ and u_- are the eigenvalues of the matrix, $\mathbf{M}'$:

$$\mathbf{M}' = \begin{pmatrix} (1 + \lambda'_{11})\gamma_1 & \lambda'_{12} \\ \lambda'_{21} & (1 + \lambda'_{22})\gamma_2 \end{pmatrix}$$

so that,

$$u_{\pm} = \tfrac{1}{2}[(1 + \lambda'_{11})\gamma_1 + (1 + \lambda'_{22})\gamma_2] \pm \tfrac{1}{2}\{[(1 + \lambda'_{11})\gamma_1$$
$$- (1 + \lambda'_{22})\gamma_2]^2 + 4\lambda'_{12}\lambda'_{21}\}^{1/2} \tag{15.10}$$

Now, the orientational correlation function $\phi(\mathbf{k}, t)$ is defined as [14, 29]

$$\phi(\mathbf{k}, t) = \sum_{m=-1}^{1} \psi_m(\mathbf{k}, t) \bigg/ \sum_{m=-1}^{1} \psi_m(k, 0) \tag{15.11}$$

where

$$\psi_m(\mathbf{k}, t) = \sum_{ij} \mu_i \mu_j \langle a_{lm}^{(i)}(-\mathbf{k}) a_{lm}^{(j)}(\mathbf{k}, t) \rangle \tag{15.12}$$

Substitution of the expressions of $a_{lm}^{(1)}(\mathbf{k}, t)$ and $a_{lm}^{(2)}(\mathbf{k}, t)$ into Eqs. (15.11) and (15.12) gives the following expression for $\phi(\mathbf{k}, t)$:

$$\phi(\mathbf{k}, t)$$
$$= \frac{[1/(s_+ - s_-)](\psi_{0+}e^{s+t} - \psi_{0-}e^{s-t}) + [2/(u_+ - u_-)](\psi_{1+}e^{u+t} - \psi_{1-}e^{u-t})}{[1/(s_+ - s_-)](\psi_{0+} - \psi_{0-}) + [2/(u_+ - u_-)](\psi_{1+} - \psi_{1-})}$$

$$\tag{15.13}$$

where

$$\psi_{0\pm} = \mu_1^2\{G_0(1, 1)[(1 + \lambda_{11})\gamma_1 - s_\mp] + G_0(1, 2)\lambda_{12}\}$$
$$+ \mu_1\mu_2\{G_0(1, 1)\lambda_{21} + G_0(1, 2)[(1 + \lambda_{22})\gamma_2 - s_\mp]\}$$
$$+ \mu_1\mu_2\{G_0(2, 1)[(1 + \lambda_{11})\gamma_1 - s_\mp] + G_0(2, 2)\lambda_{12}\}$$
$$+ \mu_2^2\{G_0(2, 1)\lambda_{21} + G_0(2, 2)[(1 + \lambda_{22})\gamma_2 - s_\mp]\} \tag{15.14a}$$

and

$$\psi_{1\pm} = \mu_1^2\{G_1(1, 1)[(1 + \lambda'_{11})\gamma_1 - u_\mp] + G_1(1, 2)\lambda'_{12}\}$$
$$+ \mu_1\mu_2\{G_1(1, 1)\lambda'_{21} + G_1(1, 2)[(1 + \lambda'_{22})\gamma_2 - u_\mp]\}$$
$$+ \mu_1\mu_2\{G_1(2, 1)[(1 + \lambda'_{11})\gamma_1 - u_\mp] + G_1(2, 2)\lambda'_{12}\}$$
$$+ \mu_2^2\{G_1(2, 1)\lambda'_{21} + G_1(2, 2)[(1 + \lambda'_{22})\gamma_2 - u_\mp]\} \tag{15.14a}$$

$G_m(i, j)$'s are the static correlation functions defined as

$$G_m(i, j) = \langle a_{lm}^{(i)}(-\mathbf{k}) a_{lm}^{(j)}(\mathbf{k}) \rangle \tag{15.15}$$

An analytic expression of $G_m(i, j)$ was derived earlier [29] and it is given by

$$G_m(i, j) = \frac{N}{4\pi} \left[\frac{\rho_0^{(i)}}{\rho_0} \delta_{ij} + \frac{\rho_0^{(i)}}{\rho_0} \frac{\rho_0^{(j)}}{4\pi} (-1)^m h^{(ij)}(11m; k) \right] \qquad (15.16)$$

$h^{(ij)}(11m; k)$ is related to $c^{(ij)}(11m; k)$ by the molecular Ornstein–Zernike relation [13]. ρ_0 is the total number density of the liquid.

One now needs the values of $c^{(ij)}(11m; k)$ for binary dipolar mixture. Exact solutions for the mean spherical model for the direct correlation function for a general n-component dipolar mixture are available [164,165]. So, within MSA, a complete analytic solution for $\phi(\mathbf{k}, t)$ is possible.

It is clear from the preceding discussion that the orientational relaxation in a binary mixture depends on many parameters. As a result it is somewhat difficult to obtain a general qualitative picture of the relaxation behavior.

In Fig. 36 the temporal behavior of $\phi(\mathbf{k}, t)$ for a binary dipolar mixture is shown at the long-wavelength limit. $\phi(\mathbf{k}, t)$ is calculated for several different values of the ratio of rotational diffusion coefficients, q, which are indicated on the figure. It is clear from the figure that relaxation becomes faster as q

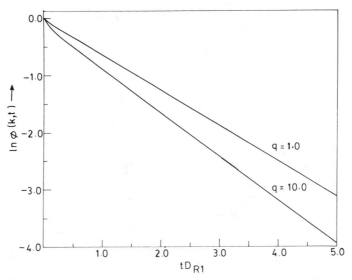

FIGURE 36. The time dependence of the dipolar correlation function, $\phi(\mathbf{k}, t)$ of a binary dipolar mixtue. $\phi(\mathbf{k}, t)$ is plotted against time for two different values of the ratio of rotational diffusion coefficients, q. The value of the wavevector, k, is zero and the sizes of the two different molecules are the same. The values of the reduced densities, $\rho_1\sigma^3$ and $\rho_2\sigma^3$ are 0.3 and 0.5, respectively. The value of the static dielectric constant is 28.0.

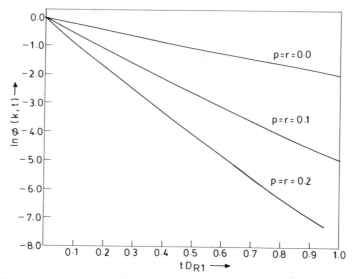

FIGURE 37. The time dependence of the dipolar correlation function $\phi(\mathbf{k}, t)$ of a binary dipolar mixture. $\phi(\mathbf{k}, t)$ is plotted against time for different values of the translational parameters p and r, which are indicated on the figure. The ratio of the rotational diffusion coefficients q is 2.0 and the sizes of the two different kinds of molecules are the same. The value of the wavevector, $k\sigma_1$ is 6.0. The values of the other parameters are the same as in Fig. 36.

is increased, which is expected, and the relaxation is not single exponential. The dependence on translational diffusion is shown in Fig. 37, where $\phi(\mathbf{k}, t)$ is plotted against time for different sets of values of the translational parameters p and r. The value of k is $6.0/\sigma$. This figure also shows that translational diffusion contributes significantly to collective orientational relaxation in this intermediate value of the wavevector k.

Now we summarize the essential features of collective orientational relaxation in a binary dipolar mixture. We have discussed the fact that the relaxation behavior depends on the values of the dipole moments of the individual components, on the values of the respective rotational and translational diffusion coefficients, and on the sizes of the dipolar molecules. Thus, the parameter space is multidimensional. Much more work is needed to unearth the essential qualitative aspects of a binary mixture.

B. The Dielectric Relaxation in a Binary Liquid

As we have discussed, dielectric relaxation studies (that is, studies of the time-dependent response to an externally applied long-wavelength time-dependent field) provide valuable information concerning the dynamics of the liquid

[166]. For this reason, a large number of experimental studies have been performed on dielectric relaxation in binary dipolar mixture [3, 167–170]. However, very little theoretical work has been done on this important problem. Here we discuss an application of the theory developed in Sections VI and XVA to dielectric relaxation in binary liquids. It has been found that the dielectric relaxation in a binary system is, in general, nonexponential. In some limits, however, one recovers a biexponential relaxation that has traditionally been used to study the dielectric relaxation in binary system. Note that a noninteracting Debye model also gives a biexponential decay. However, the relaxation times are widely different from the predictions of the Debye noninteracting model because of the intermolecular interactions. This theory can be straightforwardly extended to study dielectric relaxation in ternary and higher-order multicomponent systems.

A general expression for the frequency- (ω) and wavevector- (k) dependent dielectric function, $\varepsilon(k, \omega)$, can be obtained by using the linear response theory which has been discussed in Section VI. The experimentally observed dielectric function is given by $\varepsilon(k = 0, \omega)$. In an n-component dipolar system, the correlation function of the collective dipole moments is defined as

$$\mathbf{C}_M(\mathbf{k}, t) = \sum_{ij} \langle \delta \mathbf{P}_i(-\mathbf{k}) \delta \mathbf{P}_j(\mathbf{k}, t) \rangle \tag{15.17}$$

where $\delta \mathbf{P}_i(\mathbf{k}, t)$ is the polarization fluctuation of the i species. $\delta \mathbf{P}_i(\mathbf{k}, t)$ is related to the density fluctuation by the following relation:

$$\delta \mathbf{P}_i(\mathbf{k}, t) = \mu_i \int d\mathbf{r} e^{i\mathbf{k}\cdot\mathbf{r}} \int d\Omega \hat{\alpha}(\mathbf{\Omega}) \delta \rho^{(i)}(\mathbf{r}, \mathbf{\Omega}, t) \tag{15.18}$$

Once the polarization correlation functions are calculated, the dielectric functions can be obtained by using the same procedure as for the one-component case discussed in Section VI. As before, we assume that the equation of motion of the density fluctuation, $\delta \rho^{(i)}(\mathbf{r}, \mathbf{\Omega}, t)$, is given by the generalized Smoluchowski equation and the equation of motion for the polarization fluctuation of the i species is given by [163]

$$\frac{\partial}{\partial t} \delta \mathbf{P}_i(\mathbf{k}, t) = -(2D_{Ri} + D_{Ti}k^2)\delta \mathbf{P}_i(\mathbf{k}, t) + \frac{\rho_0^{(i)}}{3}(2D_{Ri} + D_{Ti}k^2)\sum_j \mathbf{C}_{ij}(\mathbf{k}) \cdot \delta \mathbf{P}_j(\mathbf{k}, t)$$

$$\tag{15.19}$$

where $\mathbf{C}_{ij}(\mathbf{k})$ is a tensor defined by

$$\mathbf{c}^{(ij)}(\mathbf{k}, \mathbf{\Omega}, \mathbf{\Omega}') = \mathbf{c}_{\text{iso}}^{(ij)}(\mathbf{k}) + \hat{\alpha}(\mathbf{\Omega}) \cdot \mathbf{C}_{ij}(\mathbf{k}) \cdot \hat{\alpha}(\mathbf{\Omega}') + \cdots \tag{15.20}$$

It is now straightforward to obtain the time dependence of the longitudinal and the transverse polarization correlation functions, as discussed in Ref. 163. Here we write down the final expression of the transverse part of the collective dipole moment, $C_{MT}(\mathbf{k}, t)$, which will be used to calculate $\varepsilon(\omega)$. $C_{MT}(\mathbf{k}, t)$ is given by

$$C_{MT}(\mathbf{k}, t) = \frac{4\pi}{3} \frac{1}{u_+ - u_-} \{\psi_{1+} e^{u_+ t} - \psi_{1-} e^{u_- t}\} \tag{15.21}$$

where the different quantities in Eq. (15.21) have been defined previously. In deriving Eq. (15.21), it has again been assumed that the equilibrium correlation functions are given by a linearized theory. The linearized theory can be solved analytically, as discussed elsewhere. It may be appropriate to mention here that a full nonlinear equilibrium theory would lead to a nonexponential $\varepsilon(\omega)$, whereas a linearized theory gives a biexponential form. For a not too strongly polar liquid, the linearized theory and the biexponential form should be reliable.

The preceding equations give the following expression for the long-wavelength frequency-dependent dielectric function:

$$\varepsilon(\omega) = 1 + \frac{c}{1 + i\omega\tau_1} + \frac{d}{1 + i\omega\tau_2} \tag{15.22}$$

where

$$c = \frac{(4\pi)^2 \beta}{3v} \frac{\psi_{1+}}{u_+ - u_-}, \quad d = \frac{-(4\pi)^2 \beta}{3v} \frac{\psi_{1-}}{u_+ - u_-}, \quad \tau_1 = \frac{-1}{u_+}, \quad \text{and} \quad \tau_2 = \frac{-1}{u_-} \tag{15.23}$$

$\psi_{1\pm}$, u_+, and u_- are evaluated at $k = 0$. Equation (15.22) can also be written in the following form:

$$\varepsilon(\omega) = 1 + (\varepsilon_0 - 1)\left\{\frac{g_1}{1 + i\omega\tau_1} + \frac{g_2}{1 + i\omega\tau_2}\right\} \tag{15.24}$$

where $g_1 = c/(c + d)$, $g_2 = d/(c + d)$, and $\varepsilon_0 = \varepsilon(\omega = 0)$. Equation (15.24) shows that the dielectric dispersion occurs with two relaxation times, τ_1 and τ_2. The values of these two relaxation times depend on the diffusion coefficients; on the sizes, and on the dipole moments of the constituent particles and also on the composition of the mixture. These two relaxation times are widely different from those predicted by the Debye noninteracting rotational diffusion model, which we shall discuss in the next section. A form as Eq. (15.24) is commonly known as the Bodo formula [3]. We mentioned earlier that this biexponential form is a consequence of our use of the linearized equilibrium theory of dipolar liquid.

It is interesting to note that a biexponential form of $\varepsilon(\omega)$ has traditionally been used to analyze dielectric relaxation data on binary mixture. Here one recovers this form within a microscopic theory. It is thus important to understand the approximations made in deriving Eq. (15.24). First, a Smoluchowski-equation-based description is used for the density fluctuation. We discussed in Section IV why such a description is reliable in a dense liquid. Second, the MSA model has been used for the direct correlation functions. As we discussed previously, use of any linearized theory would give rise to the same form for $\varepsilon(\omega)$. Such linearized theories are reliable for not too strongly polar liquids ($\varepsilon_0 < 60$). If a nonlinear theory, such as RHNC [13], is used, then one would get a multiexponential form for $\varepsilon(\omega)$, but then the correlation functions must be evaluated numerically. Third, the memory effects in the dielectic relaxation have been neglected. It has been shown recently [87] that in the presence of a moderate translational contribution to polarization relaxation, such memory effects may be neglected.

It is interesting to compare the microscopic theory with the noninteracting rotational diffusion model of Debye, which we shall refer to as the Debye model for binary liquids (DMBL). In this model the polarization fluctuation of each species [say, $\mathbf{P}_i(t)$] decays exponentially with the time constant given by the rotational diffusion coefficient of that species only [171]. Therefore, in this model, the time dependence of $C_{MT}(t)$ is given by

$$C_{MT}(t) = \langle P_{T1}^2 \rangle e^{-2D_{R1}t} + \langle P_{T2}^2 \rangle e^{-2D_{R2}t} \tag{15.25}$$

with

$$\langle P_{Ti}^2 \rangle = \frac{N\mu_i^2}{3} \frac{\rho_0^{(i)}}{\rho_0}$$

ρ_0 is the total number density. The dielectric function $\varepsilon(\omega)$ is given by

$$\varepsilon(\omega) = 1 + (\varepsilon_0 - 1)\left\{\frac{g_{D1}}{1 + i\omega\tau_{D1}} + \frac{g_{D2}}{1 + i\omega\tau_{D2}}\right\} \tag{15.26}$$

where $\tau_{Di} = (2D_{Ri})^{-1}$ and $g_{Di} = \rho_0^{(i)}\mu_i^2/(\sum_{i=1}^2 \rho_0^{(i)}\mu_i^2)$. Thus, in this model, when $D_{R1} = D_{R2}$, $\varepsilon(\omega)$ becomes a single exponential. However, the microscopic theory shows that even if $D_{R1} = D_{R2}$, the relaxation may be nonexponential if the dipole moments (or the sizes) of the two species are different.

As before, the numerical calculations were performed using the mean spherical approximation (MSA) to evaluate the direct correlation function components, $c^{(ij)}(110; k)$ and $c^{(ij)}(111; k)$. $\varepsilon(\omega)$ has been calculated with varying molecular sizes, dipole moments, densities, and diffusion coefficients of the two constituent components.

Figures 38–40 shows the results of the dielectric relaxation where sizes of the two types of constituent molecules are the same but their diffusion coefficients and the dipole moments are different. In Fig. 38.3, the real part of the dielectric function, $\varepsilon'(\omega)$, is plotted against the frequency, ω. The dashed line shows the predictions of the Debye noninteracting model. One can see that the dielectric relaxation is significantly modified because of the intermolecular interactions. The importance of intermolecular interactions is also evident in Fig. 39, where the imaginary part $\varepsilon''(\omega)$ is plotted against the frequency ω. This indicates that the nature of the dielectric relaxation depends critically on the details of the molecular properties of the binary mixture. The biexponential nature of the dielectric relaxation in binary mixture is clear in Fig. 40, where we have plotted the imaginary part, $\varepsilon''(\omega)$, versus the

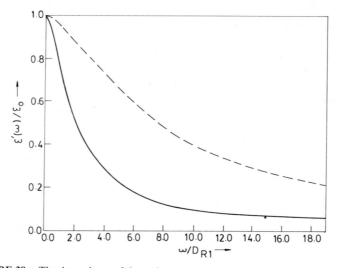

FIGURE 38. The dependence of the real part of the dielectric function, $\varepsilon'(\omega)$, on frequency, ω. The solid line shows the values of $\varepsilon'(\omega)$ obtained from the microscopic theory and the dashed line is the prediction of the noninteracting model of Debye. The sizes of the two types of molecules are the same with diameter σ. The values of the reduced densities ρ_1^* and ρ_2^* ($\rho_i^* = \rho_i \sigma_i^3$) and the ratio of the diffusion coefficients, q ($= D_{R2}/D_{R1}$), are 0.3, 0.5, and 4.0, respectively. The values of the dielectric parameters, Y_{11} and Y_{22} ($Y_{ii} = \frac{4}{9}\pi\beta\rho_i\mu_i^2$) are 0.55 and 2.35, respectively. The calculated values for the static dielectric constant (ε_0) for this set of parameter values are 30.0 for the microscopic theory. (From Ref. 163.)

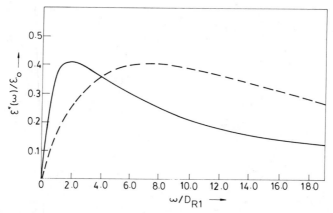

FIGURE 39. The dependence of the imaginary part of the dielectric function, $\varepsilon''(\omega)$, on the frequency, ω. As in Fig. 38, the solid line gives the values of $\varepsilon''(\omega)$ obtained from the microscopic theory and the dashed line is the prediction of the Debye noninteracting model. The sizes of the two types of molecules are the same. The values of the other parameters are the same as in Fig. 38. (From Ref. 163.)

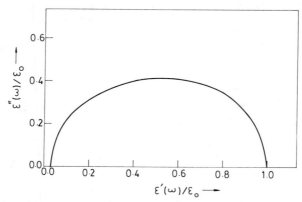

FIGURE 40. The Cole–Cole plot of the frequency-dependent dielectric function. The imaginary part of the dielectric function $\varepsilon''(\omega)$ is plotted against the real part $\varepsilon'(\omega)$ calculated at different frequencies. The values of the different parameters are the same as in Fig. 38. (From Ref. 163.)

real part, $\varepsilon'(\omega)$, calculated at different frequencies. The values of the different parameters are given in the figure caption. The biexponential nature of the dielectric relaxation is more prominent when the diffusion coefficients of the two types of molecules differ significantly.

Usually the two components of the mixture will have different molecular sizes, in addition to having different dipole moments. The effects of the

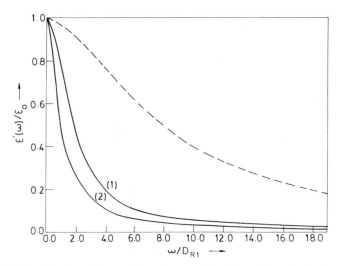

FIGURE 41. The dependence of the real part of the dielectric function on the frequency, ω. The calculated values of $\varepsilon'(\omega)$ are plotted against frequency for two different values of the size ratio, R (solid lines). As before, the dashed line is the prediction of the noninteracting model. The ratio of the rotational diffusion coefficients, q, is 4.0. The values of the other parameters $\rho_1^*, \rho_2^*, Y_{11}$, and Y_{22} are 0.5, 0.3, 0.5, and 4.5, respectively. The value of ε_0 is 60.0 for $R = 1$ and 92 for $R = 4.0$. (From Ref. 163.)

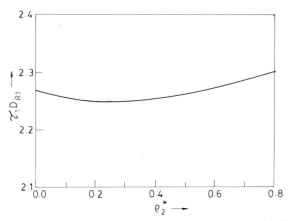

FIGURE 42. The dependence of the main relaxation time on the composition of the mixture when both the diffusion coefficients and the dipole moments of the two components are different. The calculated values of the main relaxation time (τ_1) are plotted against the density of the second component. The total density ρ^* is 0.8. The values of the different parameters, R, q, Y_{11}, and Y_{22} are 1.0, 0.48, 5.33 and 1.33, respectively. (From Ref. 163.)

differing sizes on dielectric relaxation is shown in Fig. 41, where $\varepsilon'(\omega)$ is plotted against frequency for two different values of the size ratio R. As before, the dashed line shows the results of the noninteracting model. It is seen that the relaxation is markedly different when the molecular sizes of the two constituent liquids are different. This figure also shows that the details of the molecular properties play important roles in the dielectric relaxation in binary liquids.

Figure 42 shows the effect of composition on the main relaxation time. The dipole moment of the first type of molecules is higher than that of the second type (that is, $\mu_1 > \mu_2$). The diffusion coefficients of the two types of molecules are different and the dielectric relaxation times of the two pure liquids are comparable. As we increase the proportion of the second type of molecules in the mixture, the main relaxation time first decreases, passes through a minimum, and again increases until it reaches the limiting value (the relaxation time of the pure liquid of the second type). The reason for the existence of such a minimum is the following. The dipole moment of the second type of molecules is smaller than that of first type. Thus, as we increase the proportion of the second type, the main relaxation time tends to decrease. But the diffusion coefficient of the second type of molecules is smaller than that of the first type. The relaxation time of the mixture tends to increase because of this smaller diffusion coefficient. These two opposing factors lead to the existence of a minimum in the relaxation time of the mixture. Such minimum has been found to exist in many experimental studies (e.g., in butane–butanediol system [169]).

We now summarize the essential features of this section. We have discussed a molecular theory of dielectric relaxation in a binary dipolar liquid. Within certain approximations, valid for not too strongly polar dense dipolar liquids, the dielectric relaxation is biexponential and $\varepsilon(\omega)$ is a sum of two Debye terms. In this theory, the time constants τ_1 and τ_2, and the weight factors g_1 and g_2 are functions of the microscopic structure of the liquid. Thus, it is possible to study the variations of τ_1, τ_2, g_1 and g_2 as functions of the dipole moments, the sizes, and the composition of the binary mixture. Such information is not provided by the Debye model for a binary system. The theory discussed here can explain several existing experimental results. The nonmonotonic dependence of the primary relaxation time on composition, shown in Figure 42, is an example of this.

XVI. COLLECTIVE ORIENTATIONAL RELAXATION IN THE PRESENCE OF AN EXTERNAL FIELD

The collective orientational relaxation in a dense dipolar liquid can be affected profoundly if the liquid is subjected to an external electric field. For practical reasons, it is convenient to consider two limiting cases: (i) The electric field is macroscopic where the liquid is placed between two plates of opposite charges; (ii) the electric field is microscopic where the source is an electron or an ion within the liquid. In the former case, the field contains only the $k = 0$ Fourier mode, whereas in the latter case, the field contains all the modes. For small ions (such as an electron), the intermediate-wavevector components can have a significant weight in the total field.

The effects of intense external fields on the orientational dynamics has been discussed recently by Coffey [11] within the formulation of the rotational diffusion model. His discussion was therefore limited to noninteracting single-particle dynamics and cannot be used to study the collective effects that are being considered here. The extended hydrodynamic equations presented in Section IV can be used to study the influence of the external field on the collective dynamics and will be discussed subsequently. But first let us make few general comments on the dynamics in an external field.

First, an external field distorts the equilibrium distribution and creates an nonhomogeneous state. If the external field is strong, then the nonhomogeneous state that will be created can be far from the homogeneous state in the absence of the field. The nature of the dynamics can also be quite different as one considers the fluctuations from this nonhomogeneous state. Second, the dynamical quantities, like the diffusion coefficients, of the nonhomogeneous state can be quite different from those of the homogeneous state. The equilibrium correlations, such as the direct correlation functions, in the nonhomogeneous state can also be rather different from those in the homogeneous equilibrium state. Finally, some fundamental theorems of statistical mechanics, like the fluctuation–dissipation theorem, may no longer be valid, as stressed earlier by Zwanzig [172] and discussed recently by Evans [12].

There are two entirely different classes of problems that one encounters when one considers the dynamics of the nonhomogeneous state. The first one is related to the decay (or the establishment) of an nonhomogeneous state from a homogeneous state of the system. In this case, a field may be turned on at time $t = 0$ and the establishment of the nonhomogeneous state is studied. This is the case, for example, in the solvation of a newly created ion. In the second class of problems, one is interested in the dynamics *in* the nonhomogeneous system. In this section we shall be interested in cases where the external field is time independent, so that both the homogeneous and the nonhomogeneous states are in equilibrium.

We first briefly discuss the case of the noninteracting dipolar molecules in a homogeneous external electric field. If θ denotes the angle made by the molecule with the direction of the external field, then, in the overdamped limit, the equation of motion is given by

$$\frac{\partial}{\partial t}\rho(\theta, t) = D_R \frac{\partial}{\partial \theta}\left(\frac{\partial \rho(\theta, t)}{\partial \theta} + \frac{\mu E_0}{k_B T}(\sin \theta)\rho(\theta, t)\right) \tag{16.1}$$

where μ is the dipole moment of a molecule and E_0 is the external field. The nonhomogeneous equilibrium distribution is given by the one-particle Maxwell–Boltzmann distribution

$$\rho_{eq}(\theta) = A \exp\left[\frac{\mu E_0 \cos \theta}{k_B T}\right] \tag{16.2}$$

where A is the normalization constant.

The solution of Eq. (16.1) was obtained by Debye for two special cases. In one case, the time-independent external field was kept on from $t = -\infty$ to $t = 0$, when it was turned off. Debye showed if Eq. (16.2) is linearized, then the relaxation time is given by that of the pure liquid. In the second model, the external field was assumed to be periodic in time. This can also be solved exactly in the linear approximation and connections can be made with the theory of dielectric relaxation, as discussed nicely by Coffey [11].

The situation becomes much more complicated when the external field is so strong that linearization of Eq. (16.2) is not possible. In such a situation, a simple analytic solution is no longer possible and one must have recourse to perturbative or numerical or both solutions. The results on the studies of nonlinear dielectric phenomena based on the Smoluchowski equation has been discussed by Coffey [11]. The main conclusion is that in the presence of strong field, the response is no longer linear and that the ratio of the response to stimulus now depends on the form of the applied field, and is not independent of it.

The preceding discussion is limited to the case when molecules do not interact among themselves. The situation is clearly more complicated in the presence of such intermolecular correlations. Unfortunately, little work has been done on the effects of external field on the *collective* dynamics. In the following, we briefly discuss a recently initiated approach toward understanding the collective dynamics in an nonhomogeneous system.

In general, the time-dependent potential energy because of the interaction of the liquid molecules with an external field can be written in the following form:

$$U_{ext}(t) = \int dr d\Omega u(\mathbf{r}, \mathbf{\Omega}, t)\rho(\mathbf{r}, \mathbf{\Omega}, t) \tag{16.3}$$

where $u(\mathbf{r}, \mathbf{\Omega}, t)$ is the one-body interaction potential. Let us first discuss the equilibrium part before we embark on dynamics. When the extra term due to the external field is added to the free energy functional given by Eq. (4.3) and the equilibrium density is calculated by setting the functional derivative of the free energy with respect to the density equal to zero, we obtain the following expression for the nonhomogeneous equilibrium density:

$$\rho_{eq}(\mathbf{r}, \mathbf{\Omega}) = \frac{\rho_0}{4\pi} \exp\left[-\beta V_{eff}(\mathbf{r}, \mathbf{\Omega})\right] \qquad (16.4)$$

where

$$\beta V_{eff}(\mathbf{r}, \mathbf{\Omega}) = -\int d\mathbf{r}' d\mathbf{\Omega}' c(\mathbf{r} - \mathbf{r}', \mathbf{\Omega}, \mathbf{\Omega}')[\rho_{eq}(\mathbf{r}, \mathbf{\Omega}) - \rho_0/4\pi] + \beta u(\mathbf{r}, \mathbf{\Omega}) \qquad (16.5)$$

and $u(\mathbf{r}, \mathbf{\Omega})$ is the time-independent one-body interaction potential. The last two coupled equations are to be solved simultaneously.

Let us now consider the simple case where the external field is homogeneous, so that $u(\mathbf{r}, \mathbf{\Omega}) = \mu E \cos \theta$, that is, there is no position dependence in the external field. The important point here is that in the presence of intermolecular interactions, the equilibrium density distribution is no longer given by the naive, noninteracting limit of the Maxwell–Boltzmann distribution [Eq. (16.2)], but by the generalized Boltzmann distribution [Eq. (16.4)], which depends on both the position and the orientation variables. In the case of an ion within the liquid, $u(\mathbf{r}, \mathbf{\Omega})$ consists of two parts: a harsh repulsive part at short distance from the center of the ion and a long-range electrical part. The equilibrium distribution $\rho_{eq}(\mathbf{r}, \mathbf{\Omega})$ can differ significantly from $\rho_0/4\pi$, especially at short distances from the ion.

In order to calculate $\rho_{eq}(\mathbf{r}, \mathbf{\Omega})$, it is convenient to expand it in spherical harmonics:

$$\rho_{eq}(\mathbf{r}, \mathbf{\Omega}) = \sum_{lm} a_{lm}(\mathbf{r}) Y_{lm}(\mathbf{\Omega}) \qquad (16.6)$$

Under certain approximations discussed in Ref. 31, the following expression for $a_{lm}(\mathbf{r})$ is obtained for a classical ion:

$$a_{lm}(\mathbf{r}) = \frac{\rho_0}{4\pi} g_{SR}(r) \int d\mathbf{\Omega} Y_{lm}(\mathbf{\Omega}) \exp\left[\sum_{l_1 l_2 m_1} (-1)^{m_1} Y_{l_1 m_1}(\mathbf{\Omega}) \right.$$
$$\times \int d\mathbf{r}' c(l_1 l_2 m_1; \mathbf{r} - \mathbf{r}') a_{l_2 m_1}(\mathbf{r}') - (4\pi)^{-1/2} \rho_0$$
$$\left. \times \sum_{l_1} Y_{l_1 0}(\mathbf{\Omega}) \int d\mathbf{r}' c(l_1 0 0; \mathbf{r} - \mathbf{r}') g_{SR}(r') + \beta[\boldsymbol{\mu} \cdot \mathbf{E}_0(\mathbf{r})] \right] \qquad (16.7)$$

with the following notation, $g_{SR}(r)$ is the radial distribution function due to short-range interaction between the solvent and the polar solute molecules and $\mathbf{E}_0$ is the field of the ion. This nonlinear equation can be solved numerically for $a_{lm}(r)$—a task yet to be accomplished. Several limiting forms were solved in Ref. 31, and the equilibrium polarization structure was obtained.

Now, let us turn to the dynamics in an nonhomogeneous state. Very few microscopic studies have been carried out for the transport properties in an nonhomogeneous state [46]. In fact, this problem remains both a challenge and an embarassment to the theoretical chemists and physicists. However, some progress has been made recently in the relaxation of the fluctuations in an nonhomogeneous state and also the decay of the nonhomogeneous state to the homogeneous state as the field is turned off at time $t = 0$. This is briefly discussed in the following.

Let us consider the decay of the fluctuations $\delta\rho(\mathbf{r}, \boldsymbol{\Omega}, t) = \rho(\mathbf{r}, \boldsymbol{\Omega}, t) - \rho_{eq}(\mathbf{r}, \boldsymbol{\Omega})$, where $\rho_{eq}(\mathbf{r}, \boldsymbol{\Omega})$ is given by Eq. (16.4). The linearized Smoluchowski equation for $\delta\rho(\mathbf{r}, \boldsymbol{\Omega}, t)$ in the presence of an external field is given by

$$\frac{\partial}{\partial t} \delta\rho(\mathbf{r}, \boldsymbol{\Omega}, t) = (D_R \nabla_\Omega^2 + D_T \nabla_T^2) \delta\rho(\mathbf{r}, \boldsymbol{\Omega}, t)$$

$$- D_R \nabla_\Omega \cdot \rho_{eq}(\mathbf{r}, \boldsymbol{\Omega}) \nabla_\Omega \int d\mathbf{r}' d\boldsymbol{\Omega}' c(\mathbf{r} - \mathbf{r}', \boldsymbol{\Omega}, \boldsymbol{\Omega}') \delta\rho(\mathbf{r}', \boldsymbol{\Omega}', t)$$

$$- D_T \nabla \cdot \rho_{eq}(\mathbf{r}, \boldsymbol{\Omega}) \nabla_T \int d\mathbf{r}' d\boldsymbol{\Omega}' c(\mathbf{r} - \mathbf{r}', \boldsymbol{\Omega}, \boldsymbol{\Omega}') \delta\rho(\mathbf{r}', \boldsymbol{\Omega}', t)$$

$$- \beta D_R \nabla_\Omega \cdot \rho_{eq}(\mathbf{r}, \boldsymbol{\Omega}) \nabla_\Omega [\boldsymbol{\mu} \cdot \mathbf{E}_0(\mathbf{r})] - \beta D_T \nabla_T \cdot \rho_{eq}(\mathbf{r}, \boldsymbol{\Omega}) \nabla [\boldsymbol{\mu} \cdot \mathbf{E}_0(\mathbf{r})]$$

$$(16.8)$$

For convenience, let us expand the density fluctuation $\delta\rho(\mathbf{r}, \boldsymbol{\Omega}, t)$ in spherical harmonics:

$$\delta\rho(\mathbf{r}, \boldsymbol{\Omega}, t) = \sum_{lm} A_{lm}(\mathbf{r}, t) Y_{lm}(\boldsymbol{\Omega}) \qquad (16.9)$$

Use of Eq. (16.9) in Eq. (16.8) and following the standard procedure, one can obtain a set of coupled equations for $A_{lm}(\mathbf{r}, t)$, which are given in Ref. 31. Even in the case of an ionic field, the equations are rather complicated. Unfortunately, not much theoretical work has been done on such equations. Computer simulations [173] and preliminary numerical work on Eq. (16.8)

[174] indicate that the dynamics of the decay of fluctuations in a strongly nonhomogeneous state can be considerably different from those in the pure system.

Let us now summarize the main points of this section. Although considerable work has been done on the effects of an external field on single-particle orientational dynamics at the level of simple rotational diffusion model, very little is known about the collective dynamics in a dense liquid in the presence of a strong external field. This is connected with our lack of understanding of the same in an nonhomogeneous system. This is clearly an important problem for future study.

XVII.   FUTURE PROBLEMS

It is clear from the preceding review that although considerable progress has been made in the recent years in our understanding of the collective orientational relaxation and its role in various liquid-state dynamical processes, there are still many important problems that have not been understood. In the following, we list some of the problems that we believe deserve to be studied carefully in the near future.

A.   Relaxation in a Nonhomogeneous State

A systematic study of the collective orientational relaxation in an non-homogeneous system, created by an external field, is yet to be carried out. As a result, our understanding of this problem is poor. This is despite its importance in the dynamics of solvation and the electron transfer reaction. Another important example of a nonhomogeneous system is a molecular liquid under a steady state of sheared motion, such a Couette flow. This is, of course, a nonequilibrium system and may require a quite different approach. The difficulty, at the lowest level of development, is technical in nature. As discussed in Section XVI, the equation of motion, even in the linearized form, is complicated. The main difficulty is that because of the nonhomogeneity in $\rho_{eq}(\mathbf{r}, \mathbf{\Omega})$, the advantage gained in the homogeneous system by working in Fourier space is absent here. There are two methods that are available. First, we can attempt a perturbative solution. This is the method of solution for a small nonhomogeneity. The second approach is to solve the nonhomogeneous differential equation numerically in real space. This was done by Calef and Wolynes [45], who neglected the translational modes. Clearly, considerable work is needed in either appraoch.

There are other problems of more fundamental nature that are encountered when treating strongly nonhomogeneous systems. We may need to carry out the averaging over the random forces in the nonhomogeneous system. In such a situation the transport properties will depend on the nonhomogeneity. Unfortunately, very little theoretical work has been done on this problem. This is clearly an important problem in liquid-state chemistry that deserves further study.

B. Relaxation in Hydrogen-Bonded Liquids

The theoretical calculations discussed in this review are for dipolar liquids with no specific intermolecular interaction, such as the hydrogen bonding, so, strictly speaking, they cannot be used to describe orientational relaxation of water and alcohols. Because of the importance of these liquids, a microscopic description of orientational relaxation in these liquids is desired. Recently, the dynamical properties of hydrogen-bonded liquids was reviewed by Bertolini et al. [175]. These authors have discussed the importance of the connectivity of the hydrogen-bonded liquid clusters in determining the dynamics and proposed a stochastic Liouville equation approach to treat the collective dynamics.

In principle, the extended hydrodynamic approach developed in this review should also be applicable to hydrogen-bonded liquids. The problem, however, is not trivial because the direct correlation function coefficients are not easily available. In addition to hydrogen-bonding interactions, we may need to treat the quadrupolar interactions because these liquids may also have large quadrupole moments.

Moreover, recent treatment of Stanley and Teixeira [176] suggests the presence of a "patchlike" structure consisting of a small number of molecules in water at low temperature. At low temperatures these structures can have sufficiently long lifetimes to affect the long-time relaxation properties. They are expected to influence the collective orientational relaxation also. A detailed study of these problems will be most certainly welcome.

C. Relaxation in a Supercooled Liquid

A preliminary analysis of this problem is discussed in Section XIV. The discussion was limited in scope because it considered only the dipolar spheres. In real liquids, molecular shape can play an important role in the collectivity of the orientational relaxation. Another important aspect of relaxation in the supercooled liquid is that one needs to consider the coupling between the translational and rotational dissipation that is, the term $\Gamma_{T\Omega}$ should be retained in the kinetic description. The reason is that in a supercooled liquid near its glass transition, the dissipative kernel Γ_{TT} may become very small, but $\Gamma_{\Omega\Omega}$ may remain large. So, the relaxation of spatial momentum may occur indirectly through angular momentum. A third aspect is that near the glass transition, the density relaxation may slowed down significantly, so that the nonlinear terms are important. This implies that a mode coupling theory of orientational relaxation is required. We hope to address some of these problems in near future.

D.   Relaxation of Molecules in Orientationally Disordered Crystals

The orientational relaxation in crystals whose molecules are orientationally disorderd has been a subject of continued interest [160, 177]. The orientational dynamics in such crystals can be studied by dielectric relaxation and by nmr techniques. One usually uses phenomenological models, such as the Frenkel [178] model of stepwise reorientation among the potential wells, to discuss the reorientational dynamics. We are not aware of any statistical-mechanical study of this interesting problem. One expects that an extended hydrodynamic approach, with nonhomogeneous density distribution included to account for the periodicity of the crystal, can be useful here.

E.   Orientational Relaxation in Nematic Liquid Crystals

Nematic liquid crystals are characterized by long-range orientational order, but no long-range positional order. Nematic liquid crystals are formed by charged rodlike molecules and also by ellipsoidals with large aspect (length-to-width) ratio. Because of the orientational anisotropy, the dielectric function is a tensor even in long wavelengths, with the two components $\varepsilon_\parallel(\omega)$ and $\varepsilon_\perp(\omega)$ that give polar response in the direction parallel and perpendicular to the axis of orientational order, respectively. A large number of experimental results are available in such systems. There are, however, few microscopic studies on collective orientational relaxation in such systems. The orientational dynamics in nematic liquid crystalline phases can be highly cooperative because concerted motions of nearest neighbors are needed for rotational motion of a molecule. These systems provide interesting theoretical problems for future study.

F.   Reorientational Dynamics of Molecules with an Internal Degree of Freedom

It will be interesting to study the collective orientational relaxation of molecules with internal degrees of freedom. Warchol and Vaughan [179] carried out a phenomenological analysis of orientational dynamics where molecules have free or hindered internal rotation. However, there exists no microscopic theory for this problem. The theory discussed in the previous sections can be formally extended to describe the collective orientational dynamics where molecules have one or more internal degrees of freedom. But the major difficulty arises in the calculation of the two-particle direct correlation function components for such cases. Thus, a detailed study of the equilibrium correlation functions of molecules with internal degrees of freedom is necessary to have a microscopic description of the collective orientational dynamics in those cases.

G. Polarizable Fluids

The study of collective orientational relaxation in polarizable polar fluids is another interesting problem. The importance of including the effects of polarizability lies in the fact that all real molecules are polarizable by electric fields, with both electronic and atomic motions contributing to the polarizability tensor, α. The effects of polarizability of the liquid molecules on the equilibrium correlations have been discussed by many authors [180]. For equilibrium properties the prescribed rule is that wherever the dipole moment μ appears in the rigid dipolar case, one must replace it by the actual dipole moment μ' of the molecules and wherever a function $f(\varepsilon_0)$ appears, one must replace it by $f(\varepsilon_0) - f(\varepsilon_\infty)$ [180]. This rule may not be valid for dynamics. A microscopic calculation of the collective orientational relaxation in polarizable dipolar liquids has not as yet been performed. This is clearly an important problem that deserves detailed study.

XVIII. CONCLUSION

The origin of collective dynamics in dense liquids is the strong intermolecular correlations that are inevitably present there. In atomic liquids, the consequences of the collective dynamics have been studied in great detail and several interesting phenomena such as the propagation of shear waves at intermediate wavenumbers, de Gennes' narrowing of the dynamic structure factor, total damping of sound waves at intermediate wavenumbers, and many others have been discovered. It is believed that the collective dynamics play a critical role in the anomalous rise of viscosity at low temperatures leading to the formation of the glassy state. It is becoming increasingly clear that for dense atomic liquids the slow collective relaxation at intermediate length scales has important *macroscopic* consequences.

However, in spite of the impressive advances in our understanding of the dynamics of atomic liquids, there have been few attempts to understand the collective orientational dynamics in molecular liquids, which are naturally more abundant. Just as in the spatial relaxation in atomic liquids, the collective orientational relaxation in molecular liquids is also expected to show rich and diverse behavior, some of which are discussed in this review. They include the slowing down of the longitudinal relaxation at the intermediate wavenumbers in the absence of the translational motion and also the acceleration of the same by the latter, the dramatic effect of intermolecular correlations on the solvation dynamics and on the charge transfer reactions, the reduction of the effects of dielectric friction on the dielectric relaxation and the single-particle rotation because of the translational modes and the bifurcation of $(\alpha\beta)$ relaxation times near the glass-transition temperature, and many others.

In this review, we have discussed and elaborated on an extended and generalized hydrodynamic approach to understand the collective orientational dynamics in dense dipolar liquids. The advantage of this approach is that within certain approximations it provides a detailed quantitative theory. Also, the extended hydrodynamic theory is a natural generalization of a similar theory used successfully in recent years for spatial relaxation to include orientational modes. Many of the predictions of this theory are in excellent agreement with experiments. For example, the predicted properties of dipolarons agree almost quantitatively with the computer simulation results of Pollock and Alder. The theory also provides a nice physical picture of the dynamics of solvation and of the nature of the collectivity in the orientational relaxation in a dense liquid. Because of the generality of the approach, the theory can be extended easily and applied to different problems, such as to a liquid of ellipsoidal molecules or to liquid crystals.

In Section XVII, we discussed some of the problems that need to be addressed in the near future. This list is by no means exhaustive, but we hope that it gives a glimpse of the many interesting problems that remain unexplored in the field of orientational dynamics. We believe that this field will be an exciting area of research in the coming years for both the experimentalists and the theoreticians and one can look forward to interesting new developments in the understanding of collective effects, in orientational relaxation in dense dipolar liquids.

Note added in proof

Since the preparation of this review, several interesting works on collective orientational relaxation and related topics have been reported. Wei and Patey [181] presented a detailed statistical mechanical study of orientational dynamics in molecular liquids. Two approximate schemes, the Kerr and Vineyard approximations, have been considered in this work. These authors have also calculated $\varepsilon(\omega)$ for the nonspherical model of Berne and Pecora [1] and found that in the latter model, $\varepsilon(\omega)$ can deviate significantly from the Debye result. Subsequently, Wei and Patey [181] have presented calculations of solvation dynamics using the theory developed in their earlier work [181]. The important aspect of their work is the use of the reference hypernetted chain (RHNC) approximation to evaluate the two particle direct correlation function of the liquid. It was shown that the use of RHNC for the $c(l_1 l_2 m; k)$ coefficients leads to significantly different results from the ones obtained by using MSA. However, the qualitative features, such as the divergence and the negative values of $\varepsilon_L(k)$, remain unchanged. Fried and Mukamel [183] have reported a detailed microscopic study of solvation dynamics of a classical ion. They used an approximation scheme for the dissipative kernels ($\mathbf{\Gamma}_T$ and $\mathbf{\Gamma}_\Omega$) which are different from the ones discussed here. Fried and Mukamel [183] also found the strong influence of the microscopic structure and of the translational modes of the host liquid in the dynamics of solvation. Friedman, Stell, and coworkers [184, 185] have recently presented a microscopic study of collective dynamics by using an interaction site representation of a diatomic dipolar molecule and thereby removing the point dipole assumption of the earlier theories. These authors recover the correct $k \to \infty$ limiting behavior of $\varepsilon_L(k)$ and also report several interesting studies. Maroncelli et al. [186] have reported an experimental study of solvation dynamics in methyl amide solvents. Maroncelli [187] has also reported a computer simulation study of solvation dynamics in acetonitrile.

Acknowledgments

It is a pleasure to thank Professor Graham R. Fleming, Professor Stuart A. Rice, and G. V. Vijayadamodar for collaboration and discussions on various aspects of orientational dynamics. Thanks are also due to Professor T. R. Kirkpatrick and Professor R. W. Zwanzig for discussions. This work was supported in part by the Council of Scientific and Industrial Research (INDIA) and by the Indian National Science Academy. Biman Bagchi is a recipient of the Homi Bhabha Fellowship (1989–1990).

References

1. B. Berne and R. Pecora, *Dynamic Light Scattering* (John Wiley, New York, 1976).

2. G. R. Fleming, *Chemical Applications of Ultrafast Spectoscopy* (Oxford University Press, Oxford, 1986).

3. C. G. E. Bottcher and P. Bordewijk, *Theory of Electric Polarization* (Elsevier, Amsterdam, 1979), Vol. II.

4. P. Madden and D. Kivelson, *Adv. Chem. Phys.* **LVI**, 467 (1984).

5. P. Madden and D. Kivelson, *Mol. Phys.* **30**, 1749 (1975).

6. J. P. Hansen and I. R. McDonald, *Theory of Simple Liquids* (Academic, London, 1986).

7. T. Keyes and B. M. Ladanyi, *Adv. Chem. Phys.*, **LVI**, 411 (1984).

8. C. J. Reid and M. W. Evans, in *Molecular Interactions*, H. Ratajczak and W. J. Orville-Thomas (eds.) (John Wiley, Chichester, 1982), Vol. 3.

9. R. G. Cole and G. T. Evans, *Ann. Rev. Phys. Chem.* **37**, 105 (1986).

10. W. A. Steele, *Adv. Chem. Phys.* **34**, 1 (1976).

11. W. Coffey, *Adv. Chem. Phys.* **63**, 69 (1985).

12. M. W. Evans, *Adv. Chem. Phys.* **62**, 183 (1985).

13. C. G. Grey and K. E. Gubbins, *Theory of Molecular Fluids* (Claredon, Oxford, 1984), Vol. I.

14. B. Berne, *J. Chem. Phys.* **62**, 1154 (1975).

15. E. Fatuzzo and P. R. Mason, *Proc. Phys. Soc. London* **90**, 741 (1967).

16. T.-W. Nee and R. Zwanzig, *J. Chem. Phys.* **52**, 6353 (1970).

17. A. Carrington and A. D. McLachlan, *Introduction to Magnetic Resonance* (Harper and Row, New York, 1967).

18. D. W. Oxtoby, *Adv. Chem. Phys.* **40**, 1 (1979).

19. H. Versmold, *Mol. Phys.* **37**, 201 (1986).

20. J. T. Hynes and P. G. Wolynes, *J. Chem. Phys.* **75**, 395 (1981); P. J. Stiles and G. B. Byrnes, *Chem. Phys. Lett.* **100**, 217 (1985).

21. B. Bagchi, D. W. Oxtoby, and G. R. Fleming, *Chem. Phys.* **86**, 257 (1984).

22. G. van der Zwan and J. T. Hynes, *J. Phys. Chem.* **89**, 4181 (1985).

23. B. Bagchi, *Ann. Rev. Phys. Chem.* **40**, 115 (1989).

24. M. Maroncelli, J. McInnis, and G. R. Fleming, *Science* **243**, 1674 (1989).

25. J. D. Simon, *Acct. Chem. Res.* **21**, 128 (1988).

26. P. F. Barbara and W. Jarzeba, *Acct. Chem. Res.* **21**, 195 (1988).

27. A. Chandra and B. Bagchi, *Chem. Phys. Lett.* **151**, 47 (1988).

28. B. Bagchi and A. Chandra, *J. Chem. Phys.* **90**, 7338 (1989).

29. A. Chandra and B. Bagchi, *J. Chem. Phys.* **91**, 1829 (1989).

30. A. Chandra and B. Bagchi, *J. Chem. Phys.* **91**, 2594 (1989).

31. A. Chandra and B. Bagchi, *J. Phys. Chem.* **93**, 6996 (1989).

32. C. Kalpouzos, D. McMorrow, W. T. Lotshaw, and G. A. Kenney-Wallace, *Chem. Phys. Lett.* **150**, 138 (1988).

33. D. McMorrow, W. T. Lotshaw, and G. A. Kenney-Wallace, *J. Quantum Elec.* **24**, 443 (1988).

34. P. Boon and S. Yip, *Molecular Hydrodynamics* (McGraw-Hill, New York, 1980).

35. B. Bagchi and A. Chandra, *Phys. Rev. Lett.* **64**, 455 (1990).

36. J. L. Lebowitz and J. K. Percus, *J. Math. Phys.* **4**, 116 (1963).

37. R. Kubo, *Rep. Prog. Phys.* **29**, 255 (1966).

38. H. Mori, *Prog. Theor. Phys.* **33**, 423 (1965).

39. R. Zwanzig, in *Lectures in Theoretical Physics*, W. E. Britton et al. (eds.) (Interscience, New York, 1961), Vol. III, p. 135.

40. P. C. Hohenberg and B. I. Halperin, *Rev. Mod. Phys.* **49**, 435 (1977).

41. I. de Schepper and E. G. D. Cohen, *J. Stat. Phys.* **27**, 225 (1982).

42. T. R. Kirkpatrick, *Phys. Rev. A* **32**, 3130 (1985).

43. T. R. Kirkpatrick and J. C. Nieuwoudt, *Phys. Rev. A* **33**, 2651 (1986).

44. A. Chandra and B. Bagchi, *J. Phys. Chem.* **94**, 3159 (1990).

45. D. F. Calef and P. G. Wolynes, *J. Chem. Phys.* **78**, 4145 (1983).

46. B. Bagchi, *J. Chem. Phys.* **82**, 5677 (1985); **85**, 4667 (1986).

47. B. Bagchi, *Chem. Phys. Lett.* **125**, 91 (1986).

48. B. Bagchi *Physica A* **145**, 273 (1987).

49. M. S. Wertheim, *J. Chem. Phys.* **55**, 4291 (1971).

50. P. G. de Gennes, *Physica* **25**, 825 (1959).

51. U. M. Titulaer and J. Deutch, *J. Chem. Phys.* **60**, 1502 (1974).

52. T. R. Kirkpatrick, *Phys. Rev. A* **31**, 939 (1985).

53. H. Frohlich, *Theory of Dielectrics* (Oxford University Press, Oxford, 1958).

54. W. E. vaughan, *Ann. Rev. Phys. Chem.* **30**, 103 (1979). (a) B. K. P. Scaife, *Dielectric and Related Molecular Processes* (The Chemical Society, London, 1972), Vol. I.

55. P. Colonomos and P. G. Wolynes, *J. Chem. Phys.* **71**, 2644 (1979).

56. A. L. Nichols III and D. F. Calef, J. Chem. Phys. **89**, 3783 (1988).

57. J. B. Hubbard, R. F. Kayser, and P. J. Stiles, *Chem. Phys. Lett.* **95**, 399 (1983).

58. P. J. Stiles and J. B. Hubbard, *Chem. Phys.* **84**, 431 (1984).

59. G. van der Zwan and J. T. Hynes, *Physica A* **121**, 227 (1983).

60. G. van der Zwan and J. T. Hynes, *Chem. Phys. Lett.* **101**, 367 (1983).

61. E. L. Pollock and B. J. Alder, *Ann. Rev. Phys. Chem.* **32**, 311 (1981).

62. E. L. Pollock and B. J. Alder, *Phys. Rev. Lett.* **46**, 950 (1981).

63. R. L. Fulton, *Mol. Phys.* **29**, 405 (1975).

64. M. Neumann, *Mol. Phys.* **57**, 97 (1986).

65. R. F. Loring and S. F. Mukamel, *J. Chem. Phys.* **87**, 1272 (1987).

66. R. W. Zwanzig, *J. Chem. Phys.* **38**, 2766 (1963).

67. A. Chandra and B. Bagchi, *J. Chem. Phys.* **90**, 1832 (1989).

68. A. Chandra and B. Bagchi, *J. Chem. Phys.* **91**, 3056 (1989).

69. O. V. Dolgov, D. A. Kirzhinits, and E. G. Maksimov, *Rev. Mod. Phys.* **53**, 81 (1981).

70. J. G. Kirkwood, *J. Chem. Phys.* **4**, 592 (1936); H. Frohlich, *Trans. Farad. Soc.* **44**, 238 (1948).

71. A. Chandra and B. Bagchi, *J. Chem. Phys.* (to be published).

72. J. P. Hansen and I. R. McDonald, *Phys. Rev. Lett.* **41**, 1379 (1978).

73. R. Lobo, J. E. Robinson, and S. Rodriguez, *J. Chem. Phys.* **59**, 5922 (1973).

74. G. Ascarelli, *Chem. Phys. Lett.* **39**, 23 (1976).

75. A. Chandra and B. Bagchi, *J. Chem. Phys.* **92**, 6833 (1990).

76. R. W. Zwanzig, *Phys. Rev.* **144**, 170 (1966).

77. R. W. Zwanzig, in *Statistical Mechanics: New Concepts, New Problems, New Applications,* S. A. Rice, K. F. Freed, and J. C. Light (eds.) (University of Chicago Press, Chicago, 1972).

78. G. V. Vijayadamodar, A. Chandra, and B. Bagchi, *Chem. Phys. Lett.* **161**, 413 (1989).

80. P. Madden and D. Kivelson, *J. Phys. Chem.* **86**, 4244 (1982).

81. G. A. Kenney-Wallace, S. Poona, and C. Kalpouzos, *Farad. Diss. Chem. Soc.* **85**, 185 (1988).

82. E. F. G. Templeton and G. A. Kenney-Wallace, *J. Phys. Chem.* **90**, 2896 (1986).

83. H. Mori, *Prog. Theor. Phys.* **34**, 399 (1965).

84. J. B. Hasted, S. K. Hussain, E. A. M. Frescura, and J. R. Birch, *Chem. Phys. Lett.* **118**, 622 (1985).

85. J. Barthel, K. Bachhuber, R. Buchner, and H. Hetzenauer, *Chem. Phys. Lett.* **165**, 369 (1990).

86. See, for example, R. M. Nielson, G. E. McManis, M. N. Golvin, and M. J. Weaver, *J. Phys. Chem.* **92**, 3441 (1988).

87. The calculations reported in Ref. 35 were restricted to the evaluation of the long-wavelength dielectric function, $\varepsilon(\omega)$, although wavevector and frequency-dependent relaxation times were calculated. The latter quantities also provide a description of the non-Markovian effects in the collective orientational relaxation, as discussed here. Note that these relaxation times can be used straightforwardly to study memory effects in solvation dynamics. (A. Chandra and B. Bagchi, unpublished work).

88. K. S. Cole and R. H. Cole, *J. Chem. Phys.* **9**, 341 (1941).

89. M. Evans, G. J. Evans, W. T. Coffey, and P. Grigolini, *Molecular Dynamics and Theory of Broad Band Spectroscopy* (Wiley, New York, 1982).

90. H. Farber and S. Petrucci, in *The Chemical Physics of Solvation, Part B*, R. R. Dogonadze, E. Kalman, A. A. Kornyshev, and J. Ulstrup (eds.) (Elsevier, Amsterdam, 1986), p. 433.

91. S. H. Glarum, in *Dielectric Properties and Molecular Behaviour*, N.E. Hill et al. (eds.) (Van Nostrand, London, 1969).

92. J. G. Powles, *J. Chem. Phys.* **21**, 633 (1953).

93. J. Deutch, *Farad. Sym. Chem. Soc. London* **11**, 26 (1977).

94. R. Glauber, *J. Math. Phys.* **4**, 294 (1963).

95. J. E. Shore and R. W. Zwanzig, *J. Chem. Phys.* **63**, 5445 (1975).

96. J. L. Skinner, *J. Chem. Phys.* **79**, 1955 (1983); J. Budimir and J. L. Skinner, *J. Chem. Phys.* **82**, 5232 (1985).

97. S. H. Glarum, *J. Chem. Phys.* **33**, 639 (1960).

98. P. Bordewijk, *Chem. Phys. Lett.* **32**, 592 (1975).

99. D. W. Oxtoby, *Mol. Phys.* **34**, 987 (1977).

100. B. Bagchi and A. Chandra, *J. Chem. Phys.* **93**, 1955 (1990).

101. P. G. Wolynes, *J. Chem. Phys.* **86**, 5133 (1987).

102. I. Rips, J. Klafter, and J. Jortner, *J. Chem. Phys.* **88**, 3246 (1988).

103. I. Rips, J. Klafter, and J. Jortner, *J. Chem. Phys.* **89**, 4288 (1988).

104. B. Bagchi and A. Chandra, *Proc. Indian Acad. Sci.* (*Chem. Sci.*), **100**, 353 (1988); A. Chandra and B. Bagchi, *Proc. Indian Acad. Sci.* (*Chem. Sci.*) **101**, 83 (1989).

105. B. Bagchi and A. Chandra, *Chem. Phys. Lett.* **155**, 533 (1989).

106. A. Chandra and B. Bagchi, *Chem. Phys. Lett.* **165**, 93 (1990).

107. A. Chandra and B. Bagchi, *J. Phys. Chem.* **94**, 1874 (1990).

108. A. Chandra and B. Bagchi, *J. Chem. Phys.* (to be published).

109. S.-G. Su and J. D. Simon, *J. Phys. Chem.* (to be published).

110. L. D. Zusman, *Chem. Phys.* **49**, 295 (1980).

111. I. V. Alexandrov, *Chem. Phys.* **51**, 449 (1980).

112. D. F. Calef and P. G. Wolynes, *J. Phys. Chem.* **87**, 3387 (1983).

113. J. T. Hynes, *J. Phys. Chem.* **90**, 3701 (1986).

114. H. Sumi and R. A. Marcus, *J. Chem. Phys.* **84**, 4272 (1986).

115. W. Nadler and R. A. Marcus, *J. Chem. Phys.* **86**, 3096 (1987).

116. M. Sparpaglione and S. Mukamel, *Chem. Phys.* **88**, 3263 (1988).

117. I. Rips and J. Jortner, *J. Chem. Phys.* **87**, 2090 (1987).

118. L. D. Zusman, *Chem. Phys.* **51**, 119 (1988).

119. J. S. Bader and D. Chandler, *Chem. Phys. Lett.* **157**, 501 (1989).

120. D. A. Zichi, G. Ciccotti, J. T. Hynes, and M. Ferrario, *J. Phys. Chem.* **93**, 6261 (1989).

121. T. Kakitani and N. Mataga, *J. Phys. Chem.* **90**, 993 (1986); **91**, 6277 (1987); Y. Hatano, M. Saito, T. Kakitani, and N. Mataga, *J. Phys. Chem.* **92**, 1008 (1988).

122. A. Yoshimori, T. Kakitani, Y. Enomoto, and N. Mataga, *J. Phys. Chem.* **93**, 8316 (1989).

123. M. Tachiya, *Chem. Phys. Lett.* **159**, 505 (1989).

124. E. A. Carter and J. T. Hynes, *J. Phys. Chem.* **93**, 2184 (1989).

125. J. K. Hwang and A. J. Warshel, *J. Am. Chem. Soc.* **109**, 715 (1987).

126. R. A. Kuharski, J. B. Bader, D. Chandler, M. Spirk, M. L. Klein and R. W. Impey, *J. Chem. Phys.* **89**, 3248 (1988).

127. R. M. Nielson, G. E. McManis, and M. J. Weaver, *J. Phys. Chem.* **93**, 4703 (1989).

128. G. E. McManis and M. J. Weaver, *J. Chem. Phys.* **90**, 912 (1989).

129. J. D. Simon and S. G. Su, *J. Chem. Phys.* **87**, 7016 (1987).

130. S. G. Su and J. D. Simon, *J. Chem. Phys.* **89**, 908 (1988).

131. E. M. Kosower and D. Huppert, *Ann. Rev. Phys. Chem.* **37**, 127 (1986).

132. M. A. Kahlow, W. Jarzeba, T. J. Kang, and P. F. Barbara, *J. Chem. Phys.* **90**, 151 (1989).

133. P. F. Barbara and W. Jarzeba, *Adv. Photochem.* (to be published).

134. R. A. Marcus, *J. Chem. Phys.* **24**, 966 (1956); **26**, 867 (1957).

135. R. A. Marcus, *Ann. Rev. Phys. Chem.* **15**, 155 (1964).

136. R. F. Grote and J. T. Hynes, *J. Chem. Phys.* **73**, 2715 (1980).

137. J. T. Hynes, in *The Theory of Chemical Reaction Dynamics*, M. Baer (ed.) (CRC Press, Boca Raton, FL, 1985), Vol. IV, p. 171.

138. B. Bagchi, A. Chandra, and G. R. Fleming, *J. Phys. Chem.* **94**, 5197 (1990).

139. M. A. Kahlow, T. J. Kang, and P. F. Barbara, *J. Phys. Chem.* **91**, 6952 (1987).

140. T. J. Kang et al., *J. Phys. Chem.* **92**, 6800 (1988).

141. B. Bagchi, G. R. Fleming, and D. W. Oxtoby, *J. Chem. Phys.* **78**, 7375 (1983).

142. B. Bagchi, *Intern. Rev. Phys. Chem.* **6**, 1 (1987).

143. B. Bagchi and G. R. Fleming, *J. Phys. Chem.* **94**, 9 (1990).

144. B. Bagchi, A. Chandra, and G. R. Fleming *J. Chem. Phys.* (to be published).

145. P. A. Madden, *Molecular Liquids—Dynamics and Interactions*, A. J. Barnes et al. (eds.) (Reidel, Dordrecht, 1984), p. 413.

146. D. A. Kivelson and P. A. Madden, *Ann. Rev. Phys. Chem.* **31**, 523 (1980).

147. P. A. Madden and D. J. Tildesley, *Mol. Phys.* **49**, 193 (1983).

148. D. J. Tildesley and P. A. Madden, *Mol. Phys.* **48**, 129 (1983).

149. T. I. Cox and P. A. Madden, *Mol. Phys.* **39**, 1487 (1980).

150. B. Berne and P. Pechukas, *J. Chem. Phys.* **56**, 4213 (1972).

151. C. M. Hu and R. W. Zwanzig, *J. Chem. Phys.* **60**, 4354 (1974).

152. (a) M. P. Warchol and W. Vaughan, *J. Chem. Phys.* **65**, 1374 (1976).
     (b) M. P. Allen and D. Frenkel, *Phys. Rev. Lett.* **59**, 1148 (1987).

153. A. Chandra and B. Bagchi *Physica A* **169**, 246 (1990).

154. J.-L. Colot, X.-G. Wu, H. Xu, and Baus, *Phys. Rev. A* **38**, 2022 (1988).

155. J. M. Wacrenier, C. Druon, and D. Lippens, *Mol. Phys.* **43**, 97 (1981).

156. G. D. Patterson and A. Munos-Rojas, *Ann. Rev. Phys. Chem.* **38**, 191 (1987).

157. G. P. Johari and M. Goldstein, *J. Chem. Phys.* **53**, 2372 (1970).

158. G. P. Johari and M. Goldstein, *J. Phys. Chem.* **74**, 2034 (1970).

159. G. P. Johari and M. Goldstein, *J. Chem. Phys.* **55**, 4245 (1971).

160. G. P. Johari, *Annals of N.Y. Acad. Sci.* **279**, 117 (1976).

161. D. Kivelson and S. A. Kivelson, *J. Chem. Phys.* **90**, 4476 (1989).

162. B. Bagchi, A. Chandra, and S. A. Rice *J. Chem. Phys.* (to be published).

163. A. Chandra and B. Bagchi, *J. Phys. Chem.* (to be published).

164. S. A. Adelman and J. M. Deutch, *J. Chem. Soc.* **59**, 3971 (1973).

165. D. Isbister and P. J. Bearman, *Mol. Phys.* **28**, 1297 (1974).

166. R. H. Cole, in *Molecular Liquids: Dynamics and Interactions*, A. J. Barnes, W. J. Orville-Thomas, and J. Yarwood (eds.) (Reidel, Dordrecht, 1984), p. 59.

167. D. J. Denny and R. H. Cole, *J. Chem. Phys.* **23**, 1767 (1955).

168. D. J. Denny, *J. Chem. Phys.* **30**, 1014 (1959).

169. J. Batrhel and F. Fenerleim, *J. Soln. Chem.* **13**, 393 (1984).

170. B. Gestblom, A. Elsamahy, and J. Sjoblom, *J. Soln. Chem.* **14**, 375 (1985).

171. P. Debye, *Polar Molecules* (Chemical Catalog Co., New York, 1929).

172. R. W. Zwanzig, *Lecture Notes in Physics*, **132**, 198 (1980).

173. M. Maroncelli and G. R. Fleming, *J. Chem. Phys.* **89**, 5044 (1988).

174. A. Chandra and B. Bagchi (unpublished).

175. D. Bertolini, M. Cassettari, M. Ferrario, P. Grigolini, and G. Salvetti, *Adv. Chem. Phys.* **62**, 277 (1985).

176. H. E. Stanley and J. Teixeira, *J. Chem. Phys.* **74**, 3404 (1980).

177. I. Darmon and C. Brot, *Mol. Crystals* **3**, 301 (1987).

178. J. Frenkel, *Acta Phys. Chim.* 555 **R3**, No. 123 (1935).

179. M. P. Warchol and W. E. Vaughan, *J. Chem. Phys.* **67**, 486 (1977).

180. M. S. Wertheim, *Ann. Rev. Phys. Chem.* **30**, 471 (1979).

181. D. Wei and G. N. Patey, *J. Chem. Phys.* **91**, 7113 (1989).

182. D. Wei and G. N. Patey, *J. Chem. Phys.* **93**, 1399 (1990).

183. L. E. Fried and S. Mukamel, *J. Chem. Phys.* **93**, 932 (1990).

184. Y. Zhou, H. L. Friedman, and G. Stell, *J. Chem. Phys.* **91**, 4885 (1989).

185. F. W. Raineri, Y. Zhou, H. L. Friedman, and G. Stell, *Chem. Phys.* (to be published).

186. C. F. Chapman, R. S. Fee, and M. Maroncelli, *J. Phys. Chem.* **94**, 4929 (1990).

187. M. Maroncelli, (preprint)

MECHANISTIC CLASSIFICATION OF CHEMICAL OSCILLATORS AND THE ROLE OF SPECIES

MARKUS EISWIRTH*, ALBRECHT FREUND**, and JOHN ROSS

Department of Chemistry, Stanford University, Stanford, California

CONTENTS

I. INTRODUCTION

Oscillating chemical reactions can exhibit a variety of dynamical behavior, such as limit cycles of simple or higher periodicity, mixed-mode oscillations, quasiperiodicity and chaos. Systems showing complex dynamics generally exhibit simple periodic oscillations at a different range of control parameters.

*Present address: Fritz-Haber-Institut der MPG, D-1000 Berlin 33, FRG.
**Present address: Institut für Physikalische Chemie der Universität, D-8700 Würzburg, FRG.

Advances in Chemical Physics Volume LXXX, Edited by I. Prigogine and Stuart A. Rice
ISBN 0-471-53281-9 © 1991 John Wiley & Sons, Inc.

Since the mechanistic requirements for the existence of more complicated attractors certainly depend on how these simple oscillations arise in the first place, the latter are the natural starting point for investigating a reaction mechanism. Consequently, this chapter is restricted to the discussion of simple limit cycles. Although there are several ways, known from bifurcation theory, by which stable limit cycles can arise {besides the Hopf bifurcation, these are the saddle-loop (SL), the saddle-node-homoclinic orbit (SN-hom), and the saddle-node of periodic orbits (SNP) bifurcations [1,2]}, all systems we are aware of include a supercritical Hopf bifurcation in some region of parameter space. Thus, it does not appear to be any loss of generality to restrict the study to oscillations stemming from a Hopf bifurcation, until a counterexample is found.

The goal of all mechanistic studies is first to identify the chemical nature of all species involved and second to clarify the role each species plays in the mechanism. Although there are many analytical techniques available for the identification of chemical species, their applicability depends on the details of the chemistry involved, so that it is not possible to give general procedures for the first problem. For the second one, the roles of the species must first be defined, their discernible roles examined (e.g., autocatalytic species, products, species causing a negative feedback, etc.), and their necessity established to obtain oscillations in a chemically realistic way. Next (experimentally applicable) methods have to be devised to identify the role of each species so that a mechanism can be constructed. Since quite a number of chemical oscillators are known, it follows that they have to be classified in such a way that the same definitions and identification methods of roles work within each category.

In order to avoid ambiguity in the definitions, we restrict ourselves to simple oscillators, by which we mean systems that {in the language of stoichiometric network analysis (SNA) [3]} contain only one nonweak current cycle, that is, no more than one source of instability in the essential species, (cf. Section II). This excludes coupled oscillators, but also excludes, for example, systems that contain an oscillating and a bistable subsystem. Such oscillators are probably best studied by breaking them up, if possible, and elucidating the mechanisms of the subsystems separately (which may then be simple oscillators).

We begin with an operational categorization of species into those essential and nonessential for oscillations and then give a method to reduce a mechanism to a skeleton by elimination of the nonessential species. Afterward, a classification of simple oscillators and a definition of the roles of essential species is offered, for which we use Clarke's work on stoichiometric networks [3]. All the abstract models as well as the realistic mechanisms we have studied, 25 in total (see Appendix A), can be classified according to four qualitatively distinct categories: more precisely, two categories, for one of which three subtypes are distinguished. Typical skeleton models for these are presented. Operational means are suggested by which a given unknown oscillator can be sorted into its category and the role of each species in that category can be identified. These methods may be helpful in mechanistic studies, as discussed in some detail in Section IV.

A number of ways can be tried to identify the role of species, for instance, their properties in the autonomous oscillations (amplitude, phase) can be used. In addition one can study the response of the system to external perturbations of a concentration that can be applied as a single pulse [4, 5], or a superimposed constant or periodic flow of the species [6–12]. (Time-delayed perturbations [13, 14] have been studied both experimentally and theoretically, but we have not considered this technique in the present article.)

The present study is based partly on analytical and diagrammatic techniques from stoichiometric network analysis [3] and partly on numerical simulations. The latter techniques are mainly used to obtain bifurcation diagrams in two control parameters and to calculate the effects of external perturbations, namely, changes of the stability of a focus and shifts of its steady-state concentrations (Section III). Thus, these numerical calculations provide information about the system that can, at least in part, also be obtained in an experimental situation. As shown by use of some typical examples, the assignment of a mechanism and its species to the various categories can be achieved using this information (see Section IV). There is, however, no proof that the categories found are exhaustive. Thus, there may be other oscillators (yet to be discovered) that are not classifiable with the methods given here, but may require additional categories, as well as refined means to distinguish them from each other.

The computations were performed on a VAX 3600 computer with double precision NAG routines for numerical integration and the calculation of fixed points and eigenvalues of (Jacobi) matrices. Some bifurcation diagrams were obtained with E. Doedel's AUTO package.

II.   ESSENTIAL AND NONESSENTIAL SPECIES

Chemical oscillations generally involve a large number of species, but usually only relatively few of them are required for a qualitative understanding of the oscillatory behavior (skeleton mechanisms [15, 16]). In this section a general definition of species that are nonessential for the oscillatory behavior of a particular system is given. Furthermore, experimentally applicable methods for their identification are suggested. Once identified, the non-essential species can be neglected in mechanistic studies. This approach simplifies these studies, insofar as the observed behavior can be modeled qualitatively with fewer species and the decisive nonlinearities must be contained in the differential equations of these remaining species. Some or all nonessential species can then be included in an enlarged model to improve quantitative agreement with experiment.

A supercritical Hopf bifurcation is used to define nonessential species, because all oscillating chemical systems found in the literature have this bifurcation in common. Upon variation of a control parameter (such as flow rate, temperature, or feed concentrations), a stable focus loses its stability and a stable limit cycle with continuously increasing amplitude is formed. This manifests itself in the two complex conjugate eigenvalues of the linearized system, the real parts of which change sign at the bifurcation (i.e., cross the imaginary axis from left to right) [1, 17]. The theoretical description of a Hopf bifurcation [17] in an n-variable system uses separation of slow and fast variables that set up the two-dimensional center manifold and the stable manifold of codimension two, respectively. In linear approximation these manifolds become identical to the eigenspaces spanned by the two complex conjugate and the other $(n - 2)$ real eigenvectors of the Jacobian [18] of the focus (center) at the bifurcation. When looking at the relationship among these manifolds and observable quantities, such as amplitudes and phases of oscillations, it is advantageous to use relative instead of absolute concentrations as coordinate axes. The relative concentration of a species is obtained by dividing its concentration by its stationary-state concentration or, equivalently, normalizing in such a way that the concentrations at the stable or unstable focus are all unity. For power-low (mass-action) kinetics a species exhibiting high absolute concentration change can have a very small effect on the oscillatory behavior, as long as its relative amplitude is small compared to the other oscillating species.

To a linear approximation (i.e., close to the Hopf bifurcation), there are simple relationships between the two complex conjugate eigenvectors with components $\mathrm{Re}_k \pm \mathrm{Im}_k$, which span the center eigenspace, and the relative amplitudes A_k and phases ϕ_k of the kth oscillating species: $A_k^2 \sim \mathrm{Re}_k^2 + \mathrm{Im}_k^2$ and $\tan \phi_k = \mathrm{Im}_k / \mathrm{Re}_k$. These relationships are unambiguous except for normalization. Species with high amplitude give a large contribution to the center eigenspace, while this contribution and A_k vanish if the concentration axis of the species in question is perpendicular to it. At or sufficiently close to a Hopf bifurcation, it is thus possible to construct the center eigenspace from observable quantities.

The location of the stable eigenspace can be probed close to the bifurcation by measurement of the quench vector for each species, as defined and carried out experimentally by Hynne and Sørensen [4, 5]. By rapid, pulselike, addition of a certain amount of a species at a certain phase of the oscillation,

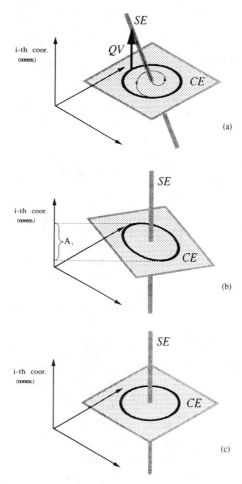

FIGURE 1.   Position of the stable (SE) and center eigenspace (CE) relative to the ith co-ordinate axes corresponding to a nonessential species, at a Hopf bifurcation for the three types of nonessential species in their limits: (a) Limiting case of a type-a nonessential species: the corresponding ith coordinate is perpendicular to CE. The quench vector (QV) is finite. A trajectory of a quench experiment is also indicated. (b) Limiting case of a type-b nonessential species: the ith coordinate axis runs parallel to SE. The oscillations in the ith coordinate show a finite amplitude A_i. (c) Limiting case of a type-c nonessential variable: the ith coordinate axis is both perpendicular to CE and parallel to SE. According QV approaches infinity, while A_i vanishes.

both of which have to be determined experimentally, the system is transferred onto the stable eigenspace, which leads to quenching of the oscillations and subsequent spiraling out of the trajectories from the unstable focus to the limit cycle (see Fig. 1a). In general, the part of phase space accessible by change of the concentration of one species from a limit cycle (a two-dimensional cylinder) and the stable eigenspace (codimension two) will intersect in two points, if at all. Only one of these can be reached by addition of the species in question. The other one is accessible by subtraction, that is, reactive removal via a selective scavenger, for example, Ag^+ to lower the concentration of Br^- [5]. The lengths of the quench vectors are, however, unique for each species. They have a minimum when the respective coordinate axis is perpendicular, and become infinite when it is parallel, to the stable manifold.

A distinction between essential and nonessential species was first put forward by Termonia and Ross, based on numerical studies of a four-variable model for glycolysis [19, 20]. They observed that when two of the species (namely, fructose-6-phosphate or phosphoenol pyruvate or both) are kept constant, the others still oscillate, albeit with changed amplitude and period, whereas the oscillations in all species are suppressed, if one of the other species is kept constant [21, 22]. These studies on glycolysis suggest that there are experimental methods to determine the minimum set of species required to explain qualitatively the oscillatory behavior, that is, when reducing the system to this minimum, the Hopf bifurcation is preserved. Quantitative properties, however, such as the locations of the oscillatory region and the bifurcating set, as well as the amplitude and the period of the oscillations, can be and usually are altered.

A. Definition of Nonessential Species

To define nonessential species, it is first of all necessary to discuss what roles they can play in an oscillating mechanism. As a start, we consider limiting cases, in which fixing the concentration of a nonessential species leaves the system not only qualitatively, but also quantitatively unchanged. In the eigenvalue problem of a system of n variables, fixing the value of the ith species means that, instead of an $n \times n$ Jacobian, only a submatrix of it has to be considered, which is obtained by deleting the ith row and column in the original matrix. The remaining eigenvalues of this reduced $(n-1) \times (n-1)$ matrix are identical to those of the original Jacobian, if all off-diagonal elements of the latter in the ith row (case a) or the ith column (case b) or both (case c) vanish. These are the three limiting cases that need to be distinguished. In such cases the matrix of eigenvectors has the same form as the Jacobian, so that it suffices to regard the Jacobian in order to learn about the topology of the system.

As was already pointed out, at a Hopf bifurcation two eigenvalues of the Jacobian are complex conjugates. One of the real eigenvalues belongs to the ith row and column representing the nonessential species and is absent in the reduced $(n-1) \times (n-1)$ matrix. Thus, at a Hopf bifurcation the topological interpretation of the preceding limiting cases a–c is particularly easy and instructive:

a. For vanishing off-diagonal elements in the ith row of the Jacobian, the ith coordinate axis is perpendicular to the center eigenspace (Fig. 1a). Consequently, the species corresponding to this axis does not oscillate at all.

b. In the case of vanishing off-diagonal elements in the ith column of the Jacobian, the ith coordinate axis is parallel to the stable eigenspace (Fig. 1b). Here the ith species has no influence on the other species, so that (constant or modulated) addition of that species has no effect on the dynamical behavior.

c. If the off-diagonal elements in both the ith row and ith column vanish, the corresponding axis is perpendicular to the center and parallel to the stable eigenspace (Fig. 1c). Such a species exhibits zero oscillation amplitude.

These situations are shown in Fig. 1 for a three-dimensional system. For more than three variables there are also more stable eigenvectors. An axis is then called parallel to the stable eigenspace if a vector pointing along the axis can be written as a linear combination of these eigenvectors.

The preceding limiting cases lead to the general definition of essential and nonessential species: Given a chemical mechanism with n species that exhibits a Hopf bifurcation; if, through continuous change of parameters, one of the limiting cases a–c can be reached for a species in such a way that the Hopf bifurcation remains—albeit shifted asymptotically to a new value of the control parameter—then the corresponding species is called nonessential of type $a, b,$ or c. A species for which none of the limiting cases can be reached in a way that the Hopf bifurcation remains is called essential.

The chemical interpretation of this mathematical definition, that is, the chemical roles of the different nonessential species in a mechanism, follows directly from the nonvanishing matrix elements in the respective limit. Type a denotes reactants that produce other (essential) species, but besides that only appear in reactions that can be neglected (and approach zero rate in the limit). Nonessential species of type b are, in the limit, inert products. If they do feed back into the system, this effect is not crucial. Species of type c react with at least one essential species or appear at least once as reactant and product. They can thus take part in several reactions, but can be constant without suppression of the oscillations. Near the limit, type c species are present in large excess (with possibly high absolute, but negligible relative amplitude of oscillation) and then correspond to the "major species" defined by Noyes et al. [23, 24] or the "external species" as defined by Clarke [3].

In the following we use simple examples first to demonstrate the preceding definition for nonessential species and second to discuss experimentally applicable methods to identify nonessential species in a system with an unknown mechanism. These examples are obtained by using the Brusselator [25] as the underlying mechanism with X and Y as essential species. The nonessential species are labeled $A-C$ according to their type. Note that $A-C$ do not correspond to the species labels normally used in the literature. Consider the following reaction mechanism occurring in an open system:

$$\to A \tag{1}$$

$$A \to X \tag{2a}$$

$$X \to A \tag{2b}$$

$$C + X \to Y \tag{3}$$

$$Y + 2X \to 3X \tag{4}$$

$$X \to B \tag{5a}$$

$$B \to X \tag{5b}$$

$$B \to \tag{6}$$

$$\to C \tag{7a}$$

$$C \to \tag{7b}$$

Processes with no species on either side are assumed to be in- or outflows. The rates of the preceding reactions are all set to unity, except the following: the rate for (2b) is written as $1/a$, for (5b) $1/b$, for (7a) $C_0 c$, and (7b) c. In this form the respective limiting cases are approached asymptotically as a, b, or c approach infinity. For example, using C_0 as control parameter, one can readily solve for the Hopf bifurcation analytically when considering only one nonessential species at a time. This means that three-variable models including X, Y, and either A, B, or C are examined, using the corresponding subset of the preceding reaction system [26]. Not surprisingly, in models obtained by fixing the concentration of either X or Y, no Hopf bifurcation is obtained: it recedes to infinity (diverging) instead of converging to a new value.

B. Identification of Nonessential Species

The preceding definition on nonessential species suggests a way to identify them by varying parameter values until one of the limiting cases is reached. This will generally not be possible in an experimental study, since it requires the variation of rate constants to infinitely large or vanishing values. It is, however, not necessary to do so; rather, it suffices that the limit exists, which can be concluded from other, experimentally applicable methods, even if it can only be incompletely approximated. We now discuss some of these methods using the Brusselator example.

For species A in the preceding example one can simply substitute a flow in X, which eliminates A altogether. This is always possible for all type a species, although a flow in several species may be required, depending on what A decays to (e.g., $A_1 + A_2 \rightarrow X + Y$). As a matter of fact, we have not found any "real" examples for type a species, though Sørensen and Hynne [5] have mentioned the possibility of a nonoscillating species with finite quench vector corresponding to a type a species at the limit (cf. Table I).

Process (5) can be made irreversible not only by suppressing (5b), but also by speeding up process (6), that is, through fast removal of B from the reactor. A real example is Br_2 in the minimal bromate oscillator, which can be removed by using an inert carrier gas like nitrogen. The higher the gas flow through the reaction mixture, the closer one comes to the limit in which Br_2 can be regarded as an inert product (its disproportionation into essential species can be neglected). Whether an effective and sufficiently selective scavenger for removal of a b type species can be found depends of course on the chemistry involved.

If oscillations can be studied under conditions where C is present in large excess, it can easily be identified as nonessential. This would be the case, if

TABLE I

Behavior of Nonessential Species as Limiting Case is Approached, Compared to an Essential Species[a]

Type	Relative Amplitude	Quench Vector	Change in Phase Shift	Perturbation Amplitude
a	$\rightarrow 0$	Finite	$\rightarrow 0$	Finite
b	Finite	$\rightarrow \infty$	$\rightarrow 0$	$\rightarrow 0$
c	$\rightarrow 0$	$\rightarrow \infty$	$\rightarrow 0$	Finite
Essential	Finite	Finite	$\sim 180°$	Finite

[a]The quantities are defined in the text.

the rate constant for (3) were considerably smaller than the others that would be compensated for by increasing C_0. If not, one can try to buffer C or, if possible, maintain a high flow of C [high rates of (7a) and (7b)] through the system independent of the other flows. The latter is, for instance, the method of choice to check whether temperature is an essential variable, because more heat can always be pumped through a reactor without affecting the flow of reactants and products (i.e., the flow of heat and mass can be decoupled by appropriate reactor design). As a rule, all species that need to be present in the initial mix for a batch oscillator are usually nonessential of type c (this includes the major species).

Some observable properties of the three kinds of nonessential species vanish or diverge in the respective limits, but far from these limits exhibit pronounced differences compared to the same properties of essential species. If these differences are found in a system, it is not necessary to approach the limit in order to use the following properties (compiled in Table I) for the identification of nonessential species:

- The relative amplitude A_{rel} in the autonomous oscillations are much smaller for nonessential species of types a and c compared with the other species. As was mentioned, it vanishes in the limit (see Fig. 1).

- Types b and c nonessential species display much higher quench vectors (QV) than the other species. In the limit they approach infinity (see Fig. 1).

- The total change of the phase shift $\Delta\Phi$ with respect to a sinusoidal perturbation in the species under consideration (taken as the distance between maxima) from one fundamental entrainment edge to the other is π for essential species [27], but is smaller for all nonessential species [22]. This reflects the fact that nonessential species in the limit (fixed values) are actually parameters and display like all other parameters zero phase shift with respect to their own perturbation.

- For type b nonessential species the relative perturbation amplitude A_{pert}, required to obtain phase locking under given conditions is much higher than those for the other species. The reason is that the coupling of type b species with the essential species vanishes as the limit is approached. Accordingly, an increasingly high and finally infinite $A_{pert.}$ is necessary to obtain entrainment.

Figure 2 shows some properties of species C from the preceding reaction scheme as the limiting case is approached ($c \to \infty$).

The differences between essential and nonessential species are often one or even several orders of magnitude, but usually not all nonessential species can be identified by the same method. Quenching experiments have recently

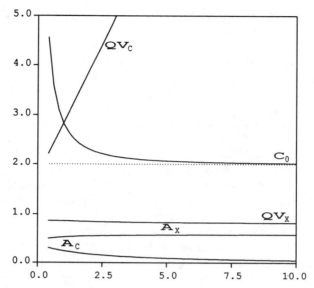

FIGURE 2. Behavior of the Brusselator reaction scheme [Eqs. (1)–(7b)] at the Hopf bifurcation as the limiting case for the nonessential species C is approached ($c \to \infty$). The bifurcation point in the dependent parameter C_0 converges to a finite value (dotted line). The quench vector QV_X of the essential species X and the relative amplitude of its oscillations A_X remain finite. In contrast the quench vector QV_C of the type c nonessential species, C goes to infinity and the corresponding relative amplitude approaches zero.

been carried out for the BZ reaction [5]. The authors found small quench vectors for Ce^{4+}, $HBrO_2$, Br^-, and Br_2. The QV for HOBr was substantially higher, while no quenching could be obtained with the organic species, like bromomalonic acid (which means that their quench vectors were too high to be achievable). From this study one can conclude that the organic acids are definitely nonessential, HOBr probably so. While Br_2 is nonessential in the FKN mechanism (see Section III. A.4), such a conclusion cannot be reached on the basis of its quench vector.

In the following description of unstable features and simple abstract models, it is assumed that the nonessential species have been eliminated. Types a and b are left out of the mechanism altogether; type c is assumed constant and included in the reaction rates, so that the processes are written as pseudoreactions [3], for instance process (3) as $X \to Y$ with an effective rate proportional to the concentration of species C. For the sake of clarity some nonessential species are shown in the discussion of realistic models, but as far as the decisive features of an oscillating mechanism are concerned they can all be eliminated or included in the parameters.

III. CLASSIFICATION OF OSCILLATORS AND ROLE OF ESSENTIAL SPECIES

The identification of the nonessential species does not say much about the mechanism of the oscillator, but simplifies the problem, since the decisive roles must be played by the remaining (essential) species. In order to define (and identify) these roles, it is required to find out what chemically realistic possibilities exist that give rise to oscillations and what kinds of species have to be present.

The importance of both positive and negative feedback in chemical oscillators has been stressed by numerous authors, for example, Tyson [28] and Franck [29, 30]. The most rigorous approach to this problem, due to Clarke, makes maximum use of stoichiometric relationships and is therefore called stoichiometric network analysis (SNA) [3, 31–33]. An important feature of SNA is that it allows qualitative statements about the stability of the stationary states of a mechanism without knowledge of the rate constants, and without computation, by applying a diagrammatic approach to stability analysis [3], which needs to be sketched briefly here.

A mechanism (= *network*) is regarded to consist of the stoichiometry and reaction orders (kinetic exponents) of all elementary reaction steps; the parameters (rate constants) are all assumed to be nonnegative. A network is stable if it does not exhibit any unstable stationary state for arbitrary parameter combinations, and unstable if there exists at least one unstable state in some region of parameter space. Both the stoichiometric coefficients and the kinetic exponents can be written in matrix form, with the row (column) index referring to the species (reaction), respectively, so that the question of network stability is closely related to a matrix stability problem [28, 34–36]. Mechanisms are conveniently represented in so-called network diagrams D_N, in which the species symbols are connected by arrows denoting reactions in such a way that the number of barbs (total number of feathers) of an arrow at a product (reactant) is equal to the stoichiometric coefficient of this product (reactant) in the respective reaction, while the number of left feathers denotes the kinetic exponent of the reactant [3] (cf. Fig. 3). By convention no feathers are shown when the stoichiometric and kinetic coefficient of a reactant are both unity [3]. The results of SNA apply equally for noninteger kinetic exponents, although these are difficult to represent graphically.

The solutions of a network for the stationary state condition are called *currents*, because of the analogy with electrical networks [3]. Their graphical representations are referred to as current diagrams D_C. The stationary concentrations of the species require that the sum over all feathers equals the sum over all barbs for every species in the D_C.

Since the set of all such currents forms a convex cone, the current cone,

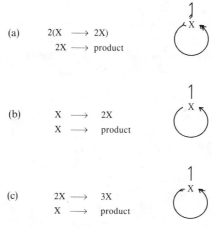

(a) $2(X \longrightarrow 2X)$
        $2X \longrightarrow$ product

(b) $X \longrightarrow 2X$
        $X \longrightarrow$ product

(c) $2X \longrightarrow 3X$
        $X \longrightarrow$ product

FIGURE 3. Examples of irreversible current 1-cycles: (a) weak $(2, 1$; exit order $= 2$, cycle order $= 1)$; (b) critical $(1, 1)$; and (c) strong $(1, 2)$. The quantities β_1, which determine the stability of the currents shown (see Appendix B), are $+2, 0$, and -1, respectively.

in the reaction velocity space [3, 34], all stationary states of a network can be parametrized as linear combinations, with nonnegative coefficients, of the finite number of vectors forming the frame of the cone (i.e., pointing along its edges), which are called *extreme currents* E_i. The extreme currents cannot be decomposed further into simpler entities without violation of the stationary-state condition. For not too large systems, the best way to find the extreme currents is usually by examining the network diagram D_N. It is modified by deleting reactions and each time multiplying stoichiometries until a current diagram D_C is obtained, which satisfies the requirement of the previous paragraph. If no further reaction can be omitted from it without violation of the stationary-state condition, it is an extreme current diagram D_{EC}. This procedure is then repeated with all possible permutations of deleted reactions yielding the complete set of extreme currents. With increasing size of the network, this method becomes more and more difficult because of the increasing number of permutations. In this case the extreme currents can also be calculated directly from the stoichiometric matrix of the complete network. (see Ref. 37 for a general algorithm.)

Clarke showed that some classes of networks are always stable, by proving the existence of a Lyapunov function, whereas certain features in a network will cause it to become unstable in some parameter region. In general, a network cannot be unstable unless it contains an autocatalytic process, called an irreversible *current cycle* in SNA (theorem V.3 in Ref. 3). In order to satisfy the stationary-state condition, an autocatalytic current cycle must have an exit reaction by which a species leaves the cycle. Current cycles are called *strong*, *critical* or *weak*, if the reaction order of the exit process is lower, equal, or higher than the cycle order (Fig. 3). Strong cycles cause an instability, weak ones always lead to stability, whereas critical current cycles can cause an instability depending on other features in the network (cf. theorems V.4–V.10 in Ref. 3).

Using results from matrix stability theory, Clarke also showed that a current is unstable, if any of certain subdeterminants β_i of the matrix formed by multiplication of the stoichiometric and the transpose of the kinetic matrix is smaller than zero (cf. Chapter V.C in Ref. 3; i refers to the number of rows and columns of the matrix for which β_i is the determinant). These β_i can be readily obtained using Clarke's diagrammatic stability analysis (see Appendix B for a worked example). To do this, one first constructs the current matrix diagram D_{CM} from the D_C of the current in question. The β_i are obtained as supergraphs consisting of sums over so-called nonoverlapping stabilizer-polygon diagrams D_{NOSP}, and are in essence a weighted sum over all (stabilizing as well as destabilizing) feedbacks occurring in the current for the species considered. These feedback loops ('sign-switched cycles') can be identified in the D_{CM} (see Ref. 3 for details). β_i will be smaller (larger) than zero, if the destabilizing (stabilizing) effects dominate. A vanishing β_i can indicate a critical current cycle (which can lead to an instability), but β_i also becomes zero for species satisfying a conservation constraint and for equilibrium currents, which consist of a single reaction and its reverse. The latter possibilities are obviously stable.

In summary, SNA shows a large number of networks to be stable, but instability can be proven, if certain features are contained in any of the currents of a network. There still, however, exists an unknown remainder of networks for which no general theorems are available regarding their stability, that is, they may be either unstable, but for different reasons than the ones shown to be unstable; or they are stable, but their stability cannot be proven with the methods used in SNA. (Clarke [3] actually gave an argument that instability is "likely" whenever the previously mentioned proof of stability using Lyapunov functions fails, but this in itself does not include restrictions on the number and type of possible unstable features.)

We therefore distinguish the following qualitatively different cases of unstable networks (depending on which feature is responsible for the instability):

1. Networks that contain a critical current cycle and a suitable, destabilizing exit reaction (CCC, Section III A).
2. Networks that contain a strong current cycle (SCC, Section III B).

3. Autocatalytic ring networks; these have two critical current cycles and are unstable if one cycle consists of at least four more species than the other (Chapter VI.F in Ref. 3).

4. Other possible, as yet unknown unstable networks.

Number 4 can obviously not be discussed. We shall also exclude category 3, because to the best of our knowledge no example of such a mechanism has yet been described in the literature (it is also strictly not a simple oscillator). Therefore in the following, two broad categories of chemical oscillators are distinguished, corresponding to entries 1 and 2. In total some 8 abstract models and 17 experimentally obtained realistic mechanisms have been considered (see Appendix A) and their unstable features all belong to these two cases.

SNA predicts unstable stationary states, but not the existence of limit cycles: all oscillating networks are unstable, but the reverse does not hold, for example, a mechanism can be bistable without exhibiting oscillations. Consequently, besides the unstable feature, other species and reactions have to be included in order to obtain oscillations. Depending on what qualitatively different possibilities there are to achieve this, subcategories have to be distinguished. There are basically two ways to tackle this problem. One can either stepwise build up simple oscillating networks from the unstable feature or analyze the ingredients of complicated realistic models. Although we have not been able to prove that the possibilities we found using the first approach, described subsequently, are exhaustive, all 17 mechanisms studied fit into the picture. Nevertheless, one has to bear in mind that there may be other basic types that have been overlooked, or other oscillating mechanisms yet to be discovered that would not belong to any of the categories to be described.

We classify chemical oscillators according to their unstable current. Although a network may have several unstable currents (see the bromate/iodide oscillator, Sections III A 4 and IV B), one of them will be dominant, and this one is used for classification. Since the oscillators are assumed to be simple (they have only one nonweak autocatalysis), it follows that in the case of several unstable currents these can only differ in some details but have to contain the same current cycle. Usually the net stoichiometric change in a reactor can be determined. This allows a decision about which currents can make a significant contribution by comparing their overall reaction with the net process known from experiment: for example, a current ending in certain products cannot be important if these products are only formed in small amounts. This argument obviously applies to stable as well as unstable currents. Since, within a given mechanism, rate constants may and frequently do differ by many orders of magnitude, only a few extreme currents are needed for a successful description of the behavior of a system, while currents involving slower reactions can be neglected [1]. This has been stated by Aguda and Clarke [38] in a "principle of simplicity," which says that a system is probably describable as a linear combination of the smallest number of extreme currents compatible with the observed dynamical behavior.

Clarke has shown (theorem V.8 in Ref. 3) that if $\beta_i = 0$ for any current, then there is an extreme current with vanishing β_i. This theorem is of considerable practical importance. On the one hand, it allows the identification of critical current cycles by examining the extreme currents only (instead of an infinite number of linear combinations); on the other hand, the task of constructing an unknown unstable current from experimentally obtained information can be simplified by using the condition that it is an extreme current. Thus, in all category 1 oscillators studied the dominant unstable current turns out to be extreme, with one exception for an enzyme model where an equilibrium needs to be included (see Section III C). For category 2 the unstable current need not be an extreme one. In all such oscillators studied, however, the unstable current can always be obtained by adding equilibrium currents to an extreme current, which still limits the number and kind of currents that have to be considered.

In the following sections we start with the unstable feature of the respective category and include more reactions and species until simple abstract oscillators are obtained. Depending on how this is done, subcategories are distinguished where appropriate. The role of the individual species in the unstable current of each category can be defined automatically from the respective unstable feature and the way the oscillator is stepwise pieced together. The resulting simple abstract oscillators serve then as prototypes for the categories and subcategories.

An abstract model, constructed to incorporate the basic qualitative features of a certain (sub-)category, ought to be made as simple as possible, containing a minimum number of species and reactions. However, all physically meaningful situations, like finite parameter changes and the suppression of feed concentrations, should lead to physically meaningful dynamics, for example, avoid the possibility that variables can approach infinity, and inclusion of sufficiently small other, higher- or lower-order, terms should not cause abrupt changes of local bifurcation structure (structural stability [1]), so that nongeneric bifurcations [40] are avoided. Similarly, since all chemical reactions are reversible at least to some extent, a model should either not change qualitatively when the reverse processes are included or allow for reversibility in the first place (inclusion of some reversibility can be regarded as a special case of adding small terms of order higher or lower than those already present in the model). The bifurcation diagram of an oscillator need not, however, be structurally stable, if a rate constant is set to zero, since this cannot be achieved experimentally. If in a model a rate can be set zero with impunity, a simpler model can be constructed that does not include this reaction in the first place.

As we will see in the following sections, the so-constructed simple prototypes cover oscillators in both closed and open systems. Operational methods are described to distinguish the different types of oscillators and the various roles of the species in them. A couple of realistic mechanisms obtained from experimental results are also discussed for each category.

A.   Category 1: Critical Current Cycles

1.   Unstable Feature

Oscillators of category 1 become unstable due to a critical current cycle (CCC) combined with a destabilizing exit reaction with another species Y that is not part of the cycle. The basic feature is shown as E_1 in Fig. 4. The CCC can consist of any number of species X_i; it may also have some fine structure (i.e., it need not consist of a single unbranched loop), as long as the conditions of the CCC theorem (V.7 in Ref. 3) are fulfilled. Some types

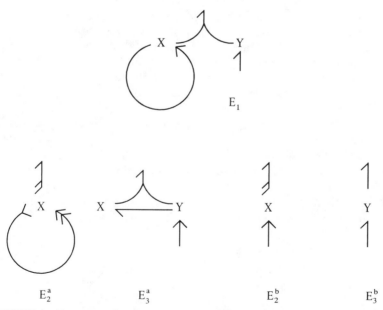

FIGURE 4.   Extreme current diagrams D_{EC} of a critical current 1-cycle with destabilizing exit reaction (E_1), and two sets of explosion-limiting stable extreme currents (E_2^a and E_3^a, respectively, E_2^b and E_3^b). The second (third) current prevents the concentration of $X(Y)$ from going to infinity.

of short circuits, however, weaken the cycle and hence destroy the instability [3]. Features such as E_1 in Fig. 4 have been proven to be always unstable, provided that no other reactions involve both Y and species from the cycle (theorem V.10 in Ref. 3). This is a sufficient but not necessary condition, so that a current can still be unstable, if it is not satisfied. Such a situation occurs, for example, in the chlorite/iodide oscillator, where I^- plays the role of Y, but also takes part in a cycle reaction. A simple calculation shows, however, that β_4 for HOI, HIO_2, HOCl, and I^- is smaller than zero for the extreme current shown in Fig. 11 (cf. Section III A 4 and Appendix B), which is therefore unstable.

Unstable currents like E_1 (Fig. 4) give rise to just an isolated saddle point with not physically meaningful explosion/extinction dynamics. In systems of nonlinear differential equations such behavior can be avoided by including (small) constant or higher-order terms or both. In SNA, at least two stable extreme currents E_2 and E_3 have to be introduced to avoid explosions [41]. Examples of such currents are included in Fig. 4, where the respective second current (E_2^a or E_2^b) prevents X and the third current (E_3^a or E_3^b) prevents Y from going to infinity. The combination of E_1, E_2^a and E_3^a (respectively, E_1, E_2^b, E_3^b) gives rise to bistability (hysteresis) in one, and a cusp bifurcation in two control parameters. In the bistable region E_2 (E_3) dominates the branch with high concentration of $X(Y)$. In realistic mechanisms there can be a large number of additional stable extreme currents, which generally involve additional species and reactions and can, in fact, be quite complicated. We do not, however, need to make any assumptions about what these currents look like, except that at least two exist which give rise to the physically required states. Two stable currents in addition to the unstable one are generally sufficient for a qualitative understanding of bistability (cf. principle of simplicity [38]).

Oversimplified oscillating models without suitable E_2 and E_3 can be constructed and may give some insight into certain features of an oscillator, but they should be handled with care, because they predict physically implausible situations, namely, infinite values of the variables or nongeneric bifurcations or both. An example of such a model, which consists of only

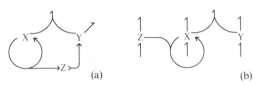

FIGURE 5.   Network diagrams for (a) the single current oscillator (for which D_N and D_{EC} are identical) and (b) the Franck model.

one extreme current (referred to as the single current oscillator of category 1), is shown in Fig. 5a. We have shown by calculations that this mechanism undergoes a supercritical Hopf bifurcation upon increasing the rate constant of the autocatalytic process. No other bifurcations occur in the model as it stands. If, however, one assumes that species Z is buffered (which is in principle always possible), there is for finite buffer capacity a direct transition from a focus to an isolated saddle point. Such a bifurcation is nongeneric in one control parameter [1]. Moreover, the concentration of X can become infinite beyond this bifurcation. The model can be extended by including other reactions resulting in a scheme similar to the Oregonator.

A somewhat more subtle problem exists in one of the models studied by Franck [30], shown in Fig. 5b. An inflow of all species, is assumed. It is physically always possible to reduce inflow concentrations to zero. If we assume zero inflow for Y, we find by calculations a single saddle-node (SN) bifurcation in the other two variables (with the possibility of explosion in one direction). If, in addition, the flow of X vanishes, a transcritical bifurcation occurs (which is again nongeneric in one parameter [1]). No such abrupt changes in the bifurcation scheme occur if a second-order term is added for X (or if the exit reaction of X is assumed to be second order in the first place).

2.   Subcategories: Types 1B and 1C

The unstable feature described in the previous section exhibits, in combination with suitable stable extreme currents, bistability stemming from a cusp. In order to obtain oscillations a third kind of species Z [42] has to be added to the model. There are two qualitatively different ways to do that, which leads to the distinction of two subcategories (Fig. 6):

Type 1B: Y is formed in a chain of reactions from the CCC via at least one intermediate Z, such that there is a negative feedback loop in the essential species of the form $a_{xx}*a_{xy}*a_{yz} < 0$ (the a_{ik} are matrix elements of the Jacobian [28]).

Type 1C: No feedback loop of the form in 1B exists. Instead, there is an essential species Z consumed by the CCC, which gives rise to a negative feedback $a_{xz}*a_{zx} < 0$.

Actually some 1C types contain in addition a negative feedback loop, different from the one defining 1B, of the form $a_{zy}*a_{yx}*a_{xz} < 0$ (see Section III A 4), but a_{zy} can also vanish, and with it the feedback loop (see for example, the Franck model). In the latter case, however, Z and Y are still correlated in a different way, since they are both contained in the flow in a fixed ratio, so that effectively a negative feedback "loop", "closed" by the flow, is still present. We shall refer to X as autocatalytic or *cycle species*,

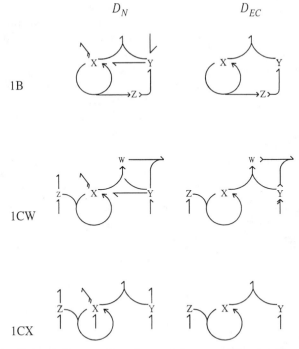

FIGURE 6. Network diagrams D_N and current diagrams of the unstable extreme currents D_{EC} of simple models for each of the three subtypes of category 1 oscillators, exhibiting a critical current cycle in X destabilized by the exit reaction with Y and a feedback species Z: $1B \equiv$ Oregonator; $1CW \equiv$ simplified NFT mechanism, and $1CX \equiv$ modified Franck model.

to Y as *exit species* and to Z as negative *feedback species*; this nomenclature gives a concise description of their roles in the unstable current.

In general, all oscillators of category 1 (except oversimplified ones like the single current oscillator) exhibit cross-shaped phase diagrams (XPDs) [41, 43, 44] when studied in two parameters, although that diagram is not necessarily detectable for all parameter combinations. This becomes evident, if, in a calculation rather than an experiment, the species Z in an oscillating state is buffered with increasing capacity. The oscillations are then more and more suppressed until the system behaves like the bistable two-variable model discussed in Section III A 1.

It follows from the preceding definitions that an oscillator of type 1C requires an inflow of at least one essential variable, namely Y, whereas in the 1B case oscillations can result without introducing Y into the reactor, since it is produced internally. For some parameter combinations, a 1B oscillator does not require any essential species to be present in the feed, that is, the trivial steady state where the concentrations of all essential species vanish ($x = y = z = 0$), can coexist with a limit cycle (which is not the case for the 1C type), but the oscillatory region need not reach down to zero inflow for all combinations of the other parameters.

In all systems studied, the distinction between 1B and 1C turns out to be identical to the experimental distinction between batch and CSTR oscillators. Note that every batch oscillator can be run in a CSTR, but not vice versa. It is conceivable, however, that a 1B oscillator has a product that poisons the oscillations. This may not cause problems in a CSTR, because the product is washed out, but could prevent the occurrence of batch oscillations, or only allow strongly damped ones. On the other hand, all flows required in a 1C oscillator may, at least in principle, be substituted by, if necessary highly selective, reactions, a situation which would result in batch oscillations. Consequently, the experimental distinction between batch and CSTR types need not always coincide with the mechanistic definition of 1B and 1C given previously, and therefore may not always give reliable information about the mechanism. It is therefore necessary to verify the tentative assignment of a batch oscillator to 1B and a flow oscillator to 1C by an independent method. This is possible through bifurcation diagrams in two parameters: although for both types an XPD is obtained in a CSTR, they differ in the relative position of the bistable and oscillatory region for certain analogous parameter combinations. Choosing the inflow concentration of Y and the speed of the autocatalysis as control parameters, the XPD is in both cases located roughly along the diagonal between the axes, but a 1B type oscillates for small, and exhibits bistability for large, values of the parameters (Fig. 7), whereas the reverse holds for a 1C oscillator (Fig. 8). Although a 1B type can oscillate without a flow in Y, this flow, or one in a species producing Y, is required to obtain the full XPD. In this case generally only two extreme currents are needed for oscillations [3]; the additional inflow of Y gives rise to (at least) one other current.

Table II compares the XPDs of the three CCC prototypes as shown in Fig. 6 in some selected combinations of bifurcation parameters (we introduce the subdivision of 1C into 1CX and 1CW later in this section). They were chosen because of their accessibility in an experimental situation: the inflow concentration of species can easily be varied and the speed of the autocatalysis can often be shifted indirectly, by changing the concentration of a nonessential species consumed by this reaction. We show in Table II the direction in which the cusp opens, that is, the location of the bistable region with respect to the oscillatory region. There is no difference between the XPDs of the two 1C prototypes (1CW and 1CX), whereas those of the 1B prototype differ in

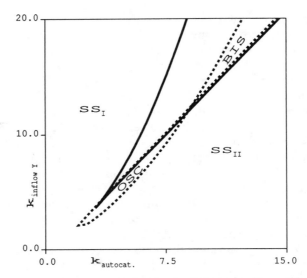

FIGURE 7. Cross-shaped phase diagram for the 1B prototype (Oregonator), obtained by using the velocity of the Y inflow ($k_{inflow} Y$) and the velocity of the autocatalysis ($k_{autocat.}$) as control parameters. The saddle-node bifurcation is shown by solid lines, the Hopf bifurcation by dotted lines. OSC denotes the oscillatory, BIS the bistable region, SS_I the steady-state region with low $[X]^{ss}$ and high $[Y]^{ss}$ and SS_{II} the one with high $[X]^{ss}$ and low $[Y]^{ss}$.

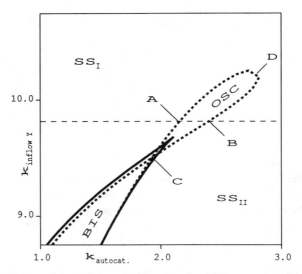

FIGURE 8. Cross-shaped phase diagram for the 1CX prototype (modified Franck oscillator); for details see Fig. 7. The labels A to D and the horizontal dashed line are referred to in the text.

TABLE II

Direction (Arrows) into which the Cusp Opens (i.e., the Location of the Bistable with Respect to the Oscillatory Region) in Cross-Shaped Phase Diagrams (XPD's) of the Three Category 1 (CCC) Prototypes in Certain Selected Combinations of Bifurcation Parameters.

Bifurcation Parameters[a]	1B Type	1CX Type	1CW Type
Y_{in} versus X_{in}	↙	↙	↙
X_{in} versus Z_{in}	↙	↘	↘
Y_{in} versus Z_{in}	b	↗	↗
X_{in} versus AC	↘	↖	↖
Y_{in} versus AC	↗	↙	↙
Z_{in} versus AC	↗	↖	↖

[a] X_{in}, Y_{in} and Z_{in} denote the velocities of the X, Y, and Z inflow. AC is the velocity of the autocatalysis.

[b] The cusp runs parallel to the parameter plane so that no XPD can be obtained.

all but one combination (Y_{inflow} vs X_{inflow}). Note that the 1B prototype shows a particular situation in the Y_{inflow} vs Z_{inflow} bifurcation diagram. The cusp runs parallel to this parameter plane so that no XPD can be obtained. This degenerated situation can easily be understood, as the only role of Z in this prototype is to produce Y. The four remaining XPDs can be used to assign a CCC mechanism under investigation to the 1B or the 1C subcategory. XPDs with two-parameter combinations not contained in Table II may also be usable to make this distinction. If this is desirable in the investigation of an experimental system, this table may easily be expanded according to the parameter combinations that are experimentally available. However, the parameter combinations of XPDs reported in the literature are very limited: With the exception of the mixed Landolt oscillators [45–49] and the Belousov–Zhabotinskii (BZ) reaction [50], the experimental and calculated XPDs reported in the literature for the examples considered use as parameters the inflow concentrations of Y (or a species producing Y) and of a nonessential species that accelerates the autocatalysis (i.e., which is consumed by the critical current cycle). In the mixed Landolt and BZ reactions the position of the bifurcation scheme is different, since other parameter combinations were used [45–50]. For the mixed Landolt (BZ), however, the oscillating region lies at higher (lower) inflow concentration of Y, which is quite in line with the fact that the BZ reaction exhibits sustained batch oscillations (1B), while the mixed Landolt system does not (1C).

Numerical simulations of 1C-type oscillators revealed that a further distinction of two subtypes is required. In some oscillators of category 1C an additional species W (*recovery species*) is essential, which is formed from the CCC and reacts with Y is the unstable current, thereby allowing the autocatalysis to recover (type 1CW). If no such essential species exists, a finite inflow concentration of X is required (1CX). For 1B oscillators no such distinction was necessary.

The network diagrams of simple abstract models belonging to the subtypes distinguished in category 1 (namely, 1B, 1CX, 1CW) are compiled in Fig. 6. In Figs. 7 and 8 we show a bifurcation scheme for the 1B and 1CX type, respectively, using the inflow concentration of Y and the rate of the autocatalytic process as parameters. The Oregonator and a modified Franck model serve as "prototypes" for 1B and 1CX oscillators, the 1CW model of Fig. 6 can be considered as a simplified NFT mechanism (for the minimal bromate oscillator [51]), from which the nonessential species (H^+, BrO_3^-, Ce^{4+}, and Br_2) have been eliminated by including them into the rate constants, and a steady-state approximation for BrO_2 has been made; this leaves $HBrO_2$ as X, Br^- as Y, Ce^{3+} as Z, and $HOBr$ as W. The stable currents included in the 1C prototypes can be seen by decomposition of the network diagrams in Fig. 6: those for the 1CW type (1CX type) are analogous to E_2^a and E_3^a (E_2^b and E_3^b) in Fig. 4.

It should be pointed out that there is no unique choice of such prototypes, rather the species can also be connected in different ways to produce the same dynamic behavior. The critical current cycle with the destabilizing exit reaction must be present, but the negative feedback loop, respectively, the explosion-limiting species Z (and the species W, if present), can connect up in different ways (see examples in Section III A 4).

Real systems are not as simple as the described abstract models. Usually there are several species X_i and the critical current cycle can have some branching in it. Also, often more than one species of type Z is present. Y is unique in most cases, but a sufficiently complicated current cycle can have more than one exit reaction with different species Y_i (see the bromate/iodide oscillator as an example, Section III A 4). 1CW oscillators with more than one essential species in the role of W have not been encountered.

3. *Identification of the Roles of Essential Species*

In the previous section we saw that three different kinds of species have to be distinguished for 1B and 1CX oscillators, while even four are needed in the case of a 1CW type, according to the roles they play in the unstable current. It follows that at least two bits of information per species are required in order to identify their roles. We now present an operational test in order to achieve this distinction of species: a constant increase or decrease of their inflow concentrations in a CSTR (shift experiments).

From the discussion in Sections III A 1 and IV A 2 it is clear that Z can in principle be distinguished from X and Y, because a cusp can be obtained in X and Y alone, with Z as a parameter (i.e., a species nonessential for bistability), while all three are essential for oscillations. Such a procedure is straightforward in computer simulations, but would hardly work in practice, because very high buffer capacities would be needed to suppress concentration changes at the jumps caused by the saddle-node (SN) bifurcations. Therefore another method is required to tell the species apart.

The species can be sorted into two subgroups as a consequence of the bistability, because one part has high and the other low concentration in a given branch of the hysteresis. Thus, species of types X and Y always reach their higher concentration on opposite branches, but this in itself does not allow absolute assignment.

Since the species have been categorized with respect to their roles in the unstable current, our efforts to find criteria for the assignment of species to these categories must concentrate on the unstable current. To do so we performed mostly numerical calculations at the Hopf and SN bifurcations, where this current must play a decisive part, because the system is about to lose stability. Only a qualitative distinction is needed, since no assumptions about the absolute rate constants are made, although the parameters have to satisfy certain inequalities to allow the instability of the network to manifest itself [3]. Therefore, it is sufficient to apply constant perturbations of the species instead of time-dependent perturbations. In an experimental situation this can be easily realized by a constant increase of the inflow rate of the species in a CSTR. Accordingly, all calculations were carried out by including the CSTR flowterms for each species into the studied system.

If a constant addition of a species is applied at a stable focus close to a Hopf bifurcation, the system can either begin to oscillate or remain stable. In the first case the system is destabilized (d), in the latter case stabilized (s) by the addition. Accordingly, if we apply a constant perturbation to a system starting at the oscillating side of a Hopf bifurcation, the oscillations may either cease (s) or remain (d). In order to ensure that the amount of addition is large enough to cross the bifurcation, the procedure is performed on both sides of the bifurcation with increasing amounts of the species added until a change (d or s) in the stability of the focus occurs.

A constant addition of a species will also shift the concentrations of all species in the mixture either to higher ($+$) or lower ($-$) values. In the present context, only the effect of the addition of each species on the stationary-state concentration of the same species needs to be considered, that is, the diagonal elements of the shift matrix. An element m_{ik} for this matrix denotes the stationary-state concentration shift of the kth species caused by a small constant addition of the ith one. Since no quantitative information is required, one has only to check, whether the concentration in question decreases (inverse regulation $-$) or increases (normal regulation $+$). The term "inverse regulation" was coined by DeKepper et al., who observed it for the first time for I_2 in the Briggs–Rauscher reaction [50, 52].

The calculations were carried out by searching for the desired bifurcation with a one-parameter continuation algorithm for fixed points (obtained by Newton's method) and solving for the eigenvalues of the Jacobian. The amount of the species added was then used as control parameter. The shift matrix and effect on the bifurcation are obtained from the difference in the fixed point and the real part of the two complex conjugate eigenvalues with and without the additions.

Analogous calculations were done at the SN bifurcations enclosing the bistable region, where it was checked, whether a jump to the other branch occurred or not and whether it led to higher or smaller concentration of the species under consideration. Here direct integration of the differential equations is the easiest way to check whether or not a jump occurs with a small constant perturbation. The calculations are easier and faster at a Hopf bifurcation, but experimentally the large jumps occurring at an SN may be more convenient. The information gained with both methods is equivalent. Therefore, only the results for shift experiments at a Hopf bifurcation will be discussed further.

If we cut through an XPD by changing only one control parameter, the oscillatory region is bounded by two opposing Hopf bifurcation points, for example, points A and B in Fig. 8 which are reached by varying $k_{autocat}$ at a constant level of $k_{inflow\ Y}$, (dashed line). Close to the crossing-over point (C in Fig. 8) the slope of the Hopf bifurcation line does not change sign. Far away from the crossing-over point the curve can be bent (near D in Fig. 8) leading to a change in the (s/d) behavior of some (or all) species so that the procedure has to be applied before such bends occur. Similarly, distinction of the species is not possible in projections of the phase diagram onto all conceivable parameter planes. From numerical studies it is concluded that an assignment can be made if the signs of the slopes of the bifurcation line in the two opposing Hopf bifurcation points are the same, for example, at the points A and B in Fig. 8. This is the case in nearly all experimentally obtained cross-shaped phase diagrams as well as those calculated from realistic mechanisms in the literature, at least not too far from the crossing-over point, but it does not hold for all possible parameter combinations (cf. some studies of the mixed Landolt reactions [45–49]). The effect on the bifurcation of a given species is opposite for the two Hopf curves, say s on one, d on the other side, but the regulation $(+/-)$ does not change. This inverted effect on the bifurcation on opposite ends of the oscillating region can also serve as a test whether the behavior determined under given conditions can be used for a valid assignment of the species.

TABLE III

Effect on the Hopf Bifurcation (s for Stabilizing, d Destabilizing)
and the Sign of the Diagonal Element of the Shift Matrix for Essential
Species in Oscillators of Category 1

Type	X	Y	Z	W
1B	s^+	d^+	d^-	—
1CW	s^+	d^+	s^-	s^+
1CX	s^+	d^+	s^-	—

The effects of the different kinds of species on the bifurcation and their regulation behavior are compiled in Table III, with the arbitrary choice of starting at the Hopf bifurcation with higher steady-state concentration of X, for example, in the XPD of Fig. 8 at point B. The autocatalytic species X_i are s^+, the exit species Y_i d^+, while the species called Z_i exhibit inverse regulation and have the same effect on the bifurcation as Y in 1B and as X in 1C oscillators. It is therefore crucial to find the species with inverse regulation to allow absolute assignment of X and Y; moreover, it must be known whether the oscillator under consideration is of type B or C (Section III A 2). The species W in 1CW oscillators exhibits the same behavior as X (s^+). This is expected since W reacts with Y in the unstable current and therefore has the opposite effect as Y.

The inverse regulation for Z comes about either because Z, indirectly by producing Y, decelerates the autocatalytic process in which it is produced (1B) or because a larger supply of Z leads to an acceleration of the autocatalysis consuming Z (1C). In nonsimple oscillators Z can be part of a second (critical or strong) current cycle, which can overcompensate the inverse regulation expected if only the first CCC is considered. Abstract reaction schemes where this effect occurs have also been studied by Franck [29, 30] (see Section IV B for an example).

With the convention to start at high X concentration, there are no species s^- in 1B nor d in 1C types. Such behavior can occur in the respective oscillators with nonessential species, but has not been found for the essential ones. For the nonessential species the shift and effect on the bifurcation may contain useful information (e.g., inverse regulation if consumed by an autocatalytic process), but often a nonessential species takes part in several different reactions and may actually change its behavior even for choices of the parameters where the essential ones do not thus change, so that no general rules can be given for nonessential species.

Inverse regulation is the key in assigning the roles of the species. It can, however, be "masked" by conservation constraints, unless the constraint is taken into account in the shift measurement. Thus, in most metal-ion-catalyzed oscillators the sum of the reduced (e.g., Mn^2, Ce^{3+}) and oxidized form (Mn^{3+}, Ce^{4+}) is conserved. A constant addition of one of them can lead to an increase in both. Instead, an increased flow in one should be compensated for by decreasing the inflow of the other by the same amount, leaving the total metal ion concentration constant. As an example, shift experiments for Ce^{4+} in the BZ reaction, which has an inflow of Ce^{3+}, have to be done by flowing in the same amount of cerium, mostly in the trivalent, but with a small content of the tetravalent state. Such conservation constraints occur for the SNB, NFT, and bromate/manganous/reductant mechanisms. Somewhat surprisingly, in the FKN mechanism Ce^{4+} does exhibit inverse regulation under the conditions studied without taking the constraint into account. This shows that the effect can, but need not be masked. Except for the cases of these metal ions, where the conservation constraint can be guessed without detailed knowledge of the mechanism, inverse regulation always takes place when expected in simple oscillators. Thus, the more complicated constraint in the SNB mechanism, where a linear combination of the concentrations of BrO_3^-, HOBr, $HBrO_2$, Br^-, and Ce^{4+} is conserved [53], does not affect the regulation behavior.

In summary, oscillators with a critical current cycle can be sorted into subcategories according to how the negative feedback loop is connected to the autocatalysis. Three (in the 1B type and 1CX type) or four (in the 1CW type) qualitatively different roles of species exist in these subcategories. Species can be assigned to their roles with the help of a constant perturbation of their inflow concentration in a CSTR.

4. Unstable Currents of Realistic Mechanisms

The rules stated in the previous section for the assignment of the roles of species have been checked not only for abstract (prototype) models, but also for a number of realistic mechanisms, which produce the main features of experimental systems. These mechanisms are compiled in Table IV (1B-) and Table V (1C-types). Their unstable (extreme) currents are shown in Figs. 9–13.

TABLE IV

1B-Oscillators: Stabilizing (s) and Destabilizing (d) Behavior as well as Normal ($+$) and Inverse ($-$) Regulation of the Essential Species at Shift Experiments Close to a Hopf Bifurcation

Model/Mechanism	Reference	Species (s^+)	Species (d^+)	Species (d^-)
Oregonator	15	X	Y	Z
SNB	93	$HBrO_2, BrO_2^-$	Br^-	Ce^{4+}
FKN	54, 55	$HBrO_2, BrO_2^-$	Br^-	Ce^{4+}
OKN	39	$HBrO_2, BrO_2^-$	Br^-	$HOBr, Br_2, Br^-$
BR	50, 57	HIO_2, IO_2^-	I^-	HOI, I_2

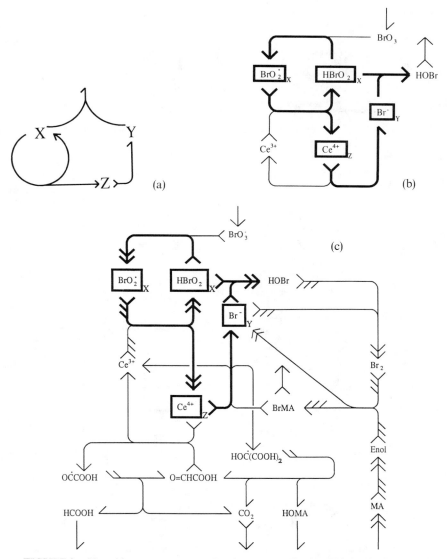

FIGURE 9. Unstable extreme currents D_{EC} for some mechanisms of the BZ reaction (1B type), where the stoichiometric factor f is assumed to be 0.5: (a) Oregonator [15]; (b) SNB mechanism [93]; (c) FKN mechanism [54, 55] (with H^+ not shown and MA ≡ malonic acid, Enol ≡ enol form of MA, BrMA ≡ bromomalonic acid, HOMA ≡ hydroxomalonic acid). In (a) all species are essential, whereas in (b) and (c) the essential species are shown in boxes and their reactions drawn in bold lines. The indices at the boxes denote the role of the species. In all diagrams flows are not distinguished from reactions to (formation by) otherwise inert products (reactants). They are shown as vertical arrows.

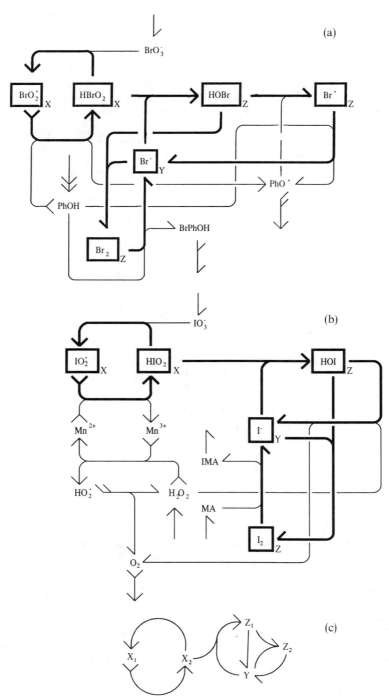

(a)

(b)

(c)

156

TABLE V

1C-Oscillators: Stabilizing (s) and Destabilizing (d) Behavior as well as Normal $(+)$ and Inverse $(-)$ Regulation of the Essential Species at Shift Experiments Close to a Hopf Bifurcation.

Model	Reference	Species (s^+)	Species (d^+)	Species (s^-)
NET	51, 58	$HBrO_2, BrO_2^{\cdot}$, $HOBr$	Br^-	Ce^{3+}
ClO_2^-/I^-	61–63	HIO_2, HOI, $HOCl$	I^-	ClO_2^-
$BrO_3/Mn^{2+}/SO_3^{2-}$	64	$HBrO_2, BrO_2^{\cdot}$, $HOBr$	Br^-, SO_3^{2-}	Mn^{2+}
BrO_3^-/I^-	66	$HBrO_2, HOBr$, IBr, HOI, HIO_2	I^-, Br^-	I_2
Franck	29, 30	X	Y	Z
Landolt L1	47, 48	H^+, I_2	SO_3^{2-}	HSO_3^-
Landolt L2	48	H^+, I_2	SO_3^{2-}	HSO_3^-, HOI, HIO_2

In these diagrams flows are not distinguished from reactions leading to inert products or formations of species from otherwise inert reactants: they are all drawn as vertical arrows without a species at one end. The nonessential species are not included in the tables, which summarize the effects of a constant perturbation close to a Hopf bifurcation of the essential species only. In the current diagrams the nonessential species have been included, unless stated otherwise. The essential species are framed and the connections between them drawn with thicker arrows, so that it can be readily seen which species have been found to be (non)essential.

Comparison of the unstable currents of the 1B oscillators in Figs. 9 and 10 shows that there are two different ways in which the negative feedback loop occurs: Z can branch off the critical current cycle (*tangent feedback*) as in the Oregonator (Fig. 9a) and SNB mechanism (Fig. 9b) or it can be formed in the exit reaction of the cycle ($X + Y$, *exit feedback*) as is the case for the OKN mechanism [39] (Fig. 10a) and the mechanism of the Briggs–Rauscher reaction of DeKepper et al. [50, 52] (Fig. 10b). Note that the shift method cannot distinguish between the two, which proves to be, however, not really necessary to construct the unstable current (cf. Section IV). Therefore only one prototype for 1B is described in Section III A 2.

FIGURE 10. Unstable extreme currents D_{EC} of 1B type mechanisms: (a) OKN mechanism [39] (Ph = phenyl); and (b) mechanism of the Briggs–Rauscher reaction [50, 57] (H^+ not shown). In both cases only the inorganic halogen species (except XO_3^-) are essential so that the two are practically identical (in the essential species). The unstable current basically looks like (c) for both. For details of the presentation of the currents (a) and (b) refer to Fig. 9.

In the currents shown in Fig. 10 there appears to be a second critical current cycle (reminiscent of the Explodator [16]) consisting of HOX and X^- (X = Br, I) of the form:

$$HOX \rightarrow X^-$$
$$X^- + HXO_2 \rightarrow 2HOX$$

This cycle is, however, short-circuited by the double-tail short circuit reaction

$$X^- + HOX \rightarrow X_2$$

and is hence weak (theorem V.7 in Ref. 3).

In the FKN mechanism (Fig. 9c), the similar current cycle formed by Br^-, HOBr, and Br_2 is weak because of an entrance reaction (BrMA $\rightarrow Br^-$) [3], but the species Br^-, Br_2, and BrMA do, in fact, form a critical current cycle, because a single-tail short circuit reaction, in this case

$$Br_2 + MA \rightarrow Br^- + BrMA$$

does not have a weakening effect as long as the cycle-tail species (Br_2) is consumed only by this reaction [3]. However, the species Br_2 and BrMA are nonessential (cf. the quench vectors [4,5], Section II) so that the FKN mechanism is still a simple oscillator. There could be conditions under which this changes, but for the rate constants of Refs. 54 and 55, we have not found feed concentrations where it does. Since the Explodator [16] postulates two critical current cycles, we regard the Oregonator (Fig. 9a) as the appropriate skeleton model for the FKN mechanism (Fig. 9c), and the Oregonator is indeed obtained when in addition to the elimination of all nonessential species, a steady-state approximation is made for $BrO_2^{\cdot}$. The exit species Br^- in the FKN mechanism is formed through both exit and tangent feedback (via HOBr, Br_2, respectively Ce^{4+}), but only the latter is essential, so that no ambiguity arises. Thus, as far as the essential species are concerned, the three unstable currents of Fig. 9 are equivalent (if X is considered to incorporate both $HBrO_2$ and $BrO_2^{\cdot}$). The BZ reaction can, however, also be carried out with oxalic acid as organic reductant [56]. In this case the tangent feedback cannot occur any more for lack of a species (like BrMA) from which Ce^{4+} could produce bromide. The corresponding model ("modified Oregonator") proposed by Field and Boyd [56] contains the same exit feedback as the OKN mechanism (Fig. 10a), and has not been included separately in Table IV nor Fig. 9. This kind of feedback loop is also contained in the "revised Oregonator" models A and B proposed by Noyes [24]. Version

C in Ref. 24 possesses a second critical current cycle. Hence, it is nonsimple and similar to the Explodator.

The OKN and the BR mechanism (Fig. 10) become identical in the essential species, if a steady-state approximation is made for Br$^\bullet$ (alternatively, I$^\bullet$ could be postulated as an intermediate in the formation of I$^-$ from HOI and H$_2$O$_2$). Although the initial mixture of reactants is very different, namely, bromate/phenol, or a similar aromatic compound, and iodate/manganous/ hydrogen peroxide/malonic acid and usually some small amount of iodine, the analogy between the two is not too surprising, since of course the chemistry of Br and I is homologous. The same reactions exhibit different rates for Br and I, but the different choice of nonessential species just gets the effective rates of the pseudoreactions in the range where oscillations occur. The very similar bromate/manganous/hypophosphite oscillator is discussed subsequently.

Oscillators of type 1C are compiled in Table V. For the 1CW systems, the species W is the last entry with s^+ behavior (HOBr, HOCl, HIO$_2$); the unstable currents are shown in Fig. 11. It is readily seen by inspection of Fig. 11 that (similar to the formation of Z in the 1B types) W can be formed in a tangent reaction (chlorite/iodide and bromate/iodide) or the exit reaction of the cycle [bromate/manganous (or cerous) oscillators].

A simplified version of the NFT mechanism [52, 58] for the minimal bromate oscillator (Fig. 11a) has already been discussed (Section III A 2), since it serves as a prototype for the 1CW subcategory. The mechanism for the bromate/chlorite/bromide system [59, 60] has an analogous unstable feature that is therefore not reproduced in Fig. 11. Calculations for this mechanism yield the same shift matrix as the 1CW prototype, but with a different species, namely, ClO$_2^-$ playing the role of Z (see Table V). The XPD in the $[\text{BrO}_3^-]_0$ vs $[\text{Br}^-]_0$ parameter plane, experimentally obtained by Orban and Epstein (Fig. 4 in Ref. 59), is expected for a 1CW-type oscillator (cf. Table II; $[\text{BrO}_3^-]_0 \equiv AC$ and $[\text{Br}^-]_0 \equiv Y_{\text{inflow}}$).

As was already pointed out in Section III A 1, the chlorite/iodide system [61, 62] does not satisfy the conditions of theorem V.10 in Ref. 3, because I$^-$($= Y$) is also consumed by the critical current cycle. Its supply does not, however, limit the autocatalysis, rather this role is played by ClO$_2^-$, so that no problems with the assignment of species arise. Besides the unstable extreme current shown in Fig. 11b there is another very similar one, in which HOCl reacts with HOI instead of I$^-$. It contains the same unstable feature and exhibits the same net stoichiometric change (see Appendix B). The ClO$_2^-$/I$^-$ system represents the minimal oscillator of a number of chlorite oscillators [60, 63], for most of which, however, the mechanisms have not yet been formulated.

A whole family of oscillators with BrO$_3^-$, Mn(II), and an inorganic reductant such as SO_3^{2-}, Sn(II), AsO_3^{3-}, N_2H_4, and I^- have been reported by Alamgir et al. [64]. The corresponding reaction with hypophosphite [65] also belongs to this group, which is closely related to the minimal bromate oscillator. We will now discuss some results of Alamgir et al. in more detail in order to demonstrate how the described methods may lead to a better understanding of the reported experimental results and also in order to show

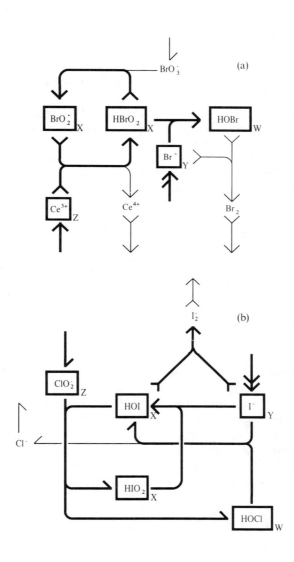

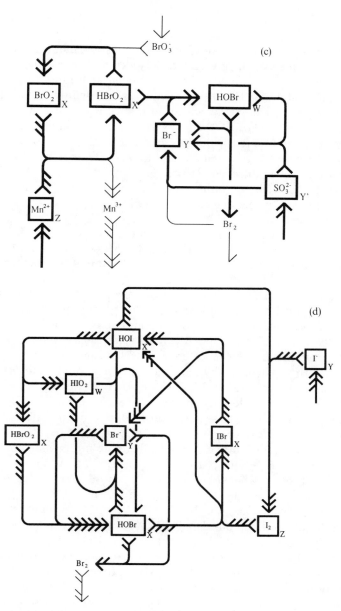

FIGURE 11. Unstable currents for some oscillators of type 1CW (H^+ not shown): (a) NFT mechanism [51, 58] for the minimal bromate oscillator, (b) simplified [61] mechanism of the chlorite/iodide oscillator (see also Ref. 62), (c) bromate/manganous/sulfite oscillator [64], and (d) bromate/iodide oscillator [66]. For details of the presentation of the currents refer to Fig. 9.

that transient batch oscillations do not necessarily indicate a 1B type oscillator.

As an example from the bromate/manganous/reductant family, only the system with sulfite as reducing agent is included in Table V and Fig. 11. The shift results were calculated with the model and rate constants given by Alamgir et al. [64]. The other members of the family (except that with hypophosphite, see below), for which no rate constants are given, however, are also expected to show d^+ behavior for the reductant species.

In addition to CSTR oscillations Alamgir et al. [64] reported a small number of transient batch oscillations with arsenite as reductant, when Br_2 was removed via a nitrogen carrier gas flow. The $BrO_3^-/Mn^{2+}/AsO_3^{3-}$ oscillator, as the oscillator with sulfite, is certainly a 1C type in the absence of a gas flow, as can be seen from the XPD in the $[AsO_3^-]_0$ vs $[BrO_3^-]_0$ parameter plane (Fig. 1 in Ref. 64: $[AsO_3^-]_0 \equiv Y_{inflow}$, $[BrO_3^-]_0 \equiv AC$ in Table II). In computer simulations we obtained a 1B type system in a CSTR, if, in addition to the removal of Br_2 due to the gas flow, we assumed that the reducing agent does not only react with HOBr and Br_2, but also reduces Mn^{3+} to Mn^{2+}. A rate of about $10^3 M^{-1} s^{-1}$ for this reaction is sufficient to generate a 1B type, with the other rates as given for the sulfite system in Ref. 64. Additional reduction of BrO_3^- by this agent is not required. The Mn species and the reductant then become nonessential, and the unstable current is basically as in the OKN and BR system, with just a minor variation in stoichiometry, because in the latter two mechanisms Br_2 forms one, in the former, two Br^-, which is why half of the Br_2 has to be removed to satisfy the stationary-state condition. The resulting current (Fig. 12) represents most likely the source of instability in the bromate/manganous/hypophosphite oscillator [65], which shows sustained instead of transient batch oscillations. As no rate constants are given in the literature [65] this could not be verified by model calculations and the system is therefore not included in Table IV. Although Mn^{3+} is a fairly strong oxidizing agent, it only reacts in a one-electron process, whereas arsenite prefers two-electron acceptors, so that the Mn^{3+} may not be able to complete with HOBr and Br_2 [64]. It appears therefore more likely that the $BrO_3^-/Mn^{2+}/AsO_3^{3-}$ oscillator with carrier gas flow should still be considered a 1C type, in which a "flow" of Br^- is produced by reduction and which quickly stops to oscillate as soon as Mn^{2+} is depleted. Alamgir et al. [64] actually reported that no further oscillations were observed when the pink color of Mn^{3+} appeared. The situation is the other way round for one-electron reductants such as $Fe(CN)_6^{4-}$, NH_2OH, which do not generate oscillations [64], presumably because the reduction of Br_2 and HOBr becomes too slow to allow the generation of a sufficiently effective negative feedback. Of course, a mixture of suitable one- and two-electron reductants should also lead to a 1B oscillator.

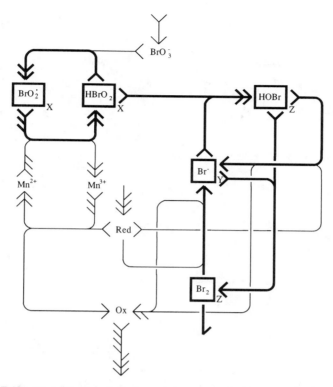

FIGURE 12. Unstable current of a bromate/manganous/reductant mechanism of type 1B. Red stands for reducing agent capable of reacting with both one- and two-electron oxidants (two different reductants could be used). Only the Br species, except BrO_3^-, are essential. The current is a minor variation of the OKN and BR mechanism in Fig. 10 [64, 65]. For details of the presentation of the currents refer to Fig. 9.

The bromate/iodide mechanism [66] possesses the most complicated unstable feature encountered so far (Fig. 11d). The critical current cycle (HOI, $HBrO_2$, HOBr, and IBr) occurs in more than one extreme current and is described in some detail in Section IV B in connection with the discussion of how one can go about constructing unstable currents.

The mixed Landolt oscillators [45–49] represent examples of 1CX types. They contain iodate or bromate with $Fe(CN)_6^{4-}$, SO_3^{2-}, and H^+. In contrast to the oscillators discussed previously, H^+ is essential, as may be concluded from the large pH changes during oscillations.

For the iodate system Luo and Epstein proposed two very similar mechanisms [48] (which are here referred to as L1 and L2). In L1 Fe(II) reduces I_2 to I^- via I_2^-; in L2 this reaction is neglected, rather Fe(II) reacts

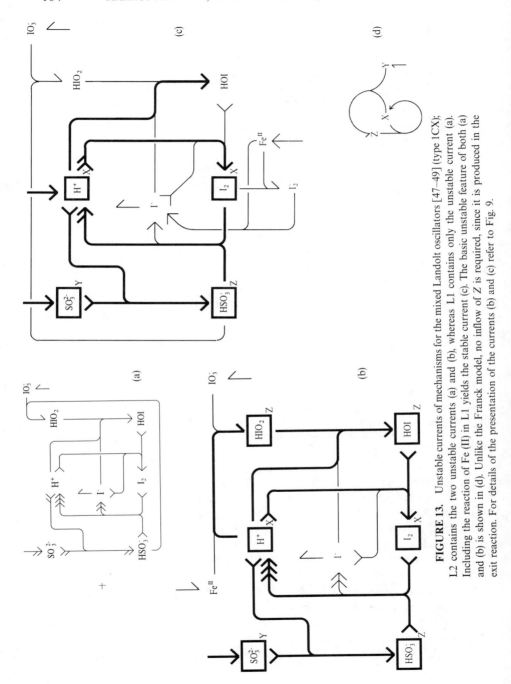

FIGURE 13. Unstable currents of mechanisms for the mixed Landolt oscillators [47–49] (type 1CX); L2 contains the two unstable currents (a) and (b), whereas L1 contains only the unstable current (a). Including the reaction of Fe (II) in L1 yields the stable current (c). The basic unstable feature of both (a) and (b) is shown in (d). Unlike the Franck model, no inflow of Z is required, since it is produced in the exit reaction. For details of the presentation of the currents (b) and (c) refer to Fig. 9.

with IO_3^- to form HIO_2. In both mechanisms H^+, I_2, SO_3^{2-}, and HSO_3^- are essential, for L2 HOI and HIO_2 are essential as well. In L1 the relative amplitudes of HOI and HIO_2 are roughly of the same order as those of the essential species and suppressing their oscillations substantially shifts the bifurcation, so that the fact that they are nonessential does not provide a straightforward way to distinguish between the mechanism. (The other nonessential species exhibit much smaller relative amplitudes.) The difference in the unstable feature is subtle. Both schemes contain the current shown in Fig. 13a (with H^+ and I_2 forming the critical current cycle, SO_3^{2-} as Y, and HSO_3^- as Z; cf. Table V). For L2 there is another current with the same unstable feature (Fig. 13b), in which IO_3^- is reduced by Fe(II) instead of HSO_3^-. No such alternative exists for L1. When the reaction of Fe(II) is introduced in L1, an entrance (flow) is required for H^+ to balance mass so that the cycle in Fig. 13c is weak (and the current stable).

Shift calculations were carried out at two different Hopf bifurcations with the use of the sulfite [ferro(II)hexacyanide] feed concentrations as parameters (the rate constants and the other inflow concentrations were taken as given in Fig. 2 of Ref. 48). Since the unstable feature is the same for both mechanisms, the shift method produces the same results for the essential species they have in common. For L2, however, the (essential) species HOI and HIO_2 are both s^- (i.e., role Z, albeit indirectly for HIO_2), whereas in L1 HIO_2 was found to be s^+, HIO s^+ for one, s^- for the other Hopf bifurcation. The nonessential species I^- was d^+ for L1, s^+ for L2 (at both bifurcations). With such distinctions one has to bear in mind, however, that the behavior of nonessential species (especially when they take part in several reactions like I^-) can change for other parameter combinations, such as other rate constants, and therefore does not provide a good qualitative criterion to discriminate between mechanisms.

There is also an interesting effect in an off-diagonal element of the shift matrix in L1. Adding I_2 decreases the H^+ concentration. In contrast, in the other oscillator mechanisms studied (including L2), an increase in one autocatalytic species always increases the concentrations of all other species on the cycle. Thus, the influence of a superimposed flow of I_2 on the pH is different in sign for L1 and L2. The effect in L1 is probably because all reactions of I^- consume H^+ and I_2 can, by reacting with Fe(II), produce I^- without simultaneously producing even more protons (as in the cycle reaction).

These differences between L1 and L2 demonstrate the limits of the proposed procedure relying only on the diagonal elements of the shift matrix. Nevertheless, it seems possible to discriminate between the two through other results than the diagonal elements of shift experiments, such as the off-diagonal elements.

The mechanism suggested for the bromate mixed Landolt oscillator [49] is qualitatively the same as L2 (exccept that the reaction of XO_3^- with X^- was neglected) and exhibits the same shift behavior.

As can be seen by inspection of Figs. 11 and 13, a similar distinction regarding the formation of the negative feedback species Z as made previously for 1B oscillators can also be made in the 1C case: Z can either be produced in the exit reaction (*exit feedback*), as in the bromate/iodide and the mixed Landolt oscillators, or can be introduced in the flow (*flow feedback*), as in the rest. A simplified unstable current for the former case is reproduced in Fig. 13d (cf. Ref. 67). An analoguous observation is made for the species W in 1CW oscillators, which forms from the cycle and is either supplied through a tangent (chlorite/iodide and bromate/iodide) or the exit reaction of the cycle (bromate/manganous/sulfite and minimal bromate systems).

Comparison of the results of shift experiments as given in Table IV with the unstable currents of the respective mechanisms shows that the assignment of the essential species is unambiguous in both directions for the 1B oscillators that is, not only do, for instance, all autocatalytic species X exhibit s^+ behavior, but also all essential s^+ species lie on the critical current cycle. Inspection of Table V and the corresponding figures shows that this is not the case for 1C oscillators. Although all species exhibit the behavior summarized in Table III, some of the s^+ species do not lie on the critical current cycle but rather are formed from this cycle and play the role of a W species. Therefore, an unambiguous assignment of all species to their roles by the results of shift experiments only is not possible for 1C type oscillators.

B.   Category 2: Strong Current Cycles

For strong current cycles (SCC) the cycle order is higher than that of the exit process; in the simplest case the cycle just contains the second-order autocatalytic reaction $2X \rightarrow 3X$ together with the exit $X \rightarrow$ (see Fig. 3c). This feature, in itself, allows for unlimited growth of X. For reasons of mass balance another species Z has to be consumed in the autocatalysis, which leads to an unstable current as in Fig. 14a. There is a negative feedback of the form $a_{xz}*a_{zx} < 0$, similar to the 1C types. The network diagram in Fig. 14a is the irreversible Sel'kov model [68], which can thus be considered as the single current oscillator of category 2. This scheme changes its bifurcation structure if the reactions are assumed to be reversible (see below) so that other, for example, equilibrium, currents have to be included for realistic models. As has been pointed out by several authors [3, 25, 69, 70] the trimolecular reaction $(2X + Z)$ is not realistic as an elementary step, but there are a number of ways to break it up into several bimolecular processes, which approximately give rise to the same overall kinetics, for example, via the

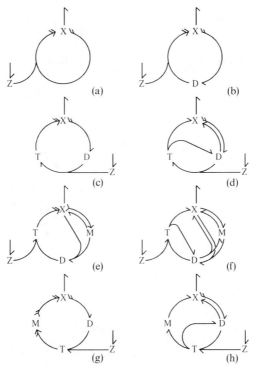

FIGURE 14. Unstable current for the Sel'kov model (a) with second-order autocatalysis and first-order exit, and its expansion into bimolecular reactions, assuming that dimerization of X is the initial process (b–h). β_i for the cycle species is -1 throughout. The currents containing a reaction and its own reverse are not extreme currents.

formation of first a dimer D and subsequently a trimer T [71]. Including a dimer, formed from $2X$, as an intermediate leads to the equivalent unstable current of Fig. 14b, which also remains qualitatively unchanged, if in addition a trimer is formed as intermediate rather than as activated complex (Fig. 14c). We expect, however, a trimer to decay into $D + X$ instead of $3X$ directly (Fig. 14d). As discussed by Clarke [3], such a premature break-up of T, which introduces a single-tail short circuit, still gives rise to a strong current cycle, if the dimerization of X is assumed to be sufficiently reversible so that a preequilibrium between X and D exists. Similarly, including other intermediates M need not destroy the strong current cycle (Figs 14e–14h). Thus, β_i for the cycle species, that is all species present except Z, is negative (-1) for all currents shown in Fig. 14. Note that the currents shown in

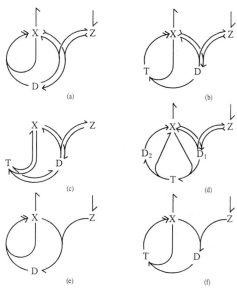

FIGURE 15. Expansion of the unstable current of Fig. 14a into bimolecular reactions, assuming an equilibrium $X + Z \rightleftharpoons D$ as initial step of the autocatalysis. Premature break-up of T requires an additional species (d). Unless the preequilibrium is fast, the dominant unstable current is of the form shown in (e) or (f), which contain a critical current cycle with Z as destabilizing exit species. β_i for the cycle species is -1 for (a), (b), and (d); zero for (e) and (f); (c) does not have an irreversible cycle. Currents containing equilibria are not extreme.

Figs. 14d–14h are not extreme, because they can be decomposed into simpler extreme currents and one or two equilibrium currents.

A dimer need not form from $2X$ in order to give rise to a second-order autocatalysis, but a similar "dimer" could be obtained in a preequilibrium from X and Z. As long as this equilibration is fast (Figs. 15a, b), the effective order of the autocatalytic process is still 2, or at least bigger than 1. However, in this case a premature break-up of T leads to an equilibrium current, as shown in Fig. 15c, unless another dimer is postulated (Fig. 15d). If the formation of D is not significantly reversible, the autocatalysis will be of first order and therefore a critical current cycle, as in the currents of Figs. 15e and 15f. This means that the category of the dominant unstable current depends on the speed of attaining the initial equilibrium.

Recently, Cook et al. [70] carried out extensive calculations on three models in which they replaced the trimolecular process in the Sel'kov model with two consecutive bimolecular steps. These models correspond to Figs. 15a, 14b, and 14d. In the model represented by Fig. 15a the authors

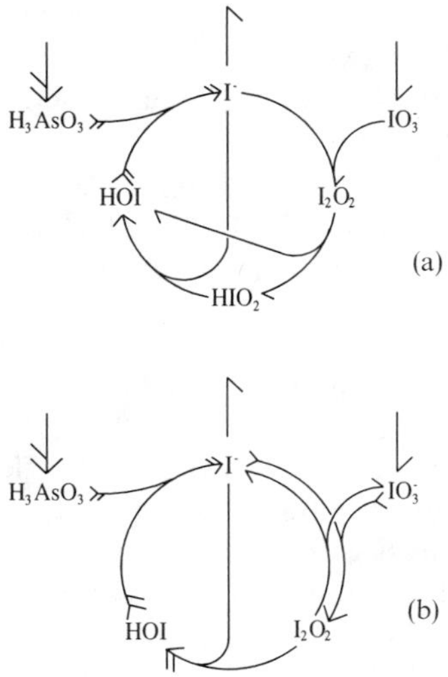

FIGURE 16. Unstable currents in the iodate/arsenous acid system [72] containing (a) a critical cycle (first-order autocatalysis) formed by I^- and HOI and (b) a strong cycle (second-order autocatalysis) formed by I^-, I_2O_2 and HOI; note that a dimer I_2O_2 is included explicitly, a (possible) "trimer" (from $I^- + I_2O_2$) is neglected; either it is very short-lived or occurs only as transition state.

included the process $D + Z \rightarrow D + X$ instead of including T explicitly, that is, T is regarded as an activated complex, not as an intermediate.

Since only two species are required to produce oscillations, strong current cycles are quite popular in abstract reaction schemes, such as the Brusselator [25] and the Sel'kov model [68], but are rare in practice, at least for isothermal homogeneous oscillators [3]. The only realistic example found has been suggested by Papsin et al. to explain the bistability in the oxidation of arsenous acid with iodate [72], but the situation in this system is more complicated because the autocatalysis in iodide can be both first and second order [72] stemming from a critical and a strong cycle (Fig. 16). In the empirical rate law [73, 74] a second-order term for iodide was indeed found so that the current in Fig. 16b probably plays a role in this reaction. Still, we have not found a realistic model of an isothermal homogeneous oscillator in which the oscillations are clearly caused by a strong current cycle.

In heterogeneous catalysis, on the other hand, strong cycles come into play quite naturally with the vacant surface sites playing the role of autocatalytic "species." The rate law of adsorption is often quadratic in the vacant sites for dissociatively adsorbing molecules, such as oxygen, which leads to a cycle order of 2, while the exit process is approximately of order

1, if the other reactant (e.g., carbon monoxide) does not need two adsorption sites. A detailed discussion of such a mechanism, that of the isothermal CO oxidation on platinum single crystals, is published elsewhere [74, 75]. The unstable current of this model, already contained in the classical Langmuir–Hinshelwood mechanism [76], is reproduced in Fig. 17a. The described argument for the occurrence of a strong current cycle is not restricted to CO oxidation, but applies whenever the adsorption rate of a molecule is of order higher than one in the vacant adsorption sites.

In thermokinetic oscillations, where temperature is an essential variable, the heat in the reactor plays the role of the cycle "species" and the process is autocatalytic if it is exothermic. The cycle "order" is exponential (> 1) for activated reactions, while the exit process, the flow of heat, is usually approximately linear (proportional to the temperature difference between the reactor and its surroundings). Consequently, an activated exothermic reaction can, through a strong current cycle (Fig. 17b), lead to thermokinetic oscillations for an appropriate reactor configuration.

The simplest oscillators of category 2, the Brusselator and the irreversible Sel'kov model, exhibit a Hopf bifurcation only. In general, the bifurcation

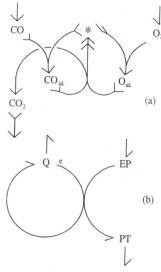

FIGURE 17.   (a) Unstable extreme current D_{EC} for the Langmuir–Hinshelwood mechanism [74–76] of catalytic CO oxidation (∗ denotes an empty site, O_{ad} and CO_{ad} adsorbed species). The gaseous species O_2, CO, and CO_2 are nonessential of type c, c, and b, respectively. The cycle formed by CO_{ad} and ∗ is weak (second-order exit), the one consisting of O_{ad} and ∗ strong (cycle order of 2). (b) Unstable extreme current D_{EC} for the nonisothermal hydrolysis of 2,3-epoxy-1-propanol EP [78]; Q stands for the amount of heat in the reactor, e denotes exponential reaction "order", the product, 1,2,3-propanetriol PT is inert.

structure of such systems is more complicated, allowing for both bistability and oscillations. Detailed diagrams have been calculated for the reversible Sel'kov model [77] (which can be regarded as prototype of category 2), the isothermal CO oxidation on Pt(110) [74, 75] and the nonisothermal hydrolysis of 2,3-epoxy-1-propanol [78]. A general analysis of thermokinetic systems in a CSTR with three parameters results in the unfolding of a Takens–Bogdanov–Cusp (TBC) point [44]. These bifurcation schemes have in common that, although they exhibit bistability and oscillations, they do not have the shape of a classical cross-shaped phase diagram. Rather, there is only one Hopf bifurcation curve instead of two opposing each other. Thus, the bifurcation scheme in two parameters can be used to distinguish categories 1 and 2 in case this cannot be guessed from the kind of system under consideration (cf. Section IV A).

Since strong current cycles are unstable by themselves without requiring an exit species Y, only two kinds of essential species need to be considered: the cycle species X and the species Z, which limits the autocatalysis of X. Consequently, only one bit is needed to assign the roles, which can be obtained from shift experiments: The species Z exhibit inverse regulation at the Hopf bifurcation, while the various X do not. The inverse regulation for Z may not always show up over the whole range of the Hopf bifurcation curve so that several trials may be required. Such shift experiments are generally out of the question in heterogeneous catalysis for the surface species (adsorption sites and adsorbates), but it is quite straightforward to measure the adsorption kinetics directly. If this is not possible, one can, for the example cited, identify the autocatalytic process by checking with a small perturbation whether the CO- or O-covered state is excitatory. This experiment was carried out recently by Fink et al. using an infrared laser pulse on a Pt(100) surface as catalyst [79]. An excursion occurred after perturbation when starting with a CO-covered surface, but not from an oxygen-rich state. This clearly shows that the reactive consumption of CO_{ad} is autocatalytic, that is, CO_{ad}, playing the role of Z, is consumed by the cycle, whereas O_{ad} is a cycle species. In this way the assignment of species can be achieved with excitation instead of shift experiments.

In thermokinetic oscillators the role of heat is obviously known and the problem reduces to finding out which essential species exhibit inverse regulation. These are the reactants in the most important activated exothermic step(s).

We have not found it necessary to distinguish subtypes in category 2, since the models considered exhibit qualitatively the same behavior. However, further classification of oscillators with a strong current cycle may become appropriate as soon as more, and possibly more complicated, examples are available.

C.   A Remark about Enzyme Reactions

The kinetics of an enzymatic process is commonly described with a steady-state approximation for the enzyme–substrate complex and the conservation constraint that the sum of the concentrations over all species containing the enzyme is constant. The resulting expressions, known as Michaelis–Menten kinetics, do not have the form of power laws. Although in such cases SNA can be applied using an effective power function defined by Clarke [3, Chapter I.C], this can lead to situations where the resulting network depends on the concentrations. It is therefore usually more convenient to include the enzymes and the enzyme–substrate complexes explicitly in an enlarged network (as is done in Ref. 38), which then exhibits mass action kinetics [80].

Some enzymatic reaction mechanisms give rise to networks with extreme currents of the same form as discussed in the previous section. As an example consider the peroxidase-catalyzed oxidation of NADH as described in the Yamazaki model [81]. The corresponding network was analyzed by Aguda and Clarke [38]. It contains an unstable current (E_5 in Ref. 38) of 1CX type. Although this reaction is usually studied in a "batch" reactor, a flow of the essential exit species oxygen (exchange with the gas phase) is required so that it is classified as 1C.

Enzymatic reactions can lead to oscillatory behavior through allosteric effects, that is, when an enzyme is activated by formation of a complex with another species (see, for example, Ref. 82). Such effects are typical of enzymatic reactions and we therefore briefly discuss how instabilities arise from this phenomenon. Two simple examples of product activation are considered, the Higgins model [83, 84] and a model for the papain-catalyzed hydrolysis of α-N-benzoyl-L-arginine [85, 86], the former (latter) of which contains an activation step with one (two) product molecules. We denote the activating species with X, the allosteric enzyme with E, the initial substrate with Z, the final (nonessential) product with P, and another enzyme (if present) with Y. The enzyme–substrate or enzyme–activator complexes are written as a combination of the species from which they are formed. If all details are included, the unstable currents are as shown in Fig. 18a and 19a, respectively.

In the Higgins model there is a critical current cycle (consisting of X, EX, and EXZ), limited by the supply of Z, with a destabilizing exit species Y. The unstable current is analogous to the Franck model, as can be seen from the reduced version shown in Fig. 18b, where the complexes are assumed to be very short-lived and the preequilibria forming the enzyme–substrate complexes are established rapidly. Figure 18a also provides the only example encountered of a category-1 oscillator where the dominant unstable current is not an extreme current. This is not a violation of theorem V.8 in Ref. 3, since

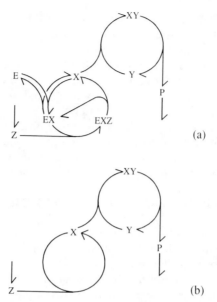

FIGURE 18. Unstable current for (a) the Higgins model [83] exhibiting a critical current cycle (X, EX, EXZ) and a simplified version (b) resembling the Franck model.

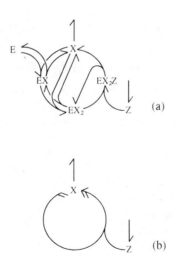

FIGURE 19. Unstable current for (a) the papain model [85, 86]; there is a strong current cycle formed by X, EX, EX_2, and EX_2Z, so that the current is similar to the Sel'kov model (b). For definition of species see text.

the extreme current obtained by leaving the equilibria out of Fig. 18a has a β_3 equal to zero if the three species included in β_3 are X, EX, and EXZ.

The papain model contains a strong current cycle (X, EX, EX_2), limited by species Z, and is in essence a more detailed bimolecular version of the Sel'kov model (cf. compare Fig. 19b to Fig. 14), which comes as no surprise since the Sel'kov model was originally suggested to describe oscillating enzyme reactions [68].

Thus, the categorization and the methods described in the preceding sections can also be applied to enzyme reactions, if one is willing to pay the price that more species have to be taken into account in SNA than variables needed for description with Michaelis–Menten approximations.

IV. APPLICATION TO MECHANISTIC STUDIES

The mechanistic classification of simple oscillators and the corresponding operational procedures described previously may be useful in studies of unknown oscillatory mechanisms, because some restrictions can be put on a reaction scheme if we determine the category of the oscillator. Identification of the roles of the involved species further restricts the number of possibilities and may in many cases allow the construction of the dominant unstable current. The role of species, however, can only be elucidated if their chemical nature has been determined by suitable analytic techniques. In the following we first outline a general procedure toward a determination of the mechanism of an unknown oscillator and then discuss some aspects in more detail using specific examples.

A. General Procedure

The classification as well as the proposed procedure are summarized in Fig. 20, in which the categories are enclosed in rectangles and successive distinctions and procedures are symbolized by encircled numbers ①–⑦. Since the procedure is limited to simple oscillators, the first question that arises is whether a system under consideration is simple or not. In step ① the system is assumed to be simple unless contradictions occur during step ②–⑥ of the proposed scheme. The vast majority of the systems investigated in this study proved to be simple (cf. Appendix A) so that the preliminary assumption of a simple oscillator may very often lead directly to a successful determination of the mechanism.

If contradictions within the proposed scheme arise, the system is either nonsimple or belongs to a new category of simple oscillators that has been overlooked so far (for an example of a nonsimple system see the Jensen oscillator in Section IV B). A systematic study of nonsimple systems and a classification of their behavior may lead to a definite answer on how to

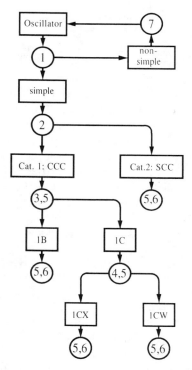

FIGURE 20. Procedure for constructing the unstable current of an unknown mechanism by assignment to one of four categories: The main categories, cat 1 ≡ critical current cycles (CCC) and cat. 2 ≡ strong current cycles (SCC). Category 1 is subdivided into various subcategories. For definition of 1B, 1C, 1CX, and 1CW refer to the text. Operational steps and decision points are represented by encircled numbers: 1 The oscillator is assumed to be simple. 2 A cross-shaped phase diagram (XPD) indicates a category 1 oscillator. The number of types of essential species found by shift experiments may also serve as a method of distinction between CCC and SCC. 3 The form of XPD's in various parameter combinations allows a distinction between 1B and 1C. Batch oscillations indicate type 1B (see Section III A 2). 4 If an inflow of a s^+ species is necessary for oscillations, the mechanism is of type 1CX; if not, the mechanism is of type 1CW (see Section III A 2). 5 Refers to the assignment of species: nonessential species are identified as described in Section II, the essential ones can be characterized by shift experiments (Section III). 6 Construction of the unstable current. 7 If contradictions to the previous assumption of a simple oscillator occur: decomposition into simpler subsystems which subsequently reenter the procedure at 1. The detailed methods of distinguishing the categories and the assignment of the species to their various roles are explained in the text.

distinguish simple and nonsimple oscillators. Such a study, however, remains subject to future investigations.

Certain types of complex dynamics, such as birhythmicity, seem to be incompatible with simple mechanisms, which suggests that a distinction may be possible from the global bifurcation structure or an examination of more complex attractors or both. If complex dynamics or contradictions within the proposed scheme indicate a nonsimple oscillator, one should try to decompose the system into simpler subsystems (⑦ in Fig. 20), for example, by choosing experimental conditions where one of the underlying unstable features clearly dominates the others. These subsystems could then be examined with the proposed procedure under the assumption that they are simple. A repetition of this decomposition step might be necessary. Afterward the separately studied subsystems can be pieced together.

Under the assumption that the system to be studied is simple, it must first be sorted into the basic categories 1 and 2, which is step ② in Fig. 20. This can often be done just on the basis of the nature of the system:

- Thermokinetic oscillators, that is, those where the temperature is an essential variable, are of category 2 (see Section III B). Whether the temperature is an essential variable can be checked by redesigning the reactor so that it can be run isothermally. The identification of the other variables can wait until step ⑤ (Fig. 20).

- In heterogeneous catalysis the adsorption kinetics of the reactants can be measured (Section III B). Similarly, in product-activated enzyme reactions the number of molecules needed for activation is experimentally accessible, allowing the distinction between category 1 and 2 systems.

- In isothermal homogeneous systems strong current cycles are rare, but can occur. Here a distinction between systems with strong and critical current cycles can be made on the basis of a detailed bifurcation scheme in two control parameters, since the bistable and oscillatory regions connect in different ways: As was already mentioned, all examples of category 1 studied exhibit a cross-shaped phase diagram; the limited number of investigated oscillators of category 2 have a bifurcation structure that corresponds to a two-dimensional section through the unfolding of a (codimension 3) Takens–Bogdanov–cusp point (TBC) [44, 74, 75, 78].

Another way to identify a strong current cycle, applied in Refs. 72 and 73 for the iodate/arsenous acid system, is to deduce the order of the autocatalytic process from empirical rate laws obtained (sometimes for subsystems) in batch, in this case the Dushman reaction [72, 73]. A strong current cycle is very likely if the cycle order is higher than one. (In principle, a critical current cycle may also result from both cycle and exit order being two, but this would require a trimolecular destabilizing exit reaction $2X + Y\rightarrow$, which is chemically unrealistic.)

For category 2 no subtypes were distinguished. The next step within that category is thus number ⑤, the assignment of species. First the nonessential ones have to be identified, as discussed in Section II; then the remaining essential ones can be assigned their role on the basis of their inverse or normal regulation as cycle species X or limiting species Z. There are only two different kinds of essential species in this category that are discernible in shift experiments, compared to three for category 1, and hence we have another way of distinguishing category 1 from 2 after the essential species and their shifts have been determined. Guidelines on how to construct the unstable current are discussed in some detail for category 1 below and apply mutatis mutandis for category 2 as well.

In the case of a category 1 system, the next step (③ in Fig. 20) is the distinction between 1B and 1C. Usually, 1B type mechanisms give rise to batch oscillations, whereas those of type 1C do not. As was pointed out in Section III A 2, this criterion need not always be reliable. Therefore, the relative positions of the bistable and oscillatory region in the cross-shaped phase diagram (see Section III A 2 and Table II) should be used as confirmation. This, however, can only be done after the assignment of the species that serve as bifurcation parameters in the XPD's in step ⑤.

Step ⑤ in Fig. 20 denotes sorting out the role of each species in the mechanism. The nonessential ones are identified first, together with their types a, b, or c, as described in Section II. The assignment of the essential species is achieved on the basis of shift experiments (Table III). The distinction between 1CX and 1CW can be made by checking which species have to be in the feed: only 1CX needs an inflow of an essential species with s^+ behavior. Some care has to be taken, since X can be produced from nonessential precursors, which is equivalent to a flow. Therefore, steps ②, a check on which inflows are required, and ⑤ have to be combined to reach a conclusion. We have not yet found a simple and generally applicable rule to distinguish X and W in the 1CW type (both are s^+ in the shift experiments).

After assignment of the species the dominant, or unique, unstable current E_1 can be constructed (⑥ in Fig. 20). Since nonessential species of type a can be substituted by flows and all reactions of type b with essential species can be neglected, only reactions among essential, and essential and type c nonessential species have to be considered. The inflows of Y and, unless it is formed internally, also of Z are present in the unstable current for 1C oscillators, but not the flow for X in 1CX, because the entrance reaction would weaken the cycle. This flow can only show up in the stable currents. Currents containing outflows of essential species are usually unimportant, because the role of essential species occurs in reaction steps prior to the outflow. Consequently, only flows of the nonessential ones (type b, possibly c) have to be taken into account to construct the dominant unstable current, but the dominating (stable) current of one branch of the hysteresis is usually governed by a flow (e.g., E_3^b in Fig. 4).

As was pointed out in Section III A 4, the essential species can connect up in different ways. Species formed from the cycle, namely, Z in 1B and W in 1CW oscillators, can be produced either by a tangent or by an exit reaction. Z in the 1C types originates from an exit process or is contained in the flow. This may suggest a further subdivision of these categories, but no general way to distinguish between them has been found. For 1C oscillators it is known whether Z is in the feed or not (i.e., produced internally), once the species have been assigned. The construction of the dominant unstable current E_1 is not expected to be ambiguous, because only a limited number of reactions are stoichiometrically possible. Furthermore, if the rate constants of some of these reactios can be determined independently, and if they are slow under the applied conditions, they can be neglected. Although, for instance, in a 1B oscillator exit and tangent feedback cannot be distinguished with the method of diagonal elements of the shift matrix, it appears very unlikely that both would be conceivable for the same set of essential species. The same argument applies mutatis mutandis for Z (and W) in 1C types. (The FKN-mechanism has both tangent and exit feedback, but the latter is nonessential.)

For 1C oscillators the assignment of species may be ambiguous (Table V), but the construction of E_1 is nevertheless expected to be unique. In case of difficulties one may examine the mechanism of the corresponding minimal oscillator first and then try to understand the more complicated systems, as was done by Epstein and co-workers in several studies (see, for example, Refs. 60, 63 and 64). For instance in the bromate/manganous/sulfite system the assignment of the exit species (d^+) is unclear (Table V), whereas in the minimal bromate oscillator (NFT) it is not.

Given the information described, the construction of the dominant unstable current should be possible unambiguously, but there can be cases where it is difficult to do so without use of some further knowledge (see, for example, the discussion of the bromate/iodide system in Section IV B).

The unstable current E_1 contains the part of the mechanism crucial for oscillations (autocatalysis and negative feedback), but a complete mechanism requires another step, namely, the determination of the other, stable currents. There are, except in the simplest models, quite a few stable currents; of these the ones dominating the stable nodes in the bistable region, called E_2 and E_3 earlier (Section III A 1), are sufficient for a qualitative understanding of the behavior [38]. From the examples in Fig. 4 we see that E_2^a and E_3^a or E_3^b (respectively, their analogues with more species) are present in most 1B and

1CW types, E_2^b and E_3^b in 1CX types, albeit with some variations. This suggests that there are only a limited number of qualitatively different possibilities; however, we have as yet no classification of all the possible stable currents nor a general method to determine them. In most cases the overall net stoichiometric change is known, or can be measured, so that, if E_1 can be constructed without using such knowledge, one can deduce information about the dominant stable currents from the difference between the overall change and the net effect of E_1 (examples are given in Section IV B). Similarly, measurement of the difference in the net turnover between the stable branches of the hysteresis can probably be used. If the stoichiometric change for these currents can be determined, this should in most cases at least allow educated guesses about what they look like. Obviously, the CSTR equations used in model calculations allow for an outflow of all species. Their construction is, however, important in order to find out indirectly which additional reactions, besides those already present in the unstable current, have to be included in the model to allow for such stable currents.

B. Examples

In this section we demonstrate with a few typical examples the application of the procedure described in Section IV A to an experimental system, show the use of SNA in addition to other chemical information and discuss some of the difficulties that might occur. In these examples we pretend not to know the mechanism but use results from numerical calculations with a given mechanism as if they were experimental data. Only properties that are expected to be easily accessible in an experimental situation are used for the interpretation. Where available in the literature, experimental results are also included in the discussion.

1. Briggs–Rauscher Reaction

As a first example we consider a mechanism for the Briggs–Rauscher reaction, taken from Refs. 50 and 57. This system consists of H_2O_2, H^+, IO_3^-, malonic acid MA, and Mn^{2+} plus usually small amounts of I_2. The example demonstrates the full scheme of steps ①–⑦ in Fig. 20 and shows in addition how sometimes even the stable currents can be obtained.

Step ①: We assume the system to be a simple oscillator.

Step ②: As the Briggs–Rauscher reaction is an isothermal homogenous oscillator, it belongs most likely to category 1. This assignment is confirmed by an XPD that is obtained in the $[I_2]_0$ vs $[IO_3^-]_0$ parameter plane.

Step ③: The above XPD shows an oscillatory region at low and a bistable region at high I_2 and IO_3^- inflow concentrations. It is thus of the type ↗ in Table II. Unless the roles of I_2 and IO_3^- are clarified, this information cannot be used to distinguish between the 1B and 1C category. As the Briggs–Rauscher reaction exhibits sustained batch oscillations, we assign it tentatively to the 1B category and proceed directly to the identification of the species ⑤, after which we will confirm the 1B type with the help of the preceding XPD.

Step ④: This step is only relevant for the 1C subcategory, since no subdivision of the 1B subcategory is defined.

Step ⑤: The first task in the identification of the species is the distinction between essential and nonessential species. Apart from the species cited previously, $IO_2^{\bullet}$, HIO_2, HOI, I^-, Mn^{3+}, iodomalonic acid (IMA), O_2 and $HO_2^{\bullet}$, are present. The short-lived radical $HO_2^{\bullet}$ is difficult to detect and is therefore excluded for the moment. The species that need to be included in the initial batch mixture are expected to be nonessential ("major species"); H_2O_2 and H^+ are present in large excess and are not treated as variables, but included in the rate constants in the first place. IMA and O_2 are inert products (type b nonessential in the limit) and are easily identified as such, because a perturbation of their concentration does not have an effect on the oscillation.

With the rate constants and inflow concentrations from Ref. 57, a Hopf bifurcation is obtained upon decreasing $[I_2]_0$ around $4.732 \times 10^{-7} M$. The relative oscillation amplitudes near this bifurcation are of the following order:

$$Mn^{3+} \gtrsim IO_2^{\bullet} \gtrsim HIO_2 > I^- \gtrsim HOI > I_2 \gg MA > IO_3^- > Mn^{2+}$$

The first three species have about the same A_{rel} (set to unity), the next two about 10^{-1}, and I_2 10^{-2}. With the same normalization MA, IO_3^- and Mn^{2+} have relative amplitudes of the order of 10^{-7}, 10^{-8}, and 10^{-10}, which clearly identifies them as nonessential (type c, cf. Table I). Although Mn^{3+} exhibits the highest relative amplitude, one can find out that it is nonessential (type b), because constant addition of Mn^{3+} requires more than 100% of its stationary-state concentration before noticeable effects on amplitude and period of the oscillation are achieved. However, addition of Mn^{3+} yields an increase in O_2 (and $HO_2^{\bullet}$) production, from which we draw a conclusion about the relative position of Mn^{3+} and O_2 in the mechanism: O_2 is "down the chain" from Mn^{3+}.

This leaves the inorganic iodine-containing species (except IO_3^-) as the only essential ones. Simulated shift experiments with these give the information that $IO_2^{\bullet}$ and HIO_2 form the critical current cycle, I^- is the exit species, and the negative feedback loop contains HOI and I_2 (Tables III and IV).

With this assignment of the species to their roles, we can verify our earlier assumption of a 1B-type oscillator: $[I_2]_0$ is equivalent to Z_{inflow} in Table II. The nonessential "major" species IO_3^- has to be consumed by the critical current cycle and thus refers to AC in Table II.

Step ⑥: The unstable current can readily be constructed from these data. Only reactions between the essential and the major species (MA, Mn^{2+}, IO_3^-, H_2O_2, and H^+) need to be considered, and we make sure that only the main products (IMA and O_2) are formed in the net result of this current. The critical cycle is straightforward to construct, consuming IO_3^- and Mn^{2+} and producing the nonessential product Mn^{3+}. The exit reaction can only be $IO_2^• + I^-$ or $HIO_2 + I^-$. The latter is faster and therefore more likely. It produces 2HOI, from which one can conclude that exit feedback takes place. The only tangentially produced species is Mn^{3+}, which is nonessential anyway. What remains is to find a way to generate I^- from HOI with I_2 as an intermediate. Since H_2O_2 and MA are present in excess, this poses no problem (Fig. 10b). The current, as it stands now, contains Mn^{3+} as a product. There is, however, hardly any of it in the outflow. It is known that Mn^{3+} generates O_2, which can reasonably come only from H_2O_2. Since the latter is a two-electron reductant, but Mn^{3+} is a one-electron oxidant, $HO_2^•$ can be postulated as an intermediate and the unstable extreme current is complete, also in the nonessential species.

Step ⑦: As no contradictions have occurred during steps ①–⑥, the initial assumption of a simple oscillator is verified and no decomposition step is necessary.

In the following we discuss how, in addition to the construction of the unstable current cycle with the procedure outlined in Section IV A and Fig. 20, in the case of the Briggs–Rauscher reaction even the stable currents of the network may be obtained.

In order to get information about the stable current one has to consider the stoichiometric change. The net effect of the unstable current is (neglecting H^+ and H_2O):

$$IO_3^- + 2H_2O_2 + MA \rightarrow IMA + 2O_2 \qquad \text{(BR)}$$

IMA and O_2 are indeed the only significant products near the Hopf bifurcation. It is therefore expected that the dominant stable extreme current has the same net stoichiometry, so that the overall stoichiometry remains unchanged. Consequently, the task is to construct such a current by leaving out some of the reactions in the unstable current (Fig. 10b) and putting in others so that the current is stable (the cycle becomes weak). A current with features such as in E_2^a (E_2^b) and E_3^a (E_3^b) in Fig. 4, which contain a quadratic

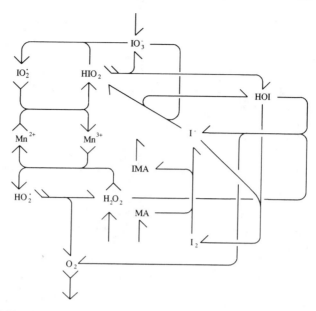

FIGURE 21. Stable current, which is most important for low flow rates during or near oscillations in the DeKepper mechanism [50] for the Briggs–Rauscher reaction. The autocatalytic cycle is weakened by an entrance as well as a quadratic exit reaction. This current has the same net stoichiometric change as the unstable one in Fig. 10.

exit and an entrance reaction to weaken the autocatalysis, suggests itself, since both reactions, namely,

$$I^- + IO_3^- \rightarrow HOI + HIO_2$$

and

$$2HIO_2 \rightarrow HOI + IO_3^-$$

are known to proceed at an appreciable rate. Indeed, inclusion of these two reactions leads to a stable current with the desired net change (Fig. 21). The reactions in the two currents are enough to account for batch and CSTR oscillations. The stable current of Fig. 21 dominates for the state with high X species, where the quadratic term in HIO_2 becomes important.

In CSTR equations outflows for all species are included and a large number of other currents are present. Basically, the network can be cut off practically anywhere and the resulting "loose ends" are assumed to flow out. However, all currents which include an outflow of a species that is present in relatively small concentration make no important contribution to the overall behavior.

In order to obtain bistability, an inflow of I_2 (or I^-) is required. This also leads to a number of additional stable currents that contain an outflow of an essential species. The simplest of these is just the flow of iodine. Since I_2 is the essential species with the highest steady-state concentration, its outflow is also more important than the others. These currents with flow in at least one essential species dominate for high I^- concentration. It is not necessary to identify explicitly any or all flow-dominated currents, since they are automatically included when the system is modeled using CSTR equations. It is thus possible not only to construct the unstable current E_1, but also the dominant stable current. No further reactions are needed so that only flows have to be included to produce bistability ("flow branch" as second stable state) in addition to oscillations in a CSTR. Other reactions can of course be tried to obtain E_2, but the condition that this current has to satisfy the overall stoichiometry (BR) excludes most conceivable combinations. Nevertheless, if several are found, rate constants have to be determined to choose between them or all have to be included.

2. Bromate/Iodide Oscillator

The example of the Briggs–Rauscher reaction demonstrates the possibility of unraveling a complete mechanism with the procedure described. In practice, not all methods needed to do so may be experimentally achievable with reasonable expenditure of effort. Our second, more complicated example is meant to demonstrate how SNA may be used in addition to other information in the development of a mechanism if the procedure outlined in Section IV A is not readily applicable. Moreover, the example shows that, although the unstable currents can probably always be constructed, one need not always achieve clear-cut conclusions for the dominant stable currents.

For the bromate/iodide oscillator, it is difficult to construct the dominant unstable current using only the shift data shown in Table V. One has to take other information into account to narrow the possible combinations of reactions that can be tried. In this case knowledge of the major overall stoichiometric processes occurring in the system [87] can be helpful; information of this kind has actually been used to piece mechanisms together in just about every study mentioned.

Experimental results [66, 88] show that in a closed system the reaction of BrO_3^- and I^- proceeds in two main stages:

$$6H^+ + 6I^- + BrO_3^- \rightarrow 3I_2 + Br^- + 3H_2O \tag{P1}$$

$$I_2 + 2BrO_3^- \rightarrow Br_2 + 2IO_3^- \tag{P2}$$

Using common knowledge about the chemistry of Br and I in addition

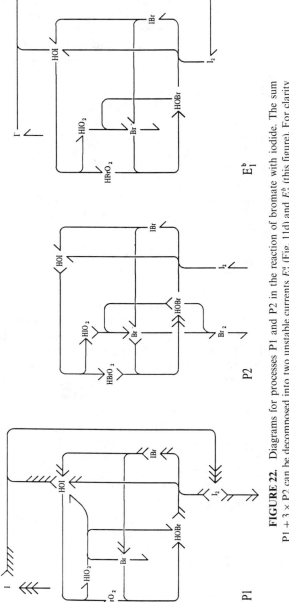

FIGURE 22. Diagrams for processes P1 and P2 in the reaction of bromate with iodide. The sum P1 + 3 × P2 can be decomposed into two unstable currents E_1^a (Fig. 11d) and E_1^b (this figure). For clarity all nonessential species except Br_2 are left out.

to shift experiments (Table V), networks for these two processes can be constructed, which is easier than trying to find a suitable unstable current for the whole oscillator directly. The networks P1 and P2 are shown in Fig. 22 without the nonessential species BrO_3^-, IO_3^-, H^+, and H_2O. The application of SNA to these networks reveals the following: P1 is indeed unstable, but P2 is not a current of the whole network at all, because it contains an inflow of I_2 that is not present in the experiments [66, 88]. Although P1 contains the unstable feature needed (critical cycle and exit reaction), it cannot be the dominant unstable current, since its products are I_2 and Br^-, whereas much more IO_3^- than I_2 and noticeably more Br_2 than Br^- are produced in the model near the Hopf bifurcation. This can be overcome by forming a new current of the form $P12 = P1 + 3 \times P2$ giving rise to the overall process:

$$6H^+ + 6I^- + 7BrO_3^- \rightarrow 6IO_3^- + 3Br_2 + Br^- + 3H_2O \qquad \text{(P12)}$$

Inspection of this current shows that it is not extreme, but can be decomposed into the one shown previously in Fig. 11d (E_1^a) and another unstable one (E_1^b) included in Fig. 22. Both currents are shown without the nonessential variables BrO_3^-, IO_3^-, H^+ and H_2O. Their net processes are

$$5I^- + 6BrO_3^- + 6H^+ \rightarrow 3Br_2 + 5IO_3^- + 3H_2O \qquad (E_1^a)$$

$$I^- + BrO_3^- \rightarrow Br^- + IO_3^- \qquad (E_1^b)$$

E_1^b contains Br^- as product, so that E_1^a is dominant, since more Br_2 forms than Br^-, but both give a finite contribution. Since the critical current cycles in E_1^a and E_1^b are identical, the system actually contains only one critical current cycle and is therefore still a simple oscillator. The cycle has a fairly complicated structure, and this is indeed the only example found where two exit species exist (I^- and Br^-) for the same cycle.

In this system it is not possible to deduce anything about the dominant stable currents from the overall process, since it is not clear exactly how large the contributions from the unstable ones are, so that the net stoichiometry of the stable ones cannot be obtained from a simple difference.

3.   Jensen Oscillator

Finally, we mention an example of a nonsimple oscillating mechanism, namely, a model suggested by Roelofs et al. [89] for the Jensen oscillator [90] [cobalt(II)-catalyzed air oxidation of benzaldehyde PhCHO], which demonstrates how contradictions in the suggested procedure (Section IV A) occur in a nonsimple system. Numerical simulations for this model were carried out for a CSTR (in contrast to Ref. 89) with a feed concentration of $2 \times 10^{-2} M$ for Co^{2+}; the oxygen pressure and the feed concentration for PhCHO as well as the total flow rate were varied.

Using O_2 pressure and the PhCHO feed concentration as parameters, we obtain four successive Hopf bifurcations, that is, two oscillatory regions separated by a stationary regime. This type of bifurcation scheme is different from those described in section III for simple oscillators of category 1 or 2. At low flow rate ($1 \times 10^{-4} s^{-1}$) oxygen exhibits normal regulation and the opposite effect on the bifurcation from the $PhCO_x$ radicals. Near another Hopf bifurcation at high flow rate ($1.25 \times 10^{-2} s^{-1}$), however, oxygen showed inverse regulation and the same effect as the mentioned radicals. Thus, the shift experiments give unclear results and do not allow well-defined assignment of the species to certain roles, rather these roles seem to be different for different bifurcations (at other parameter values).

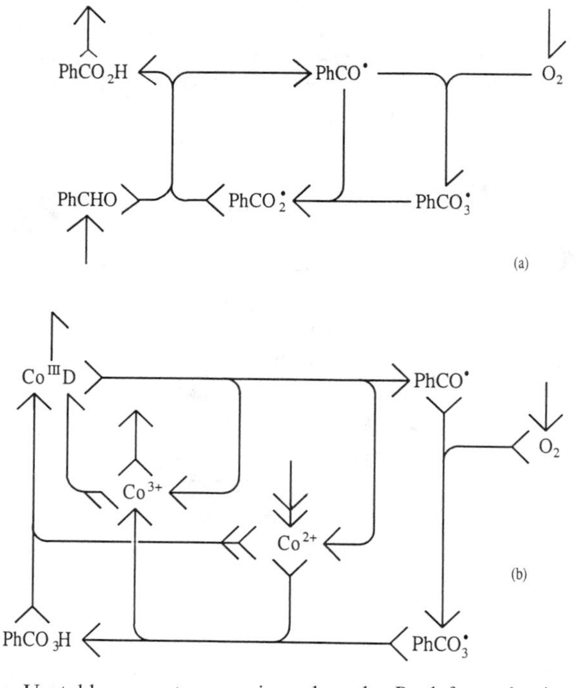

FIGURE 23. Unstable currents occurring when the Roelofs mechanism of the Jensen oscillator [89, 90] is used for simulations in a CSTR. (PhCHO is a major species; PhCOOH is an inert product.) (a) Critical current cycle with oxygen as exit species (1CX type); (b) branched strong current cycle consisting of a dinuclear Co(III)-complex [Co(III)D], PhCO·, $PhCO_3^-$, $PhCO_3H$, and Co^{3+}. Oxygen plays the role of Z (see Section III B). The cycle is strong because the cycle order of CO^{3+} is 2, while its exist order is 1.

The unstable currents derived from the model [89] are reproduced in Fig. 23. The first one (a) is of type 1CX with PhCO$^{\bullet}$ and PhCO$^{\bullet}_2$ as cycle, O$_2$ as exit, and PhCO$^{\bullet}_3$ as feedback species. This would be nearly in line with the shift behavior found at low flow rates, but no inverse regulation was obtained for PhCO$^{\bullet}_3$. The reason for this is that there is another source of instability, namely, a strong current cycle, in which PhCO$^{\bullet}_3$ is autocatalytic (among others), so that the inverse regulation that would be expected from the current in Fig. 23a alone is overcompensated. This strong cycle (Fig. 23b) is limited by the supply of oxygen, which is the reason for its inverse behavior (it plays the same role as Z in the Sel'kov model). Thus, O$_2$ has two different roles in two coexisting unstable features, so that the opposite sign in the regulation at different bifurcations is not surprising.

This example is meant to demonstrate that the interpretation of the shift data can be complicated for nonsimple oscillators (with more than one autocatalysis) and need not lead to unambiguous conclusions. It is not known, of course, whether the Jensen oscillator really contains two sources of instability. Actually, we are unaware of any experimental data supporting this, but a decision can only be made experimentally, that is, by applying the procedure proposed in Section IV A to the system. To perform the necessary CSTR experiments and to compare the resulting shift matrix and bifurcation diagrams with those predicted from the Roelofs model will be subject of future studies.

V. DISCUSSION

This chapter is to our knowledge the first attempt at a systematic classification of chemical oscillators according to mechanistic criteria. Oscillators have usually been sorted into groups [91] depending on the main species [60, 64, 66] and the nature of the system (such as bromate and chlorite oscillators, enzyme reactions, or heterogeneous systems). Although this scheme has, together with Epstein's concept of the minimal oscillator, been found useful in mechanistic studies, at least in homogeneous phase [60], it is not based on differences in the mechanisms. Stoichiometric network analysis [3] provides a way to get some order into the confusing variety of oscillators. The classification presented here is restricted to simple oscillators and, even for these, it is quite likely not complete. Thus, different unstable features may require new categories; and other subtypes of category 1 or a subdivision of category 2 may become appropriate as more and more oscillators are discovered. The classification scheme presented here may then be enlarged to include the new types.

Although operational definitions have been offered throughout this chapter, further experimental techniques to distinguish the various types as well as to identify the roles of species in them are needed, since the same method will not be practicable for all systems of interest because of possible experimental limitations. Moreover, a method independent of the assignment of species that can determine whether Z (W) in 1B (1CW) types is produced in a tangent or exit reaction is desirable. Similarly, the origin of Z in 1C oscillators (flow or exit reaction) cannot be known from the techniques suggested until the species playing the role of Z is identified.

Since only simple limit cycle oscillations are discussed here, the essential and nonessential species are defined with respect to a Hopf bifurcation. As was already pointed out in Section III A 2, the species Z in category 1 can be regarded as nonessential for the cusp (bistability), but is essential for the Hopf bifurcation. In chemically realistic mechanisms, considering just the species essential for the Hopf bifurcation, we have never found dynamical behavior more complex than simple limit cycles. However, in abstract mathematical models like the Rössler system [92], chaos can be obtained in the same three variables that are already essential for the Hopf bifurcation. This suggests the possibility of defining (non)essential species with respect to other bifurcations or other attractors, such as quasiperiodicity, mixed-mode oscillations, or chaos. As an example, we mention that the Neimark–Sacker bifurcation occurring for certain parameters in the SNB mechanism [50, 93, 94] requires HOBr as additional essential species. A systematic study of which variables are required for dynamical behavior of increasing complexity has recently been carried out with a model of isothermal CO oxidation on platinum [74, 75], in which an increasingly complicated system of differential equations was built up such that bistability, simple limit cycles, mixed-mode oscillations, and chaos were found for two, three, four respectively five variables. Although not all of these are chemical species—structural changes of the catalyst had to be included—they all correspond to effects found in experimental studies [74, 95, 96]. The number of essential species (or variables) for a certain phenomenon must be at least equal to, but has in most cases been found to be greater, than the corresponding embedding dimension. The number of variables can usually be reduced further, since for some essential ones steady-state approximations can be made; however, a systematic study of when such an approximation is possible without destroying, for example, the Hopf bifurcation has not yet been carried out. A generalization of the concept of essential and nonessential species may lead to elucidation of the minimal mechanistic requirements for dynamical behavior more complicated than the simple limit sycles discussed in the present work. Thus, for instance mixed-mode oscillations seem to be possible in simple oscillators with two negative feedback loops occurring on different time scales [74, 75]. On the other hand, we are unaware of any realistic mechanism predicting chaos that does not contain at least two nonweak current cycles. Of course, the behavior of such systems does not only depend on the nature of each cycle (or subsystem), but also on the way they are coupled together, that is, which species the two subsystems have in common. As an example, consider two oscillators coupled in such a way that an inert product of one is an essential species of the second. Such a system exhibits a bifurcation diagram like a single oscillator under the influence of a periodic perturbation of the respective essential species, an effect that has been studied extensively, both in theory and in experiment [6–12]. If, in contrast, the subsystems have essential species in common, then the bifurcation structure is different, since there is two-way coupling, and new phenomena can arise, for example, birhythmicity as reported by Alamgir and Epstein for the bromate/chlorite/iodide system [64]; the subsystems in this case, obtained by omitting one oxidant, share three essential species (cf. Table V).

We restricted ourselves to a qualitative discussion of chemical oscillators, which is also the way a mechanism (network) is defined in SNA. Understanding of an oscillator is not complete without determination of the rate constants, which usually have to be measured in independent experiments. On the other hand, inequalities can be derived with SNA [3], which the parameters have to satisfy in order to allow a system to become unstable. Similarly, conditions for the parameters can be obtained for bifurcation points, for example, a Hurwitz determinant has to vanish at a Hopf and a certain polynomial at an SN bifurcation [3, 33], which allow the determination of some parameters in terms of others. Although such a procedure does not yield rate constants directly, since the computations in SNA are carried out using so-called convex parameters [3], these parameters can be transformed into the more familiar kinetic parameters. Obviously, no such quantitative aspects can be addressed until the (qualitative) mechanism has been established, which is why the latter problem was examined in the present work, while systematic investigation of quantitative applications need to be the subject of future studies.

VI. SUMMARY

We presented an operational classification of chemical oscillators in open and closed systems, and a categorization of the species in these oscillators. This was achieved partially by analytical but mostly by numerical investigations within the framework of stoichiometric network analysis and bifurcation theory. We confined ourselves to simple oscillators, that is, systems that contain only one nonweak current cycle (and therefore only one source of instability). The classification of oscillators is based on this basic unstable feature (i.e., their autocatalysis) and the type of the required negative feedback loop. Two main categories were distinguished depending on whether the autocatalytic process is a critical (category 1) or strong (category 2) current cycle. Category 1 was subdivided into three types (1B, 1CX and 1CW), which exhibit qualitative differences in the nature of the negative feedback. The suggested operational method to assign an unknown system to its category uses the bifurcation structure in two control parameters and the role of species in the system. The species were primarily categorized into those essential and nonessential for a Hopf bifurcation. Three different kinds of nonessential species were defined and several ways to identify them described. Depending on the category of the mechanism two to four classes of essential species needed to be distinguished: species on the current cycle (X), species causing a negative feedback (Z), species that destabilize a critical current cycle (Y, in category 1 only), and species allowing the critical current cycle to recover by reacting with Y (W, in 1CW only). These roles can be identified using shift experiments. About 25 simple oscillators, abstract models and realistic mechanisms, were investigated and classified in this way. All fitted into the four categories. The proposed procedure may be useful in mechanistic studies, as demonstrated with the help of typical examples, where the numerical results of model calculations were treated in analogy to experimental data. Thus the procedure may contribute to the solution of the "inverse problem", i.e., to find a generally applicable way to develop a mechanism for an oscillating system from experimental results.

APPENDIX A

DIRECTORY OF OSCILLATORS

The following directory gives an overview over the abstract models and realistic mechanisms investigated in the present paper, the categories to which they belong (only for simple oscillators) and the section(s) in which they are discussed.

I. SIMPLE OSCILLATORS

Abstract Models	Category	Discussed in Section(s)
Oregonator [15]	1B	III A 2, III A 4
Revised oregonator versions A and B [24]	1B	III A 4
Franck model [29, 30]	1CX	III A 2, III C
Sel'kov model and its variations [68, 77]	2	III B, III C, IV B
Brusselator [25]	2	II, III B
Exothermic reaction in a CSTR [44]	2	III B
Higgins model [83, 84]	1CX	III C
Papain model [85, 86]	2	III C

Furthermore, the single current oscillator, an oscillator with a nongeneric bifurcation scenario, is mentioned in Section III A 1.

Realistic Mechanisms	Category	Discussed in Section(s)
FKN mechanism [54, 55]	1B	II, III A 3, IV A 4, IV A
SNB mechanism [93]	1B	III A 3, III A 4, V
OKN mechanism [39]	1B	III A 4
NFT mechanism [51, 58]	1CW	III A 2, III A 3, III A 4, IV A
BZ with oxalic acid ("modified oregonator") [56]	1B	III A 4
$BrO_3^-/Mn^{2+}/SO_3^{2-}$ reaction [64]	1CW	III A 4, IV A
$BrO_3^-/Mn^{2+}/$hypophosphite reaction [65]	1B	III A 4
$BrO_3^-/Br^-/ClO_2^-$ reaction [69, 60]	1CW	III A 4
BrO_3^-/I^- reaction [66]	1CW	II, III A 2, III A 4, IV B
ClO_2^-/I^- reaction [61–63]	1CW	III A 1, III A 4

Briggs–Rauscher reaction [50, 52]	1B	III A 3, III A 4, IV B
Mixed Landolt reaction L1 [47, 48]	1CX	III A 2, III A 3, IV A 4
Mixed Landolt reaction L2 [49]	1CX	III A 2, III A 3, III A 4
BrO_3^- mixed Landolt oscillator [49]	1CX	III A 4
Nonisothermal hydrolysis of 2, 3 epoxy-1-propanol [78]	2	III B, IV A
Isothermal CO oxidation on Pt(110) [74, 75]	2	III B, V
Peroxidase catalysed NADH oxidation [81]	1CX	III B

II.   Nonsimple Oscillators

System	**Discussed in Section(s)**
Two further Franck models [29, 30]	III A 3
Explodator [16]	III A 4
Revised oregonator version C [24]	III A 4
Jensen oscillator [89, 90]	IV B

APPENDIX B

DIAGRAMMATIC STABILITY ANALYSIS

The stability of a current is determined by certain subdeterminants β_i of the product of the stoichiometric matrix and the transpose of the kinetic matrix, where the index i refers to the numbers of species included. A current is

- unstable, if any β_i for any subset of species is smaller than zero,
- stable, if all β_i's for all combinations of species are nonnegative.

The general procedure of how these quantities are obtained is described in Ref. 3. In the following we outline this procedure using a specific example, namely, an unstable extreme current from the chlorite/iodide oscillator shown in Fig. 24a. From this current we select the three species HOI, HIO_2 and I^- as species 1, 2, and 3, respectively, and their reactions (marked 1 to 4 in Fig. 24a). For the kth reaction we denote the number of left feathers of the lth species with l_{kl}, the total number of feathers with f_{kl} and the number of barbs with b_{kl}. In our example the corresponding matrices L, F, and B are:

$$L = \begin{pmatrix} 1 & 0 & 0 \\ 0 & 1 & 1 \\ 1 & 0 & 1 \\ 1 & 0 & 0 \end{pmatrix} \qquad F = \begin{pmatrix} 1 & 0 & 0 \\ 0 & 2 & 2 \\ 2 & 0 & 2 \\ 1 & 0 & 0 \end{pmatrix} \qquad B = \begin{pmatrix} 0 & 1 & 0 \\ 4 & 0 & 0 \\ 0 & 0 & 0 \\ 0 & 1 & 0 \end{pmatrix}$$

The first step is the construction of the current matrix diagram D_{CM}. To do this one computes the sum over all k of $l_{kl} * f_{kl}$ for every species l and writes the resulting number in front of the species symbol. For the two species (1 and 2) in β_2 in Fig. 24b $\sum_{k=1}^{4} l_{k1} * f_{k1} = 1 + 0 + 2 + 1 = 4$ yields 4HOI and $\sum_{k=1}^{4} l_{k2} * f_{k2} = 0 + 2 + 0 + 0 = 2$ results in $2HIO_2$. These sums are referred to as *stabilizers* if positive and *destabilizers* if negative. Destabilizers can occur if 1-cycles are present, but there are none in this example.

If a species Y produces another species X in a reaction k, this reaction is called a *promotion* of X. A reaction m in which X reacts with Y and is thereby consumed is refered to as an *inhibition* of X by Y. The reaction is also an inhibition of Y by X. Next, such promotions and inhibitions are drawn according to the following rules:

- Promotions of Y by X are drawn as full arrows with $l_{kX} * b_{kY}$ barbs.
- Inhibitions of Y by X are drawn as dashed arrows with $l_{mX} * f_{mY}$ barbs from X to Y, inhibitions of X by Y as dashed arrows with $l_{mY} * f_{mX}$ barbs from Y to X.

- If several arrows occur between the same two species in the same direction, they are added algebraically, where barbs at full and dashed arrows compensate each other.

It is not necessary to construct the full D_{CM} in order to obtain a certain β_i, rather only those species for which β_i is to be calculated need to be

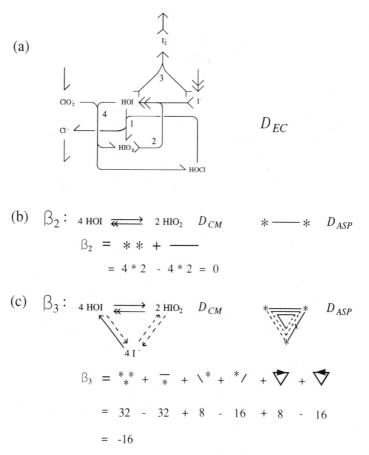

(a)

D_{EC}

(b) β_2: $4\ HOI \rightleftharpoons 2\ HIO_2$ D_{CM} $* \text{———} *$ D_{ASP}

$\beta_2 = ** + \text{———}$

$= 4 * 2 - 4 * 2 = 0$

(c) β_3: $4\ HOI \rightleftharpoons 2\ HIO_2$ D_{CM} D_{ASP}

$4\ I^-$

$\beta_3 = \overset{*}{\underset{*}{*}} + \overline{\underset{*}{}} + \backslash^* + {}^*/ + \bigtriangledown + \bigtriangledown$

$= 32 - 32 + 8 - 16 + 8 - 16$

$= -16$

FIGURE 24.   Diagramatic stability analysis of an unstable current (a) from the clorite/iodide oscillator. Shown are the *current matrix* diagrams D_{CM}, the *all stabilizer and polygon* diagrams D_{ASP}, and the supergraphs—the sums of *nonoverlapping stabilizer and polygon* diagram D_{NOSP}—yielding the subdeterminants β_i that determine the stability of the current. (b) Subdeterminant β_2 of two species (HOI and HIO$_2$). (c) Subdeterminant β_3 of three species (HOI, HIO$_2$, and I$^-$).

included. Thus, in Fig. 24b only that part of D_{CM} incorporating HOI and HIO$_2$ is shown: HOI produces HIO$_2$ in two different first-order reactions, resulting in a promotion with two barbs, HIO$_2$ reacts to 4HOI in a single first-order process, which gives rise to a promotion with four barbs. No inhibitions occur in this part of the current matrix diagram since the two species under consideration do not react with each other.

It is possible to obtain β_i directly from the D_{CM}. However, especially in complicated cases, it is more convenient first to construct a diagram of *all stabilizers and polygons* D_{ASP}, where the (de)stabilizers are symbolized by stars and feedback cycles drawn as polygons. The D_{ASP} can then be decomposed into *nonoverlapping stabilizer and polygon* diagrams D_{NOSP}, that is, diagrams which contain each species under consideration exactly once either in a (de)stabilizer or in a feedback cycle. The corresponding β_i is obtained as a sum (*supergraph*) over all nonvanishing D_{NOSP}'s taking into account that positive (negative) feedback loops are destabilizing (stabilizing) and therefore enter with negative (positive) sign (*sign-switched cycles* [3]). Therefore, only destabilizers (not present in our example) and polygons with 0, 2, 4, ... dashed edges or inhibitions yield negative terms in the supergraph. In the case of β_2 for HOI and HIO$_2$ there are two stabilizers and one positive 2-cycle in the D_{ASP} (Fig. 24b). Consequently, two D_{NOSP}'s have to be taken into account, namely, the product of the stabilizers ($2*4 = 8$) and the 2-cycle. The number of barbs of this 2-cycle in the D_{CM} are also multiplied yielding 8 again, but with negative sign because of the 0 dashed edges of the corresponding polygon. This leads to a β_2 of zero and to the conclusion that HOI and HIO$_2$ form a critical current cycle. (They do not exhibit a conservation constraint of the form $c_1[\text{HOI}] + c_2[\text{HIO}_2] = \text{const.}$) This CCC can be destabilized by an exit reaction with an additional species. However, theorem V.10 from Ref. 3 (see Section III A 1) does not apply, since the exit species in question, I$^-$ also enters in a tangent reaction of the cycle (HIO$_2$ + I$^-$). Therefore, β_3 for these species has to be worked out explicitly to judge the stability of the current. The required parts of the D_{CM} and the corresponding D_{ASP} are shown in Fig. 24c. The latter contains all three possibilities of 2-cycles: besides the positive one consisting of two promotions between HOI and HIO$_2$, there is a negative 2-cycle between I$^-$ and HOI as well as another positive one which results from the two inhibitions between I$^-$ and HIO$_2$ (competitive positive feedback shown as a crossed full line). In addition there are two different 3-cycles (inner triangles of the D_{ASP}) differing in orientation and sign: a negative feedback (I$^- \to$ HOI $\to$ HIO$_2 \to$ I$^-$) with an odd, and a positive one (I$^- \to$ HOI $\to$ HIO$_2 \to$ I$^-$) with an even number of inhibitions. In the supergraph occur six D_{NOSP}'s: one for the product of all stabilizers, three combinations of a stabilizer and a 2-cycle, as well as the two 3-cycles. As shown in Fig. 24c, β_3 for HOI, HIO$_2$, and I$^-$ is negative, indicating that the CCC is destabilized by the exit reaction and the extreme current is therefore unstable.

The β_i's can also be obtained through formal matrix operations (see Ref. 3), but with such an approach the feedback loops occurring in the network are not as evident as in the diagrammatic analysis. The matrix formalism is, however, the method of choice when using a computer.

Acknowledgments

The authors are much indebted to Bruce L. Clarke for his hospitality, a detailed introduction to stoichiometric network analysis, and many helpful discussions. We thank Raima Larter for useful questions and suggestions. Financial support by Deutsche Forschungsgemeinschaft (DFG) is greatly appreciated. This work was supported in part by the National Science Foundation and the Air Force Office of Scientific Research.

References

1. J. M. T. Thompson and H. B. Stewart, *Nonlinear Dynamics and Chaos*, Wiley, Chichester, 1987.
2. J. Guckenheimer and P. Holmes, *Nonlinear Oscillations, Dynamical Systems, and Bifurcations of Vector Fields*, Springer, New York, 1986.
3. B. L. Clarke, *Adv. Chem. Phys.* **43**, 1 (1980).
4. F. Hynne and P. G. Sørensen, *J. Phys. Chem* **91**, 6573 (1987).
5. P. G. Sørensen and F. Hynne, *J. Phys. Chem.* **93**, 5467 (1989).
6. F. W. Schneider, *Ann. Rev. Phys. Chem.* **36**, 347 (1985).
7. P. Rehmus and J. Ross, in *Oscillations and Traveling Waves in Chemical Systems*, R. J. Field and M. Burger, Eds., Wiley Interscience, New York, 1983, pp. 287–332.
8. W. Vance and J. Ross, *J. Chem. Phys* (to be published).
9. M. Eiswirth and G. Ertl, *Phys. Rev. Lett.* **60**, 1526 (1988); *Appl. Phys. A* **47**, 91 (1988).
10. M. Eiswirth, P. Möller, and G. Ertl, *Surf. Sci* **208**, 13 (1989); *J. Vac. Sci. Technol. A* **7**, 1882 (1989).
11. F. Buchholtz, A. Freund, and F. W. Schneider, in *Temporal Order,* L. Rensing and N. J. Jaeger, Eds., Springer, Berlin, 1985, pp. 116–121.
12. A. Freund, F. Buchholtz, and F. W. Schneider, *Ber. Bunsenges. Phys. Chem.* **89**, 637 (1985).
13. J. Weiner, F. W. Schneider, and K. Bar-Eli, *J. Chem. Phys.* **93**, 2704 (1989).
14. T. Chevalier and J. Ross (unpublished).
15. R. J. Field and R. M. Noyes, *J. Chem. Phys.* **60**, 1877 (1974).
16. Z. Noszticzius, H. Farkas, and Z. A. Schelly, *J. Chem. Phys.* **80**, 6062 (1984).
17. J. Carr, *Applications of Center Manifold Theory*, Springer, New York, 1981.
18. The Jacobi matrix contains the partial derivatives of the kinetic equations with respect to state variables.
19. Y. Termonia and J. Ross, *Proc. Nat. Acad. Sci (U.S.)* **78**, 2952 (1981).
20. Y. Termonia and J. Ross, *Proc. Nat. Acad. Sci. (U.S.)* **78**, 3563 (1981).
21. The simple criterion [19, 20] to identify nonessential species through their phase shift with respect to external periodic perturbations unfortunately is not generalizable: the reported

zero phase shift at resonance for essential species seems to be a special property of the glycolysis model.

22. Y. Termonia and J. Ross, *Proc. Nat. Acad. Sci (U.S.)* **79**, 2878 (1982).

23. K. R. Sharma and R. M. Noyes, *J. Am. Chem. Soc.* **98**, 4345 (1976).

24. R. M. Noyes, *J. Chem. Phys.* **80**, 6071 (1984).

25. J. J. Tyson, *J. Chem. Phys.* **58**, 3919 (1972).

26. The form of the Jacobian in the limit is obvious for the subsets with A and B. In the case of C the diagonal element diverges with c approaching infinity. The variable C can be renormalized leading to a finite diagonal and vanishing off-diagonal elements.

27. W. S. Loud, *Q. J. Appl. Math.* **22**, 222 (1967).

28. J. J. Tyson, *J. Chem. Phys.* **62**, 1010 (1975).

29. U. F. Franck, *Ber. Bunsenges. Phys. Chem.* **84**, 334 (1980).

30. U. F. Franck, in *Temporal Order*, L. Rensing and N. I. Jaeger, Eds., Springer, Berlin, 1985, pp. 2–12.

31. B. L. Clarke, *J. Chem. Phys* **75**, 4970 (1981).

32. B. L. Clarke, in *Chemical Applications of Topology and Graph Theory*, R. B. Kind, Ed., Elsevier, Amsterdam, 1983, pp. 322–357.

33. B. L. Clarke, *Cell Biophys.* **12**, 237 (1988).

34. J. Quirk and R. Ruppert, *Rev. Econ. Stud.* **32**, 311 (1965).

35. C. Jeffries, V. Klee, and P. Van den Driessche, *Can. J. Math.* **29**, 315 (1977).

36. G. Hadley, *Linear Programming*, Addison-Wesley, Reading, MA, 1962.

37. B. von Hohenbalken, B. L. Clarke, and J. E. Lewis, *J. Comp. Appl. Math.* **19**, 231 (1987).

38. B. D. Aguda and B. L. Clarke, *J. Chem. Phys.* **87**, 3461 (1987).

39. M. Orbán, E. Körös, and R. M. Noyes, *J. Phys. Chem.* **83**, 3056 (1979); P. Herbine and R. J. Field, *J. Phys. Chem.* **84**, 1330 (1980).

40. A bifurcation is called nongeneric in n parameters if at least $n + 1$ parameters need to be fixed in order to locate the bifurcation in the parameter space.

41. J. Boissonade and P. De Kepper, *J. Chem. Phys.* **84**, 501 (1980).

42. In principle, while a limit cycle can be embedded in only two dimensions, in the case of a CCC system two species are not sufficient to obtain oscillations.

43. P. De Kepper and J. Boissonade, *J. Chem. Phys.* **75**, 189 (1981).

44. J. Guckenheimer, *Physica* **20D**, 1 (1986).

45. E. C. Edblom, M. Orbán, and I. R. Epstein, *J. Am. Chem. Soc.* **108**, 2826 (1986).

46. V. Gáspár and K. Showalter, *J. Am. Chem. Soc.* **109**, 4869 (1987).

47. E. C. Edblom, L. Györyi, M. Orbán, and I. R. Esptein, *J. Am. Chem. Soc.* **109**, 4876 (1987).

48. Y. Luo and I. R. Epstein, *J. Phys. Chem.* **93**, 1398 (1989).

49. E. C. Edblom, Y. Luo, M. Orbán, K. Kustin, and I. R. Epstein, *J. Phys. Chem.* **93**, 2722 (1989).

50. P. De Kepper and J. Boissonade, in *Oscillations and Traveling Waves in Chemical Systems*, R. J. Field and M. Burger, Eds., Wiley Interscience, New York, 1983, pp. 223–256.

51. R. M. Noyes, R. J. Field, and R. C. Thompson, *J. Am. Chem. Soc.* **93**, 7315 (1971).

52. P. De Kepper and J. Boissonade, *Compt. Rend. Scéances Acad. Sci. Ser. C* **286**, 437 (1987).

53. D. Barkley, J. Ringland, and J. S. Turner, *J. Chem. Phys.* **87**, 3812 (1987).

54. R. M. Noyes, R. J. Field, and E. Körös, *J. Am. Chem. Soc.* **94**, 1394 (1972).

55. R. J. Field, E. Körös, and R. M. Noyes, *J. Am. Chem. Soc.* **94**, 8649 (1972).

56. R. J. Field and P. M. Boyd, *J. Phys. Chem.* **89**, 3707 (1985).

57. P. De Kepper and I. R. Epstein, *J. Am. Chem. Soc.* **104**, 49 (1982).

58. W. Geiseler, *Ber. Bunsenges. Phys. Chem* **86**, 721 (1982); *J. Phys. Chem.* **86**, 4394 (1982).

59. M. Orbán and I. R. Epstein, *J. Phys. Chem.* **87**, 3212 (1983).

60. I. R. Epstein and M. Orbán, in *Oscillations and Traveling Waves in Chemical Systems*, R. J. Field and M. Burger, Eds., Wiley Interscience, New York, 1983, pp. 257–286.

61. O. Citri and I. R. Epstein, *J. Phys. Chem.* **91**, 6034 (1987).

62. I. R. Epstein and K. Kustin, *J. Phys. Chem.* **89**, 2275 (1985).

63. M. Orbán, C. Dateo, P. De Kepper, and I. R. Epstein, *J. Am. Chem. Soc.* **104**, 5911 (1982).

64. M. Alamgir, M. Orbán, and I. R. Epstein, *J. Chem. Phys.* **87**, 3725 (1983).

65. L. Adamčíková and P. Ševčík, *Int. J. Chem. Kin.* **14**, 735 (1982).

66. O. Citri and I. R. Epstein, *J. Am. Chem. Soc.* **108**, 357 (1986).

67. R.-S. Li, *J. Chem. Soc. Faraday Trans. 2* **84**, 737 (1988).

68. E. E. Sel'kov, *Eur. J. Biochem.* **4**, 79 (1968).

69. J. J. Tyson and J. C. Light, *J. Chem. Phys.* **59**, 4164 (1973).

70. G. B. Cook, P. Gray, D. G. Knapp, and S. K. Scott, *J. Phys. Chem.* **93**, 2749 (1989).

71. The terms dimer and trimer are used in the loose sense that D or T are capable of breaking up into two or three smaller pieces, which are identical to X or can be transformed into X, but other essential or nonessential species can be involved so that the formulas for D or T need not be multiples of the formula of X.

72. G. A. Papsin, A. Hanna, and K. Showalter, *J. Phys. Chem.* **85**, 2575 (1981).

73. H. A. Liebhafsky and G. M. Roe, *Int. J. Chem. Kinet.* **11**, 693 (1979).

74. M. Eiswirth, K. Krischer, and G. Ertl, *Appl. Phys. A* **51**, 79 (1990).

75. K. Krischer, Ph.D. Thesis, Freie Universität Berlin, Berlin (1990).

76. T. Engel and G. Ertl, *Adv. Catal.* **28**, 1 (1979).

77. P. H. Richter, P. Rehmus, and J. Ross, *Progr. Theor. Phys.* **66**, 385 (1981).

78. W. Vance and J. Ross, *J. Chem. Phys.* **88**, 5536 (1988).

79. T. Fink, R. Imbihl, and G. Ertl, *J. Chem. Phys.* **91**, 5002 (1989).

80. For some reviews on enzyme oscillators see A. Goldbeter and S. R. Caplan, *Ann. Rev. Biophys. Chem.* **5**, 449 (1976); P. E. Rapp, *J. Exp. Biol.* **81**, 217 (1979); **81**, 281 (1979); R. Larter, *Chem. Rev.* **90**, 355 (1990).

81. K. Yokota and I. Yamazaki, *Biochemistry* **16**, 1913 (1977).

82. B. Hess and A. Boiteux, *Ann. Rev. Biochem.* **40**, 237 (1971).

83. J. Higgins, *Ind. Eng. Chem.* **59**, 5 (1967).

84. Y. Termonia and J. Ross, *J. Chem. Phys.* **74**, 2339 (1981).

85. S. R. Caplan, A. Naparstek, and N. J. Zabusky, *Nature* **245**, 364 (1973).

86. H. S. Hahn, A. N. Nitzan, P. Ortoleva, and J. Ross, *Proc. Nat. Acad. Sci. (U.S.)* **71**, 4067 (1974).

87. R. M. Noyes, in *Synergetics—Far from Equilibrium*, A. Pacault and Ch. Vidal, Eds., Springer, Berlin, 1979.

88. D. E. C. King and M. W. Lister, *Can. J. Chem.* **46**, 279 (1968).

89. M. G. Roelofs, E. Wasserman, J. H. Jensen, and A. E. Nader, *J. Am. Chem. Soc.* **105**, 6329 (1983).

90. J. H. Jensen, M. G. Roelofs, and E. Wasserman, in *Nonequilibrium Dynamics in Chemical Systems*, Ch. Vidal and A. Pacault, Eds., Springer, Berlin, 1984, p. 35.

91. O. Gurel and D. Gurel, *Oscillations in Chemical Reactions*, Springer, Berlin, 1983.

92. O. E. Rössler, *Phys. Lett.* **57A**, 397 (1976).

93. K. Showalter, R. M. Noyes, and K. Bar-Eli, *J. Chem. Phys.* **69**, 2514 (1978).

94. J. Rinzel and J. B. Schwartz, *J. Chem. Phys.* **80**, 5610 (1984).

95. M. Eiswirth, P. Möller, K. Wetzl, R. Imbihl, and G. Ertl, *J. Chem. Phys.* **90**, 510 (1988).

96. S. Ladas, R. Imbihl, and G. Ertl, *Surf. Sci.* **197**, 153 (1984); **198**, 42 (1988).

COMPLEX SCALING AND DYNAMICAL PROCESSES IN AMORPHOUS CONDENSED MATTER

C. A. CHATZIDIMITRIOU-DREISMANN

Iwan N. Stranski-Institut für Physikalische und Theoretische Chemie
Technische Universität Berlin
Berlin, FRG

CONTENTS

Advances in Chemical Physics Volume LXXX, Edited by I. Prigogine and Stuart A. Rice
ISBN 0-471-53281-9 © 1991 John Wiley & Sons, Inc.

I. INTRODUCTION

A. General Introductory Remarks

The aims of this chapter are; (1) to show that the extension of the complex scaling method into quantum statistics represents a powerful approach to dynamical processes in condensed matter, (2) to demonstrate that this theory yields novel information, which is directly comparable to real experiments, and (3) to make this approach popular among experimentalists.

1. On the Complex Scaling Method (CSM)

The operators in conventional quantum mechanics, which correspond with physical observables, are as a rule self-adjoint. Therefore they have real eigenvalues, *provided* that the corresponding eigenfunctions fulfill the standard *boundary conditions*. This holds true especially for the Hamiltonian H of a quantum system: its energy spectrum is always real.

However, already in 1927 Dirac [1,2] showed the following: If one considers a system consisting of a scattering center and incoming and outgoing waves, and if one properly changes the boundary conditions so that they can fit to this scattering problem, then the system may exhibit certain *resonant states* (or resonances) with a finite lifetime corresponding to "complex energy" eigenvalues

$$\mathscr{E}_k \equiv E_k - i\varepsilon_k$$

with E_k and ε_k being real. This "proper change" of the conventional boundary conditions was recognized to be associated with a certain mathematical extension of the standard quantum-mechanical formalism. Furthermore, experience showed that these complex energies are related with physical resonant states.

This idea (or, better, conjecture) was brought to a firm mathematical basis, but not before 1971, by the fundamental investigations of Balslev and Combes [3,4] and others (see Refs. 5–10). Namely, it was shown that the preceding complex eigenvalues may be calculated by means of the so-called complex scaling method (CSM)—also called complex coordinate method or complex dilation theory—in which the ordinary coordinates $\mathbf{r}$ are replaced by complex coordinates

$$\mathbf{r}' = \eta\mathbf{r}$$

where η is a complex "scaling factor." One of the main features of the CSM is that, if one puts

$$\eta = |\eta|e^{i\theta}$$

then the continuous parts of the spectrum of the original Hamiltonian H become rotated about their respective thresholds (or staring points) with an angle equal to -2θ out in the complex energy plane; additionally, this rotation reveals certain new discrete complex eigenvalues $\mathscr{E}_k$, which correspond to physical resonances. Since this approach is computationally simple, it has already become very popular in current atomic and molecular physics; for review articles and examples, see Refs. 11–18. The mathematical foundation and further extension of CSM constitute an active field of modern research, see Refs. 6, 7 and 18. Further details on CSM are presented in Section II.

2.   On CSM and Quantum Statistics of Condensed Matter

The theoretical part of this chapter considers the extension of CSM into the field of statistical mechanics of condensed systems (and not the CSM per se).

From the preceding introductory remarks one can see that up to now, the complex scaling method has been applied on the so-called Hamiltonian level (or operator level) of quantum theory, only.

Recently, the possibility of an extension of CSM into the Liouville (or superoperator) level was envisaged by Brändas and Dreismann [19, 20]. The starting point of our investigations was the possibility of an *intrinsic* physical connection between the CSM and the conceptual basis of Prigogine's novel theory of microscopic irreversibility; see Refs. 21–27. Our theoretical investigations are based on

(a) the CSM of ordinary quantum mechanics and

(b) the density matrix theory for *fermionic* systems, which is mainly due to Coleman [28–31], and they led to

(c) the first "extension" of the CSM of ordinary quantum mechanics into the canonical ensemble formalism of quantum statistics [20].

An unexpected result of this theoretical work is the following: By analyzing the fermionic second-order density matrix $\Gamma^{(2)}$, in connection with the canonical ensemble formalism of statistical mechanics, we proved that the associated CSM-transformed second-order density matrix $\Gamma^{(2)c}$ may contain submatrices γ that cannot be diagonalized under any similarity transformation (being unitary or not).

The physical meaning of this theoretical result is that these quantities γ, which concern quantum correlations, represent a new kind of "indivisible units" (or structures), owing to the fact that the well-known probabilistic interpretation for the diagonal elements γ_{ii} ($i = 1, \ldots, s$) is now strictly forbidden.

These units are called *coherent-dissipative structures*, because they are short-lived and short-ranged and exhibit a finite dimension s in the space of state functions (that is, the corresponding Hilbert space). Their physical significance and connection with the "standard" coherent states of quantum theory (like the BCS states of superconductivity [32]) as well as with Prigogine's dissipative structures of phenomenological irreversible thermodynamics [33, 34] has been recognized.

3.   Comparison with Experiments

Very recently, we succeeded with the application of the general CSM theory of quantum correlations to different and concrete microdynamical processes in condensed amorphous systems [35–41]. The predictive power of the general theory may be demonstrated by its explicit experimental context as well as by the formulation of quantitative predictions concerning certain new experiments, some of them in progress now; see Section VI.

These investigations strongly indicate that quantum correlations play an important role in the dynamics of condensed matter, and—this being the crucial point—that they can be connected directly with experimental results. To put it in another way, the asserted specific connection between theory and experiment is intrinsically related with quantum correlations and—at the same time—independent of the explicit form of the actual Hamiltonian. (This somehow "surprising" statement will be clarified in Section VI, in connection with concrete experimental results.)

B.   On the Physical Basis of Coherent-Dissipative Structures: CSM and Microscopic Irreversibility

Our approach proposes a treatment of the spontaneous emergence of coherent-dissipative structures in amorphous condensed systems (like a liquid or a paramagnetic spin system) from first principles [20]. In the context of kinetic equations and, more general, of other phenomenological theories, the appearance of coherent (or cooperative, synergetic) effects is nowadays a common feature. Within the *microscopic* level of description of natural phenomena, however, the appearance of these structures must be considered to be surprising, owing to the well-known reversible character of the fundamental equations of motion (classical as well as quantal). In fact, one expect even intuitively that time reversible mechanical laws cannot account for the appearance of qualitatively new forms of matter, as, for example, highly organized forms of matter, originating from a disordered form, for example, a liquid being in the so-called thermal equilibrium.

1.   Classical and Quantum Correlations

The matter under thermal equilibrium conditions, however, is not totally "disordered", even in the sense of classical statistical mechanics. As standard statistical mechanics tells us, there exist correlation patterns that obey the BBGKY (Bogoliubov–Born–Green–Kirkwood–Yvon) hierarchy (classical or quantal) that may have a very large size. These patterns appear due to the thermal motion of the molecules (or atoms), and thus correlations are continuously created and destroyed. The "distribution" of these correlation patterns is represented by the reduced distribution functions that fulfill the BBGKY hierarchy; see Ref. 42.

A microscopic disturbance of the system (e.g., an enforced motion of an ion due to an external electric field acting on an ionic solution) that may occur in the equilibrated macroscopic system can be traced as follows: All the correlation patterns that contain the disturbed quantum system (for example, the considered ion) become disturbed and thus they begin to "relax", showing a time dependence differing from the time dependence they had at thermal equilibrium; that is, during the actual relaxation time τ_{rel} of the disturbed quantum system (for example, the ion, in the preceding example), the motions of all the other quantum systems (other ions or molecules or both in the surroundings) that belong to the same correlation pattern are affected too. In other terms, the dynamical behavior of all these quantum systems has been changed, even if they do not interact with each other directly, and although only one of them has been initially disturbed.

One should remember, however, that the BBGKY hierarchy is still valid, although some of the reduced distribution functions have now become explicitly time dependent [42, 43].

It should be noted that the preceding considerations are also valid in classical statistical mechanics. Thus, they ought to be looked upon as illustrative. From the beginnings of quantum theory, however, it was well known that there also exists a novel "kind" of correlation having no classical analog. In this context one often speaks about the "holistic nature" of quantum mechanics. These correlations correspond to nonlocal quantum entanglements and are often called Einstein–Podolsky–Rosen (EPR) correlations [44]. At least after the famous Aspect experiment [45] it became obvious that EPR correlations play a fundamental role in physical processes. The universal relevance of quantum correlations for dynamical processes in *condensed* matter is theoretically well established, but thus far there has been no clearcut experiment—at, say, standard experimental conditions of temperature and pressure—demonstrating their presence.

The importance of these correlations in the physical context of spectroscopic experiments in condensed matter has been studied in the framework of traditional projection-operator (or Zwanzig–Mori) methods [46–48].

2.   Microscopic Irreversibility

The appearance of irreversible phenomena is known to be in conflict with the well-established microscopic mechanical theories because the time symmetry is "broken" in nature. The apparent "contradiction" between "reversible" microscopic dynamics (classical as well as quantal) and "irreversible" macroscopic thermodynamical processes still remains one of the most intriguing and challenging problems of modern science [21]. The presently well-established dynamical laws of physics are exactly CPT invariant [49], and the time-reversal asymmetry inferred from some observed CP violations is extremely small. Nevertheless, most (macroscopic) physical processes we observe in our daily lives display a gross time asymmetry. In this context one sometimes speaks about "arrows of time" [50]. Most prominent among these arrows is the "thermodynamic arrow" represented by the law of increase of entropy. Another is given by the retardation of electromagnetic radiation. Additionally, there is also the obvious fact that our perceptions of the past and the future are totally different.

One often argues, however, that the (almost exact) time symmetry of the basic physical laws is not really in any conflict with the observed irreversibility, namely, one says that time-symmetrical equations of motion also admit time-asymmetrical solutions. Especially the thermodynamic arrow of time has repeatedly been attributed to "special" initial conditions for a system under consideration. For example, Boltzmann had already stressed at the close of the last century that the one-sidedness of the change of the *H* function "lies uniquely and solely in the initial conditions"; see Ref. 51.

Despite these considerations, however, many scientists expect (or hope) that a more fundamental origin of the observed irreversible phenomena eventually will be discovered [21].

In this context let us mention that, during the last two decades, one often meets the conjecture (if not the thesis) that there may exist a "master time arrow" being due to the "special initial conditions" of the universe, from which all the others could follow. In other words, the different observed aspects of irreversibility are suspected to be related with the expansion of the universe. This "monistic" viewpoint is furthermore considered to be supported by the "big bang" cosmological model and the observed background radiation.

The aforementioned contradiction between the irreversible phenomena and the reversible physical laws represents by no means only a "philosophical" problem, since, for example, this is experienced by any chemist considering the time development of a chemical reaction. Many scientists are therefore not willing to accept Einstein's view that " irreversibility is only an illusion, a subjective impression, coming from exceptional initial conditions" [21, p. 203]. Furthermore, there is also need to "disentangle" questions concerning the preceding expansion of the universe from those concerning physical, chemical, and biological processes "on earth". Therefore, the question arises: Can irreversibility be founded on the microscopic level of description of natural phenomena?

It will become clear in the following that this question is intimately connected with the appearance of coherent-dissipative structures as given by the general theoretical framework of this chapter. In Ref. 20 we discussed these aspects in some detail and showed how and why the general CSM approach may be considered to be an extension of the standard quantum-mechanical formulation.

In this context one should mention the pioneering work of Prigogine and co-workers concerning the microscopic foundation of time and entropy as observables represented by superoperators (for more references see Refs. 22–25, and 27). It is clear that this novel work contrasts with the "Boltzmann–Gibbs" traditional viewpoint, that is, that the entropy is some functional of the distribution function (or the density operator) in phase space.

The formalism underlying our work on CSM and statistical mechanics does indeed make contact with the problem of the precise "microscopic foundation" of irreversibility. This is a simple consequence of the property attributed to the complex scaling method as being a well-defined mathematical theory for resonance phenomena, that is, being devised to deal with irreversible phenomena (decay of molecular quantum states, and so on) starting from first principles.

However, each attempt to introduce irreversibility on the microscopic level is inevitably coupled to the question: What are the direct and experimentally accessible new theoretical predictions that cannot satisfactorily be predicted and explained via standard "reversible" microscopic theory, be it classical or quantum mechanical? Formulated more emphatically, what would be the decisive reason (this should of course be intimately connected with the experimental situation) that would force us to accept that a "deep revision" of the well-established time reversible theory is inevitable? It is well known that the majority of the scientific community believes that, thus far, these questions have not been answered convincingly.

Motivated by the preceding remarks, we made efforts to explore explicitly the connection between the coherent-dissipative structures as derived within our CSM approach to quantum correlations and some real experiments in condensed matter. The different results achieved thus far (see Section VI) should also be considered as a strong indication of the physical significance of microscopic irreversibility.

3.   *Quantum Correlations and Coherent-Dissipative Structures*

A recent extension [20] of the complex scaling method (CSM) to the quantum statistics of disordered condensed systems, which is conceptually connected with Prigogine's subdynamics, predicted a new manifestation of EPR correlations, the so-called coherent-dissipative structures. These structures are shown to be quantum mechanical in nature, without a classical analog. The general theory [20] revealed the possibility that "self-organized" structures may appear, after "excitation" of some of the elementary quantum systems that constitute the macroscopic condensed system.

It should also be pointed out that the size of the coherent-dissipative structures is primarily not determined by the range of the intermolecular interactions (as, for example, the classical and thus illustrative viewpoint might suggest). Moreover, it holds that the appearance of these structures represents an intrinsically *dynamical* phenomenon that cannot be described with "standard" formalism, that is, the time-independent Schrödinger equation. In fact, the quantization conditions that lead to the appearance of coherent-dissipative structures are directly connected with the lifetimes that the CSM generically assigns to the quantum states (that is, the resonances) constituting these structures. In a more physical language we may say that in a space consisting of states with infinite lifetime, there cannot exist coherent-dissipative structures.

In this context, it is interesting to mention some similarities between coherent-dissipative structures and the well-known *dissipative structures* of the Brussels school [33, 34], which appear in the phenomenological thermo-dynamics of systems far from equilibrium. In both cases, self-organized forms of matter occur only after some "external disturbance" of the (initially) equilibrated, unstructured macroscopic system; these forms, moreover, die out, if the "external disturbance" stops. Furthermore, their size cannot fall below a critical value.

C.   Outline

Hitherto, in Section I, some general remarks on the formalism and the physical context of our CSM approach were considered.

In Section II, the conventional complex scaling method (that is, on the operator level) is considered and its fundamental character is pointed out. An intuitive introduction to CSM and a qualitative discussion of the main results of the Balslev–Combes theorem are present in Section II A. Certain mathematical details and the latest developments of the CSM theory are treated in Section II B, following the example of Ref. 20. The Reid–Brändas corollary concerning a specific complex symmetric matrix is formulated (and proved in Appendix A), because of its particular importance for the theory.

In Section III, some subdynamical aspects of the CSM are formulated in a general setting, for the purposes of the present theory. For example, the CSM transformation of the canonical density operator of quantum statistics is studied. The underlying point of view is that typical irreversible processes, like scattering [25] and decay, are intrinsically connected with "microscopic irreversibility" (see Section I B) as being displayed explicitly in subdynamics of the Brussels school or the resonances being exposed after complex scaling or both.

In Section IV, the fundamentals of Coleman's theory of the fermionic second-order density operator are presented. Here one can also see how Yang's off-diagonal long-range order (ODLRO) appears in the "box part" of the fermionic density matrix.

In Section V, the extension of CSM to quantum statistics, that is, the theory of coherent-dissipative structures is studied, in connection with previous investigations [20] on this topic. Complete analytic details are given in Section V. B. It is argued that the considered CSM theory of quantum correlations represents a novel synthesis of the three formalisms (that is, complex scaling, canonical ensemble, and fermionic density matrix) treated in the previous sections.

In Section V C, different topics on the physical interpretation (and significance) of the theoretical results are treated in some detail. Of particular importance for the understanding of the general formalism and results may be the discussions concerning the connection of coherent-dissipative structures with (1) the BCS states of superconductivity, (2) Prigogine's dissipative structures of phenomenological irreversible thermodynamics, and (3) Yang's ODLRO. These discussions strongly indicate that coherent-dissipative structures represent a novel kind of "spontaneously created order" in amorphous matter, on the microscopic level.

For readers who are not interested in the complete theoretical and mathematical details of Sections II–V, a short outline of the general formalism and a schematic derivation of the main results are given in Appendix B. This can also serve as a "theoretical introduction" for experimentalists being interested in applications exclusively.

In Section VI, applications of the general CSM theory to concrete dynamical processes in (amorphous) condensed matter are studied in detail. In Section VI A it is pointed out that the CSM approach under consideration should be regarded as a new, general theoretical treatment of quantum correlations in condensed matter. This is because the results obtained so far are independent of the specific properties of the respective system Hamiltonian. The comparison with experimental results concern the following fields: (A) proton transfer in water, (B) anomalous decrease of H^+ conductance in H_2O/D_2O mixtures; (C) ionic conductance of molten alkali chlorides; (D) spin waves of paramagnetic Gd at high temperatures; (E) correlations in Cu–O superconductors. The results are noteworthy because they represent a quantitative comparison with experiments without any adjustable parameters.

II. THE COMPLEX SCALING METHOD

A. Introductory Remarks

In conventional spectroscopy of (isolated) atoms and molecules, the concept of stationary states is of fundamental importance, since the spectra experimentally observed are interpreted as being the result of transitions between such states of the system. It is well known, furthermore, that well-defined stationary states correspond to the bound states of standard quantum mechanics [52].

In most realistic cases, however, and especially in the case of dynamical processes in condensed matter, one is confronted with unbounded states and continuous spectra. In the corresponding physical context, the term "stationary state" looses its physical meaning; confer the detailed discussion of this important point by Landau and Lifshitz [54]. The generalized eigenvectors associated with the continuous (part of the) spectrum are not normalizable, and therefore they fall outside the common "world" of Hilbert space. (At this stage, the modern mathematical theory of the rigged Hilbert spaces [53] should be mentioned, which allows one to treat generalized eigenvectors rigorously. This theory is not considered in this chapter.)

The distinction between proper and generalized eigenstates is clear cut with respect to mathematics. The experiments, however, often exhibit the presence of phenomena "in the continuum", which can be interpreted as being due to the appearance of so-called *resonant* (or metastable, quasibound) states corresponding to certain energies of the system. The time evolution of such states is characterized physically by a localization (in coordinate space) of a wave-packet for a certain time interval and by an exponential decay of the wave packet for longer time periods.

Thus, it becomes clear that a "natural" description of such processes should be made in a time-dependent framework. On the other hand, at least the huge number of existing computer programs dealing with the time-independent Schrödinger equation works as a stimulus for research on methods, which could be able to handle resonant states in a time-independent framework rigorously. Indeed, it was shown that this is possible, if one allows the introduction of more general boundary conditions into the formalism.

Historically, these ideas already existed just after the discovery of quantum mechanics; for example, Gamow's theory of the radioactive decay [55] shows the occurrence of solutions of the Schrödinger equation with complex energies $\mathscr{E}_k \equiv E_k - i\varepsilon_k$, by using "physically meaningful" (although mathematically not well-defined) boundary conditions.

There exist two main definitions for resonances. The first one is based on the properties of the resolvent

$$R(z) = \frac{1}{z - H}$$

where H is the system Hamiltonian. The complex resonance energies may be defined to be given by certain singularities occurring in R when properly continued onto the second Riemann sheet of the (complex) energy plane. More physical insight gives the second definition of a resonance, which relies on the scattering process and imposes the following very specific boundary conditions on the resonance wavefunction: This function should be a purely outgoing wave in the resonance channel, that is, it should satisfy the Siegert boundary condition [56]

$$\Psi(r) \to e^{ikr} \quad \text{for} \quad r \to \infty$$

For the two-body system, the equivalence of the two definitions can be proved. For many-body systems, one assumes that this correspondence exists.

The aforementioned analytic continuation of the resolvent and exposition of resonances in the complex energy plane can be achieved by using the complex scaling (or rotation, dilation) techniques. The latter have become very popular in the study of resonant phenomena in many-body systems; see Refs. 10–18.

The CSM for the many-particle system was put on a firm mathematical basis through the Balslev–Combes theorem [3,4]. This method is based on the analytic continuation of the unitary group of scalings, which can be defined through its action on the wavefunctions:

$$U(\eta)\Psi(x_1, x_2, \ldots, x_N) = \eta^{3(N-1)/2}\Psi(\eta x_1, \eta x_2, \ldots, \eta x_N)$$

where $\eta = |\eta| \exp(i\theta)$ is a complex number. The specific form of the exponent on the right-hand side of this equation can be regarded to correspond to the separation of the center-of-mass degrees of freedom. For more mathematical details and references, see the following discussion. Here, let us just illustrate the main results of the Balslev–Combes theorem.

Let $H = T + V$ be the Hamiltonian. The corresponding CSM transformed Hamiltonian is formally given by $H^c \equiv U(\eta)HU^{-1}(\eta)$. If the potential V is dilation analytic (that is, allows for an analytic continuation into the complex plane) and Δ-compact (that is, not too strong as compared with the kinetic energy operator T), then the defined complex scaled Hamiltonian exhibits the following features:

1. The branch cuts extending from each threshold of the continuum spectrum will rotate about their respective threshold into the complex plane by an angle -2θ.

2. The spectrum of the well-known bound states is not affected, that is, it remains unaltered.

3. The rotated continua (or cuts) may expose resonances (which are said to belong to higher Riemann sheets). These complex eigenvalues are stable under variations of θ, as long as they are not "covered up" by a rotated continuum. These points are illustrated furthermore by the Figure 1.

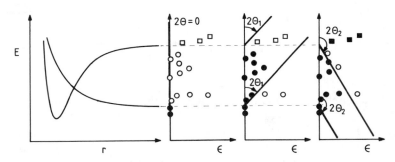

FIGURE 1. Illustration of the effects of complex scaling of a two-body system, where the nuclear motion in a two-channel scattering problem is considered. Solid symbols (circles and squares) represent the discrete energy eigenvalues that can be calculated. Open symbols represent hidden (or covered) resonances. Thick lines represent the channel energy continua. On the left are shown the potentials. Moving to the right, one sees the rotation of the energy continua (with an angle $2\theta_1$) about their thresholds and the new discrete eingenvalues being exposed. The latter are now calculable, because the corresponding resonance eigenstates are square integrable after complex scaling. If the rotation angle is increased ($2\theta_2$), one can uncover some new resonances, but some of the previously exposed ones are now "absorbed" from one rotated continuum. This figure is adapted from Ref. 16.

B. CSM, Unbounded Similarity Transformations, and Complex Symmetric Forms

1. Details of CSM

Since the method of complex scaling, or complex rotation, plays such a fundamental role in our present study, we will give a short review of some of its more puzzling aspects. It is not improper to consider the deformation of the coordinates into complex-valued degrees of freedom as an attempt to analytically continue quantum mechanics beyond its conventional domain of a time reversible microscopic formulation into a valid theory of a more general nature, where irreversibility is naturally embedded in the dynamics. The meaning of this somewhat "cryptic" statement will be clarified in the following sections. The following mathematical details of this subsection are based on the presentation of Ref. 20.

Let us first introduce some mathematical concepts needed for the actual extension. Although our technique emanates from the pioneering work of Balslev and Combes [3], Simon [5], and van Winter [8] we will here follow the domain characterization of Brändas [9] in order to arrive at a transformation theory that not only contains standard quantum mechanics, but also leads to the desired extension, that is, including mappings onto a contractive (semigroup) evolution [57], and the concomitant approach to equilibrium. As we will see in the following sections, this has fundamental consequences also for the properties of the CSM-transformed canonical density operator.

In nonrelativistic quantum mechanics, it is customary to introduce the N-body Hamiltonian as

$$H = T + V$$

where T is the kinetic energy operator and V is the interaction potential. Since H is an unbounded operator, albeit bounded from below, we need to restrict the domain somewhat. Let $\mathscr{H}$ denote the Hilbert space, complete with respect to the standard norm of square integrability, that is,

$$\Phi \in \mathscr{H} \quad \text{with} \quad \| \Phi \|_{L^2} < \infty$$

Then the domain of H is given by

$$\mathscr{D}(H) = \{ \Phi \in \mathscr{H} \,|\, H\Phi \in \mathscr{H} \} \tag{2.1}$$

We will basically consider interactions such that the unboundedness of H arises entirely from the kinetic energy operator T. The potential V will be further specified subsequently, however at the present stage it will be sufficient to treat the case with $\mathscr{D}(H) = \mathscr{D}(T)$.

A key quantity in the CSM formulation is the scaling operator

$$U(\theta) = \exp(iA\theta) \tag{2.2}$$

where

$$A = \frac{1}{2} \sum_{k=1}^{N} (\mathbf{p}_k \mathbf{x}_k + \mathbf{x}_k \mathbf{p}_k) \tag{2.3}$$

is the generator of the scaling transformation, θ is a parameter, which may be real or complex, and $\mathbf{x}_k$ and $\mathbf{p}_k$ are the coordinate and momentum operators of particle k. If $\theta \in R$, then U is a unitary operator defining the dilation group, with $\mathscr{D}(U) = \mathscr{H}$, which effectuates the scaling

$$U(\theta)\Phi(x_1, \ldots, x_N) = \exp\left(\frac{3N}{2}\theta\right)\Phi(e^{\theta}x_1, \ldots, e^{\theta}x_N) \tag{2.4}$$

By considering complex scalings, that is,

$$U(\eta) \quad \text{with} \quad \eta = |\eta|e^{i\theta}$$

(note that our parameter in U now refers to the whole analytic parameter η rather than the dilation group parameter θ) Balslev and Combes [3] proved that for certain classes of Hamiltonians, the so-called dilation analytic Hamiltonians, the continuous spectrum changed under the complex deformation and became rotated down -2θ in the complex energy plane, thereby opening up sectors on the "unphysical" Riemann sheet, where possible finite-dimensional "resonance eigenvalues" would appear. They also demonstrated the nonexistence of singularly continuous spectra in these cases, as well as invariance of the exposed spectrum (including bound states) to variations of θ.

Without going into too many mathematical details, we will thus consider

$$\eta = |\eta|e^{i\theta} \quad \text{with} \quad 0 \leqslant \theta < \theta_0$$

Generally, θ_0 depends on the potential, for instance, in the Coulomb case no limit is invoked, while in some other cases, as will be seen, it is natural to have $\theta_0 \leqslant \pi/4 - \delta$. We also introduce Ω as

$$\Omega = \{\eta \mid \theta_0 > |\arg(\eta)|\}. \tag{2.5}$$

Furthermore, we make the decomposition $\Omega = \Omega^+ \cup \Omega^- \cup R$, where the real axis $R = R^+ \cup R^- \cup \{0\}$ partitions Ω into its upper and lower parts in the complex plane. In what follows we will assume that the interaction V is a sum of two-body interactions V_{ij} such that V_{ij} is Δ_{ij}-compact [3] in $L^2(R^3)$ and furthermore that $V_{ij}(\eta) = U_{ij}(\eta)V_{ij}U_{ij}^{-1}(\eta), \eta \in R^+$, has a compact analytic extension to Ω^+. The definition of the dilatation analytic family of operators $H(\eta)$ is then given by

$$H(\eta) = U(\eta)HU^{-1}(\eta), \quad \eta \in R^+ \tag{2.6a}$$

and

$$H(\eta) = \eta^{-2}T + V(\eta), \quad \eta \in \Omega^+ \tag{2.6b}$$

In the previous expression it is notable that $H(\eta)$ is obtained in two steps. First via a unitary transformation to a scaled representation and thereafter followed by an analytic continuation to Ω^+. Although this is a mathematically rigorous procedure, it is preferable, as we will see, to consider the similarity transformation $U(\eta)$, with $\eta \in \Omega$, directly [20]. Here it should be emphasized that $U(\eta)$ is an *unbounded* operator. We will then introduce the domain $\mathcal{N}(\Omega)$ as the following subset, called Nelson class [58]:

$$\mathcal{N}(\Omega) = \{\Phi \,|\, \Phi \in \mathcal{H},\, U(\eta)\Phi \in \mathcal{H}, \eta \in \Omega\} \tag{2.7}$$

We furthermore complete the space with respect to the norm

$$\int_{-\theta_0}^{+\theta_0} \| U(\eta)\Phi \|_{L^2}^2 \, d\theta = \| \Phi \|_{N_{\theta_0}}^2 \tag{2.8}$$

Since we also want the first and second partial derivatives to satisfy (2.8), it is natural to define the spaces

$$\mathcal{N}_{\theta_0}^{(i)}, \quad i = 0, 1, 2$$

analogously.

We are now in the position to make an alternative definition of the analytic family $H(\eta)$ (see Ref. 9), that is,

$$H(\eta) = U(\eta)HU^{-1}(\eta); \quad \mathcal{D}(U H U^{-1}) = \mathcal{N}_{\theta_0}^{(2)} \tag{2.9a}$$

and

$$H(\eta) = \eta^{-2}T + V(\eta); \quad \mathcal{D}(U H U^{-1}) \longrightarrow \mathcal{D}(T) \tag{2.9b}$$

Note that $\mathcal{N}_{\theta_0}^{(i)}$ $(i = 0, 1, 2)$ are Hilbert spaces, which allow $H(\eta)$ to be interpreted as a similarity transformation of the self-adjoint unscaled operator H. We also note that $U(\eta)$ exhibits the "star-unitary" property (see also Refs. 21 and 22)

$$U^{\dagger}(\eta^*) = U^{-1}(\eta) \tag{2.10}$$

This relation can be easily proven from

$$\langle \Phi | \Psi \rangle = \langle \Phi(\eta^*) | \Psi(\eta) \rangle = \langle \Phi | U^{\dagger}(\eta^*) U(\eta) | \Psi \rangle \tag{2.11}$$

which holds for all $\Phi, \Psi \in \mathcal{N}_{\theta_0}$.

In the definition presented previously, leading of course to the same analytic family $H(\eta)$, we can see that the two steps involved in Eq. (2.9) are different in comparison to Eq. (2.6). Here the first step consists of restricting $\mathcal{H}$ to a smaller domain $\mathcal{N}_{\theta_0}^{(2)}$ for which the scaling $U(\eta)$ is defined for all η such that $\arg(\eta) < \theta_0$. After making the complex rotation, that is, changing the parameter θ in e^{θ} from real to complex, one completes the Nelson subset, which is dense in $\mathcal{H}$, to $\mathcal{D}(T)$. This means that completion is made with respect to the "standard" L^2-norm for the functions and its first and second partial derivatives.

Before ending this subsection it should be mentioned that the present definition is particularly useful in connection with time evolution and associated Lyapunov converters [57]. Other uses of this formulation will be considered below.

2. Similarity Transformations and Complex Symmetric Forms

One can understand the abstract transformations of the previous subsection in a very simple way. Consider for example, the integral

$$I(b) = \int_0^{+\infty} e^{-br} \, dr = \frac{1}{b} \tag{2.12}$$

whose simple analytic form trivially allows analytic continuation to negative values of b. In the latter case the integral is not convergent even if $I(b) = 1/b$ for all $b \neq 0$. If $b \neq 0$ is complex (or negative) so that the integrand in (2.12) does not vanish for $r \to \infty$, then one may consider the trick of defining a complex integration path will θ sufficiently large so that

$$I = \lim_{R \to \infty} \int_0^{Re^{i\theta}} e^{-br} \, dr = \frac{1}{b} \tag{2.13}$$

The complex path can hence be used to explicitly compute a numerical value of the integral I even when its analytical form is unknown.

When evaluating matrix elements, using a complex deformation as in (2.13), a word of warning should be issued. For instance, if one does not realize that the spaces $N_{\theta_0}^{(i)}$, $i = 0, 1, 2$, have a finer topology than $\mathscr{H}$, then paradoxical situations may occur. Consider, for example, the resolvent

$$(H - \lambda I)^{-1}$$

which for $\lambda^* \neq \lambda$ is a bounded operator in $\mathscr{H}$. *This is not true in* $\mathscr{N}_{\theta_0}$. We will exemplify this as follows [59]:

In momentum space we obtain with $H_0 = k^2$, $\Phi = \pi^{-1/4} e^{-k^2/2}$, $\eta = |\eta| e^{i\theta}$, $\theta_0 \leqslant \pi/4 - \delta$ the following result:

$$(U(\eta)\Phi) = \pi^{-1/4} \frac{1}{\sqrt{\eta}} e^{-k^2/2\eta^2} \qquad (2.14)$$

and

$$\| U(\eta)\Phi \|_{L^2}^2 = \frac{1}{\sqrt{\cos(2\theta)}} \qquad (2.15a)$$

$$\| U(\eta)\Phi \|_{N_{\theta_0}}^2 = \int_0^{\pi/2} \frac{d\theta}{\sqrt{\cos\theta}} < \infty \qquad (2.15b)$$

Since $0 < \theta_0 \leqslant \pi/4 - \delta$, then $k^n e^{-k^2/2} \in \mathscr{N}_{\theta_0}$ for all n, and hence $\Phi \in \mathscr{N}_{\theta_0}^{(\infty)}$. Now it is easy to show that if $-2\theta_0 < \arg\lambda < 0$, then

$$\chi = (H_0 - \lambda I)^{-1} e^{-k^2/2} \notin \mathscr{N}_{\theta_0}$$

Thus one obtains

$$(U(\eta)\chi)(k) = \frac{1}{\sqrt{\eta}} (\eta^{-2} k^2 - \lambda)^{-1} e^{-k^2/2\eta^2} \qquad (2.16)$$

and

$$\| \chi \|_{L^2}^2 = 2(\cos 2\theta)^{3/2} \int_0^\infty \frac{e^{-k^2} dk}{(k^2 - \lambda^* \cos 2\theta e^{-2i\theta})(k^2 - \lambda \cos 2\theta e^{2i\theta})}$$

with: $\lambda \cos 2\theta^{2i\theta} = k_0^2 e^{2i\varepsilon}$; $k_0 = |\lambda|^{1/2} (\cos 2\theta)^{1/2}$

$$= 2(\cos 2\theta)^{3/2} \int_0^\infty \frac{e^{-k^2} dk}{|k + k_0 e^{i\varepsilon}|^2 (k - k_0 e^{i\varepsilon})(k - k_0 e^{-i\varepsilon})} \qquad (2.17)$$

It is obvious from (2.17) that the divergence of the integral as $\varepsilon \to 0$ determines whether $\| \chi \|_{N_{\theta_0}}^2$ is finite or not. For small ε one finds that

$$\| \chi \|_L^2 \sim \frac{e^{-k^2}}{4k_0^2} \int_{-\delta}^{\delta} \frac{dx}{(x - i\varepsilon k_0)(x + i\varepsilon k_0)}$$

$$\sim \pi \frac{e^{-k_0^2}}{4k_0^3 |\varepsilon|} \tag{2.18}$$

which, since $\int_{-\delta}^{\delta} d\varepsilon / |\varepsilon| = \infty$, implies that $\chi \notin \mathcal{N}_{\theta_0}$.

This simple example [59] clarifies the following somewhat surprising situation. In accordance with (2.11) we write the following matrix (2.19) of H, utilizing the relations (2.9), (2.10), and Φ, $\Psi \in \mathcal{N}_{N_{\theta_0}}^{(2)}$, so that

$$\langle \Phi | H | \Psi \rangle = \langle \Phi(\eta^*) | H(\eta) | \Psi(\eta) \rangle \tag{2.19}$$

holds true. It is important to realize that

$$H(\eta) \Psi(\eta) = U(\eta) H U^{-1}(\eta) U(\eta) \Psi = U(\eta)(H\Psi)$$

are all valid operations since $\Psi \in \mathcal{N}_{N_{\theta_0}}^{(2)}$ implies $H\Psi \in \mathcal{N}_{\theta_0}$ and hence (2.19) reduces to (2.11).

The invariance (2.19) shows that the matrix element is not altered by the complex deformation, if Ψ and Φ are in the appropriate domain. As (2.12) and (2.13) indicate, however, the scaled integral to the right in (2.19) may have a finite value for a suitable η even when the unscaled integral to the left in (2.19) may not exist. Hence, the right-hand side of (2.19) exists for a more general class of functions than those that are L^2 integrable, or more precisely, for those that belong to $\mathcal{N}_{\theta_0}^{(2)}$. Thus we see that complex scaling allows us to extend quantum mechanics beyond its conventional domain in that it assigns a well-defined meaning to vectors and matrix elements with respect to operators (resolvents), requiring in analogy with (2.19) the existence of the scaled representation for some nonreal η.

Before we continue, we want to emphasize that we have implicitly assumed that $V\Psi \in \mathcal{N}_{\theta_0}$ for $\Psi \in \mathcal{N}_{\theta_0}^{(2)}$. Even if this may be a restriction on V in general, one can circumvent this problem by considering more general deformations, like exterior scaling [5]. These are, however, technical points that will not alter our general considerations and conclusions as well as the physical interpretation.

With the preceding development in mind, it is now easy to avoid the following paradoxical situation. If we employ (2.19) with H replaced by the resolvent operator, one may ask whether the following relation with $\eta \in \Omega^+$, $\lambda^* \neq \lambda$:

$$\langle \Phi | (H - \lambda I)^{-1} | \Psi \rangle = \langle \Phi(\eta^*) | (H(\eta) - \lambda I)^{-1} | \Psi(\eta) \rangle \qquad (2.20)$$

holds for all Φ, $\Psi \in \mathcal{N}_{\theta_0}$. The puzzle consists in that the left-hand side of (2.20) is always finite for λ complex, while the right-hand side may become infinite if λ is a complex eigenvalue of the complex-rotated operator $H(\eta)$, which is a possibility according to the Balslev–Combes theorem [3]. Now we can return to our analysis carried out in (2.18). For instance, with the choice $H = H_0 = k^2$, $\Psi = \Phi$, and $-2\theta_0 < \arg \lambda < 0$, it was demonstrated that $\chi = (H - \lambda I)^{-1} \Phi \notin \mathcal{N}_{\theta_0}$. Therefore, (2.11) (will χ replacing Ψ) is *not* applicable and, *a fortiori*, the equality (2.20) is *not* valid. Nevertheless, the scaled matrix element in (2.20) exists if, for example, $2\theta_0 > \arg \lambda > 0$, and it is equal to the unscaled integral and more importantly has a meromorphic continuation into the sector given by $-2\theta_0 < \arg(\lambda) < 0$.

For simplicity we have displayed a simple situation with one threshold at zero, but the generalization to the general many-body situation should be obvious.

It is also clear that we have considered the Hilbert spaces $\mathcal{N}_{\theta_0}^{(i)}$ ($i = 0, 1, 2$), in order to find convenient domains for the unbounded (in $\mathcal{H}$) similetude $U(\eta)$, $\eta \in \Omega$. After appropriate deformations completion with respect to the standard L^2 norm is made. In this manner we arrive at the formal eigenvalue relation

$$H(\eta) \Psi(\eta) = \mathcal{E}(\eta) \Psi(\eta), \quad \eta \in \Omega^+ \quad \text{and} \quad \eta = |\eta| e^{i\theta} \qquad (2.21)$$

with $\arg (\eta)$ sufficiently large to uncover the resonance $\mathcal{E}$. The conjugate of equation (2.21) becomes

$$\overline{H(\eta)} \, \overline{\Psi(\eta)} = \overline{\mathcal{E}(\eta)} \, \overline{\Psi(\eta)}, \quad \eta \in \Omega^- \qquad (2.22)$$

with the involution $\overline{A(\eta)} = A^*(\eta^*)$ being introduced to include also (complex) optical potentials. Obviously (2.21) and (2.22) in conjunction with (2.19) motivates the construction

$$\langle \overline{\Psi(\eta^*)} | H(\eta) | \Psi(\eta) \rangle = \mathcal{E}(\eta) \langle \overline{\Psi(\eta^*)} | \Psi(\eta) \rangle \qquad (2.23)$$

from which stationary variational principles can be derived in almost the same fashion as in ordinary quantum mechanics [60, 61], with the important distinction that the extremum property of the principle has been lost. Note also that if H satisfies the relation $H^* = H$, then there is no restriction *from* *the variational point of view* to assume that $\Psi^* = \Psi$ [62] and then the construction (2.23) defining a trial $\mathscr{E}(\eta)$ based on a trial $\Psi(\eta)$ and corresponding $\overline{\Psi(\eta^*)} = \Psi(\eta)^*$ becomes complex symmetric. However, in many cases it may be useful to analyze the wavefunction in terms of nonreal spherical harmonics. One might then leave the angular momentum eigenfunctions as they are and consider each angular block in a complex symmetric representation even if the corresponding term in the wavefunction representation contains the nonreal component $Y_{l,m}(\vartheta, \varphi)$.

Since η sometimes is used primarily as a numerical convergence factor, we may replace the explicit η-dependence for the quantity A by replacing $A(\eta)$ by A^c. Hence, we write

$$H^c \Psi^c = \mathscr{E} \Psi^c \tag{2.21a}$$

and for the complex conjugate equation

$$H^{c*} \Psi^{c*} = \mathscr{E}^* \Psi^{c*} \tag{2.22a}$$

and

$$\langle \Psi^{c*} | H^c \Psi^c \rangle = \mathscr{E} \langle \Psi^{c*} | \Psi^c \rangle \tag{2.23a}$$

Even if this simple notation is very appealing in that almost any standard quantum-mechanical technique can be taken over provided it is appropriately modified, one should note that it results in a formulation that goes beyond conventional quantum mechanics "on the real axis." The most direct consequence is, as was mentioned, the appearance of complex resonance eigenvalues and associated Gamow vectors [55]. A closer study of the full generalized spectral properties of the complex deformed problem [63] shows that these eigenstates essentially deflate the (generalized) spectral density, giving in an asymptotic sense a decomposition of the continuum into resonances and background. This has the important consequence that each Gamow vector represents a well-defined section of the continuous spectrum associated with the unscaled self-adjoint problem. In other words, one can say that

the Gamow representation condenses an infinite-dimensional Hilbert space associated with a particular spectral part of the continuum into a finite-dimensional linear space of suitable Gamow vectors.

This "condensation" will play a fundamental role in a following section of this chapter. The prize for this convenience is the appearance of so-called Jordan blocks, see, for example, the corollary of Reid and Brändas [64].

If the Hamiltonian H in (2.21) and (2.22) has the property that it is real for $\eta \in R^+$, then it follows, as we have pointed out, that the construction (2.23) will lead to complex symmetric forms. Furthermore it is a general theorem that any matrix can be brought to a complex symmetric form by a similarity transformation, see, for example, Ref. 65. Even if this is always possible in general, it may not be the most convenient construction in every problem. Nevertheless, it will be an important representation in our general CSM treatment of quantum correlations to be presented in the following sections.

3.   The Reid–Brändas Corollary

The specific complex symmetric matrix $\mathbf{q}$ with elements

$$q_{kl} = \left(\delta_{kl} - \frac{1}{s} \right) \exp \left(i\pi \, \frac{k+l-2}{s} \right) \tag{2.24}$$

where

$$1 \leqslant k, l \leqslant s \tag{2.25}$$

is of particular interest for the theory of this article; namely, for the general formalism to be presented in Section V, the following algebraic corollary, which is due to Reid and Brändas [64], is of basic importance:

The Jordan block represented by the $s \times s$ matrix (s is a positive integer)

$$\mathbf{C}_s(0) = \begin{pmatrix} 0 & 1 & 0 & \cdots & 0 \\ \vdots & \ddots & \ddots & & \vdots \\ \vdots & & \ddots & \ddots & 0 \\ \vdots & & & \ddots & 1 \\ 0 & \cdots & \cdots & \cdots & 0 \end{pmatrix} \tag{2.26}$$

is similar to the $s \times s$ matrix $\mathbf{q}$ defined by (2.24). A short proof [64] of this corollary may be found in Appendix A.

It should be mentioned that the "range" of the integer variables k and l is not unique. This follows from the details of the proof, namely the explicit form (2.24) of a matrix $\mathbf{q}$ being similar to a Jordan block can be rewritten as

$$q_{kl} = \left(\delta_{kl} - \frac{1}{s} \right) \omega^{k+l-2} \tag{2.27}$$

with $\omega \equiv \exp(i\pi/s)$, or

$$\omega^s = -1 \tag{2.28}$$

From the details of the proof, however, it becomes obvious that the condition

(2.28) is the only crucial one used in the derivations. Thus, the replacement

$$\omega \to \omega' \equiv \omega \exp\left(i\frac{\pi}{s} 2K \right) \tag{2.29}$$

where K is an arbitrary (negative or positive) integer, preserves the desired "character" of the matrix form (2.24). To put it another way, the matrix elements of $\mathbf{q}$ being under consideration are given by

$$q_{kl} = \left(\delta_{kl} - \frac{1}{s} \right) \exp\left(i\pi \frac{(k+K)+(l+K)-2}{s} \right) \tag{2.24a}$$

where K is an arbitrary integer.

Thus we have found that the $s \times s$ matrix $\mathbf{q}$ corresponds to the special Jordan block $\mathbf{C}_s(0)$ with Segré characteristic s and trace zero. This just means that there does not exist any similarity transformation (for example, unitary, star-unitary, or orthogonal) that is able to diagonalize the matrix $\mathbf{q}$. Since the more general form of a Jordan block corresponding to $\mathbf{C}_s(0)$ is given by

$$\mathbf{C}_s(\lambda) = \lambda \cdot \mathbf{1} + \mathbf{C}_s(0)$$

where $\mathbf{1}$ is the unit $s \times s$ matrix and λ is a (real) constant, we see that $\mathbf{C}_s(0)$, and *a fortiori* $\mathbf{q}$, have eigenvalues and traces (of all orders) equal to zero.

These remarks will play an important role in the physical interpretation of the general formal results of the theory of coherent-dissipative structures; see Section V.

III. SUBDYNAMICS IN THE LIGHT OF CSM

In order to set the frame for a dilatation analytic extension of a subdynamical superoperator formulation, we will introduce the following quantities. As will be seen, this leads directly to the relevant evolution superoperator and the corresponding Boltzmann superoperator [20].

To allow different choices of representation we define the superoperator $\hat{P}$ by

$$\hat{P} = A| \rangle\langle | + | \rangle\langle |B \tag{3.1}$$

where the choice of A and B will be made below and $| \rangle\langle |$ is an abbreviation for the ket-bra components [66] of the density operator. With this notation it is straightforward to show that

$$e^{\hat{P}} = e^A | \rangle \langle | e^B \tag{3.2a}$$

and

$$\partial \rho = dA\rho + \rho d\mathbf{B} \tag{3.2b}$$

There are basically two choices of A, B, and $\hat{P}$ that correspond to physically meaningful subdynamical formulations.

The first choice of operators concerns the time evolution of isolated systems, where the Hamiltonian is well defined, as studied by Obcemea and Brändas [26]. With $A = -(i/\hbar)Ht = B^\dagger$ and $\hat{P} = -i\hat{L}t$ (where $\hat{L}$ is the Liouville superoperator [22, 42, 66]), one obtains

$$e^{-i\hat{L}t} | \rangle \langle | = |e^{-(i/\hbar)Ht}\rangle\langle e^{-(i/\hbar)Ht}| \tag{3.3a}$$

and furthermore

$$i\frac{\partial \varrho}{\partial t} = \hat{L}\varrho \tag{3.3b}$$

which is the Liouville–von Neumann equation of motion. After complex scaling and using the construction of Ref. 26 for the density operator, one has

$$\varrho_{kl}^c \equiv |\Psi_k^c\rangle\langle\Psi_l^c| \tag{3.4}$$

where

$$H^c\Psi_k^c = \mathscr{E}_k\Psi_k^c \tag{3.5}$$

and

$$\mathscr{E}_k = E_k - i\varepsilon_k \tag{3.6}$$

From the last three equation it follows that the eigenvalue relation

$$e^{-i\hat{L}^c t}\varrho_{kl}^c = e^{\{-(i/\hbar)(E_k - E_l) - (1/\hbar)(\varepsilon_k + \varepsilon_l)\}t}\varrho_{kl}^c \tag{3.7}$$

holds. The choice (3.4) is good in that (3.7) contains energy differences (as in conventional quantum mechanics), while the widths are added. Note, however, that ϱ_{kk} in (3.4) (that is, for $k = l$) is *not* a projector, because

$$(\varrho_{kk}^c)^2 \neq \varrho_{kk}^c$$

which makes the interpretation of ϱ^c fundamentally different as compared to the corresponding one of ϱ before CSM transformation; see also the following discussion.

The second choice [20] concerns the canonical density operator of quantum statistics, which is extensively considered in Section V. With the standard abbreviation $\beta = 1/k_B T$, where k_B is the Boltzmann constant and T is the absolute temperature, one chooses the representation $\hat{P} = -(\beta/2)\hat{L}_T$ and $A = -\beta H = B$. Then it holds

$$-\frac{\partial \rho}{\partial \beta} = \hat{L}_T \rho \tag{3.8}$$

The ket-bra operators defining the appropriate dilated density operator [20] are now given by

$$\rho^c_{kl} \equiv |\Psi^c_k\rangle\langle\Psi^{c*}_l| \tag{3.9}$$

and the following eigenvalue relation holds:

$$e^{-(\beta/2)\hat{L}^c_T}|\Psi^c_k\rangle\langle\Psi^{c*}_l| = e^{-(\beta/2)\{(E_k + E_l) - i(\varepsilon_k + \varepsilon_l)\}}|\Psi^c_k\rangle\langle\Psi^{c*}_l| \tag{3.10}$$

It should be observed that—in contrast to the previous choice—here the "diagonal" ket-bra operator

$$\rho^c_{kk} = |\Psi^c_k\rangle\langle\Psi^{c*}_k|$$

does represent a *projector*, that is,

$$(\rho^c_{kk})^2 = \rho^c_{kk}$$

although it is not self-adjoint,

$$(\rho^c_{kk})^\dagger \neq \rho^c_{kk}$$

Such quantities are called *skew* projectors [66, 67]. In addition to this feature we also see that both widths and energies are added in (3.10). One can see easily that this formulation yields the standard Boltzmann factor for the diagonal elements of the density operator in the limiting case of vanishing imaginary parts of the resonances, that is,

$$e^{-(\beta/2)\hat{L}^c_T}\rho^c_{kk} = e^{-\beta E_k}\rho^c_{kk} \quad \text{for } \varepsilon_k = 0 \tag{3.11}$$

Therefore

$$e^{-(\beta/2)\hat{L}_T^c}$$

may be called the (CSM-transformed) Boltzmann superoperator at temperature T.

It should be observed that $\hat{L}$ in (3.3b) is defined by

$$\hat{L} = \frac{1}{\hbar}(H| \rangle\langle |-| \rangle\langle |H) \qquad (3.12)$$

while $\hat{L}_T$ in (3.8) is given by

$$\hat{L}_T = \frac{1}{\hbar}(H| \rangle\langle |+| \rangle\langle |H) = \frac{2}{\hbar}\mathcal{H} \qquad (3.13)$$

$\mathcal{H}$ is called the energy superoperator [24, 25]; see Section V. We also see the difference between ϱ and ρ in the two preceding formulations. In the undilated (that is, conventional) formalism they are of course identical.

The usefulness of the analytically continued projector, in connection with variational principles and related virial theorems, has been demonstrated by Brändas and co-workers [68].

As we have discussed in some detail [20], complex scaling leads to a certain asymptotic decoupling of the localized resonance state from the background. This is also very easy to visualize from the physical point of view. By scaling the coordinates so that the appropriate outgoing Gamow waves, fulfilling the Siegert condition [56], become square integrable, the environment will be shielded from the system in a manner precisely given by the deformation. The incorporation of complex scaling into the present superoperator picture therefore gives in a physical as well as in a mathematical sense a valid subdynamical picture. The consequences of this point of view will be explored in the following.

Although complex scaling exhibits the previously mentioned localization properties, it is important to remember that irreducible nondiagonal structures, so-called Jordan blocks, may appear (in contrast to the conventional formulation) "on the real axis". Here this will be a desired "complication," since in fact it is associated with a new correlation pattern of quantum-statistical origin.

It is important to realize that these irreducible structures occur in the second-order density operator and not on the Schrödinger level. It was immediately observed (see, for example, Refs. 69 and 70) that Jordan blocks could appear in connection with complex scaling. It was further pointed out by Moiseyev and Friedland [71] that there always exists (if the kinetic and potential operators do not commute) at least one rotation angle θ so that the complex rotated matrix Hamiltonian exhibits a Jordan block with Segré characteristic larger than unity.

Another line of thought emanates from the suggestion of Marcus (see, for example, Ref. 72) that many avoided crossings in the energy plane are an indication of "quantum chaos". That such situations can occur is not unusual, see, for example, recent studies of the so-called Zel'dovich phenomenon [73]. In connection with this discussion the conjecture arises that the present occurrence of irreducible blocks in the density matrix, as will be studied and analyzed, could contribute to the clarification of the idea of quantum chaos [74], if it exists at all as a meaningful concept.

IV. COHERENCE, ODLRO, AND FERMIONIC DENSITY MATRIX

A. Introductory Remarks

In this section we will review some of the definitions and formulas of the theory of density matrices, which are needed for the general theory of coherent-dissipative structures presented in Section V. We will try to make this section self-contained, so that the most relevant aspects of the present approach becomes easy to comprehend in its main features.

The mathematical analysis of the structure of fermionic density operators of quantum mechanics was established by Coleman [28, 29]. Yang [77] and Coleman [30, 31] showed that the corresponding reduced second-order density operator $\Gamma^{(2)}$ is of considerable physical importance, in particular for the description of macroscopic quantum phenomena like superfluidity and superconductivity. Particularly noteworthy is Yang's finding that the assumption of ODLRO (off-diagonal long-range order; see. Section VC) in $\Gamma^{(2)}$ explains flux quantization in superconducting rings and the ODLRO is accompanied by a large eigenvalue of this operator. Another important result, in this context, is given by Blatt's proof [90] that the antisymmetrized geminal power (AGP) function of N fermions is the "N-particle projection" of the BCS function [32] of superconductivity.

For our CSM theory of quantum correlations, the so-called "large box part" of Coleman's "box and tail" representation of the fermionic second-order reduced density matrix (in the "extreme case", see following discussion) is of importance [20], and especially the so-called "small box part" of this matrix, which—as a matter of fact—did not attract any closer consideration in the scientific literature thus far.

For a didactical presentation of the basic concepts of Coleman's theory and an explicit example, see Ref. 75.

B. Box-and-Tail Form and Eigenvalues of the Fermionic $\Gamma^{(2)}$

The mathematical details of this subsection are based on the presentation of Ref. 20.

Let us start with some definitions. For N fermions, the reduced density operator (also called, matrix) of order p is defined as follows, see, for example, Löwdin [76]:

$$\Gamma^{(p)}(x_1 \cdots x_p | x_1' \cdots x_p')$$

$$= \binom{N}{p} \int \psi^*(x_1 \cdots x_p, x_{p+1} \cdots x_N) \psi(x_1' \cdots x_p', x_{p+1} \cdots x_N) dx_{p+1} \cdots dx_N \quad (4.1)$$

In literature one finds several similar definitions with different normalizations:

$$\text{Löwdin [76]: } \Gamma^{(p)}; \quad \text{Tr } \Gamma^{(p)} = \binom{N}{p} \quad (4.2)$$

$$\text{Yang [77]: } \rho^{(p)}; \quad \text{Tr } \rho^{(p)} = p! \binom{N}{p} \quad (4.3)$$

$$\text{Coleman [28]: } D^p; \quad \text{Tr } D^p = 1 \quad (4.4)$$

In the following we will use the Löwdin normalization.

The term geminal means a two-particle (pair) function. Since the N-particle projection of the BCS function [32] is an *antisymmetrized geminal power* (AGP), we will review the fundamentals of the AGP state. The interest in antisymmetrized geminal powers was related to the following two basic facts:

1. The two-particle density operator $\Gamma^{(2)}(x_1, x_2 | x_1', x_2')$ has a simple matrix representation, the "Box and Tail" [28, 29].
2. The AGP ansatz is the ground state for the polarization propagator [78, 79].

(There is also the generalization to GAGP states that we will not need in the following.) In addition to the usefulness of the polarization propagator technique in chemistry, we will also mention the importance of the *extreme* case (see subsequently and Section V), which is a consequence of the assumption of equal *a priori* probability in the microcanonical ensemble as given by equilibrium (and quasiequilibrium) thermodynamics.

In this chapter we will focus on the first simplifying feature. Consider a geminal of ranks s expanded in a basis of $r = 2s$ orthonormal spin orbitals $(2s \geqslant N)$,

$$g(1, 2) = \sum_{i=1}^{s} g_i |\phi_i, \phi_{i+s}\rangle \quad (4.5)$$

where $|\phi_i, \phi_{i+s}\rangle$ is (normalized) Slater determinant of the spin orbitals ϕ_i and ϕ_{i+s}. From g, one may construct the AGP as (using $N = 2m$ fermion states) the following antisymmetrized product:

$$|g^{N/2}\rangle = [S_{N/2}]^{-1} \sum_{\substack{j_\tau < j_{\tau+1} \\ 1 \leqslant \tau \leqslant N/2}} g_{j_1} g_{j_2} \cdots g_{j_{N/2}} |\phi_{j_1} \phi_{j_1+s} \cdots \phi_{j_{N/2}} \phi_{j_{N/2}+s}\rangle \quad (4.6)$$

where the normalization integral $S_{N/2}$ is given by the symmetric sum

$$S_{N/2} = \sum_{\substack{j_\tau < j_{\tau+1} \\ 1 \leqslant \tau \leqslant N/2}} n_{j_1} n_{j_2} \cdots n_{j_{N/2}} \quad (4.7)$$

and $n_j = |g_j|^2$. The one-particle reduced density operator $\Gamma^{(1)}(g^{N/2})$ associated with the AGP is then written

$$\Gamma^{(1)}(g^{N/2}) = \sum_{j=1}^{s} \lambda_j^{(1)} \{|\phi_j\rangle\langle\phi_j| + |\phi_{j+s}\rangle\langle\phi_{j+s}|\} \quad (4.8)$$

with

$$\text{Tr } \Gamma^{(1)}(g^{N/2}) = N \quad (4.9)$$

The eigenvalues $\lambda_i^{(1)}$ of $\Gamma^{(1)}(g)$ and $\Gamma^{(1)}(g^{N/2})$ are doubly degenerate (corresponding to natural spin orbitals (NSO) ϕ_i and ϕ_{i+s}), and given by

$$\lambda_i^{(1)} = [S_{N/2}]^{-1} n_i \frac{\partial S_{N/2}}{\partial n_i} \quad (4.10)$$

Before we continue, one should point out that $\Gamma^{(p)}(g), p = 1, 2$, are different although related to another. For instance, one has the relations

$$\Gamma^{(2)}(g) = |g\rangle\langle g| \quad (4.11)$$

$$\text{Tr } \Gamma^{(2)}(g) = 1 = \sum_{i=1}^{s} n_i \quad (4.12)$$

and also

$$\Gamma^{(1)}(g) = \sum_{j=1}^{s} n_j \{|\phi_j\rangle\langle\phi_j| + |\phi_{j+s}\rangle\langle\phi_{j+s}|\}$$

$$\text{Tr } \Gamma^{(1)}(g) = 2$$

Therefore, while $\Gamma^{(1)}(g)$ and $\Gamma^{(1)}(g^{(N/2)})$ have the same NSO's, their occupation numbers are different as shown in (4.10) and following (4.7).

Although $|g\rangle$ is a natural geminal of $\Gamma^{(2)}(g)$, it is not in general an eigengeminal to $\Gamma^{(2)}(g^{N/2})$ (only in the extreme case, see the subsequent discussion). However, the representation of $\Gamma^{(2)}(g^{N/2})$ in the basis

$$\{\{|\phi_i, \phi_{i+s}\rangle | 1 \leqslant i \leqslant s\}, \{|\phi_i, \phi_j\rangle | 1 \leqslant i \leqslant j \leqslant 2s; i + s \neq j\}\}$$

assumes a very simple "box and tail" form

$$
\begin{pmatrix}
b_{11} & b_{12} & \cdots & b_{1s} & 0 & 0 & \cdots & 0 & 0 \\
b_{21} & b_{22} & \cdots & b_{2s} & 0 & 0 & \cdots & 0 & 0 \\
\vdots & \vdots & \ddots & \vdots & \vdots & \vdots & & \vdots & \vdots \\
b_{s1} & b_{s2} & \cdots & b_{ss} & 0 & 0 & \cdots & 0 & 0 \\
0 & 0 & \cdots & 0 & t_{12} & 0 & \cdots & 0 & 0 \\
0 & 0 & \cdots & 0 & 0 & t_{13} & \cdots & 0 & 0 \\
\vdots & \vdots & & \vdots & \vdots & \vdots & \ddots & \vdots & \vdots \\
0 & 0 & \cdots & 0 & 0 & 0 & \cdots & t_{(2s-2)2s} & 0 \\
0 & 0 & \cdots & 0 & 0 & 0 & \cdots & 0 & t_{(2s-1)2s}
\end{pmatrix} \tag{4.13}
$$

The "box" **B**, which has dimension s, has the elements

$$b_{j,j+s;k,k+s} \equiv b_{jk} = [S_{N/2}]^{-1} g_j g_k^* \frac{\partial^2 S_{N/2+1}}{\partial n_j \partial n_k}, \quad 1 \leqslant j \neq k \leqslant s \tag{4.14}$$

and

$$b_{j,j+s;j,j+s} \equiv b_{jj} = [S_{N/2}]^{-1} n_j \frac{\partial S_{N/2}}{\partial n_j} = \lambda_j^{(1)}, \quad 1 \leqslant j \leqslant s \tag{4.15}$$

where $S_{N/2+1}$ in (4.14) is the symmetric function of order $N/2 + 1$. The "tail" **T**, which has the dimension $2s(s-1)$, has the elements

$$
\begin{aligned}
t_{ij} &= [S_{N/2}]^{-1} n_i n_j \frac{\partial^2 S_{N/2}}{\partial n_i \partial n_j} \\
&= \lambda_i^{(1)} \lambda_j^{(1)} + \frac{1}{2}\left\{ n_j \frac{\partial \lambda_i^{(1)}}{\partial n_j} + n_i \frac{\partial \lambda_j^{(1)}}{\partial n_i}\right\}
\end{aligned} \tag{4.16}
$$

where

$$1 \leqslant i \leqslant j \leqslant 2s; \quad i + s \neq j$$

Instead of constructing the general BCS function [32], where N goes to infinity, we will take a closer look at the finite N case. The exact form of $\Gamma^{(2)}(g^{N/2})$ as reviewed here was thoroughly studied by Coleman [28, 29] and, furthermore, investigated for the extreme case [29]. The upper bound on $\lambda^{(2)}$ was obtained by Yang [77] (for N even) and the existence of a large eigenvalue shown to be a consequence of the assumption of ODLRO [77]. The general case of N even and odd was studied by Sasaki [80] and the upper limit to $\lambda^{(2)}$ found to be

$$\lambda_i^{(2)} = \begin{cases} \leqslant (N-1)/2 & \text{if } N \text{ odd (Sasaki)} \\ < N/2 & \text{if } N \text{ even (Yang)} \end{cases} \tag{4.17}$$

Coleman introduced the following definition: $g^{N/2}$ (N even) is said to be *extreme* if all occupation numbers n_j of $\Gamma^{(1)}(g)$ are equal. Using (4.10), (4.14), (4.15), and (4.16) the "box and tail" becomes in the extreme case:

$$b_{jk} = b = \lambda^{(1)}\left(1 - \frac{N-2}{2(s-1)}\right) = \frac{N(2s-N)}{4s(s-1)}, \quad 1 \leqslant j \neq k \leqslant s \tag{4.18a}$$

$$b_{jj} = \lambda^{(1)} = \frac{N}{2s}, \quad 1 \leqslant j \leqslant s \tag{4.18b}$$

$$t_{ij} = \frac{N(N-2)}{4s(s-1)} = \lambda^{(1)} - b, \quad 1 \leqslant i \leqslant j \leqslant 2s; i + s \neq j \tag{4.18c}$$

The characteristic equation for the "nondiagonal" part becomes

$$(\lambda^{(1)} + (s-1)b - \lambda^{(2)})(\lambda^{(1)} - b - \lambda^{(2)})^{(s-1)} = 0 \tag{4.19}$$

with the "large eigenvalue" given by

$$\lambda_L^{(2)} \equiv \lambda_1^{(2)} = \lambda^{(1)} + (s-1)b = \frac{N}{2} - \frac{N(N-2)}{4s} \tag{4.20}$$

and an $(s-1)$ degenerate (and small) eigenvalue

$$\lambda_S^{(2)} \equiv \lambda_j^{(2)} = \lambda^{(1)} - b = \frac{N(N-2)}{4s(s-1)}, \quad \text{with } j = 2, \dots, s \tag{4.21}$$

For reasons of a simplification of the nomenclature that is extensively used in the following two sections, this "small" eigenvalue is labeled with an "S" (instead of a "D").

Hence the "box and tail" becomes in the extreme case a diagonal matrix with one eigenvalue $\lambda_1^{(2)} = \lambda_L^{(2)}$ nondegenerate and the remaining $(2s+1)(s-1)$ eigenvalues $\lambda_j^{(2)} = \lambda_s^{(2)}$, $j \neq 1$, degenerate.

Note that $s = N/2$ gives one degenerate eigenvalue 1 of order $\binom{N}{2}$. This is the *Hartree–Fock* independent particle model. In this case $\Gamma^{(2)}(g^{N/2})$ has a diagonal representation in the basis chosen here. Normally we will assume that $2s > N$. Note also that $\lambda_L^{(2)} = 0$ for $s = N/2 - 1$, which gives $\lambda_S^{(2)} = N/(N-4)$. (This is obviously unphysical, but the possibility of such a condensation off the real axis exists in the CSM framework.)

In the extreme case $\Gamma^{(2)}(g^{N/2})$ thus writes

$$\Gamma^{(2)}(g^{N/2}) = \frac{N}{2s} \sum_{i=1}^{s} |\phi_i, \phi_{i+s}\rangle\langle\phi_i, \phi_{i+s}| + \frac{N(2s-N)}{4s(s-1)} \sum_{i \neq j} |\phi_i, \phi_{i+s}\rangle\langle\phi_j, \phi_{j+s}|$$

$$+ \frac{N(N-2)}{4s(s-1)} \sum_{\substack{i<j \\ i+s \neq j}} |\phi_i, \phi_j\rangle\langle\phi_i, \phi_j|$$

$$= \Gamma_L^{(2)} + \Gamma_D^{(2)} \tag{4.22}$$

where

$$\Gamma_L^{(2)} = |\Phi_L\rangle\lambda_L^{(2)}\langle\Phi_L|; \quad \Phi_L = \sum_{i=1}^{s} \frac{1}{\sqrt{s}} |\phi_i, \phi_{i+s}\rangle = g \tag{4.23}$$

with

$$\lambda_L^{(2)} = \frac{N}{2} - \frac{N(N-2)}{4s}$$

and

$$\Gamma_D^{(2)} = \lambda_S^{(2)} \left\{ \sum_{i,j} |\phi_i, \phi_{i+s}\rangle \left(\delta_{ij} - \frac{1}{s} \right) \langle\phi_j, \phi_{j+s}| + \sum_{\substack{i<j \\ i+s \neq j}} |\phi_i, \phi_j\rangle\langle\phi_i, \phi_j| \right\} \tag{4.24}$$

with

$$\lambda_S^{(2)} = \frac{N(N-2)}{4s(s-1)}$$

see Eqs. (4.20) and (4.21)

It is easily seen from relations (4.23) that, if the basis $|\phi_i, \phi_{i+s}\rangle$ is localized in some sense, then the *extreme* geminal g is delocalized. Hence, in the limit of large s this delocalization does not vanish because of the properties of the off-diagonal matrix elements. This limit is related to the existence of a large eigenvalue, proportional to the number of particles, which also means that the off-diagonal parts of $\Gamma^{(2)}$ remain finite in the thermodynamic limit, that is, $N \to \infty$, $V_N \to \infty$ and N/V_N being finite. For more details on the thermodynamic limit "lim therm" see, for example, Girardeau [81].

As we will see below, the degenerate eigenvalue $\lambda_S^{(2)}$ approaches 0 for large s, and this will lead to an interesting observation regarding the remaining part of the "box" after subtraction of the large "macroscopic" component corresponding to $\Gamma_L^{(2)}$.

From (4.22) one obtains as a check

$$\text{Tr } \Gamma^{(2)}(g^{N/2}) = \frac{N}{2} + 2s(s-1)\frac{N(N-2)}{4s(s-1)} = \binom{N}{2} \qquad (4.25)$$

and from (4.23) and (4.24) that

$$\text{Tr } \Gamma^{(2)}(g^{N/2}) = \text{Tr } \Gamma_L^{(2)} + \text{Tr } \Gamma_D^{(2)} = \frac{N}{2} - \frac{N(N-2)}{4s} + (2s+1)(s-1)\frac{N(N-2)}{4s(s-1)} = \binom{N}{2} \qquad (4.26)$$

Before continuing we note that, if the rank s of the geminal g becomes infinite, $\lambda_L^{(2)}$ approaches $N/2$ and the degenerate eigenvalue $\lambda_S^{(2)}$ goes to zero. The degeneracy also becomes infinite, but in such a way that the trace relations (4.25) and (4.26) should hold in the limit.

The results quoted here can be strictly formulated into a theorem proved by Coleman [29]:

The geminal g is an eigenfunction of $\Gamma^{(2)}(g^{N/2})$ with a nonvanishing eigenvalue if and only if g is of extreme type, that is, the eigenvalues of $\Gamma^{(1)}(g)$ are all equal.

Since Yang's assumption of ODLRO [77] leads to a large eigenvalue of the second-order reduced density matrix, as well as magnetic flux quantization and the Meissner effect, it is obvious that the extreme condition for g is a necessary condition for ODLRO. However, one may wonder whether this is a sufficient condition also, that is, whether the extreme condition for g is sufficient for ODLRO. As we will see in following sections, another type of off-diagonal order will also appear although not corresponding to the macroscopically large eigenvalue.

Additional to the preceding analysis we will now evaluate one- and two-body expectation values in the extreme-type situation. For an one-body observable $H = \sum_{i=1}^{N} \hat{H}_i$ one obtains

$$\text{Tr}\{\hat{H}\Gamma^{(1)}(g^{N/2})\} = \frac{N}{2s}\sum_{i=1}^{s}\{\langle\phi_i|\hat{H}|\phi_i\rangle + \langle\phi_{i+s}|\hat{H}|\phi_{i+s}\rangle\} \qquad (4.27)$$

If ϕ_i and ϕ_{i+s} have identical spatial parts but opposite spins and if $\hat{H}$ is spin independent, this relation reduces to

$$\text{Tr}\{\hat{H}\Gamma^{(1)}(g^{N/2})\} = \frac{N}{2}\langle\phi|\hat{H}|\phi\rangle \qquad (4.28)$$

For a two-particle property $V = \sum_{i<j}\hat{V}_{ij}$ one obtains similarly

$$\text{Tr}\{\hat{V}\Gamma^{(2)}(g^{N/2})\} = \text{Tr}\{\hat{V}\Gamma_L^{(2)}\} + \text{Tr}\{\hat{V}\Gamma_D^{(2)}\} \qquad (4.29)$$

with

$$\text{Tr}\{\hat{V}\Gamma_L^{(2)}\} = \frac{1}{s}\left(\frac{N}{2} - \frac{N(N-2)}{4s}\right)\sum_{i=1}^{s}\sum_{j=1}^{s}\langle\phi_i,\phi_{i+s}|\hat{V}|\phi_j,\phi_{j+s}\rangle$$

$$= \lambda_L^{(2)}\langle g|\hat{V}|g\rangle \qquad (4.30a)$$

and

$$\text{Tr}\{\hat{V}\Gamma_D^{(2)}\} = \frac{N(N-2)}{4s(s-1)}\left\{\sum_{i=1}^{s}\langle\phi_i,\phi_{i+s}|\hat{V}|\phi_i,\phi_{i+s}\rangle - \langle g|\hat{V}|g\rangle\right.$$

$$\left. + \sum_{\substack{i<j\\i+s\neq j}}\langle\phi_i,\phi_j|\hat{V}|\phi_i,\phi_j\rangle\right\} \qquad (4.30b)$$

The preceding results are applicable to the general extreme case, when there exists a large eigenvalue as well as in other cases of interest. In particular, we will see in the next section how ODLRO may appear from a CSM transformation of the first part of equation (4.24), that is, the "small box part" of $\Gamma^{(2)}$ defined by

$$\Gamma_S^{(2)} \equiv \lambda_S^{(2)}\sum_{i,j}|\phi_i,\phi_{i+s}\rangle\left(\delta_{ij} - \frac{1}{s}\right)\langle\phi_j,\phi_{j+s}| \qquad (4.31)$$

In contrast to the "large box part" $\Gamma_L^{(2)}$, this part of the fermionic density matrix was considered to be without any particular importance thus far.

V. COMPLEX SCALING AND STATISTICAL MECHANICS: COHERENT-DISSIPATIVE STRUCTURES

A. Introductory Remarks

Our CSM-approach to quantum correlations in condensed systems [20] was motivated by the conjecture that there may exist an intrinsic interrelation between the CSM of quantum mechanics and the conceptual basis of Prigogine's subdynamics. This conjecture is supported by the observation that, in the framework of both theories, the existence of decaying states represents a basic fact following from first principles. This is clearly in contrast to conventional theory, which treats decaying states as "approximations" and considers rather the concept of "stationary state" as the fundamental one.

Our theoretical approach [20] represents a novel combination of the following three topics: (A) the CSM of ordinary quantum mechanics, (B) the fermionic density matrix, and (C) the canonical ensemble of quantum statistics. The most remarkable theoretical result (which has direct experimental consequences, see Section VI) is given by the following: One can prove that, under specific conditions, the CSM-transformed and "thermalized" second-order density matrix $\Gamma^{(2)}$ has no diagonal representation. In other words, it contains submatrices γ that represent a new kind of "indivisible units", which we called coherent-dissipative structures. These structures are due to quantum interference between resonances being exposed after complex dilation; furthermore, they are short-lived and exhibit a finite "size" in the space of state functions.

Coherent-dissipative structures are intrinsically connected with the complex delated (or scaled) small box part $\Gamma_s^{(2)}$, Eq. (4.31), of the fermionic $\Gamma^{(2)}$. Their existence follows from first principles of the microscopic theory.

B. Formalism of Coherent-Dissipative Structures

1. Canonical Density Operator and Complex Scaling

In this section we implement the concept of the canonical ensemble into the formalism. This is, for our purposes, physically necessary because condensed quantum systems cannot be properly treated as an ensemble of isolated small systems. It should also be observed that the concept of the canonical density operator is confined, in a strict sense, to physical systems being in thermal equilibrium [42]. Nevertheless, this concept has also found a wide applicability to systems being in quasiequilibrium and, more general, in a steady state. For the density operator ρ we have in general

$$\rho = \sum_{k,l} |\Psi_k\rangle \rho_{kl} \langle \Psi_l| \qquad (5.1)$$

where $\{|\Psi_k\rangle\}$ is a set of appropriate (many-particle) states. In the special case where these states are eigenfunctions of the system Hamiltonian H, one defines the *canonical* density operator through the well-known formula

$$\rho^{\mathrm{d}}_{\mathrm{can}} = \frac{1}{Z}\sum_k |\Psi_k\rangle e^{-\beta E_k} \langle \Psi_k| \tag{5.2}$$

where $\beta = 1/k_B T$ (k_B is the Boltzmann constant) and Z is an appropriate normalization constant. (The superscript d stays for "diagonal"; see following discussion.) In the conventional representation, that is, before applying CSM, H is of course self-adjoint and E_k is real. Thus, one sees immediately that the characteristic Boltzmann factor $e^{-\beta E_k}$ can be obtained by the action of the superoperator.

$$\frac{1}{Z}e^{-\beta\mathscr{H}}, \quad \text{with} \quad \mathscr{H} = \tfrac{1}{2}\hat{L}_T$$

see (3.13), on the operator $\rho_{kk} = |\Psi_k\rangle\langle\Psi_k|$, as the following equations show, see, for example, (4.10),

$$e^{-\beta\mathscr{H}}\rho_{kk} = e^{-\beta E_k}\rho_{kk} \tag{5.3}$$

In this equation, $\mathscr{H}$ is the energy superoperator, see, for example, Refs. 21, 24, and 66. This quantity is a direct generalization of the conventional Hamiltonian. For the proof of (5.3) see Section III, where it is simply seen that the superoperator $\mathscr{H}$ is also self-adjoint, since H is self-adjoint. The obvious physical meaning of the superoperator in (5.3) is that it produces the well-known Boltzmann factors. The factorization property displayed in Section III and shown in (5.3) can also be applied to the general ket-bra operators [66] $\rho_{kl} = |\Psi_k\rangle\langle\Psi_l|$, $k \neq l$, in a consistent manner producing pure quantum-mechanical off-diagonal contributions to the density operator, see Eq. (5.1), that is, correlations of pure quantum-mechanical character (see Refs. 79–84). We thus obtain

$$e^{-\beta\mathscr{H}}\rho_{kl} = e^{-(\beta/2)(E_k + E_l)}\rho_{kl} \tag{5.4}$$

from which one obtains the off-diagonal terms that correspond to (5.2), if one divides the exponential factor in (5.4) by Z. From this it follows that:

(i) the density operator is self-adjoint, that is $\rho^\dagger = \rho$,
(ii) the diagonal elements ($k = l$) are proportional to $e^{-\beta E_k}$.

Let us summarize the situation in the following way and let

$$\rho_m \equiv \sum_{k,l} |\Psi_k\rangle c_{kl} \langle \Psi_l| \tag{5.5}$$

with

$$c_{kk} \equiv \text{const, \quad for all } k \tag{5.6a}$$

and

$$c_{kl} = c_{ik}^* \quad (l \neq k) \tag{5.6b}$$

be a "nonthermalized" density operator. (Clearly, this operator corresponds to a microcanonical-type ensemble.) The canonical operator ρ_{can} is then given by

$$\rho_{\text{can}} = \frac{1}{Z} e^{-\beta \mathcal{H}} \rho_m \tag{5.7}$$

It follows that by disregarding the off-diagonal elements of ρ_{can} one obtains the very well-known "diagonal" part as given in Eq. (5.2), which—per definition—contains no information about quantum correlations among different states.

Until now no complex scaling has been considered. In the following we proceed to the extension of the preceding results in the framework of complex scaled quantities. This is easily done here, because of the simple analytic form of the "Boltzmann" superoperator. Thus, one needs in principle only to know how to scale the Hamiltonian H. In the following we make use of certain results presented in Sections II and III, which are based on results of Brändas [9] and Löwdin [10]:

1. The complex scaled Hamiltonian H^c is given by a similarity transformation

$$H^c \equiv UHU^{-1} \tag{5.8}$$

where the operator U is the *unbounded* dilation operator fulfilling the condition

$$(U^\dagger)^{-1} = U^* \tag{5.9}$$

in the appropriate domain; see Section II.

2. The Hamiltonian H^c is complex symmetric in the same domain, that is,

$$(H^c)^\dagger = (H^c)^* \tag{5.10}$$

This property is of particular importance in the context under consideration, because it allows us to construct the complex scaled canonical operator ρ^c_{can} in a very simple way. Owing to the simple analytic form of the exponential operator, one finds immediately

$$[e^{-\beta H}]^c = e^{-\beta H^c} \tag{5.11}$$

Using (3.5) and (3.6), where

$$E_k, \varepsilon_k = \text{real} \tag{5.12a}$$

and

$$\varepsilon_k \geqslant 0 \tag{5.12b}$$

one finds

$$e^{-(\beta/2)H^c}|\Psi_k^c\rangle\langle\Psi_l^{c*}|e^{-(\beta/2)H^c} \equiv \omega_{kl}|\Psi_k^c\rangle\langle\Psi_l^{c*}| \tag{5.13}$$

with ω_{kl} given by

$$\omega_{kl} = e^{-(\beta/2)\{(E_k+E_l)-i(\varepsilon_k+\varepsilon_l)\}} \tag{5.14}$$

see Eq. (3.10). This proves that the density operator being here under consideration

$$\rho^c \equiv \sum_{k,l}|\Psi_k^c\rangle\omega_{kl}\langle\Psi_l^{c*}| \tag{5.15}$$

is trivially complex symmetric, that is,

$$\rho^{c\dagger} = \rho^{c*} \tag{5.16}$$

and this is in agreement with (5.10), since ρ^c is per definition an operator function of H^c (see also Ref. 68).

We also see easily that all the results concerning the complex scaled quantities coalesce to the well-known "unscaled" results in the limit

$$\varepsilon_k \to 0 \tag{5.17}$$

if one uses the accompanying formal transitions

$$H^c \to H \tag{5.18}$$

and

$$\Psi_k^c \to \Psi_k \tag{5.19}$$

This remark may be considered as an additional check that the preceding considerations concerning the complex scaled extension of the canonical density operator are physically meaningful.

Finally we will formulate the analog of Eq. (5.7) in the CSM representation. The density operator of a microcanonical ensemble ρ_m^c may be represented as

$$\rho_m^c = \sum_{k,l} |\Psi_k^c\rangle c_{kl} \langle \Psi_l^{c*}| \tag{5.20}$$

with $c_{kk} = $ const, all k, and $c_{kl} = c_{lk}$, which indeed is a complex symmetric operator. The corresponding complex scaled canonical density operator ρ_{can}^c is then given by the relation

$$\rho_{can}^c = \frac{1}{Z} e^{-(\beta/2)H^c} \rho_m^c e^{-(\beta/2)H^c} \tag{5.21}$$

From the preceding consideration one finds that this operator is also complex symmetric. From (5.24) and (5.20) we obtain

$$\rho_{can}^c = \frac{1}{Z} \sum_{k,l} c_{kl}\omega_{kl} |\Psi_k^c\rangle \langle \Psi_l^{c*}| \tag{5.22}$$

where ω_{kl} is given by (5.14), which gives the explicit form of the canonical operator needed in the derivations that follow.

2. Coherent-Dissipative Structures on the Microscopic Level of Description

We proceed now to present the formal conditions that accompany the spontaneous creation of microscopic coherent-dissipative structures. The physical discussion of this formal derivation will be presented in the next section; for further illustration within the experimental context of this theory, we refer to Section VI.

The three main formulas that are needed in the following are given by (2.24a), (4.31), and (5.22):

$$q_{kl} = \left(\delta_{kl} - \frac{1}{s}\right) \exp\left(i\pi \frac{(k+K)+(l+K)-2}{s}\right) \tag{2.24a}$$

$$\Gamma_S^{(2)} = \lambda_S^{(2)} \sum_{k,l} \left(\delta_{kl} - \frac{1}{s}\right) |k, k+s\rangle\langle l, l+s| \tag{4.31}$$

and

$$\rho^c_{can} = \frac{1}{Z}\sum_{k,l} c_{kl}\omega_{kl}|\Psi^c_k\rangle\langle\Psi^{c*}_l| \qquad (5.22)$$

A comparison of Eq. (4.31) with (5.22) shows the following formal correspondence: Applying our CSM approach to the "small box part" (4.31) and utilizing the formal replacement

$$|k, k + s\rangle \rightarrow |(k, k + s)^c\rangle$$

one sees that the right-hand side (rhs) of Eq. (4.31) is formally a special case of the complex scaled microcanonical-type operator ρ^c_m with the coefficients

$$c_{kl} \equiv \lambda^{(2)}_S\left(\delta_{kl} - \frac{1}{s}\right) \qquad (5.23)$$

In what follows, the absolute values of the "nondiagonal" numbers $c_{kl}, k \neq l$ (representing quantum correlations, see also subsequent discussion) do not obey any restriction, and thus—if desired—they may also be considered to be arbitrarily small. We can immediately see that $\Gamma^{(2)c}_S$ is also complex symmetric, as one expects.

We are now ready to ask what happens to $\Gamma^{(2)c}_S$ if the systems of interest are subject to the canonical ensemble formalism of statistical mechanics [20]. With the considerations of the previous section in mind, we just make use of Eq. (5.21). The "part of the ensemble correlations" represented by the CSM-transformed small box part $\Gamma^{(2)c}_S$ is then subject to the transformation

$$\gamma \equiv \frac{1}{Z}e^{-(\beta/2)H^c}\Gamma^{(2)c}_S e^{-(\beta/2)H^c} \qquad (5.24)$$

Here, H^c represents the complex scaled second-order reduced Hamiltonian [88]. Using the explicit result (5.22) and the additional physical assumption (corresponding to a mathematical restriction) that the here considered real energies of all the pairs $|(k, k + s)^c\rangle$ are equal, that is,

$$E_k = E \quad (k = 1, \ldots, s) \qquad (5.25)$$

we immediately obtain

$$\gamma \equiv \sum_{k,l}\gamma_{kl}|(k, k + s)^c\rangle\langle(l, l + s)^{c*}| \qquad (5.26)$$

with

$$\gamma_{kl} = \frac{\lambda_S^{(2)}}{Z} e^{-\beta E} \left(\delta_{kl} - \frac{1}{s} \right) e^{i(\beta/2)(\varepsilon_k + \varepsilon_l)} \qquad (5.27)$$

We thus make the following crucial observation. The matrix elements γ_{kl} of the operator γ, Eq. (5.26), are formally very similar to the matrix elements of the matrix $\mathbf{q}$, Eq. (2.24a). More precisely, if we require the validity of the *quantization conditions*

$$\pi \frac{k + K - 1}{s} = \frac{\beta}{2} \varepsilon_k \quad (k = 1, 2, \ldots, s; K \text{ being an arbitrary integer}) \qquad (5.28)$$

we obtain the important equality

$$\gamma_{kl} = \frac{1}{Z} \lambda_S^{(2)} e^{-\beta E} q_{kl} \equiv \text{const} \times q_{kl} \qquad (5.29)$$

In other terms, if the "widths" ε_k and the energies E_k of the complex scaled pairs $|(k, k + s)^c\rangle$ fulfill the conditions (5.28) and (5.25), then the matrix elements of the canonical density operator γ constitute a Jordan block $\mathbf{C}_s(0)$ (due to the Reid–Brändas corollary, see Section II B 3). The density operator γ represents the *coherent-dissipative structures*, that is, the cooperative (synergetic) phenomenon of interest.

3. Minimal Size of Coherent-Dissipative Structures: Derivation and Physical Interpretation

In this subsection we proceed in the completion of the formalism by determining the "size" (in the space of state functions under consideration) the coherent-dissipative structures can have. This is a crucial task, because of its importance of the direct contact between the theory and real experiments, the latter being considered in Section VI. In the previous subsection we found the quantization conditions (5.28) that are necessary for the appearance of Jordan blocks of Segré characteristic s. As was shown previously, these blocks are not diagonizable, and—as a consequence—the complex dilated pairs $|(k, k + s)^c\rangle$ constituting them appear to "act cooperatively"; see also the subsequent discussion.

Here the following point should be observed. Any choice concerning the value of K affects the magnitude of the lhs of the quantization conditions (5.28). At this stage we may proceed to the specification of the *physically meaningful range* of K through the following considerations.

First, we already know from the general theory of CSM that the "widths"

ε_k, Eq. (5.23), should be *positive real* quantities. This physically means that one is dealing with decaying states. This implies the requirement that the lhs of (5.28) has to be positive, which leads to the restriction

$$K \geqslant 1 \qquad (5.30)$$

Second, we require that no lifetime

$$\tau_k = \frac{\hbar}{2\varepsilon_k} \qquad (5.31)$$

of the considered paired states should be greater than a characteristic *relaxation time* τ_{rel} of a quantum system, which can be studied in a specific experiment by standard methods. (This point is extensively illustrated by the experiments considered in Section VI.) This requirement thus physically means that all the paired states participating to the considered Jordan block will "disappear"—because they decay—after a time of the order of the experimentally measured time interval τ_{rel}. Or, in other terms, the "lifetime" of the coherent-dissipative structure is expected to be not larger than the quantum relaxation process creating this structure; see also subsequent discussion.

This physical requirement is formally represented through the relation

$$\tau_{rel} = \tau_1 \equiv \frac{\hbar}{2\varepsilon_1} = \frac{\hbar\beta}{4\pi} \frac{s}{K} \qquad (5.32)$$

which follows from Eq. (5.28) immediately. Furthermore, we already know from (5.28) that $K \geqslant 1$. Thus, we arrive at the following crucial point. The further restriction

$$K = 1 \qquad (5.33)$$

of the relation (5.30) also guarantees that the Segré characteristic of the considered Jordan block becomes *minimal*: one easily sees that any other choice (that is, $K > 1$), yields a large numerical value for s, since s appears to be proportional to K due to (5.32). Remember that the Segré characteristic of the considered Jordan block coincides with the number of the pairs constituting the Jordan block.

Let us summarize the preceding results. The quantization conditions that lead to the creation of Jordan blocks may be formulated as

$$\tfrac{1}{2}\beta\varepsilon_k = \pi\frac{k}{s}, \quad k = 1,\ldots,s \qquad (5.34)$$

or, alternatively,

$$\tau_k = \frac{\hbar\beta}{4\pi}\frac{s}{k}, \quad k = 1,\ldots,s \qquad (5.35)$$

see Eq. (5.28), (5.31), and (5.33). The additional requirement

$$\tau_{rel} = \frac{\hbar\beta}{4\pi}\frac{s}{1} \qquad (5.36)$$

guarantees that—physically speaking—all the "members" (that is, pairs) of the Jordan block, and therefore also the Jordan block itself, decay within a time of the order of the (measurable) relaxation time of the microscopic quantum systems under consideration. Also, we have seen that the same conditions which led us to the Eq. (5.34)–(5.36) determine, at the same time, the minimal "size" s_{min} the Jordan block must exhibit:

$$s_{min} = \frac{4\pi}{\hbar\beta}\tau_{rel} = \frac{4\pi k_B T}{h}\tau_{rel} \qquad (5.37)$$

This remark concludes the desired derivations. Further considerations and the physical context of this general theory are presented below.

C. Spontaneous Creation of Coherent-Dissipative Structures: Physical Considerations

In this subsection, additional remarks concerning the physical content and interpretation of the general CSM theory are presented. It will become clear that many formal concepts and steps in the preceding derivations can be interpreted physically in a natural way—as far as condensed matter processes are considered.

Some remarks on the differences between the physical nature of these structures and the well-known superconducting BCS states (which, of course, represent typical coherent structures) may be of particular interest, since the latter states are related with the "large box part" $\Gamma_L^{(2)}$ of the complete $\Gamma^{(2)}$.

The following discussions are also intended to give further physical insight into the general CSM theory of quantum correlations as well as to illustrate its physical significance, before we proceed to concrete experimental applications.

1. Microdynamical Processes and the Basis Set $\mathscr{S}$

As we intend to apply the above formalism to microdynamical processes in condensed matter, we will interpret the units

$$|\phi_k, \phi_{k+s}\rangle \equiv |k, k+s\rangle$$

appearing in the box part of the second-order density matrix $\Gamma^{(2)}(g^{N/2})$ in the following manner; $|k\rangle$ and $|k+s\rangle$ represent two possible quantum states of a microscopic quantum system being associated with the specific dynamical process under consideration. Or, more simply, these kets could be considered to represent some proper quantum states of a microscopic system "before" and "after" a microdynamical process "happens", respectively.

One should observe that these two states must be considered to be *entangled* (even with more states of the other quantum systems of the surroundings), at least for times of the order of the "relaxation time" of the considered microscopic system; see the books of d'Espagnat [83], Primas [82], and Wheeler and Zurek [84].

Thus, it appears to be natural to represent a microdynamical process in condensed systems through a unit (or "paired state" $|k, k+s\rangle$), which plainly reflexes this entanglement "during the dynamical process." Therefore, it is physically meaningful to represent the dynamical process under consideration by the set of pairs:

$$\mathscr{S} = \{|k, k+s\rangle | k = 1, \ldots, s\}. \tag{5.38}$$

Once again we like to stress that this "cooperation" between elementary quantum systems follows from first principles of quantum mechanics and CSM (see the previous derivation), and that it is entirely due to the nondiagonal part of the density operator of the condensed system. As is well known, these correlations are of pure quantum-mechanical nature, thus having no classical analog, see Ref. 77 and subsequent discussion.

Oversimplifying and for illustration one could also say that, "during the time" the process "takes place", the state of an elementary quantum system has to be described by some "combination" of $|k\rangle$ and $|k+s\rangle$.

2. Fermionic Degrees of Freedom and the Units $|k, k+s\rangle$

The general theory considers fermionic degrees of freedom that are involved in the build-up of the pairs $|k, k+s\rangle$; see Section IV. The entanglement of the two-particle states represented by $|k, k+s\rangle$ is intrinsically connected with the fact that many-body correlations and interactions always exist in condensed matter. These in general cause the first- and second-order density matrices to exhibit nonintegral one- and two-particle occupation numbers, respectively. In any case it is clear that the degrees of freedom connected with the considered dynamical process of the microscopic quantum system have manifestly fermionic origin, even though they are coupled (or paired) into (quasi-) bosonic degrees of freedom. Hence, it is obvious that the unit

$|k, k + s\rangle$, which is associated with two fermionic degrees of freedom, should be viewed as a boson-type degree of freedom, and referring to the number of degrees of freedom to follow, we always mean the latter entity.

Let us also point out the—more or less obvious—fact that the properties of the density matrix $\Gamma^{(2)}$, which are presented in Section IV, are valid because of the fermionic character of the "one-particle" states $|k\rangle$ and not (as incorrectly might be believed) because these states may be occupied by distinguishable, different particles that happen to be fermions. The latter point of view is intuitively motivated, for example, by the build-up of Cooper pairs from "individual" electrons in the BCS theory of superconductivity; see subsequent discussion. However, the actual occupation number of a specific $|k, k + s\rangle$ is without interest here. To avoid confusion, we would like to emphasize that the concept of "paired degrees of freedom", as represented by $|k, k + s\rangle$, is not identical with the (intuitive) concept of "particle pairs."

3.   The Degeneracy Condition $E_k = E$

In (5.25) we have used the physical condition $E_k = $ const, for all $k = 1, ..., s$. This means that the (real) energies of all the units that constitute the aforementioned Jordan block are equal. Moreover, it suffices to assume that

$$|E_k - E| \leqslant \Delta E, \quad k = 1, ..., s \tag{5.39}$$

This condition is, of course, more realistic, and it easily permits us to recognize its physical meaning: In any realistic experiment, the energy of the detected processes can only be measured up to a finite accuracy, which is given by the so-called spectral resolution of the apparatus. It should be observed that all the detected microdynamical processes that fulfill condition (5.39) are then indistinguishable in the sense that there is no possibility to attribute each of them to some individual quantum system (for example, molecule) of the considered specimen; for examples see Section VI.

4.   Geminals of Extreme Type

The microdynamical process fulfilling the condition (5.39) are indistinguishable, as discussed previously. Hence it appears to be "natural" to assume that the $N/2$ considered microdynamical processes are equally "distributed" over all the possible units $|k, k + s\rangle$. This assumption is in line with the basic ideas of statistical mechanics of equilibrium (and quasiequilibrium); it corresponds with the best "unbiased choice" that conforms with Eq. (5.39). We should point out, however, that this remark may illustrate the "extremity" of the AGP function $|g^{(N/2)}\rangle$ in the present physical context; but it does not explain (or give a clear physical insight into) the fact that the considered coherent-dissipative structures are connected—as presented in the subsection

B—with functions of that type. It may be interesting to note that almost the same remark holds true for the case of superconductivity, see Refs. 28 and 30.

5. *Finite Rank of* $\Gamma^{(2)}$ *and the Associated Dissipative Phenomena*

In the derivations presented in Section V B, we considered the case where the index s is finite, exclusively. In other words, the dimension of the physical space under consideration is assumed to be finite. Therefore, one might wonder why this restriction does not prevent the treatment of irreversible (or dissipative) phenomena as those being under consideration in this chapter. Indeed, as has been pointed out by many authors (see Prigogine [21] and references cited therein) a necessary condition to break the reversible character of the fundamental equations of motion (Schrödinger or Liouville–von Neumann equation) is that the physical system under consideration has infinitely many degrees of freedom.

The clarification of this apparent "contradiction" is more or less trivial, and it is given by the physical meaning of a resonance, in connection with its mathematical specification in the framework of CSM: namely it is a well-known fact that an infinite-dimensional space of some conventional quantum representation is representable by a finite-dimensional space after the appropriate complex scaling; cf. Reed and Simon [6]. For illustration, let us consider a scattering process. An isolated resonance being revealed after complex scaling, under certain conditions is found to be associated with a real, measured "bump" in the observed continuous energy spectrum of the scattering system. Of course, the mathematically precise "resonance" is not identical with experimental quantities. Nevertheless, experience shows that the considered correspondence between mathematical "resonances" and experimental "scattering bumps" works extremely well, this being also one reason for the popularity of CSM.

In the subsection B we showed how coherent-dissipative structures can appear after complex scaling of the density operator. Remember that, prior to the application of the complex scaling, no cooperative phenomenon (of the kind of the considered coherent-dissipative structures) has been reported. These remarks are also in line with the fact that we considered the canonical ensemble associated with the (second-order) density matrix. This means physically that any degree of freedom of the canonical ensemble should be considered to be "coupled" with infinitely many degrees of freedom of the thermal bath. (Remember that the use of the canonical ensemble formalism presupposes the existence of a thermal bath, which is assumed to be weakly coupled with any system of the ensemble [42].) In the light of the CSM, one can say that the "lifetimes" associated with the imaginary energies $i\varepsilon_k$, Eqs. (3.6) and (5.12b), are due to those "couplings."

The preceding comments may also be helpful in order to justify the occurrence of a contractive evolution operator $\exp(-i\hat{L}^c t)$, that is, a Lyapunov converter, after complex scaling of the Liouville–von Neumann equation [26].

6. Missing Probabilistic Interpretation

We found in the previous subsection that the CSM-transformed part $\gamma \equiv (\Gamma^{(2)}_{S,\text{can}})^c$ of the quantum correlated canonical (or "thermalized") second-order density matrix, Eq. (5.24), is similar to a Jordan block, if the appropriate "quantization" conditions are fulfilled. As is well known, however, a Jordan block is characterized by its nondiagonalizability [65, 89]. Hence, there does not exist any unitary, star-unitary, restricted similarity, or another similarity transformation that diagonalizes γ. Thus the units, or pairs, that constitute—after CSM transformation—the considered part γ of the full density matrix $\Gamma^{(2)}$ act cooperatively and create a new "unit" which we called a coherent-dissipative structure. In other terms we may say that—in the case under consideration—all the units $|k, k + s)^c\rangle$ for $k = 1, \ldots, s$, coalesce and act "as a whole" or as an "indivisible unit".

Here, and in the following parts of the chapter, we will leave out for simplification the upper index "c" on the complex scaled quantities, if there is no danger of confusion.

In more physical terms we may say that, in the considered case, one cannot attribute a probability to the "particular" microdynamical process represented by $|k, k + s\rangle\langle(k, k + s)^*|$ just because of the fact that now this particular process does not exist (and this holds for all $k = 1, \ldots, s$). Namely the fact that

(i) $\operatorname{Tr} \gamma = 0$, and
(ii) that physically meaningful occupation probabilities must be nonnegative, $p_k \geqslant 0$,

imply $p_k \equiv 0$, for all k.

7. Lifetime and Spatial Extension of Coherent-Dissipative Structures

It was mentioned that the units $|k, k + s\rangle$ (before and after complex scaling) are considered to describe cooperatively a possible microdynamical process "between" the "one-body" states $|k\rangle$ and $|k, k + s\rangle$ of a quantum system. All these units ($k = 1, \ldots, s$) have the same (real) energy, $|E_k - E| \leqslant \Delta E$. Thus, all these units are multiplied by the same Boltzmann factor, in the formalism of the canonical ensemble; see Section V B. Now, one quantum system (or more of them) may appear to be delocalized over the space of states

$$\mathscr{S}_E \equiv \{|k, k + s\rangle| \quad E_k = E; \quad k = 1, \ldots, s\} \tag{5.40}$$

As is well known, the complex scaling method assigns "widths" ε_k to that units. These widths are considered to determine the "lifetimes" (or decay times) of the elements of $\mathscr{S}_E$ through the relation $\tau_k = \hbar/2\varepsilon_k$. These lifetimes, however, cannot be identified with possible real decay (or relaxation) times of an excited quantum system, because, in the present case, all the elements of $\mathscr{S}_E$ act cooperatively, and thus the whole Jordan block represents the preceding decay of a single quantum system. This also means that there does not exist any physical interpretation of an *individual* τ_k in terms of measurable quantities. The requirement $\tau_{\mathrm{rel}} = \tau_1$, Eq. (5.32), which can be reformulated as $\tau_{\mathrm{rel}} \geqslant \tau_k$ (for all $k = 1, \ldots, s$), may be interpreted in physical terms as the condition that the coherent-dissipative structure associated with the Jordan block over $\mathscr{S}_E$ does decay with the time constant τ_{rel}.

The actual *spatial dimensions* (or size) of the coherent-dissipative structures have not been determined thus far, because this problem cannot be solved with the aid of the general formalism of quantum *correlations* exclusively. As a matter of fact, the treatment of this problem must take the relevant Hamiltonian (that is, the actual mechanism) of the specific considered physical system into account explicitly. However, the following investigations will show that coherent-dissipative structures are always spatially restricted quantities. For illustration of this important point and some examples of such estimates, see Section VI.

8.   Temperature Dependence

The main formula (5.37) exhibits the following remarkable feature. Since $s_{\min}$ is proportional to T, an increasing temperature seems to favor the extension of the specific quantum correlations under consideration. Of course, the effect of increasing T on the relaxation time τ_{rel} is in general quite the opposite. In this context it should be observed that τ_{rel} is intrinsically connected with the effective (or relevant) Hamiltonian of the quantum systems, rather than with "correlation patterns" to which they may participate.

These remarks hold true in the case of low temperatures, too. In the limit $T \to 0$, however, one should observe that the coherent part $\Gamma_L^{(2)}$ becomes dominant. This part of the complete density matrix $\Gamma^{(2)}$ represents a pure case and not a mixture [83], and it exhibits the well-known (infinite-lived) quantum correlations of standard quantum mechanics.

9.   Possible Connection with Dissipative Structures

At this stage we would like to mention that there seems to be a similarity between the coherent-dissipative structures and the very well known *dissipative* structures of Prigogine and co-workers: Both structures cannot have a "size" smaller than a critical one; both cannot continue to exist, if "external energy supply" stops. On the other hand, it should be stressed that

Prigogine's dissipative structures are concepts of phenomenological thermo-dynamics, whereas the coherent-dissipative structures are concepts of a microscopic theory. Thus, the following speculation may be mentioned: The formalism of coherent-dissipative structures may represent the framework in which the phenomenological dissipative structures can be established "from first principles."

10. Differences to BCS States of Superconductivity

Further clarification of the physical content of the preceding theory may now be achieved by pointing out some conceptual difference between our formalism and the well-known BCS theory (or better, model) of super-conductivity.

It has been proved by Blatt [90] and repeatedly pointed out by Coleman [28–30] that the BCS ground-state ansatz is "equivalent" to an AGP ansatz. As is well known, the formation of Cooper pairs of electrons is accompanied by a negative energy, which—in conventional treatments—is considered to be the "main necessary physical condition" for the formation of the superconducting state. But, as Coleman [28] and Yang [77] have pointed out, the most fundamental aspects of the phenomenon of superconductivity ought to be presented mainly in connection with the wavefunction of this state. As an example, Coleman found that AGP's of extreme type do represent the cases where "coherent pairing" is most intense. In this context, moreover, it is interesting to observe that the explicit form of the interaction Hamiltonian does not appear in the fundamental papers of Yang and Coleman cited previously. (The reason for this is solely that the Hamiltonian is not needed in detail, in order to describe many of the main aspects of superconductivity!)

The preceding remarks may be used to stress the following point: In the case of superconductivity, the ground state is represented by the part $\Gamma_L^{(2)}$, Eq. (5.9), of the total density matrix, where the geminal $|g\rangle$ is extreme. This physical state is clearly associated with a pure quantum state that, additionally, has the lowest possible energy of all the states of the system. In contrast to this, a coherent-dissipative structure is not represented by a pure quantum state (and therefore we do not call it a "state" but a "structure"). Furthermore, the "pairing" $|k, k + s\rangle$ has here a totally different physical meaning (see preceding discussions), and it is associated with a positive (real) energy E. This shows that the coherent part $\Gamma_L^{(2)}$ is associated with a very high energy (being of the order of $E\lambda_L^{(2)}$), and thus there is a vanishing probability at quasiequilibrium for the occupancy of this state. Remember that, in the case of superconductivity, quite the "opposite" holds true.

Thus, the condensed system may occupy the "fully incoherent" part $\Gamma_{Tail}^{(2)}$ (which is of no interest in the present context), or the "small-box part" $\Gamma_S^{(2)}$, which, *after complex scaling*, exhibits a qualitatively new form of cooperative (or synergetic) behavior, which is formally represented by the Jordan block of order s. Also note that the appearance of the Jordan block is intrinsically connected with the specific phase factors $\exp\{-i\beta(\varepsilon_k + \varepsilon_l)/2\}$, see Eqs. (5.14), which the CSM introduces in the formalism of the second-order density matrix.

In more physical terms, one may say that the spontaneous creation of coherent-dissipative structures is due to the dynamical process of quantum transitions, which, as already has been mentioned, do disturb the equilibrium (quantum or classical) correlation patterns of the quantum systems in the condensed macroscopic system. Clearly, these structures are not caused by "noise", and they ought to be considered to belong to the field of quantum dynamics, rather than to the theory of stochastic processes.

It should also be pointed out that the coherent-dissipative structures always have a *finite lifetime*. In contrast to this fact, a superconducting state has an infinite lifetime.

Another (rather formal) feature that might be worth mentioning is the following: In the case of BCS states, the occupation number of any involved pair state is equal to one. In contrast to this, in the present case there is no specification of the expectation value of the occupancy of the units that create the Jordan block.

From the preceding considerations it follows that the coherent-dissipative structures and the superconducting states represent *qualitatively different* organized forms of matter. Furthermore, it should be concluded that the process of creation of coherent-dissipative structures, as discussed previously, differs substantially from the physical processes leading to superconducting or superfluid states. However, also these structures cannot be considered to be localized in the coordinate space.

11. Connection with ODLRO

As we mentioned in Section IV, the extreme condition on the AGP under consideration is a necessary condition for the appearance of Yang's concept of off-diagonal long-range order (ODLRO). We have also shown in Section V B that the appearance of coherent-dissipative structures is intrinsically connected with the existence of *off-diagonal* terms in the thermalized and complex dilated $\Gamma_s^{(2)}$. Therefore, and in order to prevent possible confusion and misunderstanding, it is necessary to stress that coherent-dissipative structures are intimately (although not in the "BCS sense") connected with Yang's ODLRO with respect to the fact that the corresponding density operator cannot be diagonalized.

In many steps of the previous derivations, however, we have tried to illustrate the entity "coherent-dissipative structure" using the term "correlation pattern". In this context, we would like to stress that no intuitively appealing explanation (that is, an explanation in classical mechanical terms) of the emergence of coherent-dissipative structures is possible: As Yang [77] points out: "Since off-diagonal elements [of the density matrix] have no classical analog, the off-diagonal long-range order ... is a quantum phenomenon not describable in classical mechanical terms."

VI. APPLICATIONS TO DYNAMICAL PROCESSES IN CONDENSED MATTER

A. Remarks on "Correlations" and "Mechanisms"

The general CSM theory of quantum correlations, as developed in previous sections, will now be applied to concrete physical processes. Before we proceed to this problem, however, we must take precautions against a source of confusion and misunderstanding, which—as experience shows—appears frequently. This remark concerns the physical significance of the concept of quantum (or EPR) correlations, not only in the context of "theory", but also in the context of "experiment".

Nowadays, of course, there is no doubt of the fundamental character of such correlations. This became obvious during, let us say, the last two decades, with the aid of many important experiments that clearly demonstrated the violation of Bell's inequalities, the best known of these experiments being due to Aspect [45]. These experiments do not concern processes in condensed matter (or, more general, in many-body systems). However, it follows from the principles of modern quantum theory that the correlations under consideration exist everywhere, although in most cases it is very difficult (or even impossible—due to technical limitations) to detect them directly.

The latter remark touches the aforementioned "source of confusion". Let us now consider experiments in condensed matter and stress the important point by asking the following questions:

1. Are quantum correlations necessary for the interpretation of experiments? Probably, many scientists would answer in the affirmative.

2. Are quantitative explanations (or even predictions) of concrete experimental outcomes possible, which are based on the notion of quantum correlations explicitly, but—at the same time—do not make use of the specific form of the interaction mechanism between the microscopic quantum systems? Probably, many scientists would answer in the negative, although the opposite is correct.

For illustration, let us consider the conventional (that is, low-T_c) superconductivity. The interpretation of this effect on the fundamental (or microscopic) level requires the knowledge of the specific interaction between pairs of electrons, that is, of the coupling mechanism leading to the

Cooper-pair formation. (This coupling is given by the so-called Fröhlich or phonon-mediated mechanism.)

However, the formation of the superconducting state cannot be proved on the basis of this mechanism alone. Or, in other words, the appropriate coupling mechanism (of electron pairs) represents a necessary condition for the formation, but not a sufficient one. This is easily recognized by considering the famous BCS theory. The (physical) superconducting state is here represented by the (mathematical) BCS function. Many physical properties of the superconductors are explained extremely well on the basis of the concrete, explicit form of the BCS function, but without using explicitly the specific form of the Fröhlich mechanism. To put it another way, the basic importance of the BCS theory (for low-T_c superconductivity) would remain to a great extent unaffected in a (fictitious) case, where there would be another mechanism responsible for electron pair formation.

From the physical point of view, this remark is by no means surprising, because of the fact that superconductivity is a quantum interference effect on a macroscopic scale, where a huge number of microscopic quantum systems (here Cooper pairs) act cooperatively, thus revealing the "strange" aspects of the microscopic quantum world to our senses. Certainly, the physical nature of this "macroscopic interference" differs significantly from the physical nature of the "pair formation mechanism"—the former term has to do with "correlations" (or also "wavefunctions") and the latter one with "operators" (representing "mechanical interactions").

One can say that quantum-mechanical interactions and quantum correlations represent two aspects of physical reality. For our mind, however, there is an apparent asymmetry: The former has classical analog, whereas the latter has not; see Refs. 77 and 79–85.

In the following parts of this section, correlation aspects—and not interaction mechanisms—of certain concrete physical processes are considered. The physical information being envisaged is thus due to quantum correlations. This is important to stress, because the considered processes are in condensed matter and at "high" temperature; it is often believed that in such cases quantum effects should be "smeared out" by the thermal motion. This is true in the sense that one cannot observe here discrete energy eigenvalues. But quantum correlations may become "detectable" during sufficiently short time intervals.

Since we are dealing with (possibly existing) correlation effects, the specific interaction Hamiltonians are not needed explicitly. This means that the results to be obtained do not depend on the specific forms of these Hamiltonians. For illustration, let us mention two examples:

1. The quantitative relation concerning activation energies of proton transfer reactions in water, presented in Section VI B, is independent

of the question concerning the specific form in which a proton "really exists" in water: H^+, or H_3O^+, or $H_5O_2^+$, and so on.

2. The derived "scaling law" of the very recently observed spin-ordered domains in paramagnetic Gd at temperatures far above T_c, presented in Section VI F, is independent of the actual spin–spin interaction mechanism in this system.

The results to be presented in the following strongly indicate that quantum correlations play a significant role in dynamical processes in condensed matter.

Anticipating the following topics, it may be helpful to make some remarks on the "common scheme" which the applications of the preceding CSM theory exhibit. The general *ansatz* for the actual delocalization (in the geometric space) of microscopic quantum objects is given by

$$\Xi_X = F_X(\hat{H}_{\text{eff}}, \ldots) W_X^{dB} s_{\min, X} \qquad (6.1)$$

The symbols have the following meanings: Ξ represents (i) a transport coefficient or (ii) the geometrical size of a coherent-dissipative structure; X is the specific microscopic quantum system; $s_{\min, X}$ is the "size" (in the space of state functions) of the structure as given by the main formula (5.37); W_X^{dB} is the conventional thermal de Broglie wavelength of a quantum system X; and, finally, $F_X(\hat{H}_{\text{eff}}, \ldots)$ is a functional of the "effective" or "relevant" Hamiltonian $\hat{H}_{\text{eff}}$ being proper for the dynamics of X in the condensed system. F may depend on some external parameters, too.

In the light of the preceding remarks, one should mention that in this *ansatz* the specific "mechanism" being responsible for the microdynamics of X does appear. However, it will become clear that its actual form is often not needed.

And some remarks on the used nomenclature: Hamiltonians, $\hat{H}$, always carry a "hat," in order not to confuse them with protons, H. The thermal de Broglie wavelength is denoted with W^{dB}, because the symbols λ and Λ now represent ionic and molar conductances, as usual in many experimental contexts.

B. Proton Transfer and Proton Delocalization in Water and Aqueous Solutions

1. Introductory Remarks

One of the oldest and most fundamental problems in the physical chemistry of water is the evaluation of the rate constants characterizing the following processes [91]:

$$H_3O^+ + H_2O \xrightleftharpoons{k_1, E_1} H_2O + H_3O^+ \qquad (6.2a)$$

$$OH^- + H_2O \xrightleftharpoons{k_2, E_2} H_2O + OH^- \qquad (6.2b)$$

These processes play an important role in many biological processes, too. With the pioneering work of Meiboom [92] it has been proved that one can measure these reaction rates, k_i, and activation energies, E_i (with $i = 1, 2$) by NMR spectroscopic methods; see following discussion. The study of proton transfer reactions in water is also of importance for the understanding of the excess (or anomalous) conductivities (or mobilities) of the hydronium (H_3O^+) and hydroxyl ions in water and aqueous solutions. Furthermore, in aqueous solutions of acids and bases, fast proton transfer in the water is often considered to be involved as part of the actual reaction scheme; see the classic work of Eigen [93].

Recently Hertz [94] presented a detailed analysis of a series of different experimental methods (NMR, X-ray and neutron scattering, and so on), which—in principle—should be able to detect the H^+ (or the H_3O^+) ion in aqueous solutions *directly*. This analysis, in connection with an extensive discussion of some corresponding experimental results, revealed that none of the considered experimental investigations was thus far able to prove the *real existence* of the so-called H^+ particle in aqueous solution [94].

In all the traditional (and thus well-established) theories of the ionic solutions, however, the entity H^+ is postulated to exist and to correspond to some fast "moving" or "jumping" particle (a proton). In this context let us just mention the very well-known Grotthus mechanism, a traditional model that is believed to explain classically the high excess conductivity of H^+ (and OH^-) in aqueous solutions; see Ref. 95 for a detailed discussion. Nevertheless, the aforementioned exhaustive analysis [94] yields the result that the object usually defined to be the H^+ (or H_3O^+) thus far cannot be considered to represent a particle in the conventional sense. This finding leads Hertz to the conclusion: "...what we call H^+ ion in aqueous solutions is really a dynamical property of the solution."

At this stage, it is important to point out that the analysis of Ref. 94 is carried out entirely within classical mechanics. In this framework, of course, there are no delocalization effects like those being typical for quantum-mechanical processes, and thus the aforementioned conclusion is clearly remarkable. Motivated by the preceding remarks and Ref. 94, it appeared worthwhile to investigate some of the main aspects of the microdynamical behavior of the entity "H^+ ion" with the aid of modern quantum-mechanical theory of condensed matter and, in particular, the CSM combined with the canonical ensemble formalism of quantum statistics; see previous sections. This framework constituted the theoretical basis for the present analysis [35].

Before we proceed to the theoretical treatment of the physical context under consideration, let us briefly stress one reason that makes the quantum-mechanical framework for the description of the H^+ ion necessary. In order to show this, it suffices to observe that the thermal de Broglie wavelength W_H^{dB} of the "quasifree" proton at $T = 300$ K is about 1 Å. Additionally, a very well-known (static) criterion [96, 97] that must be fulfilled before classical mechanics can be applied, requires the following: The mean nearest-neighbor separation R between an H^+ particle and the surrounding water molecules must be much larger than its (thermal) de Broglie wavelength, that is, $R \gg 1$ Å in the present case. Without going into more details, it is clear that this condition cannot be fulfilled in water (or in other condensed systems).

It should also be observed that remarks concerning the "effective mass" of a proton in water cannot change this conclusion significantly: as the well-known high value of the (excess) mobility of H^+ indicates, the effective mass one should attribute to this ion can not have the extremely high value necessary to fulfill the condition $W_{H^+}^{dB} \ll R$.

In this context, it may be interesting to note that there exists an additional *dynamical* criterion for the applicability of classical mechanics, which is even more restrictive. It requires that "the motion of the particles in the liquid should be observed only on a time scale much longer than $\hbar/k_B T$"; for example see Ref. 96. p. 3. It is obvious that this criterion follows from the fundamental energy–time uncertainty relation of quantum mechanics. At room temperature one finds $\hbar/k_B T \approx 2.6 \times 10^{-14}$ s. Some further reasoning shows explicitly that the concepts of "classical particle" and "classical trajectory"—in the considered physical context—strongly violate this dynamical criterion. (For details and numerical examples, see Ref. 98.)

2. H^+ Conductance and Proton Transfer Reactions in the Light of CSM and Coherent-Dissipative Structures

In this section, we present the new CSM approach [35] to the dynamics of the H^+ and OH^- transport in water and aqueous solutions, which connects the following topics and/or measurable quantities:

(a) our CSM theory of quantum correlations and the associated coherent-dissipative structures;

(b) the ionic conductances λ_{H^+} and λ_{OH^-};

(c) the rate constants k_i and activation energies $E_i (i = 1, 2)$ of the acid- and base-catalyzed proton transfer (or exchange) reactions in water, as represented by Eqs. (6.2a) and (6.2b).

Let us start just by noting the following: From the quantum-mechanical viewpoint, and in connection with the aforementioned criterion, it appears that the H constituents forming the H^+ ions are indistinguishable from those belonging to the water molecules being in the vicinity of the ions. (It is surprising that this quantal point of view is even in line with a modern classical analysis of the H^+ conductance mechanism in water [see 94, 95, and 99 and references therein].) In more illustrative (and thus classical) terms, one may say the following: The quantity $W_{H^+}^{dB}$ is, at room temperature, large enough so that one almost always will find water protons in a distance of the order of $W_{H^+}^{dB}$ around each H^+. This leads to the typical delocalization or interference effects or both being characteristic for the quantum theory.

Some of our theoretical results can now be applied—more or less—straightforwardly to the present physical context. We proceed as follows:

(A) A proton of the solution can be considered to be in the state "H^+ ion," represented by

$$|H_i^+\rangle \text{ or simply } |i\rangle \quad (i = 1, \dots, s) \tag{6.3}$$

or in the state "H-water", represented by

$$|H_{i+s}^W\rangle \text{ or simply } |i+s\rangle \quad (i = 1, \dots, s) \tag{6.4}$$

The relaxation process of interest is (classically) represented by the proton transfer reactions (6.2a) and (6.2b). Let us now consider (6.2a). The corresponding proton transfer process can be represented schematically as

$$|H_i^+\rangle \rightleftharpoons |H_{i+s}^W\rangle \text{ or } |i\rangle \rightleftharpoons |i+s\rangle \tag{6.5}$$

Thus, an individual process of the kind (6.5) should be associated with a pair of these "one-body" states, that is, with

$$|H_i^+, H_{i+s}^W\rangle \text{ or } |i, i+s\rangle \quad (i = 1, \dots, s) \tag{6.6}$$

[Let us mention here that the states (6.5) are of fermionic character, whereas the paired states (6.6) can reasonably be considered to be of bosonic character; see Section V.] This "pairing" is physically meaningful, because it is caused by the quantum correlations that intrinsically connect the simple states (6.5) participating in the considered relaxation process. In Schrödinger's terms, one can say that the two simple states appearing in (6.6) are *entangled*; see Refs. 82 and 83. These quantum correlations have a lifetime of the order of the (conventional) relaxation time τ_{rel} of the process under consideration, see Section V and Ref. 20. In the present context, τ_{rel} can be regarded to represent the lifetime of the proton in the state H^+-ion [94]. By standard theory one has

$$\frac{1}{\tau_{rel,H^+}} = k_1[H_2O] \tag{6.7}$$

see Refs. 92 and 100. An analogous formula holds for the process (6.2b).

(B) As was mentioned, our theory shows that these quantum correlations may cause a novel kind of short-lived coherence (or interference) effect, which corresponds to the formation of so-called coherent-dissipative structures in an amorphous condensed system. Here, let us mention that these "structures" represent a cooperative behavior of many degrees of freedom (and, as a rule, also of particles) that is caused by the thermal motion. These "structures" are also of dynamic origin, less specified than a coherent superposition of stationary quantum states, but more organized than the state associated with the completely uncorrelated canonical ensemble; see also Ref. 20.

A coherent-dissipative structure corresponds to a *finite* number s of intrinsically connected (paired) states $|i, i + s\rangle$. The mathematical representation of these structures is given by *Jordan blocks* of order s in the density matrix of the condensed system, and not (as the usual coherence phenomena) by wavefunctions (or representations in terms of stationary states). The important point, however, is that there exists a lower bound s_{min} for s, that is, $s \geqslant s_{min}$, which has been determined by the theory. As was derived in Section V, one has

$$s_{min} = \frac{4\pi kT}{\hbar} \tau_{rel} \tag{6.8}$$

For example, for $T = 300\,K$ and $\tau_{rel} \approx 2 \times 10^{-12}\,s$ (as in the present physical context, see subsequent discussion) one obtains $s_{min} \approx 1000$. This means that (not less than) 1000 of the physical states (6.6) may now interfere and constitute a coherent-dissipative structure. This means that these states become intrinsically connected and act "as a whole" (or as an indivisible unit) during a time interval Δt of the order of τ_{rel}. During Δt, every one of the individual constituents (6.6) of the coherent-dissipative structure loses its individuality, and therefore also its physical significance [20].

(C) Owing to the preceding considerations, one can also state that each H^+ is delocalized over a (high-dimensional) physical space given by

$$\mathscr{S} = \{|i, i + s\rangle | \quad i = 1, \ldots, s\} \tag{6.9}$$

According to previous work [20] we will assume in what follows that $s = s_{min}$.

The aforementioned loss of individuality of the elements of $\mathscr{S}$ thus prevents the detection of an elementary relaxation process (6.5) and, *a fortiori, an individual* H^+ or H_3O^+; see next section.

This leads us to propose the following "picture" for the H^+ conductance in water—in the light of CSM: An H^+ particle is considered to be delocalized over a spatial area occupied by "its" coherent-dissipative structure, that is, by the "structure" that is created by the dynamics of the relaxation process (6.5) of H^+ (see Fig. 3 in Ref. 95). Let this area be characterized by a linear dimension of magnitude d. As was mentioned, this coherent-dissipative structure decays after $\Delta t \approx \tau_{rel}$. If, now, an electric field is applied to the water specimen, then it is reasonable to expect the H^+ ion to "reappear" with larger probability in a position with lower potential energy, after its coherent-dissipative structure dies out. In more illustrative terms: The H^+ ion seems to exist and to follow the field gradient in the conventional manner, as it appears to "jump" over a linear dimension d. The high extra conductivity of H^+ in water is a consequence of the preceding delocalization (and, consequently, of the corresponding "jumping") of the H^+. This "picture," of course, has some intuitive similarity with the standard Grotthus mechanism. (The discussion of this point lies outside the scope of the present article.) Let us mention here that the correlated reorientational motions of some water molecules, which are characteristic for the Grotthus mechanism, do not appear in the present CSM treatment.

The mechanism of the OH^- conductance in water is very similar to that of H^+, that is, the ion can be considered to be in the state OH^--*ion* or in the state OH^--*water*, see Eqs. (6.3) and (6.4). Note that even if both mechanisms, see the classical reaction schemes (6.2a) and (6.2b), are caused by a proton transfer reaction it is the OH^- ions that transport charge and mass to the anode, see Refs. 95 and 101.

(*D*) The physical assumptions (which have the logical status of axioms in the derivation) leading to a quantitative connection between (i) the ionic conductivities λ of H^+ and OH^- and (ii) the proton transfer reaction rates k_i of Eqs. (6.2a) and (6.2b) are the following:

First, the high ionic conductivities under consideration are assumed to be related with the degree of quantum delocalization as given by the corresponding coherent-dissipative structures, see Eq. (6.1). In the present case we have:

$$\lambda_X = F_X(\hat{H}_{eff}, T)W_X^{dB}s_{min, X} \quad (X = H^+, OH^-) \tag{6.10}$$

The linear dependence on the thermal de Broglie wavelength appears as natural, since W^{dB} describes the minimal delocalization of a quantum object as given by standard quantum mechanics. The *linear* dependence on the "size" of the coherent-dissipative structures s_{min}, however, does represent a strong physical hypothesis, which may be tested through comparison with experiments; see subsequent discussion. (Nevertheless, the reader should realize that the arbitrary introduction of some "fractal dimensionality" in (6.10) would be tantamount to the introduction of a fitting parameter.) The functional F depends on $\hat{H}_{eff}$, the relevant (or effective) Hamiltonian [21, 42] for the microdynamical treatment of the proton transfer processes (6.2a) and (6.2b). Since $\tau_{rel} \approx 2\,\text{ps}$, as was mentioned, and thus

$$\tau_{rel} \gg \hbar/k_B T \approx 2.6 \times 10^{-14}\,\text{s} \quad (\text{at } 300\,\text{K})$$

we may expect that $\hat{H}_{eff}$ depends on the degrees of freedom of a large number N of water molecules surrounding each ion. In more plain (that is, classical) terms, one could say that τ_{rel} is large enough—with respect to the quantum-uncertainty time $\hbar/k_B T$—so that a local relaxation process may create long-ranged correlation patterns in its surroundings. Furthermore, $\hat{H}_{eff}$ must be considered to depend on T (and possibly other external parameters) too, because also N is expected to be temperature dependent.

Second, in the present context we may also assume that

$$F_{H^+} = F_{OH^-} \tag{6.11}$$

because of the fact that, in both cases, the greatest part of the system with Hamiltonian $\hat{H}_{eff}$ consists of water molecules, that is, of the same compound. It should be pointed out that Eq. (6.11) represents a physical assumption, which is based on the extension of quantum correlations around each "relaxing center", and thus it is of approximative character. Its validity (or physical significance), however, can be shown with the aid of its experimental consequences, see subsequent discussion.

With the aid of these physical considerations, we obtain from (6.10) and (6.11):

$$\frac{\lambda_{H^+}}{\lambda_{OH^-}} = \frac{d_{H^+}}{d_{OH^-}} = \frac{W_{H^+}^{dB}}{W_{OH^-}^{dB}} \cdot \frac{\tau_{rel,H^+}}{\tau_{rel,OH^-}}$$

Furthermore, with the identity

$$W_X^{dB} = \hbar \sqrt{\frac{2\pi}{m_X k_B T}}$$

and Eq. (6.7) it follows immediately

$$\frac{k_1}{k_2} = \sqrt{\frac{m_{OH}}{m_{H^+}}} \frac{\lambda_{OH^-}}{\lambda_{H^+}} \tag{6.12}$$

which is our main result [35]. To our knowledge, this formula did not exist in the scientific literature. It should be observed that this formula contains measurable quantities exclusively.

(E) The theoretical considerations presented can now easily be extended in order to make a prediction about the activation energies of the proton transfer processes (6.2a) and (6.2b). In the literature it is often assumed that these two processes have a temperature dependence of the Arrhenius-type, that is,

$$k_i = C_i e^{-E_i/RT} \quad (i = 1, 2) \tag{6.13}$$

In the present context, this should be considered as an approximation (or a numerical procedure) in order to fit the NMR data, see Refs. 100–103 and 111. This means that the predictions of our theory do not depend on the explicit form of the temperature dependence of the reaction rates k_i. Thus, the further use of Eq. (6.13) does not imply Arrhenius-like behavior for the ionic conductivities.

Inserting these two relations into the main result (6.12), one immediately obtains the relation:

$$\log\left(\frac{\lambda_{H^+}}{\lambda_{OH^-}}\right) = C + \frac{E_1 - E_2}{RT} \tag{6.14}$$

This represents a second new formula that is derived with the aid of our CSM theory, and which can be directly compared with thus far existing experimental results.

(F) Before the derived novel formulas can be compared with experiment, however, the following point should be observed. In electrochemistry one usually relates the proton transfer rates k_1 and k_2 to the *excess* conductivities (or mobilities)

$$\lambda_{H^+}^e = \lambda_{H^+} - \lambda_{X^+} \tag{6.15}$$

(with $X^+ = K^+$ or Na^+) and

$$\lambda_{OH^-}^e = \lambda_{OH^-} - \lambda_{Cl^-} \tag{6.16}$$

rather than to the "full" conductances λ_{H^+} and λ_{OH^-}; see for example, Refs. 92, 100, 102, and 104. This means in other terms that the "normal" contributions to the considered ionic mobilities—which are associated with the conventional "sizes" of the hydronium and the hydroxyl ions—have to be subtracted. One usually uses Cl^- as the reference anion, and K^+ or Na^+ [92, 100] as the reference cation.

Therefore, the derived new formulas (6.12) and (6.14) of our CSM treatment have to be written more correctly as follows (see Ref. 37):

$$\frac{k_1}{k_2} = \sqrt{\frac{m_{OH^-}}{m_{H^+}} \cdot \frac{\lambda^e_{OH^-}}{\lambda^e_{H^+}}} \qquad (6.17)$$

$$\log\left(\frac{\lambda^e_{H^+}}{\lambda^e_{OH^-}}\right) = C + \frac{E_1 - E_2}{RT} \qquad (6.18)$$

With the aid of the very precise data of Ref. 105 for the ionic conductivities we obtain the predicted values [37]

$$\frac{k_1}{k_2} \approx 1.75 \quad \text{at } T = 25\,^\circ\text{C} \qquad (6.19)$$

and

$$E_1 - E_2 \approx \begin{cases} +2.1\,\text{kJ/mol} & \text{for} \quad X^+ = K^+ \\ +1.9\,\text{kJ/mol} & \text{for} \quad X^+ = Na^+ \end{cases} \qquad (6.20)$$

which follows from the graphical representation of Fig. 2. The positive sign of this difference is in accordance with Ref. 35, where the complete (instead of the excess) ionic conductivities are considered.

One may also observe that the graph in this figure is slightly "curved." The high precision of the conductivity data [105], however, implies that this finding is not due to experimental or numerical errors. [In this connection please remember the approximative character of (6.11) and (6.13) mentioned previously.]

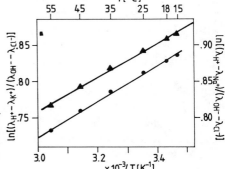

FIGURE 2. Graphical representation of Eq. (6.18). Conductivity data are from Ref. 105. Graph with circles (triangles) corresponds to the left (right) ordinate, with reference cation K^+ (Na^+). See Ref. 37.

3. On the Classical Theory of H^+ Conductance and Proton Transfer Reactions

Before we proceed to the examination of the previously derived predictions, and in order to point out their surprising features, it is instructive to present here the classical mechanical treatment of the connection between the quantities considered. This classical treatment, which is due to Eigen [91] and Meiboom [92, 100], is based on the well-established equations of Nernst

$$\lambda^e = qD/k_B T \tag{6.21}$$

and Einstein

$$D = \langle x^2 \rangle / 6\tau_{rel} \tag{6.22}$$

The notations are as follows: q is the elementary charge, D is a diffusion coefficient describing charge transport due to proton transfers, τ_{rel} is the average lifetime of a H_3O^+ (or OH^-) ion, and $\langle x^2 \rangle$ is the average of the square of the charge displacement accompanying a proton transfer. (In a simple model, one may identify $\langle x \rangle$ with the mean distance between two oxygen atoms of adjacent water molecules; see Ref. 92.)

From these equations and (6.7),

$$\frac{1}{\tau_{rel,H^+}} = k_1 [H_2O]$$

one derives the relation

$$\lambda_{H^+}^e T = Ck_1 \tag{6.23}$$

where C is a temperature-independent constant. A corresponding equation holds true with respect to OH^-. Thus, one obtains

$$\frac{k_1}{k_2} = \frac{\lambda_{H^+}^e}{\lambda_{OH^-}^e} \tag{6.24}$$

From Eq. (6.23) and the aforementioned Arrhenius-type ansatz for the rate constants it follows

$$\log\left(\frac{\lambda_{H^+}^e}{\lambda_{OH^-}^e}\right) = C - \frac{E_1 - E_2}{RT} \tag{6.25}$$

which is clearly in disagreement with the corresponding quantal relation (6.18).

The aforementioned precise data of Ref. 105 for the ionic conductances in water yield the classically predicted value

$$\frac{k_1}{k_2} \approx 2.35 \quad \text{at } T = 25\,^\circ\text{C} \tag{6.26}$$

and

$$E_1 - E_2 \approx -2.0\,\text{kJ/mol} \tag{6.27}$$

for the considered difference of the activation energies [100].

Comparison of the corresponding quantum and classical formulas exhibit surprising differences. As an example, one may compare (6.18) with (6.25), which obviously predict quite the opposite result.

4.   Comparison with Experimental Results

It was mentioned that the reaction rates of (6.2a) and (6.2b) can be measured by different NMR methods. These measurements are based on the following phenomenon: When observing the NMR relaxation times T_1 and T_2 for both protons and ^{17}O nuclei (spin $\frac{5}{2}$) in water containing some ^{17}O, it is found that T_1 is nearly pH-independent while the transverse relaxation time T_2 is not [92, 106]. This effect can be explained qualitatively by the fact that the proton and the ^{17}O are spin–spin coupled. In the domain of slow proton exchange, this is, around pH $\approx$ 7, the shape of both the proton and ^{17}O signals (showing widths proportional to $1/T_2$) are partially collapsed multiplets that are obtained when proton transfer is faster than molecular

TABLE I

Rate Constants ($\times 10^9\,\text{dm}^3\,\text{mol}^{-1}\,\text{s}^{-1}$) and Activation Energies (kJ mol^{-1}) of the Proton Transfer Reactions (6.2a) and (6.2b)

Reference	$k_1{}^a$	k_2	E_1	E_2	k_1/k_2	$E_1 - E_2$
Meiboom [92]	7.1*	3.8			1.8_6	
Loewenstein and Szöke [102]	7.3*	5.5	10.9	19.7	1.3_3	−8.8
Luz and Meiboom [100]			10.0	8.8		+1.25
Glick and Tewari [103]	5.3*	4.5	10.9	11.3	1.1_7	−0.4
Rabideau and Hecht [106]	5.5*	4.6			1.2	
Turner [110]	9.2*	3.1			3.0	
Hertz et al. [107]	6.4	4.5			1.4	
Halle and Karlström [108]	7.1	3.4			2.1	
Average					1.7_2	

$^a k_1$ values marked with an asterisk are transformed to the "normalization" corresponding to Eq. (6.28).

reorientation processes in acid or basic solutions. The measured effect on proton relaxation appears to be nearly proportional to the ^{17}O concentration in the water, this being the reason for using ^{17}O-enriched probes in many experiments.

During the last three decades k_1 and k_2 have been measured repeatedly (mainly around 25 °C) by NMR methods. The corresponding results can be found in Table I. It is interesting to observe that the determined values of these quantities, as well as their ratio, exhibit a considerable scattering, which—following the opinion of some experts in the field—remains still unexplained.

The thus far published experimental data for the difference $E_1 - E_2$ exhibit even a larger scattering: $-8.8\,kJ/mol$ in Ref. 102; $-0.4\,kJ/mol$ in Ref. 103; $+1.25\,kJ/mol$ in Ref. 100; see Table 1. The reasons for this considerable "disagreement" are not known precisely (see the corresponding remark in Ref. 100).

Before we compare the aforementioned theoretical predictions with these experiments, the following detail should be observed. As an example, let us consider the ^{17}O-NMR measurement. In this case one writes the relation

$$\frac{1}{\tau_{\text{rel}}} = k_1 c_{H^+} + k_2 c_{OH^-}$$

$$= k_1 c_H + k_2 \frac{K_W}{c_{H^+}} \tag{6.28}$$

where K_W is the dissociation constant for water; see for example, the recent references 107 and 108. τ_{rel} is the time interval over which a given ^{17}O nuclear magnetic moment experiences a specific value of magnetic energy. Thus this time may be considered to represent the mean lifetime of a proton in the water molecule.

In former publications, however, a factor $\frac{2}{3}$ occurs in front of k_1 in the above formula. This appears naturally within a classical-mechanical picture of the H_3O^+ ion and the exchange of a H^+ (considered as a classical particle) with surrounding water molecules. This factor $\frac{2}{3}$ may be regarded—in this picture—as the probability that reaction (6.2a) will result in the exchange of one of the water protons, because there is a $\frac{1}{3}$ chance that the incoming and outgoing proton will be the same. The situation is similar also in the case of H-NMR measurements. (See Ref. 92 for a detailed discussion.)

In this chapter, the "normalization" of Eq. (6.28) is used, since our quantal treatment does not allow us to accept the physical significance of the preceding classical "distinguishability" of the protons. Therefore, the k_1 values of some former references have been "renormalized" according Eq. (6.28) (see the values marked with a star in Table I).

It should now be observed that the quantitative predictions (6.19) and (6.20) being based on the CSM theory of quantum correlations differ from the classically predicted values (6.26) and (6.27). Besides, the CSM-predicted value (≈ 1.75) for k_1/k_2 at $T = 25\,°C$ appears to be almost equal to the average of the corresponding experimental values. However, the considerable "scattering" of the latter (standard deviation: ca. 0.65) does not permit to refute the classical prediction (≈ 2.35). Additionally, the aforementioned different possibilities for the "normalization" of k_1 let the considered quotient appear even less suitable with respect to the envisaged decision between the two predictions.

The difference $E_1 - E_2$ of the two activation energies is clearly a more suitable quantity for comparison with experiments. Namely, the predicted values (6.20) and (6.27) differ significantly. They also appear to be almost independent of the choice of the reference cation, although the conductances of Na^+ and K^+ differ significantly. Moreover—this being a crucial point—these predictions are independent of the "normalization" of the rate constant k_1. The thus far existing three experimental results (see Table I) do not agree sufficiently. In this context it should be mentioned that the traditional (classical as well as quantal) theories of proton transfer in water (cf. Refs. 102 and 109) favor a negative difference $E_1 - E_2$, which is clearly in contradiction to our CSM-prediction (6.20), Refs. 35 and 37. To our knowledge, there is no theory—up to the present one—predicting a positive value for the difference $E_1 - E_2$.

It might be interesting to note that, from the viewpoint of the present theoretical treatment, the most "accurate" experimental values for k_1/k_2 as well as $E_1 - E_2$ were thus far given by Meiboom [92, 100], see Table I.

The preceding findings clearly indicate that there is a need for more reliable measurements of k_i and E_i ($i = 1, 2$). The derived contradictions between the classical and quantum-statistical points of view initiated, about two years ago, new high precision NMR experiments for the correct measurement of the difference $E_1 - E_2$ and the quotient k_1/k_2 over the whole pH range between 4 and 10 (in the laboratory of H. G. Hertz, Karlsruhe). The first results obtained rule out the large "negative difference" of Ref. 102 definitely. Moreover, the most recent experiments [111] (which are still in progress) in the temperature range $T = 10–65\,°C$ give the result

$$E_1 - E_2 \approx + 1.8\,kJ/mol \qquad (6.29)$$

which seems to agree with the quantum-correlations prediction (6.20), and obviously disagrees with the classical mechanical prediction (6.27).

5. Summary and Additional Remarks

The ionic conductances (or mobilities) of H^+ and OH^- in water and, more general, in aqueous solutions, have been analyzed by our theory [20] that combines the so-called CSM with the canonical ensemble formalism of statistical mechanics. The main formula (6.8), which characterizes the "size" of the coherent-dissipative structures [20], has been used in order to connect

 (i) the kinetic constants characterizing the proton transfer reaction rates of (6.2a) and (6.2b), with
 (ii) the ionic conductivities of H^+ and OH^-

The derived main formula (6.12) contains no fitting parameter, and thus it can be compared with existing experimental results immediately.

The new experiments mentioned will help us to obtain a deeper insight into the dynamics of proton transfer and, *a fortiori*, of H-bonding [112].

Some more special points concerning the underlying general CSM theory may also be mentioned, for reasons of illustration:

It would be interesting to apply our theory to the case of D_2O, too. The *bosonic* character of D^+ let us expect that considerable "deviations" from the preceding results (concerning the fermionic H^+) should become apparent. Indeed, some experimental results on D^+ transfer in D_2O (Ref. 108) seem to be in line with this qualitative expectation. It should be observed, however, that D-NMR experiments are much more difficult to carry out than the corresponding H-NMR experiments (owing to the smallness of the appropriate spin–spin coupling constant in the former case). In view of this fact and, at the same time, of the considerable scattering of experimental data even in the case of the "easier" H-NMR experiments, there seems to be at present little chance of obtaining reliable and reproducible data for D^+-transfer in D_2O.

The present analysis of the proton transfer mechanism in water with the aid of CSM thus gives the physical rationale for using γ as described above. It is then obvious that the possibility of finding a Jordan block rests on a coupling between two possible degenerate quantum states of the same system via time-dependent correlations, although, as was pointed out, the amorphous system in other respects exhibits no particular symmetry or long-range order. Hence one could say that CSM introduces a coupling between a *finite* number of degrees of freedom appearing in the canonical (reduced) density matrix and *infinitely* many degrees of freedom of the environment or thermal bath. More details on this aspect are given in the previous sections of the article.

The quantity s_{min} determines the minimal "size" that the Jordan block must exhibit, and it represents the site of the new "unit" that we call a coherent-dissipative structure. As we discussed in Section V, one cannot attribute the usual probability interpretation to the individual elementary relaxation processes contributing to a coherent-dissipative structure. There is also the question whether the description refers to electronic or nuclear motion or both. Generally speaking the dynamical features of the nuclei are given by the potentials associated with the corresponding electronic distributions. Hence, there is necessarily a correlation between the nuclear and electronic degrees of freedom, which is playing a vital role in the conductance measurement affecting both charge and mass [101].

C.   Experimental Evidence for H^+ and D^+ Delocalization in H_2O/D_2O Mixtures

1.   A New Experiment for the Detection of Proton Delocalization: Anomalous Decrease of H^+ Mobility in H_2O/D_2O Mixtures

Further investigations on the H^+ (and D^+) conductance, based on the CSM theory presented in the previous subsection, have predicted an *anomalous decrease* of this measurable quantity in H_2O/D_2O mixtures [37]. This prediction was surprising, because, as will be discussed, it is in contradiction to every known electrochemical theory or model. (An equivalent statement holds for the OH^- conductivity, too.)

The physical considerations leading to this prediction are as follows:

As already mentioned in Section VI B, quantum correlations between "protons" in aqueous H^+ solutions may be expected even within conventional quantum theory, namely due to the large thermal de Broglie wavelength of a quasifree H^+ that is about $1\,\text{Å}$ at room temperature. (See also the corresponding remarks concerning the "effective mass" of H^+ in water.) In more concrete terms, the underlying physical idea is that H^+ is delocalized and EPR correlated with water protons of its surroundings, "belonging" to H_2O or HDO molecules. This physical picture is clearly in contrast to the viewpoint taken by quantum chemistry and molecular dynamics, where protons are considered as classical particles being subject to the Born–Oppenheimer approximation.

The following point is now crucial [37]. If the well-known high H^+ conductance, λ_{H^+}, in liquid water is caused by the assumed specific quantum interference effects (that is, the coherent-dissipative structures), then there must be an anomalous decrease of λ_{H^+} in H_2O/D_2O mixtures due to the so-called mass and spin superselection rules (see Ref. 82). In these mixtures, namely, the possible quantum interference between appropriate protonic states becomes disrupted by deuterons ("belonging" to D_2O, HDO, or D^+

ions and) being "near" or "between" the considered protons. For exactly the same reasons we also may expect an anomalous decrease of the D^+ conductance in H_2O/D_2O (cf., however, Ref. 101).

A crude estimation of the decrease of λ_{H^+} in an 1:1 mixture H_2O/D_2O showed that the predicted effect should be of the order of some percent of the corresponding conductivity in pure water [37].

[The oversimplified, illustrative model leading to this estimate is the following: First, let us consider the water molecules as classical bodies, and a volume V of liquid water. Let V_O and V_H be the volume parts being occupied by oxygen and hydrogen atoms, respectively. Then $V = V_O + V_H$. Second, in the 1:1 mixture of H_2O/D_2O one has correspondigly $V = V_O + V_H^* + V_D^*$ with $V_H^* = V_D^*$. The coherent-dissipative structure around an H^+ may be imagined as a sphere of radius r, in the case of pure H_2O, and r^*, in the mixture. The disruption of the structures by some D atoms in the mixture leads to the following simple relation:

$$\frac{r^*}{r} = \left(\frac{V_O + V_H^*}{V_O + V_H}\right)^{1/3} \equiv \left(\frac{V - V_D^*}{V}\right)^{1/3}$$

Furthermore, and in line with the general theory, the H^+ mobility is assumed to be proportional to the spatial size r or r^* of the coherent-dissipative structures. Thus, the model gives the following values:

1. For, let say, $V_O : V_H = 1:1$ one has $r^*/r \approx 0.91$ and thus an "anomalous decrease" of the H^+ mobility (or conductance) of about -9%.

2. For, let say, $V_O : V_H = 4:1$ one has $r^*/r \approx 0.97$ and thus an "anomalous decrease" of about -3%.

See the experimental results, Eqs. (6.32a) and (6.32b).]

In order to test this prediction of our CSM theory [35], an adequate experiment has been conceived and carried out very recently [39]. In this experiment we measured molar conductivities, Λ, of different HCl/DCl solutions in H_2O/D_2O mixtures. The experimental setup, the results, and their analysis are discussed in the following.

It will be demonstrated in the following parts of this subsection that the experimental results clearly confirm the predicted effect. This finding may have far reaching consequences for the further understanding of the microstructure and some microdynamical processes in water and aqueous solutions (like the dynamics of H-bonding), because of the revealed quantum-mechanical character of nuclear degrees of freedom participating significantly in the physical properties of these systems.

2. Experimental Details

Owing to the possible importance of the effect under consideration, certain details concerning the experimental procedure and the precision of the measurements [39] should be mentioned.

For solutions in normal water, deionized and distilled H_2O with a specific conductance of about $1 \times 10^{-6} \, S \, cm^{-1}$ was used. A stock solution of HCl in H_2O with a molar concentration of slightly above $0.5 \, mol \, dm^{-3}$ was obtained by diluting concentrated aqueous HCl. The exact concentration was determined by potentiometric titration of H^+ against NaOH and of Cl^- against $AgNO_3$. By adding H_2O, the concentration of this solution was then adjusted exactly to $0.5 \, mol \, dm^{-3}$. The conductance of the adjusted stock solution agreed to within the present experimental error ($\pm 0.05\%$) with the value interpolated from the data of Stokes [113]. More dilute solutions were obtained by weight dilution of the stock. Density values of Ref. 114 were used in the calculations.

Solutions of DCl in D_2O were prepared using the same procedure. The D_2O used (D content > 99.99 at%) had a specific conductance less than $2 \times 10^{-6} \, S \, cm^{-1}$. The used DCl was obtained as a 36 wt% solution in D_2O (D content > 99.9 at%). In the calculations the density of pure D_2O was taken from Kell's compilation [115], and the partial molar volume of HCl was assumed to be the same in H_2O and D_2O, which allowed one to convert concentration units from wt% to molarity. This assumption was proved by representative density measurements.

Solutions of KCl in H_2O and D_2O with desired concentrations (see subsequent discussion) were prepared.

Mixed solutions of the desired D content (that is, D atom fractions $X_D = 0.25$, 0.5, and 0.75) were prepared by mixing HCl/H_2O solutions with DCl/D_2O solutions of the same molarity. The same procedure was applied for obtaining solutions of KCl in H_2O/D_2O mixtures. The small excess volume upon mixing was assumed to be the same as in salt-free H_2O/D_2O mixtures [116].

Electrical conductances were measured with a Wayne–Kerr B-905 bridge as described elsewhere [113]. Cells were of the design described in Robinson and Stokes' monograph [105] with cell constants ranging from 30 to $300 \, cm^{-1}$. Duplicate measurements in cells with different cell constants agreed to better than $\pm 0.05\%$. Measurements were performed at four frequencies, f, between 400 and 10,000 Hz, and if necessary, minor frequency corrections (less than 0.05%) were performed by extrapolation of the resistance against the inverse frequency ($f \to 0$). The overall experimental accuracy of the conductance measurements is $\pm 0.05\%$. All data presented below are based on Jones–Bradshaw standards [105], converted from $(int \, \Omega)^{-1} cm^{-1}$ to $(absolute \, \Omega)^{-1} \, cm^{-1}$, that is, $S \, cm^{-1}$.

For density measurements, an "Anton Paar" vibrating tube densimeter was used. The corresponding accuracy was $\pm 2 \times 10^{-5} \, \text{g cm}^{-3}$.

3. Experimental Results

Table II presents the experimental results for the molar conductance Λ of HCl/DCl in H_2O/D_2O mixtures at 25 °C. The deuterium content of the solutions is characterized by the mole fraction X_D. C is the molar concentration of HCl/DCl. More digits than significant have been retained in this table as the "internal consistency" of the data is higher than their absolute accuracy. We determined the maximum error of the data to be smaller than $\pm 0.1\%$.

We consider first the results for the conductances in pure H_2O and D_2O, respectively. By forming the ratio of the conductances at $X_D = 0$ and 1, respectively, one finds an average factor $\Lambda(\text{HCl}/H_2O)/\Lambda(\text{DCl}/D_2O) = 1.364$, which—within the present experimental precision—is independent of the acid concentration. Frivold et al. [117] report the value 1.362 for this factor. Other (presumably less accurate) data can be found in the compilation of Gmelin [118]. By extrapolating we obtain a limiting conductance, Λ^0, of DCl in D_2O at 25 °C of 312.7 s cm^2 mol^{-1}, which, with the limiting

TABLE II
Molar Conductances of HCl/DCl in H_2O/D_2O^a

C	Λ				
(mol dm^{-3})	$X_D = 0$	0.25	0.5	0.75	1
0.50	361.07	325.68	296.27	276.13	264.90
0.40	367.36	330.78	301.51	280.92	269.16
0.30	374.10	336.86	307.06	286.15	274.11
0.20	381.71	343.85	313.31	292.01	279.96
0.10	391.49	352.70	321.66	299.51	286.92
0.05	399.30	359.72	328.41	305.30	292.70
0.02	407.31	366.92	335.00	311.61	298.62
0.01	412.06	371.04	338.56	315.18	302.15
$C \to 0$	426.3	384.5	350.5	326.2	312.7
$\Delta\Lambda(X_D, C \to 0)$	—	−13.4	−19.0	−14.9	—
$\Delta^{\text{rel}}(X_D, C \to 0)$	—	−3.4%	−5.1%	−4.4%	—

^{a}Molar conductances (in S cm^2 mol^{-1}) of HCl/DCl in H_2O/D_2O mixtures at 25 °C as a function of solvent composition (expressed by the atom fraction X_D of D) and of acid concentration C (in mol dm^{-3}). $\Delta\Lambda(X_D, C \to 0)$ is the deviation of the molar conductance at infinite dilution of the acid from the value calculated by linear interpolation between the limiting conductances of the two pure solutions (with $X_D = 0$ and $X_D = 1$). $\Delta^{\text{rel}}(X_D, C \to 0)$ is the corresponding relative deviation in percent; see Ref. 39.

conductance of $62.83 \, \text{S cm}^2 \, \text{mol}^{-1}$ for Cl^- in D_2O, yields a limiting conductance of the D^+ ion in D_2O of $249.9 \, \text{S cm}^2 \, \text{mol}^{-1}$. This compares well with the value reported by Longsworth and McInnes [119] (that is, $250.0 \, \text{S cm}^2 \, \text{mol}^{-1}$) for a mixture containing 99.35% D_2O.

To estimate the magnitude of the excess conductance of D^+, the preceding value may be compared with the corresponding one for K^+ in D_2O, which is $61.40 \, \text{S cm}^2 \, \text{mol}^{-1}$, Ref. 120. The ionic conductance λ_{H^+} of H^+ in H_2O at 25 °C is $350.0 \, \text{S cm}^2 \, \text{mol}^{-1}$, Ref. 105. The conductances of the alkali metal ions in D_2O differ from those in H_2O by about a factor between 1.19 and 1.21 [120], which is close to, but not identical with, the ratio between the viscosities of D_2O and H_2O, the latter being equal to 1.23 [121]. Based on these ratios and the well-known conductance of H^+ in H_2O, one could expect a value between 285 and $295 \, \text{S cm}^2 \, \text{mol}^{-1}$ for the D^+ conductance in D_2O. This is markedly higher than the preceding experimentally observed value ($= 249.9 \, \text{S cm}^2 \, \text{mol}^{-1}$). Note that the latter value is still four times larger that the limiting conductance of K^+ in D_2O.

Our main interest, however, is focused on the ionic conductances of H_2O/D_2O mixtures. For reasons to be seen below we define

(i) the deviation of the measured conductance of a mixture, $\Lambda(X_D, C)$, from that being determined by linear interpolation between the values of the two pure solutions ($X_D = 0$ and 1), $\Lambda_{lin}(X_D, C)$, by the equation

$$\Delta\Lambda(X_D, C) = \Lambda(X_D, C) - \Lambda_{lin}(X_D, C)$$
$$= \Lambda(X_D, C) - [(1 - X_D)\Lambda(0, C) + X_D\Lambda(1, C)] \qquad (6.30)$$

as well as

(ii) the corresponding relative deviation in percent given by

$$\Delta^{rel}(X_D, C) \equiv 100\Delta\Lambda(X_D, C)/\Lambda_{lin}(X_D, C) \qquad (6.31)$$

The analysis of the experimental results shows that $\Delta^{rel}(X_D, C)$ is independent of the acid concentration, so that convenient extrapolations can be made to obtain the limiting conductances of HCl/DCl in H_2O/D_2O mixtures, which are also given in Table II. These results are in good agreement with corresponding data at $0.05 \, \text{mol dm}^{-3}$ being reported in Ref. 119. Other data, for example, see Ref. 118, are also in qualitative agreement with the present experimental findings [39].

Figure 3 shows the resulting conductances at infinite dilution, Λ^0, plotted against the mole fraction X_D. It is seen that at intermediate solvent compositions the curve lies distinctly below the straight line connecting the limiting values in pure H_2O and D_2O. The relative deviation at the equimolar

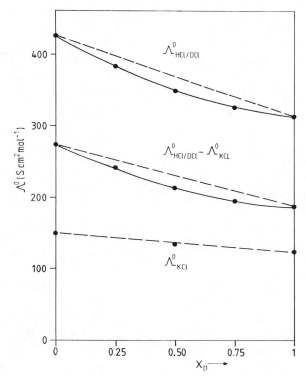

FIGURE 3. Molar conductances of HCl/DCl, $\Lambda^0_{\text{HCl/DCl}}$, and of KCl, Λ^0_{KCl}, in H_2O/D_2O mixtures at infinite dilution and $T = 25\,^\circ\text{C}$ as a function of the mole fraction X_D of D. Also shown in the excess conductance as determined from the difference between the data for HCl/DCl and KCl. For details see the text and Ref. 39.

solvent composition is

$$\Delta^{\text{rel}}(0.5, C \to 0) \approx (-5.1 \pm 0.1)\% \qquad (6.32a)$$

The small experimental error of $\pm 0.1\%$ strongly supports the significance of this result. The deviations $\Delta\Lambda(X_D, C \to 0)$ and $\Delta^{\text{rel}}(X_D, C \to 0)$ are given in Table II. It also appears that in the D_2O-rich region these deviations are slightly larger than in the H_2O-rich regime, but the corresponding effects are at the limits of the present experimental accuracy.

It should be pointed out that the considered relative decrease at the equimolar solvent composition, $\Delta^{\text{rel}}(0.5)$, is always about -5%, that is, independently of the solute concentration C. (This result follows immediately

TABLE III

Molar Conductances of KCl in H_2O/D_2O Mixtures at 25 °C[a]

C	Λ		
(mol dm^{-3})	$X_D = 0$	0.5	1
0.1	129.02	117.31	107.48
0.05	133.43	121.23	110.73
0.01	141.33	128.23	117.19
$C \to 0$	150.0	136.06	124.3
$\Delta\Lambda(X_D, C \to 0)$	—	−1.09	—
$\Delta^{rel}(X_D, C \to 0)$	—	−0.8%	—

[a]The notations correspond to those in Table II.

from the data of Table II.) This finding supports—at least indirectly—the high accuracy of the presented experimental data, and it indicates the possible importance of the effect with respect to other dynamical processes in water and aqueous solutions.

The conductivity data for KCl in H_2O/D_2O mixtures are presented in Table III and Fig. 3. Our results are in good agreement with the well-known [120] values of the conductance of KCl in pure H_2O and pure D_2O. In an equimolar H_2O/D_2O mixture (that is, at $X_D = 0.5$) the observed deviation of the limiting conductance Λ^0_{KCl} from the corresponding linearly interpolated value is only slightly negative, that is, about −0.8%. This result corresponds excellently

1. to the only slightly positive deviation (of about +0.95%) in the viscosity [121] of this mixture and
2. to the small deviation of self-diffusion coefficients [122].

It should also be observed that the considered almost linear dependence of the molar conductance of KCl on X_D appears to be independent of the concentration C; see Table III.

In the light of the considerations of Section VI B, however, one should investigate the excess molar conductance, Λ_{excess}, which is conventionally given by the difference of the data obtained for HCl/DCl and KCl. (Note that this is a widely used procedure in electrochemistry.) The resulting curve for infinite dilution is also included in Fig. 3. The excess conductance at $X_D = 0.5$ is now −7.7% smaller than that determined from linear interpolation between the corresponding conductances in pure H_2O and pure D_2O, that is,

$$\Delta^{rel}_{excess}(0.5, C \to 0) \approx (-7.7 \pm 0.1)\% \qquad (6.32b)$$

For illustration, the following point may also be observed: The "disrupted quantum interference" mentioned previously is easily seen to become most effective at $X_D = 0.5$, thus leading to a maximum of the predicted anomalous decrease of the H^+/D^+ conductance at $X_D = 0.5$. The experimental data confirm this expectation, too.

In conclusion, we found that the conductances of KCl solutions in H_2O/D_2O mixtures are completely in accord with standard electrochemical theory. But, at the same time, it is the almost

- linear dependence of Λ_{KCl}^0 on the mole fraction X_D

in connection with the

- linear dependence of the fluidity of H_2O/D_2O on X_D [121]

that proves that the observed decrease $\Delta\Lambda^0$ of the molar conductance of HCl/DCl is a specific property of the H^+ and D^+ ions and the H_2O/D_2O solvent.

Therefore, it follows that the considered effect cannot be interpreted in terms of standard (or classical) electrochemistry. At the same time, the presented experimental results are in line with the underlying physical picture [35] of H^+ delocalization in water and clearly confirm the prediction [37] of the anomalous decrease of the H^+/D^+ conductance in H_2O/D_2O mixtures.

It should be noted that the preceding experimental findings suggest that this effect may play an important role in the dynamics of H^+ transport and H bonding formation dynamics [112] in further physical, chemical, and biological systems.

D. Further Proposed Experiments on H^+ Delocalization

1. H^+ Conductivity of $H_2{}^{16}O/H_2{}^{18}O$ Mixtures

The measurement of the ionic conductivity λ_{H^+} of H^+ in mixtures of $H_2{}^{16}O$ with $H_2{}^{18}O$ (or in $H_2{}^{16}O/H_2{}^{17}O$ mixtures) would be interesting in the present context, for the following reasons.

Thus far, the possible delocalization of *protons* in aqueous solutions has been considered. In the well-known model of Eigen and de Maeyer [91, 93] for the proton and hydroxyl ionic conductances in water, however, the ionic complexes $H_9O_4^+$ and $H_7O_4^-$ play a crucial role. Thus, one might be inclined to extend the preceding considerations—in the light of the successful experiments described in the previous section—by asking if even quantum delocalization of *water molecules* in these ions may appear. If this holds true, then the aforementioned conductivity measurements will exhibit an "anomalous" decrease of λ_{H^+}; see Fig. 4. Namely, the reasoning is almost the same as in the previous experimental context (see the previous section),

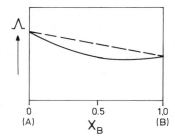

FIGURE 4. Schematic representation of HCl (or NaOH) molar conductivity in water mixtures A/B. Here, $A = H_2{}^{16}O$, and $B = H_2{}^{18}O$ or $= H_2{}^{17}O$. Brocken line: Conductance as expected by standard theories. Solid line: Anomalous decrease of conductance, if quantum correlations of the discussed nature between water molecules (and not only protons) would be physically significant. see Ref. 37.

and it is based on the fact that possible quantum states of the "molecules" $H_2{}^{16}O$ and $H_2{}^{18}O$ will be subject to the mass superselection rule that prevents their mutual interference in the isotopic mixtures under consideration. Because of this, the spatial extension of the coherent-dissipative structures around the ionic complexes $H_9O_4^+$ (or $H_7O_4^-$) is then expected to exhibit a corresponding decrease in the considered mixtures, thus leading to an anomalous decrease of λ_{H^+}, too [37].

The preceding physical considerations also imply that this decrease could be expected to be of the same magnitude as the corresponding one already measured in H_2O/D_2O mixtures.

2.   On Light Scattering Experiments

The proposed experiment [37] to be discussed in this section is of a different type. It was motivated by the following considerations.

It has been proved in Section V (and in Ref. 20) that coherent-dissipative structures correspond to Jordan blocks γ with

$$\text{Tr}\,\gamma = 0$$

This mathematical result has an important physical consequence. It implies that the measurable average

$$\langle E \rangle = \frac{\text{Tr}(\Gamma E)}{\text{Tr}\,\Gamma} \tag{6.33}$$

(where Γ is the full second-order density matrix, after proper thermalization and complex scaling) over an appropriate physical quantity E is expected to contain no contribution from the Jordan blocks γ, because of the facts that (1) the latter are not normalizable and (2) it holds trivially

$$\langle E \rangle < \infty \tag{6.34}$$

Now let **E** represent the electric field component of a continuous or pulsed laser field acting on an appropriate H^+ or OH^- aqueous solution. If coherent dissipative structures (with a lifetime of about 2 ps [35]) do really exist in the solutions, then Eq. (6.34) implies physically that the laser light field must be "expelled" from the spatial domains being occupied by these structures. In more formal terms, we should assume the validity of the equation

$$Tr(\gamma E) = 0 \qquad (6.35)$$

in order to be able to fulfill the necessary physical condition (6.34). It follows furthermore that the corresponding part of the scattered light should contain a component that exhibits non-Poissonian statistics.

This may be illustrated and specified a little more by the following concrete experimental situation. Let us assume that one is able to measure the cross correlation between the scattered light intensity in two *antiparallel directions*. The cross correlation is then expected to exhibit an enhancement of "simultaneously" scattered photons [37]. This enhancement is expected to be due to light scattering on coherent-dissipative structures. Namely, one easily sees that the aforementioned simultaneous antiparallel scattering is—more or less—the simplest way to fulfill the condition (6.35) over the spatial domains being occupied by coherent-dissipative structures. (An experiment of this type is now under preparation in Berlin.)

E. Quantum Correlations in Molten Alkali Chlorides

1. Introductory Remarks

Recently the physical reasons for the high electric conductivities of molten alkali halides have been considered and further analyzed (also in connection with silver halide melts); see Refs. 123 and 124 and references cited therein. In particular, it has been pointed out that these conductivities are at least 10^5 times larger than those of the corresponding crystals, and that no one of the well-known "free-volume" theories of the melts can explain this fact. Recent additional conductivity measurements under high pressure confirmed this statement exactly; see Ref. 123 for details and further references. Furthermore, it has been concluded that "the absence of long range order is the property of the molten state which makes charge transport possible. The elementary step then should be a concerted motion of particles belonging to different coordination shells" [123].

Motivated by this analysis, as well as further considerations in that reference, we recently applied our CSM theory of quantum correlations to this physical problem [36]. This theoretical work has a strong conceptual connection with the CSM treatment of H^+ and OH^- conductivity in aqueous solutions presented in the Sections VI B and VI C. The ionic melts are described by means of coherent-dissipative structures. These structures—in the present physical context—are assumed to be caused by quantum correlations of delocalized ions in the melt [36].

Within the CSM treatment, a novel relation concerning the considered molar conductivities is derived, and its connection with experimental results as well as "practical" importance is demonstrated.

2. Coherent-Dissipative Structures and Ionic Conductance of Molten Salts

In the light of the preceding remarks, it appears natural to apply our quantum theory of coherent-dissipative structures [20] to the present physical problem. As in the case of the H^+ and OH^- ionic mobilities in aqueous solutions (see preceding discussion), the theoretical treatment is based on the main formula (5.37), that is,

$$S_{min} = \frac{4\pi k_B T}{\hbar} \tau_{rel}$$

which gives the minimal dimensionality of a coherent-dissipative structure in the physical space of paired states

$$|k, k+s\rangle \quad (k=1,\ldots,s)$$

see Section V. In the present physical context, the "single" states $|k\rangle$ and $|k+s\rangle$ represent two possible states of a system consisting of one ion X ($= Li^+$, Na^+, K^+, Rb^+, Cl^-) taken together with the relevant part of its environment in the melt. The underlying microdynamical process, exhibiting the characteristic relaxation time τ_{rel}, consists—plainly spoken—in the temporal change of the ionic correlation patterns to which X participates, due to the thermal motion:

$$|k\rangle \rightarrow |k+s\rangle$$

Here it is clear that τ_{rel} may also be considered to represent the characteristic decay time of *local charge fluctuations* in the melt [123]. The paired state $|k, k+s\rangle$ expresses then the (almost always) existing quantum correlations between the displayed individual "single" states.

Now the application of the preceding theory becomes straightforward [36]. For the "ionic conductance" λ_X of an ion X in the melt we make the same *ansatz*, Eq. (6.1), as in the case of H^+ mobility in water, namely,

$$\lambda_X = F_X(\hat{H}_{eff}, T, p) W_X^{dB} s_{min,X} \quad (X = \text{cation}, Cl^-) \qquad (6.36)$$

For the relaxation time τ_{rel} appearing in s_{min} we may assume

$$\tau_{X^+} = \tau_{Cl^-} \equiv \tau_{XCl} \tag{6.37a}$$

and make the assumption

$$\tau_{XCl} \equiv \tau_{rel} \tag{6.37b}$$

for each melt (LiCl, NaCl, KCl, RbCl, CsCl). The physical reason for this is given by the fact that the microdynamical behavior of both the cations X^+ and anions Cl^- has the same physical origin, namely, the aforementioned temporal development of the local charge fluctuations (around these ions) in the melt. Of course, τ_{rel} is also expected to depend on temperature and pressure, since the considered charge fluctuations are intrinsically connected with the aforementioned correlation patterns [36] to which the ions belong.

For the same reasons it is physically justified to assume the equality

$$F_{X^+} = F_{Cl^-} \equiv F_{XCl}, \tag{6.38}$$

since the relevant (or effective) Hamiltonian for the different ions is here the same. Owing to the fact that $\hat{H}_{eff}$—in the present physical context—is mainly determined by the strong and long-ranged Coulomb interactions, and less by the different weaker "corrections terms", it may be expected that the functionals F_{XCl} do not strongly depend on the specific melts XCl (X = cation). Of course, Eqs. (6.37) and (6.38) hold approximatively.

The contributions λ_{X^+} and λ_{Cl^-} of the two ionic species to the molar conductivity Λ_{XCl} of the molten salt XCl (X = Li, Na, K, Rb, Cs) are directly dependent on the existence of the aforementioned coherent-dissipative structures.

The individual quantities λ_X, however, have no well-defined physical meaning, since they cannot be measured separately. Of course the identity

$$\Lambda_{XCl} = \lambda_{X^+} + \lambda_{Cl^-} \tag{6.39}$$

holds per definition (where X = Li, Na, K, Rb, Cs). Furthermore, with the aid of Eqs. (6.36–6.38) one immediately obtains:

$$\Lambda_{XCl} = F_{XCl}(W_{X^+}^{dB} + W_{Cl^-}^{dB}) \frac{4\pi k_B T}{\hbar} \tau_{XCl}$$

$$= F_{XCl} \sqrt{32\pi^3 k_B T} \left[\frac{1}{\sqrt{m_{X^+}}} + \frac{1}{\sqrt{m_{Cl^-}}} \right] \tau_{XCl}$$

which can be rewritten as

$$\Lambda_{XCl}^* \equiv \frac{\Lambda_{XCl}}{(m_{X^+})^{-1/2} + (m_{Cl^-})^{-1/2}}$$

$$= C_{XCl}(T,p)\tau_{XCl} \qquad (6.40)$$

The result (6.40) connects the measurable molar conductivities with the characteristic relaxation times of the charge fluctuations of the melts, the latter being considered to represent the physical reason for the spontaneous creation of the short-lived coherent-dissipative structures in the amorphous melt.

3. Comparison with Experimental Results

The preceding main formula can now be further analyzed in connection with existing precise experimental results on the molar conductivity of molten alkali chlorides [125, 126] in their common temperature range, that is, $T = 800$—$900\,°C$. Now the left-hand side of Eq. (6.40) can be determined.

For $T = 800\,°C$, the results presented in Table IV reveal the following remarkable finding:

All the considered *scaled molar conductivities* Λ_{XCl}^* exhibit very similar magnitudes, the corresponding maximal variations of these values being equal or less than 4%.

This surprising finding is clearly confirmed by the accurate conductivity

TABLE IV
On the Constancy of the Scaled Molar Conductivities Λ_{XCl}^*, Eq. (6.41), of alkali chlorides at $T = 800\,°C^a$

	M [g mol^{-1}]	κ [Ω^{-1} cm^{-1}]	ρ [g cm^{-3}]	Λ_{XCl} [Ω^{-1} cm^2 mol^{-1}]	Λ_{XCl}^* [Ω^{-1} cm^2 mol^{-1} (g/mol)$^{1/2}$]
LiCl	42.4	6.7	1.43	198.5 (197.5)	362 (360.5)
NaCl	58.45	3.5	1.55	132.2 (134.2)	350 (356)
KCl	74.55	2.2$_2$	1.48	111.5 (110.5)	339 (337)
RbCl	120.92	1.8	2.17$_5$	100.1	363
			Mean value		353.5 ± 3%

^{a}Conductivity and density data are taken from ref. 125; data in parentheses refer to Ref. 126.

TABLE V
On the Constancy of the Scaled Molar Conductivities Λ^*_{XCl} at $T = 900\,°C^a$

	κ $[\Omega^{-1}\,cm^{-1}]$	ρ $[g\,cm^{-3}]$	Λ_{XCl} $[\Omega^{-1}\,cm^2\,mol^{-1}]$	Λ^*_{XCl} $[\Omega^{-1}\,cm^2\,mol^{-1}\,(g/mol)^{1/2}]$
LiCl	6.9_2	1.39	$211._6$	$386._5$
NaCl	3.8_4	1.50	$149._6$	397
KCl	2.4_6	1.41	$129._9$	396
RbCl	1.9_2	2.09	$111._4$	$403._5$
			Mean value	$396 \pm 2\%$

^{a}Conductivity and density data are taken from Ref. 125.

and density data [125] for the considered alkali chlorides at $T = 900\,°C$; see Table V. Here the scaled molar conductivities exhibit a similar small variation with standard deviation of 2%.

As the experimental accuracy and the reading errors for the molar conductivities (from the data of Refs. 125 and 126) are of the same order of magnitude, the conclusion is permitted that

$$\Lambda^*_{XCl} = \text{const} \quad \text{for all } X\text{Cl} \tag{6.41}$$

The preceding remark on the approximative independence of the functionals F_{XCl} with respect to the specific salt XCl, that is,

$$F_{XCl} \approx F, \quad \text{for all } X\text{Cl} \tag{6.42}$$

implies then the relation

$$\tau_{XCl} \approx \text{const} \equiv \tau \quad (\text{all } X\text{Cl}) \tag{6.43}$$

(with $X = $ Li, Na, K, Rb), that is, that the aforementioned charge fluctuations exhibit (at any given temperature) the same relaxation time τ in all melts.

Furthermore, an inspection of the interpolation formulas of the handbook [127] for the molar conductivities considered, at the above temperatures, reconfirms the result (6.41); see Table VI. It should be observed that the common temperature range of these interpolated molar conductivities almost coincides with the considered range $T = 800$—$900\,°C$ (the corresponding data are contained in Tables I.B.1 and III.B.2 of Ref. 127). Note also that the variations of the numerical values of the present scaled molar conductivities (standard deviation = 3%) are comparable with the corresponding variations presented in Tables IV and V.

TABLE VI

Approximative Values of Molar Conductivities and Scaled Molar Conductivities of Alkali Chlorides at $T = 800\,°C$ and $T = 900\,°C^a$

	Λ_{XCl} $[\Omega^{-1}\,cm^2\,mol^{-1}]$		Λ^*_{XCl} $[\Omega^{-1}\,cm^2\,mol^{-1}\,(g/mol)^{1/2}]$	
	$T = 800\,°C$	$T = 900\,°C$	$T = 800\,°C$	$T = 900\,°C$
LiCl	197.5	214.1	361	391
NaCl	134.0	151.0	356	401
KCl	110.5	126.6	337	386
RbCl	95.8	114.2	347	413.5
		Mean values	$350 \pm 3\%$	$398 \pm 3\%$
CsCl	100.3	123.1	394	487
	Deviations from mean values		$+ 13\%$	$+ 22\%$

^{a}Data are calculated with the interpolation formulas of Ref. 127, Table III.B.1. (The maximal common temperature range for the validity of these interpolation formulas is about $T = 800–900\,°C$; see Table I.B.1 and III.B.2 of Ref. 127.)

The aforementioned interpolation formulas [127] yield also (approximative) data for the molar conductivity of CsCl at the temperature range under consideration. These data seem to deviate significantly from the constancy represented by Eq. (6.41), and this deviation seems to grow up with increasing temperature; see Table VI. A possible reason for this deviation is discussed below.

4. Additional Remarks

The remarkable result derived previously may be reformulated in a slightly different way, to point out its probable "practical" significance: Assuming that the molar conductivity of one of the salts (say, for LiCl) is known from experiment at some given temperature T, then the above result (6.41) permits to predict the experimental values of the remaining three molar conductivities (of NaCl, KCl, and RbCl) at the same temperature T within an accuracy of about 3–4%.

The constancy revealed deserves some more comments.

In this context, the following steps of the derivations should be emphasized: The main assumptions (postulates, axioms) in the derivation of Eq. (6.41), that is, formulas (6.36)–(6.39), were motivated physically, and they are identical with the corresponding ones of Section VI B concerning the H^+ and OH^- mobility in solutions. The assumptions concerning F and τ_{rel} have been physically justified through the present strong Coulombic interactions in the ionic melts. The further derivation of the main formula (6.41) is then straightforward.

The variations of the values of Λ^*_{XCl} (for $X = $ Li, Na, K, Rb) around their mean value, at some given temperature, are not greater than the uncertainties connected with the experimental data for molar conductivities. Thus, the revealed constancy strongly indicates that there may be a common "mechanism" that governs the ionic conductivities in all melts under consideration.

The new "scaling law" of the molar conductivities, as shown in Eq. (6.41), has—to our knowledge—not appeared before in the scientific literature [36]. This scaling has neither a formal nor a conceptual resemblance to the so-called "law of corresponding states" for fused salts [128]. In this context it should also be observed that the corresponding deviations from the predicted constancies of some "scaled" physicochemical quantities for alkali chlorides (as presented in Ref. 128) are significantly greater than the deviations shown in Tables IV–VI.

Of particular interest may be the remark that, in contrast with [128], our CSM treatment reveals no "abnormal" behavior for the "light" Li^+ ion in the melt; see the preceding tables. On the other hand, the scaled molar conductivity of CsCl apparently does not obey the constancy (6.41); see Table VI. In this context, however, it might be interesting to observe that the crystalline phase of CsCl is bcc, in contrast to the other four alkali chlorides that form fcc lattices at room temperature. These two facts seem to indicate that the "geometrical order," as reflected in these two lattice structures, could be (on the microscopic time scale) still "persistent" in the two respective *microstructures* of the melts. This microstructural difference clearly determines also different relevant Hamiltonians $\hat{H}_{eff}$, thus affecting the approximate constancy of the factor F appearing in formula (6.42).

Comparison of the present with the "equivalent" application of the general CSM theory presented in Section VI B shows a strong resemblance, formally as well as conceptually. In both treatments the considered ions are assumed to represent delocalized quantum objects undergoing concerted motions. In other terms, one can also say that the considered ions are not "particles" in the conventional (that is, classical-mechanical) sense, but rather that they intrinsically participate into dynamically induced "cooperative structures" (or correlation patterns) of the matter. As was previously pointed out, this short-lived cooperative behavior between many degrees of freedom (represented by coherent-dissipative structures) becomes spontaneously created by the thermal motion.

As was mentioned, the Coulombic interactions dominate the interionic interactions in the melt. Since they are long ranged, there is no surprise that also the coopperative behavior represented by the coherent-dissipative structures may become long ranged, too. This leads to the aforementioned large molar conductivities (see the introductory remarks) as well as to the preceding confirmation of our theoretical predictions, which are based on the assumption that coherent-dissipative structures do really exist in the system.

The relation (6.43) indicates that the relaxation times (and, more general, the microdynamical behavior) of the charge fluctuations in the considered melts are *independent of the masses* of the participating ions. This finding, furthermore, once more supports the viewpoint that the process under consideration is mainly due to quantum delocalization effects, rather than due to "fast moving particles" in the sense of classical mechanics.

As was mentioned, the relaxation process causing the creation of coherent-dissipative structures in the melts is given by the charge fluctuations that spontaneously appear around ions, these fluctuations being caused by the temporal behavior of the interionic correlation patterns to which the considered ions are participating. In this context, it may be interesting to mention the well-known differences between the autocorrelation functions of the ionic velocity and current being typical for ionic melts, which are considered to be related to charge density fluctuations [123]. These fluctuations also appear to have an effect on the viscosity of molten salts; see Ref. 123.

F.   Quantum Correlations in the Spin Dynamics of Paramagnetic Gd at High Temperatures

1.   Introductory Remarks on the Physical Context

The existence of propagating spin waves in magnetic systems above the Curie temperature T_c and their relation to magnetic short-range order is of current interest; see Refs. 129 and 130 and references cited therein. In particular, neutron inelastic scattering [129] and muon spin rotation [130, 132] experiments give new insight into this important physical process. Recently, it has been shown that many systems (for example, rare-earth aluminides [130], EuO, EuS, and Gd [129, 131, 132]) clearly prove the existence of spin waves above T_c. Of particular importance for our following treatment are the very recent and precise experimental results [129] of neutron scattering on Gd (along the $\langle 110 \rangle$ direction), which clearly demonstrate a crossover from spin diffusion at small scattering wavevector q to damped spin–wave behavior at large q. A surprising result consists in the existence of static spin correlations even at $T = 850$ K, that is, well above the Curie point $T_c = 293$ K. It was also concluded [129] that, in Gd, it is not the thermal breakup of spin-pair correlations that destroys the spin waves for $T > T_c$.

These remarkable experimental findings have been considered [40] in the context of our general CSM theory of quantum correlations in disordered condensed matter. Explicit connection of the general theory with the most recent experimental inelastic neutron scattering results reveals that the size of the quantum correlated domains in the considered system is independent of temperature between $T = 320$ K and $T = 850$ K. Furthermore, the analysis strongly indicates that the size of the magnetically ordered regimes is independent of T in the considered range, too.

2. Coherent-Dissipative Structures and Spin-Ordered Regions

Very recently we applied for the first time our CSM theory of quantum correlations to the physical context of spin-ordered regions in the paramagnetic temperature range of magnetic materials [40]. This treatment is intended to reveal general features of (possibly existing) quantum correlations in the "amorphous" paramagnetic spin system, which also means that the explicit form of the actual Hamiltonian appears to be of less importance in the present context. As it will be shown, the following results strongly indicate the existence of the specific quantum correlations (as represented by coherent-dissipative structures) in the paramagnetic Gd, even far above the Curie temperature.

It might also be helpful to mention that the present approach is very similar to the previous applications, and, in particular, that it is intrinsically connected with the previous treatment of ionic conductance of molten salts.

The starting point of the theoretical considerations [40] is given by the general formula (5.37),

$$.s_{\min} = \frac{4\pi k_B T}{\hbar} \tau_{\text{rel}} \tag{6.44}$$

which gives the minimal dimensionality of a coherent-dissipative structure in the function space of paired states

$$|k, k + s\rangle \quad (k = 1, \dots, s)$$

In the present physical context, the "single states" $|k\rangle$ and $|k + s\rangle$ are considered to represent two possible states of an individual spin system. The preceding paired states represent the quantum (or EPR) correlations between these "single states," which are created due to the thermal motion and the interactions between the elementary quantum systems. The minimal "size" $s_{\min}$ represents the minimal number of (paired) degrees of freedom, which are necessary for the spontaneous creation of a coherent-dissipative structure, with lifetime τ_{rel}.

As was mentioned in Section VI A, the crucial relation for the comparison of the theory with experimental results is given by the general formula (6.1), which reads here

$$d_{\min} = F(\hat{H}_{\text{eff}})W_{\text{sys}}^{dB}s_{\min} \qquad (6.45)$$

$d_{\min}$ represents the spatial dimension of a coherent-dissipative structure. This quantity depends linearly on the thermal de Broglie wavelength

$$W_{\text{sys}}^{dB} = \hbar\sqrt{2\pi/k_B T m_{\text{sys}}}$$

where m_{sys} is the (effective) mass of an elementary spin system, and on the functional F, which relies on the effective (or relevant) Hamiltonian $\hat{H}_{\text{eff}}$ as well as other thermodynamical variables (like temperature and external magnetic field). It should be stressed that this *ansatz* coincides with formulas (6.10) and (6.36) of the applications presented above.

With the aid of the last three equations one obtains straightforwardly

$$d_{\min} = \text{const } F(\hat{H}_{\text{eff}})\sqrt{T}\,\tau_{\text{rel}} \qquad (6.46)$$

where the explicit form of the constant appearing here is not needed furthermore. In connection with the remarks presented in Section VI A, one may say that the term $F(\hat{H}_{\text{eff}})$ is due to the specific mechanism of the present problem, whereas the quantity $\sqrt{T}\,\tau_{\text{rel}}$ is due to quantum correlations.

3. Comparison with Neutron Scattering Results

In the following, connection of the main formula (6.46) with the recent experimental results on Gd presented by Cable and Nicklow [129] is made. In this reference, the size of spin-ordered regions above $T_c = 293$ K, which are a prerequisite for the appearance of short-wavelength spin waves, is qualitatively represented by the inverse of the reduced crossover scattering wavevector ζ_c of inelastically scattered neutrons. The experimental results (as being presented in the insert of Fig. 6 of Ref. 129) exhibit a slight, nearly linear increase of ζ_c with temperature above T_c. To be more specific: ζ_c increases only about 50% between 320 K and 850 K, thus implying a corresponding slight decrease of the size of the existing magnetically ordered domains.

This surprising observation is clearly in contrast with the predictions of standard theories of phase transitions; see Refs. 129 and 130 for details and further references.

The present CSM approach is based on the assumption that coherent-dissipative structures may properly describe the magnetic short-range-ordered regimes of Gd above T_c, that is, in the temperature region, where the ferromagnetic long-range order is clearly broken. The derived simple formula (6.46) allows the following detailed theoretical treatment [40] of the experimental data:

(A) From Fig. 6 of Ref. 129 we extract the numerical values of the crossover wavevector ζ_c at the considered five different temperatures ($T = 293$ K, 320 K, 440 K, 586 K, and 850 K).

(B) The inelastic neutron scattering results being parametrized in Table I and II of Ref. 129 [on the basis of the so-called damped harmonic oscillator (DHO) model] allow us to determine the values of the damping parameter β_q corresponding to the crossover wavevector $\zeta_c(T)$ at the temperatures given; see Table VII. Note that Tables I and II of Ref. 129 represent parametrizations of two different experimental series, which are based on different mean intensities I_0 of the scattered neutron beams.

(C) To apply our theoretical result (6.46) we need explicitly the quantity τ_{rel}, that is, the characteristic relaxation time of the quantum systems that contribute to the creation of a coherent-dissipative structure. In the present physical context, it is reasonable to associate the inverse of the damping parameter β_q with this relaxation time, as β_q represents the lifetime (or decay time) of the magnetically ordered clusters. This identification is completely in line with the physical model (that is, the DHO model) being used in [129].

With this choice we calculate the quantum correlations term $\sqrt{T/\beta_q}$ as desired by Eq. (6.46); the corresponding numerical values are shown in Table VII.

(One might also be inclined to associate the characteristic relaxation time

TABLE VII
Inelastic Neutron Scattering Data for Gd of Ref. 129[a]

$T(K)$	ζ_c	β_q	$\sqrt{T/\beta_q}$
293	0.114	0.67	25.5
(293)		(0.745)	(23.0)
(320)	0.177	(1.16)	(15.4)
440	0.184	1.325	15.8
586	0.228	1.59	15.2
850	0.290	1.79	16.3

[a]Magnetic data for Gd are from Ref. 129, Table I, for $Q = (\zeta, \zeta, 2)$. T: temperature; ζ_c: reduced crossover wavevector (in units of $4\pi/a$); β_q: damping parameter corresponding to ζ_c (in THz). [Data in parentheses are taken from Table II of that reference, for $Q = (\zeta, \zeta, 2)$.] See also Ref. 40.

τ_{rel} of the theory with the inverse of the characteristic frequency ω_q at the crossover wavevector. This choice, however, would yield no additional physical information due to the "crossover condition" $\beta_q = 0.707\omega_q$ of the DHO model [129].)

(D) Both the aforementioned different intensities underlying the parametrizations of the Tables I and II of Ref. 129 contain data at $T_c = 293$ K. This allows us to determine two numerical values for the desired quantity $\sqrt{T}/\beta_q$ at T_c. The data at $T = 320$ K (cf. our Table VII) are extracted from Table II of Ref. 129.

4.   Results and Additional Remarks

This treatment of the experimental data reveals a remarkable result that may be shortly stated by the following:

The experimental neutron scattering results show that the quantity

$$\sqrt{T}\tau_{rel} \tag{6.47}$$

is independent of the temperature. This quantity is due to quantum correlations of general character and—in the light of our theory—is connected with the size d_{min} of the magnetically ordered regions above T_c.

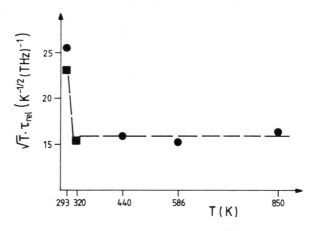

FIGURE 5. Temperature dependence of the quantity $\sqrt{T}\,\tau_{rel}$ that is associated with the extension of the quantum correlations in the paramagnetic Gd spin system. The constancy of this quantity above T_c is clearly shown. The only small "scattering" of the data points between 320 K and 850 K around their mean value is presumably due to the high precision experimental results of Ref. 129. Circles (squares) correspond to the data of Table I (II) of Ref. 129. For details, see the text and Ref. 40.

This novel constancy is shown in Table VII and graphically presented in Fig. 5. To our knowledge, this "scaling law" did not exist so far in the scientific literature [40].

Let us now consider this finding in more detail. The value of $\sqrt{T}\,\tau_{\rm rel}$ at the transition temperature $T_c = 293$ K is clearly larger than that at all other temperatures, and this is obviously not due to numerical uncertainties of the calculation. It also appears that the two values of this quantity differ. In our opinion, however, this difference seems not to be related with "experimental errors," since near T_c the considered quantity is expected—owing to standard theory of phase transitions—to vary significantly. Thus, it seems that this difference is due to a small temperature difference in the two experimental series.

In the light of the general CSM theory, the preceding result also reveals the remarkable point that the effective Hamiltonian $\hat{H}_{\rm eff}$ (describing the dynamics of the magnetically ordered regions during their lifetimes) may depend at most slightly on the temperature. To see this, one may compare the aforementioned temperature dependence of the crossover wavevector with the expression of $d_{\rm min}$ and the preceding finding. As already mentioned above, the linear spatial dimension of a magnetically ordered region, $D_{\rm spin}$, is of the order ζ_c^{-1} and, at the same time, is represented by the theoretical quantity $d_{\rm min}$.

This comparison immediately shows that the value of the functional $F(\hat{H}_{\rm eff})$ may decrease only slightly with increasing temperature (that is, about 50% in the temperature region considered). In this context, it should be reminded that $\hat{H}_{\rm eff}$ is a specific quantity of a concrete physical system and, therefore, it cannot be determined by the general CSM theory of quantum correlations.

The revealed constancy of the considered quantity $\sqrt{T}\,\tau_{\rm rel}$ in the temperature range between 320 K and 850 K, in connection with the small "scattering" of the data, see Fig. 5, indicates that the underlying experimental errors [129] are small (that is, about $\pm 4\%$).

The preceding observations suggest the following additional possibility for an interpretation of the considered experimental data.

The crossover wavevector ζ_c represents, of course, an approximate measure for the actual size of the magnetically ordered regions $D_{\rm spin}$ under consideration. This thus implies that the slight temperature dependence of this quantity (as shown in Fig. 6 of Ref. 129) does not strictly prove that $D_{\rm spin}$ does really decrease with increasing temperature. In connection with the revealed constancy of $\sqrt{T}\,\tau_{\rm rel}$, this remark suggests that the functional $F(\hat{H}_{\rm eff})$ may be a T-independent quantity, which furthermore supports the conjecture that the actual size $d_{\rm min}$ may then be independent of the temperature, too.

In other words, the present theoretical analysis of the experimental results of Ref. 129, which is based on our theory of coherent-dissipative structures and therefore assumes the validity of the relation

$$D_{\text{spin}} \approx d_{\text{min}} \qquad (6.48)$$

suggests that the actual size D_{spin} of the considered magnetically ordered regimes remains constant between 320 K and 850 K—which is a remarkable finding.

It should be observed that the preceding theoretical analysis of the experimental data, which is free of fitting parameters, does not use any specific feature of the underlying spin–spin interaction mechanism or, more general, of the system Hamiltonian $\hat{H}_{\text{eff}}$. Therefore, the derived T independence of the quantum correlation factor $\sqrt{T}\, \tau_{\text{rel}}$ (and the suggested T independence of the magnetically ordered domains) should be considered to be due to the general character of quantum correlations in condensed matter as being revealed in the framework of the general CSM theory.

G.   Quantum Correlation Effects in High-T_c Superconductivity

1.   Introductory Remarks on the Physical Context

In this subsection, certain aspects and experimental results on the topic "high-T_c superconductivity" are considered, in the light of the above general CSM theory. The following point should be emphasized here. The present treatment (as well as the general theory) does not intend to reveal the microscopic mechanism of pair formation in the high-T_c superconductors, which represents the aim of many new theories [133, 134]. In contrast, the present approach deals with quantum *correlations* appearing explicitly in the thermalized and CSM-transformed fermionic density operator $\Gamma^{(2)}$, and their experimental manifestation. To put it another way, these correlations, as it will become obvious, permit the extraction of physical information that is *independent* of the (still unknown) precise nature of the high-T_c pair-formation mechanism. In this connection, please compare with the considerations of Section VI A.

Since the discovery of high-temperature superconductivity, one of the most disputed issues is whether the new superconducting state is of conventional type or not. There exist many experimental results, however, that cannot be understood with the aid of the "weak-coupling" limit of the BCS theory [32]. Two of them are the following:

(A) the large energy gap $2\Delta \approx (7\text{–}8)\, k_B T_c$ of $YBa_2Cu_3O_7$ and $Bi_2Sr_2CaCu_2O_8$ superconductors [135, 136, 137, 138], and

(B) the "universal" linear relation [139] between T_c and n_s/m^* (carrier density over effective mass) that very recently has been found to hold for many different compounds of copper-oxide superconductors, see Fig. 6;

see also references cited in these papers. The well-defined experimental conditions underlying these precise infrared-reflectivity [135], uv-photoemission [136, 137], electron energy-loss [138], nuclear quadrupole resonance [141], and muon-spin-relaxation [139] measurements make the aforementioned results (A and B) reliable, so that they should be considered to constitute experimental data which must be explained by any successful theory of high-T_c superconductivity.

Recently, we proceeded to the explanation of these experimental data (and some additional ones) with the aid of our general CSM formalism of quantum correlations [41]. In the following it is demonstrated that our general CSM transformation of the fermionic second-order reduced density matrix represents a suitable framework for the analysis of these effects. Moreover, quantitative comparison with these experimental results on high-T_c copper-oxide superconductors is made explicitly. In particular, the following results are considered in detail:

1. The contribution of quantum correlations to the energy gap is found to be $6.28 \, k_B T_c$; this is in satisfactory agreement with very recent and high precision experiments [130–133, 139], and other meausrements.

2. An explanation of the universal linear relation between T_c and n_s/m^* (carrier density over effective mass) being observed in all copper-oxide superconductors by Uemura et al. [139] follows "from first principles" of the theory.

3. The experimental evidence for a considerable energy dissipation "within" the energy gap of the copper-oxide superconductors [130–132] is explained from first principles of the general formalism.

2. Summary of Formulas

In order to simplify the presentation and to point out the crucial theoretical points, some formulas derived in the previous three sections are summarized here.

The second-order reduced density matrix for a fermionic system in the so-called extreme case [28, 29], that is, when all the eigenvalues of the first-order reduced density matrix are equal (see Section IV), is given by

$$\Gamma^{(2)}(g^{N/2}) = \Gamma_L^{(2)} + \Gamma_S^{(2)} + \Gamma_{\text{Tail}} \tag{6.49}$$

(This partitioning appears to be in line with the standard description of Penrose and Onsager [140] and Yang's concept of ODLRO [77]; see Refs. 41

and 144.) Remember that

$$\Gamma_L^{(2)} = |g\rangle \lambda_L \langle g|, \quad \lambda_L = \frac{N}{2} - \frac{N(N-2)}{4s} \tag{6.50}$$

(where N is the number of fermions) and

$$\Gamma_S^{(2)} = \lambda_S \sum_{i,j} (\delta_{ij} - 1/s)|i, i+s\rangle\langle j, j+s|, \quad \lambda_S = \frac{N(N-2)}{4s(s-1)} \tag{6.51}$$

where $g^{N/2}$ is an antisymmetrized geminal power (AGP). The uncorrelated contribution Γ_{Tail} is not needed in the following derivations. The geminal g (of rank s) is expanded in a (localized) basis of $m = 2s$ orbitals (with $2s > N$), that is,

$$g(1,2) = \sum_{i=1}^{s} g_i |i, i+s\rangle \tag{6.52}$$

where $|i, i+s\rangle$ is a (normalized) Slater determinant of spin orbitals i and $i + s$. More explicitly, the geminal will have the form (for even N)

$$|g^{N/2}\rangle = \mathscr{A}_N\{g(1,2)g(3,4)\cdots g(N-1,N)\} \tag{6.53}$$

where $\mathscr{A}_N$ represents the conventional antisymmetrization operator.

The eigenvalues λ_L and λ_S are coupled through the simple relation

$$\text{Tr}\{\Gamma^{(2)}\} = N/2 = \lambda_L + (s-1)\lambda_S, \tag{6.54}$$

as it follows easily from Eqs. (6.49)–(6.51); for full details, see Section IV. It is also important to note that λ_S may be nonvanishing even in the case of a large λ_L. [This will become clear after the discussion of the experimental results $(C-E)$.]

3. Carrier Dependence of T_c and Energy Gap in the Light of the CSM Theory

For the present treatment of quantum correlations in the superconductor we do not need—as was already pointed out—the explicit form of the pair forming mechanism, $\mathscr{V}_{12}$. Here it is sufficient just to assume that it is representable by a two-body effective interaction Hamiltonian. The energy gap 2Δ of the superconductor being in a physical state represented by the density matrix $\Gamma_L^{(2)}$ is then expressed by the formula

$$2\Delta = \text{Tr}(\mathscr{V}_{12}\Gamma_L^{(2)}) \tag{6.55}$$

and with the aid of Eq. (6.50) we obtain

$$2\Delta = \lambda_L \mathrm{Tr}(\mathcal{V}_{12}|g\rangle\langle g|) \tag{6.56}$$

In this formula only the coherent part of the density operator $\Gamma^{(2)}$ appears since the supercurrent is known to be representable by a macroscopic pure state (that is, a state vector $|g\rangle$ in the present case).

At this stage, the CSM transformation of the general theory is applied on Eq. (6.56). In the following, the effect of this transformation will be denoted by a "tilde". It has been shown how specific Jordan blocks occur into the thermalized and CSM-transformed part $\tilde{\Gamma}_S^{(2)}$ of the complete density matrix—after appropriate quantization conditions concerning the resonances being exposed after complex scaling. These Jordan blocks have the minimum "size"

$$s_{\min} = \frac{4\pi k_B T}{\hbar}\tau_{\mathrm{rel}} \tag{6.57}$$

see Eq. (5.37).

Here, however, we are considering the corresponding transformation of the coherent part $\Gamma_L^{(2)}$ of the density matrix, Eq. (6.50); this term represents in standard theory the superconducting state. By this transformation one obtains easily a matrix $\tilde{\Gamma}_L^{(2)}$ with elements proportional to

$$\exp\left[i\frac{\pi}{s}(k+l)\right]$$

This matrix is shown to have the rank one and the trace zero. Furthermore, one sees that the square of $\tilde{\Gamma}_L^{(2)}$ is the zero matrix, which implies that $\tilde{\Gamma}_L^{(2)}$ contains Jordan blocks of order $s = 2$.

Our general CSM theory of quantum correlations associates with these Jordan blocks the characteristic energy $\hbar/\tau_{\mathrm{rel}}$, and using the general result (6.57) one has (with $s = 2$)

$$\frac{\hbar}{\tau_{\mathrm{rel}}} = 2\pi k_B T \tag{6.58}$$

For T being equal to the critical temperature T_c of the superconductor, this characteristic energy may be associated on physical grounds with the gap energy 2Δ, which is intrinsically connected to $\tilde{\Gamma}_L^{(2)}$ through the CSM transformed Eq. (6.55). Thus we obtain for $T = T_c$ the relation

$$2\Delta = \frac{\hbar}{\tau_{rel}} = 2\pi k_B T_c \tag{6.59}$$

and, furthermore, with the aid of (6.55),

$$T_c = \frac{1}{2\pi k_B} \langle g | \mathscr{V}_{12} | g \rangle \lambda_L \tag{6.60}$$

with

$$\lambda_L = \frac{N}{2} - (s-1)\lambda_S = \frac{N}{2} - \frac{N(N-2)}{4s} \tag{6.61}$$

see Eq. (6.54). The simple formulas (6.59) and (6.60) represent our main results [41, 144].

4. Energy Gap

These theoretical formulas allow for a straightforward connection with the experimental results $(A-B)$ mentioned previously. The condition (6.60) for the collapse of the supercurrent [as represented by the destruction of the wavefunction (6.52) and the simultaneous creation of a Jordan block of order 2 over the thermalized and CSM transformed Γ_L] yields the numerical value

$$2\Delta \approx 6.28 k_B T_c$$

see Eq. (6.59), for the energy gap, which obviously differs from the standard BCS result, which is about $3.5 k_B T_c$. At the same time, our result appears to approach the aforementioned experimental result (A) for copper-oxide superconductors.

Although there are some controversies regarding the correct value for the energy gap in the new superconductors, the most recent and precise experimental results show that the actual value for 2Δ is clearly greater than the corresponding standard BCS result. The energy gap in the super-conducting a-b plane of high-T_c Cu-O superconductors appears to be

$$\frac{2\Delta}{k_B T_c} \approx 5-8 \tag{6.62}$$

see Refs. 135–139. This result should be considered as well established nowadays. To our knowledge, no other theory could so far predict quantitatively the observed large energy gap from first principles, that is, without making use of "fitting parameters".

[In this context, let us also mention the following speculation: The gap anisotropy of $YBa_2Cu_3O_7$ and other related materials is considerable: 2Δ is $\approx 8k_BT_c$ along the a and b crystalographic axes and $\approx 3k_BT_c$ along the c axis. These values give for the mean value of 2Δ averaged over all three axes: $2\Delta_{av} \approx 6.3\,k_BT_c$. This numerical value "coincides" with our theoretical result. Of course, the preceding correlation treatment does not make use of the specific crystal asymmetry of the new superconductors. This property must however be reflected in the form of the relevant Hamiltonian.]

5.   Uemura's Universal Linear Relationship

The observed universal linear relation (B) mentioned previously is in obvious agreement with our result (6.60) and (6.61), since T_c is predicted to be proportional to the number of carries $N/2$, for small carrier concentration.

Indeed, a small carrier concentration corresponds theoretically with the limiting case $N^2 \ll s$. We immediately obtain

$$T_c = \frac{1}{2\pi k_B}\langle g|\mathscr{V}_{12}|g\rangle\frac{N}{2} \qquad (6.63)$$

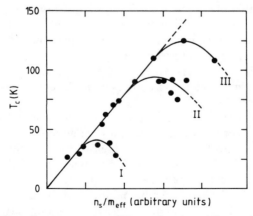

FIGURE 6.   Schematic representation of Uemura's universal linear relation [139] between transition temperature and carrier (hole) density over effective mass. Roman numbers give the number of Cu–O layers of the different copper-oxide superconductors. The parabolae shown represent schematically the predicted "quadratic decrease" as discussed in the text; see Eqs. (6.60) and (6.61), and Refs. 41, 144.

This theoretical prediction agrees excellently with the experimentally well-established *universal linear relation* of Uemura et al., see Ref. 139 and Fig. 6. Up to now, to our knowledge, no other theory has been able to explain this effect sufficiently, this remark being in agreement with the corresponding remarks of Uemura et al. [139].

In this context it should be observed that the quantity "effective mass" represents a physical concept that takes into account some specific aspects of the actual Hamiltonian, and thus it is not immediately connected with the concept of quantum correlations between states; see Section VI A. This is of course the reason why m^* does not appear explicitly in the mathematical part of the preceding theoretical treatment.

6. *Dissipation within the Energy Gap*

We now turn our attention to some additional experimental observations that are connected with the aforementioned two effects (A and B) and which clearly indicate the relevance of the small box part Γ_S in the present physical context.

It has been repeatedly observed, by different spectroscopic methods, that there appears a considerable "energy dissipation" or "background" in the energy gap domain of the superconductor. For example, Collins et al. [135] pointed out that there exists a considerable infrared absorption, that is, nonzero conductivity $\sigma(\omega)$, below $500\,cm^{-1}$ (which corresponds to about $8k_B T_c$) in the superconducting state, at $T = 45\,K$. Because of the reproducibility of this effect and the high quality of the samples used, it was concluded that this unexpected absorption is intrinsic to $YBa_2Cu_3O_7$ and may arise from a "contribution to $\sigma(\omega)$ that does not have the $500\,cm^{-1}$ gap" [135]. In this connection, it is also interesting to mentioned the observed small depression in the conductivity (below $500\,cm^{-1}$) at $T = 105\,K$, that is, well above the critical temperature ($T_c \approx 92\,K$). This effect [135] is reminiscent of the corresponding much larger depression in the conductivity at $T = 45\,K$; see Fig. 1(c) of Ref. 135.

Similar observations have been made and conclusions have been drawn with respect to other experiments, too, for example, high-resolution photo-electron spectroscopy [136, 137] and high-resolution electron-energy-loss spectroscopy [138]. The experimental investigations of Refs. 136 and 137 clearly show that the measured uv-photoemission spectra of single-crystal specimens of $Bi_2Sr_2CaCu_2O_8$ exhibit a finite intensity in the gap that cannot be attributed to the instrumental resolution. This intensity was tentatively ascribed to states belonging mainly to "nonsuperconducting layers" [136].

In the light of the presented theory of quantum correlations, this effect is of considerable importance, since it is intrinsically connected with the fraction of energy-dissipating carriers being represented by the "incoherent" but still quantum-correlated part Γ_S of the complete density matrix, Eq. (6.49). This part, as was mentioned, is expected to be of importance even in the presence of the coherent part Γ_L that describes the supercurrent. From the viewpoint of the theory, the intensity of the observed "background" in the gap is expected to be proportional to the degenerate small eigenvalue λ_S being associated with Γ_S.

Detailed experimental results on this effect, and in particular the knowledge of its temperature dependence, could be decisive for the determination of the fraction "coherent/incoherent carriers" (that is, λ_L/λ_S) and *a fortiori* for the further development of the present CSM theory.

7. Additional Implications of the Theory

There are also some additional remarkable predictions of the preceding general CSM theory, which are immediately connected with the thus far considered experiments.

First, for larger values of $N/2$ one obtains a general *negative quadratic deviation* from Uemura's linear dependence treated previously, which is due to the quadratic term appearing in the right-hand side of formula (6.61). Remember that this term was neglected in the case of the small carrier concentration considered.

This predicted negative quadratic deviation of the "linear dependence" is schematically presented in Fig. 6. This is obviously in qualitative agreement with the extensive experimental data of Ref. 139. Indeed, this deviation seems to be a universal feature of the presented data, but thus far it was considered just as a "disturbing accompanying effect" being due to higher carrier concentration. However, as Uemura et al. mention [139], there is no clear explanation for the suppression of T_c in this region of carrier concentration.

Second, our theoretical approach predicts also a *sudden breakdown* of the superconducting state at higher carrier concentration, that is, in the limit

$$s = 2N \qquad (6.64)$$

This case corresponds to the well known Hartree–Fock limit of independent particles; see Section IV. Remember, also, that the general restriction $2N \leqslant s$ must hold, owing to the fermionic character of the considered quantum systems, that is, the electrons. A corresponding effect may have been seen in experiment; see Refs. 139 and 144.

The general result (6.60) allows us to make the following new quantitative prediction. The carrier concentration $(N/2)_{max}$ giving the maximum transition temperature T_c^{max} for each group of Cu-O superconductors (see Fig. 6) is predicted to be half of the value of the carrier concentration $(N/2)_{break}$ at the breakdown limit (6.64), that is,

$$(N/2)_{max} = 0.5(N/2)_{break} \tag{6.65}$$

This relation is easily proved by differentiating Eq. (6.60) with respect to N and putting the derivative equal to zero, which leads to the condition $s + 1 = N$ for the maximum of T_c. Comparison of this value with the Hartree—Fock limit (6.64) yields the prediction (6.65).

Thus, from our theoretical viewpoint, there is an urgent need for the detailed experimental exploration of the predicted "quadratic decrease" and "breakdown" effects under consideration. These experiments would permit us to determine the numerical value of the Hilbert space dimension s, which appears as a parameter in the general treatment of quantum correlations. Furthermore, this would enable us to study new aspects of the breakdown mechanism of the supercurrent.

In conclusion, we found strong evidence that certain very recently discovered effects concerning the high-T_c copper-oxide superconductors are intrinsically connected with the special kind of quantum correlations being predicted by our general CSM formalism. The quantitative agreement of the theoretical predictions with these experiments represents further evidence for the important role that quantum correlations may play in the considered physical context.

VII. OUTLOOK

As Prigogine and co-workers have emphasized [21], the entity "density operator" has to be considered as the primary concept in dynamics, whereas the "wave fuction" should be considered as a quantity of approximative character, that is, as an idealization. In the physical context of condensed matter processes, this may become clear easily—at least if one wants to go beyond the usual "coarse graining", "time smoothing", and other similar approximations.

Another point of interest is given by the fact that, in condensed matter, the physical meaning of the concept "stationary state" is lost. This is beautifully demonstrated by Landau and Lifshitz in their classic textbook [54]. With respect to isolated (and relatively small) molecules, of course, stationary states do have physical significance.

By the study of the Liouville–von Neumann equation (for condensed systems), one is confronted with the crucial role that correlations play in physical processes. The fundamental character of the concept of correlations has been clearly demonstrated by the Brussels school; see Ref. 42. Here, the concept of subdynamics arises. In contrast to several widely used descriptions based on the so-called Zwanzig–Mori projection-operator methods, the projection operator Π is not arbitrary, but uniquely determined by the dynamics [42]. As a consequence, the concept of *microscopic* irreversibility is revealed to be of fundamental character. Going a step further, the concept of the star-unitary transformation is introduced, and the corresponding transformed density operator $\tilde{\rho}$ appears to undergo an irreversible time evolution; see Refs. 21–25.

It is this "nonarbitrariness" that indicates that there may exist an intrinsic connection between the subdynamics (of statistical mechanics) and the Balslev–Combes mathematical theory of complex scaling [3]: namely, the resonances that complex scaling (or dilation) reveals are intrinsic to the Hamiltonian. Thus, the physically well-established resonant (decaying, metastable) states have been put on a firm mathematical basis. In other words, these states exhibit irreversibility "from first principles."

Our extension of the CSM into quantum statistics [20] further emphasizes the fundamental role that quantum correlations play. Indeed, if some very specific quantization conditions concerning the exposed resonances are fulfilled, then the thermalized and CSM-transformed fermionic $\Gamma^{(2)}$ has no diagonal representation. This suprising result means physically that there is now no way to "transform away" quantum correlations. This stresses the physical significance of Yang's concept of ODLRO [77] even more. These persisting correlations constitute the coherent-dissipative structures.

In future work we intend to study in detail the interdependence of the theories of complex scaling, quantum statistics, and subdynamics, trying to understand (from first principles) the fundamentals of irreversibility and self-organization in amorphous condensed matter.

Some current work in this direction seems to show the remarkable feature that *bosonic* systems may also exhibit quantum correlations of the nature considered previously, if certain more restrictive conditions can be fulfilled.

Another problem concerns the question whether the preceding theory can be extended into the framework of the *grand* canonical ensemble. Certain preliminary investigations show that a new kind of "condensation" can appear: a macroscopic condensation (that is, of all the particles) into one huge Jordan block, if the temperature is sufficiently low.

The physical relevance of the preceding CSM theory of quantum correlations was demonstrated by the applications presented in Section VI. The theory allows one to derive qualitative new results, some of them even being in contradiction to corresponding well-established theories. This may illustrate the predictive power of the CSM in connection with statistical mechanics.

Current work concerns the further development of our treatment (see Section VI F) of magnetically ordered domains in paramagnetic regimes of

Gd (and other materials) far above the Curie temperature [143]. The CSM theory also initiated certain novel experiments, some of them being now in progress (see Section VI). As an example, let us mention the experiments concerning the proton delocalization in water and the nature of H bond (see Sections VI B and VI C).

APPENDIX A

In this appendix we consider the proof of the Reid–Brändas corollary [64]. The following theorem holds:

Theorem

Any matrix can be brought to complex symmetric form by a similarity transformation.

The theorem is seldom mentioned although it is more or less implicit in linear algebra treatments, see, for example, Löwdin [89]. For a more direct proof, see Gantmacher [65], where it is explicitly shown that every square matrix is similar to a symmetric matrix and that every (complex) symmetric matrix is (complex) orthogonally similar to an explicitly given normal form.

The following proof—to some extent—parallels the Gantmacher construction, the explicit symmetric form is however quite different. It should be noted that the connection between the latter and the off-diagonal long-range order (ODLRO) in the sense of Yang [77] plays a fundamental role in the theory of coherent-dissipative structures.

The proof of the theorem [64] reads as follows.

Since any matrix is similar to a block-diagonal matrix in which the submatrices along the main diagonal are Jordan blocks [65], it suffices to show that any Jordan block can be transformed into a complex symmetric matrix. Moreover, since

$$\mathbf{C}_n(\lambda) = \begin{pmatrix} \lambda & 1 & \cdots & \cdots & 0 & 0 \\ 0 & \lambda & \ddots & \cdots & 0 & 0 \\ \vdots & \vdots & \ddots & & \vdots & \vdots \\ \vdots & \vdots & & \ddots & \ddots & \vdots \\ 0 & 0 & \cdots & \cdots & \lambda & 1 \\ 0 & 0 & \cdots & \cdots & 0 & \lambda \end{pmatrix}$$

can be written as

$$\mathbf{C}_n(\lambda) = \lambda \mathbf{1} + \mathbf{C}_n(0)$$

we may deal only with matrices of the form $\mathbf{C}_n(0)$. Given a matrix $\mathbf{A}$, we need to show that there is a matrix $\mathbf{B}$ such that $\mathbf{B}^{-1}\mathbf{A}\mathbf{B}$ is a symmetric matrix. This means that

$$\mathbf{B}^{-1}\mathbf{A}\mathbf{B} = (\mathbf{B}^{-1}\mathbf{A}\mathbf{B})^T = \mathbf{B}^T\mathbf{A}^T(\mathbf{B}^{-1})^T = \mathbf{B}^T\mathbf{A}^T(\mathbf{B}^T)^{-1} \qquad (A.1)$$

299

Multiplying this equation from the left by $\mathbf{B}$ and from the right by $\mathbf{B}^T$ gives

$$\mathbf{ABB}^T = \mathbf{BB}^T\mathbf{A}^T$$

Since $\mathbf{BB}^T$ is a symmetric matrix—call it $\mathbf{S}$—we can write this as

$$\mathbf{AS} = \mathbf{SA}^T = \mathbf{S}^T\mathbf{A}^T = (\mathbf{AS})^T$$

and so $\mathbf{AS}$ is also symmetric. Thus, if we can find a nonsingular symmetric matrix $\mathbf{S}$, factorizable into the form $\mathbf{BB}^T$, such that $\mathbf{AS}$ is also symmetric, then the transformation (A.1) can be carried out.

In the following we consider the special case where $\mathbf{A}$ is a Jordan block $\mathbf{C}_n(0)$, so that $a_{ij} = \delta_{i+1,j}$. In this case we have

$$(AS)_{ij} = \sum_{k=1}^{n} a_{ik}s_{kj} = \sum_{k=1}^{n} \delta_{i+1,}s_{kj} = \begin{cases} s_{i+1,j} & i < n \\ 0 & i = n \end{cases}$$

Then, using the fact that both $\mathbf{AS}$ and $\mathbf{S}$ are symmetric, we find

$$s_{i+1,j} = s_{j+1,i} = s_{i,j+1} \quad (i \neq n \neq j) \tag{A.2}$$

while

$$s_{i+1,n} = 0 \tag{A.3}$$

Equation (A.2) shows that the elements of $\mathbf{S}$ depend only on the sum of the indices; that is, $\mathbf{S}$ is persymmetric. By (A.3), the only nonzero element in the last column of $\mathbf{S}$ is s_{1n}, and so all elements below and to the right of the secondary diagonal are zero. Those on the secondary diagonal must not vanish, since $\mathbf{S}$ must not be singular, while those above and to the left are arbitrary except for the requirement of persymmetry.

These requirements can be satisfied as follows: Let be

$$\omega \equiv \exp(\pi i/n)$$

and

$$\mathbf{B} = \begin{pmatrix} 1 & \omega & \omega^2 & \cdots & \omega^{n-1} \\ 1 & \omega^3 & \omega^6 & \cdots & \omega^{3(n-1)} \\ \cdots & \cdots & \cdots & \cdots & \cdots \\ 1 & \omega^{2n-1} & \omega^{2(2n-1)} & \cdots & \omega^{(n-1)(2n-1)} \end{pmatrix}$$

That is,

$$b_{kl} = \omega^{(2k-1)(l-1)}$$

Then

$$(BB^T)_{rs} = \sum_{l=1}^{n} b_{rl}b_{sl} = \sum_{l=1}^{n} \omega^{2(r+s-1)(l-1)} = \frac{1 - \omega^{2n(r+s-1)}}{1 - \omega^{2(r+s-1)}} \qquad (A.4)$$

This depends only on $r + s$, leading to the required persymmetry. The numerator in Eq. (A.4) is zero for all r and s, but the denominator is zero only if $r + s = n + 1$; that is, for the secondary diagonal. In this case each term of the sum is unity. Thus S has the form

$$\mathbf{S} = \begin{pmatrix} 0 & 0 & \cdots & 0 & n \\ 0 & 0 & \cdots & n & 0 \\ \cdots & \cdots & \cdots & \cdots & \cdots \\ n & 0 & \cdots & 0 & 0 \end{pmatrix}$$

As a specific example, for $n = 4$, $\omega = \sqrt{2}(1 + i)$, and $\mathbf{A} = \mathbf{C}_4(0)$, one has

$$\mathbf{B}^{-1}\mathbf{AB} = \frac{1}{4} \begin{pmatrix} 3 & -\omega & -i & \omega^* \\ -\omega & 3i & \omega^* & 1 \\ -i & \omega^* & -3 & \omega \\ \omega^* & 1 & \omega & -3i \end{pmatrix}$$

Observation of this and other examples suggests that $\mathbf{B}^{-1}\mathbf{AB}$ is the sum of a persymmetric matrix and a diagonal matrix.

This statement can be proved is the following way: If we denote the elements of $\mathbf{B}$ and its inverse by b_{kl} and α_{kl}, respectively, and those $\mathbf{B}^{-1}\mathbf{C}_n(0)\mathbf{B}$ by q_{kl}, we have

$$\sum_{k=1}^{n} \alpha_{jk}b_{kl} = \sum_{k=1}^{n} \alpha_{jk}\omega^{(2k-1)(l-1)} = \delta_{jl}$$

and

$$q_{jl} = \sum_{k=1}^{n-1} \alpha_{jk}b_{k+1,l} = \sum_{k=1}^{n-1} \alpha_{jk}\omega^{(2k+1)(l-1)}$$

Multiplying the first by $\omega^{2(l-1)}$ and comparing with the second gives

$$q_{jl} = \delta_{jl}\omega^{2(l-1)} - \alpha_{jn}\omega^{(2n+1)(l-1)} = \delta_{jl}\omega^{j+l-2} - \alpha_{jn}\omega^{l-1} \qquad (A.5)$$

since $\omega^{2n} = 1$ by definition. Since **B** is the matrix associated with a Vandermonde determinant, we can use the standard expression for the inverse of this type of matrix [142] to find

$$\alpha_{jn} = \frac{(-1)^{j+n}\sigma_{n-j}(\omega, \omega^3, \ldots, \omega^{2n-3})}{\prod_{r=1}^{n-1}(\omega^{2n-1} - \omega^{2r-1})} \qquad (A.6)$$

where σ_{n-j} is an elementary symmetric polynomial. The latter may be generated by any of the expressions

$$\prod_{i=1}^{n}(x + \lambda_i) = \sum_{i=0}^{n} \sigma_{n-i}x^i$$

or

$$\prod_{i=1}^{n}(1 + \lambda_i t) = \sum_{i=0}^{n} \sigma_i t^i$$

where the symmetric functions should carry the arguments λ_i, for $i = 1, \ldots, n$. Equation (A.6) can be evaluated by noting that its arguments, together with $\omega^{2n-1} (= 1/\omega)$, are the roots of $x^n + 1 = 0$. Writing $-1/\omega^n$ for 1 and dividing out the undesired root $1/\omega$, we find

$$x^{n-1} + \frac{x^{n-2}}{\omega} + \frac{x^{n-3}}{\omega^2} + \cdots + \frac{1}{\omega^{n-1}} = 0 \qquad (A.7)$$

Then the relation between the coefficients of an algebraic equation and the symmetric functions of its roots leads to the result that the numerator of α_{jn} is $1/\omega^{n-j} = -\omega^j$, since $\omega^n = -1$.

From the construction of Eq. (A.7) it is clear that the denominator in the expression for α_{jn} is the value that the left side takes when $x = \omega^{2n-1} = 1/\omega$. But these substitutions reduce each term to $-\omega$, so that the denominator is $-n\omega$, and $\alpha_{jn} = \omega^{j-1}/n$. Equation (A.5) then yields

$$q_{jl} = \omega^{j+l-2}(\delta_{jl} - 1/n)$$

$$= \exp\left(i\pi\frac{j+l-2}{n}\right)(\delta_{jl} - 1/n)$$

which represents the desired result. The transformation with the matrix $\mathbf{B}$ as defined previously is, of course, not unique, since any (complex) orthogonal transformation following this one will preserve the symmetry.

The matrix $\mathbf{q}$ constructed previously looks rather "innocent", since it can be viewed as an unitary matrix with each element perturbed by $1/n$. Even if this perturbation is small (for large n) the consequences are of fundamental importance as $\mathbf{q}$ by construction corresponds to a Jordan block with Segré characteristic n. It also may be interesting to note that such matrix representations are not unfamiliar in the area of complex deformations of general N-particle Hamiltonians.

APPENDIX B

In this appendix, a short outline of the CSM theory presented in Sections II–V is given.

(a) The algebraic corollary due to Reid and Brändas states the following: The $s \times s$ matrix

$$\mathbf{C}_s(0) = \begin{pmatrix} 0 & 1 & 0 & \cdots & 0 \\ \vdots & \ddots & \ddots & & \vdots \\ & & \ddots & \ddots & 0 \\ \vdots & & & \ddots & 1 \\ 0 & \cdots & & \cdots & 0 \end{pmatrix} \qquad (B.1)$$

is similar to the $s \times s$ matrix $\mathbf{q}$ with matrix elements

$$q_{kl} = \left(\delta_{kl} - \frac{1}{s} \right) \exp\left(i\pi \frac{k+l-2}{s} \right) \qquad (B.2)$$

where

$$1 \leqslant k, l \leqslant s \qquad (B.3)$$

$\mathbf{C}_s(0)$ represents a Jordan block. Therefore, no similarity transformation can diagonalize it.

(b) An intuitive description of the physical significance of the second-order density operator $\Gamma^{(2)}$ is given by the following remarks. Let us consider an amorphous condensed system, like a liquid or a melt. It exhibits no particular symmetry or measurable long-range order. This system may be viewed as being composed of many "identical copies" of a sufficiently smaller sub-system. (It is not necessary to specify the size of the subsystem.) The possible physical states of a subsystem (at time t) are assumed to be represented by a set

$$S_t = \{ |k\rangle \,|\, k = 1, \ldots, s \} \qquad (B.4)$$

of state functions $|k\rangle$, with s being sufficient large, that is, $s \gg N$, where N is the number of physical particles. Owing to the interactions between the different subsystems, the elements of S_t vary with time going on. After a time interval δt they constitute another set $S_{t+\delta t}$ being denoted by

$$S_{t+\delta t} = \{ |k+s\rangle \,|\, k = 1, \ldots, s \} \qquad (B.5)$$

The following point is crucial: If δt is sufficiently short, the possible states $|k\rangle$ and $|k+s\rangle$ are connected through quantum correlations (often called EPR correlations). In many cases, the existence of these correlations is easily shown to follow directly from the standard energy–time uncertainty relation. (For an explicit example concerning the molecular motion in liquids, see Ref. 98.)

Thus, in the considered case, the relevant physical situation of a subsystem during a short time interval δt ought to be associated with the set of correlated (or "paired") states $|k, k+s\rangle$, that is, with

$$\mathscr{S} = \{|k, k+s\rangle \; | \; k = 1, \ldots, s\} \tag{B.6}$$

(c) The theoretical entity describing the preceding quantum entanglement of physical states in the level of "two-particle-" (better, two-states-) correlations is the second-order reduced density matrix $\Gamma^{(2)}$. For fermionic systems, many fundamental results on $\Gamma^{(2)}$ have been achieved, a considerable part of them being proved by Coleman (see Section IV).

Coleman considered the $\Gamma^{(2)}$ defined with respect to the AGP function $|g^{(N/2)}\rangle$, where $N/2$ is the number of "paired fermions," $|g\rangle$ is a geminal of rank s, with $s > N/2$,

$$|g(1,2)\rangle = \sum_{i=1}^{s} g_i |\phi_i, \phi_{i+s}\rangle \tag{B.7}$$

and the AGP ("antisymmetrized geminal power") is defined by

$$|g^{N/2}\rangle = \mathscr{A}_N\{g(1,2)g(3,4)\cdots g(N-1,N)\} \tag{B.8}$$

$\mathscr{A}_N$ being the conventional antisymmetrization operator. This AGP function is called extreme if the eigenvalues of the corresponding first-order reduced matrix $\Gamma^{(1)}(g)$ are equal. In this case, all the wave amplitutes g_i in (B.7) are equal.

It was then proved that $\Gamma^{(2)}(g^{N/2})$ exhibits a simple "box and tail" matrix form, if it is represented in the basis

$$\{\{|i, i+s\rangle, \; 1 \leqslant i \leqslant s\}, \{|i, j\rangle, \; j \neq i+s, \; 1 \leqslant i, j \leqslant 2s\}\} \tag{B.9}$$

In the following one considers extreme AGP's exclusively. In this case $|g\rangle$ is an eigenfunction of the matrix $\Gamma^{(2)}(g^{N/2})$ corresponding to the large eigenvalue $\lambda_L^{(2)}$ of the matrix $\Gamma^{(2)}(g^{N/2})$. The remaining eigenvalues are equal and constitute a $(2s+1)(s-1)$ degenerate eigenvalue $\Gamma_S^{(2)}$, which is very small, if $2s/N \gg 1$. The matrix $\Gamma^{(2)}(g^{N/2})$ can therefore be decomposed as follows:

$$\Gamma^{(2)} = \Gamma^{(2)}(g^{N/2}) = \Gamma^{(2)}_{\text{Box}} + \Gamma^{(2)}_{\text{Tail}}$$
$$= (\Gamma^{(2)}_L + \Gamma^{(2)}_S) + \Gamma^{(2)}_{\text{Tail}} \tag{B.10}$$

where the "large box part" is identical with

$$\Gamma^{(2)}_L = |g\rangle \lambda^{(2)}_L \langle g|, \quad |g\rangle = \frac{1}{\sqrt{s}}\sum_i |\phi_i, \phi_{i+s}\rangle \tag{B.11}$$

the "small box part" is defined by

$$\Gamma^{(2)}_S = \lambda^{(2)}_S \sum_{i,j} |\phi_i, \phi_{i+s}\rangle (\delta_{ij} - 1/s)\langle \phi_j, \phi_{j+s}| \tag{B.12}$$

and the completely uncorrelated "tail part" is given by

$$\Gamma^{(2)}_{\text{Tail}} = \lambda^{(2)}_S \left\{ \sum_{\substack{i<j \\ i+s\neq j}} |\phi_i, \phi_j\rangle \langle \phi_i, \phi_j| \right\} \tag{B.13}$$

(Full details are presented in Section IV.)

For our purposes, the relevant part of this $\Gamma^{(2)}$ is the small box part $\Gamma^{(2)}_S$, Eq. (B.12), where the constant $\lambda^{(2)}_S$ depends on the (finite) number of particles N and the accompanying (finite number of) fermionic degrees of freedom $(s > N)$,

$$\lambda^{(2)}_S = \frac{N(N-2)}{4s(s-1)}$$

(d) In order to apply the preceding density matrix formalism to a physical microdynamical process exhibiting irreversibility, the focus is on the resonance picture of unstable states [20]. The "part of the ensemble-correlations" represented by (B.12) is then subjected to the "thermalization" transformation (with $\beta = 1/k_B T$)

$$\gamma \equiv \frac{1}{Z} e^{-(\beta/2)H^c} \Gamma^{(2)c}_S e^{-(\beta/2)H^c} \tag{B.14}$$

This transformation makes the connection with the canonical ensemble formalism of statistical mechanics. Here, H^c represents the appropriate complex scaled second-order reduced Hamiltonian and Z is the appropriate normalization factor. The notation $(\ldots)^c$ refers to complex scaled quantities.

From our CSM investigations it follows, considering the additional natural assumption that the (real) energies of all the pairs $|(k, k + s)^c\rangle$ are equal, that is,

$$E_k = E \quad (k = 1, 2, \ldots, s) \tag{B.15}$$

that

$$\gamma \equiv \sum_{k,l} \gamma_{kl} |(k, k + s)^c\rangle\langle(l, l + s)^{c*}| \tag{B.16}$$

with

$$\gamma_{kl} = \frac{\lambda_s^{(2)}}{Z} e^{-\beta E} \left(\delta_{kl} - \frac{1}{s} \right) e^{i(\beta/2)(\varepsilon_k + \varepsilon_l)} \tag{B.17}$$

Here the familiar Boltzmann factor appears explicitly.

The general CSM formalism relates the so-called widths ε_k to the corresponding lifetimes τ_k of the resonances in the usual fashion, that is,

$$\varepsilon_k = \hbar/2\tau_k \tag{B.18}$$

At this stage one can observe the following. The matrix elements γ_{kl} of the operator γ are proportional to the matrix elements of the matrix q, Eq. (B.2), if one requires the validity of the quantization conditions

$$\pi \frac{k}{s} = \frac{\beta}{2} \varepsilon_k \quad (k = 1, 2, \ldots, s) \tag{B.19}$$

In this case, one has the equality

$$\gamma_{kl} = \frac{1}{Z} \lambda_s^{(2)} e^{-\beta E} q_{kl} \equiv \text{const} \times q_{kl} \tag{B.20}$$

In other words, if the "widths" ε_k and the energies E_k of the complex scaled pairs $|(k, k + s)^c\rangle$ fulfill the conditions (B.15) and (B.19), then the matrix elements of the density operator γ constitute a Jordan block similar to $\mathbf{C}_s(0)$, Eq. (B.1). Thus, this operator γ has no diagonal representation. This implies furthermore that all the states constituting γ act as an indivisible unit. The probabilistic interpretation of the diagonal elements γ_{kk} is lost here.

The density operator γ gives the mathematical representation of the coherent-dissipative structures, that is, the cooperative phenomenon of interest.

(e) In Section V, it is demonstrated that the degeneracy condition (B.15) has a natural physical interpretation. The same holds true for the "paired states" mentioned previously, which are assumed to be equally distributed over all the possible correlated states $|k, k + s\rangle$. This assumption corresponds to the best unbiased choice being in line with the basic ideas of equilibrium (and quasiequilibrium) statistical mechanics.

Although it may be beyond the present discussion to comment on the relevance of the choice of a strong AGP-pairing component in the total wavefunction (as Coleman suggests, see Section IV and Ref. 20 for details), one should realize that the AGP function leads to the preceding simple box and tail form for the second-order reduced density matrix. The box part of this matrix is not wavefunction-representable.

The analysis of Section V gives the physical reasons or motivations or both for using the operator γ, as described previously. It then becomes obvious that Jordan blocks associated with the quantization condition (B.19) correspond with a short-lived "cooperative behavior" between possible degenerate quantum states of one and the same physical subsystem via time-dependent correlations, although, as pointed out previously, the whole system (for example, a liquid or a melt) in other respects exhibits no particular symmetry or long-range order. Hence, CSM introduces a novel quantum correlation between

(i) a finite number of degrees of freedom (of fermionic character) represented by the thermalized canonical (reduced) density matrix of a subsystem, and

(ii) infinitely many degrees of freedom of the environment or thermal bath, where the latter—in some sense—stabilizes the associated coherent-dissipative structures for a short time interval (τ_{rel}, see following paragraph).

From the quantization conditions (B.19) and some physical reasoning, one finally derives the result that the minimal "dimension" s_{min} of γ is given by

$$s_{min} = \frac{4\pi k_B T}{\hbar}\tau_{rel} \qquad (B.21)$$

τ_{rel} represents the relaxation time (or lifetime) characterizing the specific microdynamical process of one subsystem. Thus, s_{min} determines the minimal "size" (in the space of state functions) that the Jordan block at least must have, and at the same time it defines the new "unit" that we call a coherent-dissipative structure. Since all the paired states $|(k, k + s)^c\rangle$ coalesce and act as a whole, it is not longer possible to apply the usual probability interpretation to each of them. In this context it is interesting to observe that the trace of γ vanishes identically, as, for example, one sees from Eq. (B.1).

Acknowledgments

The author is indebted to E. Brändas (Uppsala), G. Hertz (Karlsruhe), E. Karlsson (Uppsala, Stockholm), and H. Weingärtner (Karlsruhe) for many elucidating discussions and other active support concerning several parts of this work, during the last three years. The author wishes to express his gratitude to E. Brändas and G. Hertz for fruitful collaboration in the framework of a common research program on coherent-dissipative structures. This work is, in part, supported by the Commission of the European Communities (SCIENCE-Program), the Deutsche Forschungsgemeinschaft, and the Fonds der Chemischen Industrie.

References

1. P. A. M. Dirac, *Proc. Roy. Soc. (London) A* **114**, 243 (1927).

2. P. A. M. Dirac, *The Principles of Quantum Mechanics*, 4th ed., Clarendon, Oxford, 1984.

3. E. Balslev and J. M. Combes, *Commun. Math. Phys.* **22**, 280 (1971).

4. J. Aguilar and J. M. Combes, *Commun. Math. Phys.* **22**, 269 (1971).

5. B. Simon, *Commun. Math. Phys.* **27**, 1 (1972); *Ann. Math.* **97**, 247 (1973).

6. M. Reed and B. Simon, *Methods of Modern Theoretical Physics*, Academic Press, New York, 1978, Vol. IV.

7. See *Int. J. Quant. Chem.* **14**, (4) (1978), and **31** (5) (1987), which are devoted entirely to complex scaling.

8. C. van Winter, *J. Math. Anal.* **47**, 633 (1974).

9. E. Brändas, *Int. J. Quantum Chem.* **S20**, 119 (1986).

10. P.-O. Löwdin, *Adv. Quantum Chem.* **19**, 87 (1988).

11. W. P. Reinhardt, *Ann. Rev. Phys. Chem.* **33**, 223 (1982).

12. B. R. Junker, *Adv. At. Mol. Phys.* **18**, 207 (1982).

13. Y. K. Ho, *Phys. Rep. C* **99**, 1 (1983).

14. P. Krylstedt, N. Elander, and E. Brändas, *J. Phys. B: At. Mol. Opt. Phys.* **22**, 1623 (1989).

15. P.-O. Löwdin, P. Froelich, and M. Mishra, *Int. J. Quantum Chem.* **36**, 93 (1989).

16. M. Rittby, Resonant Phenomena in Atomic Collosions, PhD thesis, University of Stockholm, Research Institute of Physics, 1985.

17. T. Maniv, E. Engdahl, and N. Moiseyev, *J. Chem. Phys.* **86**, 1048 (1987).

18. E. J. Brändas and N. Elander (Eds.), *Resonances—The Unifying Route towards the Formulation of Dynamical Processes*, Lecture Notes in Physics, Springer, Berlin, 1989, Vol. 325.

19. C. A. Chatzidimitriou-Dreismann and E. J. Brändas, *Ber. Bunsenges. Phys. Chem.* **92**, 549 (1988).

20. E. J. Brändas and C. A. Chatzidimitriou-Dreismann, in Ref. 18, pp. 485–540.

21. I. Prigogine, *From Being To Becoming*, Freeman, San Francisco, 1980.

22. I. Prigogine, C. George, F. Henin, and L. Rosenfeld, *Chem. Scr.* **4**, 5 (1973).

23. I. Prigogine, *Nature* **246**, 67 (1973).

24. C. George, F. Henin, F. Mayné, and I. Prigogine, *Hadronic J.* **1**, 520 (1978).

25. C. George, F. Mayné, and I. Prigogine, *Adv. Chem. Phys.* **61**, 223 (1985).

26. Ch. Obcemea and E. Brändas, *Ann. Phys.* **151**, 383 (1983).

27. C. A. Chatzidimitriou-Dreismann, *Int. J. Quantum Chem. Symp.* **19**, 369 (1986); *Int. J. Quantum Chem.* **23**, 1505 (1983).

28. A. J. Coleman, *Rev. Mod. Phys.* **35**, 668 (1963).

29. A. J. Coleman, *J. Math. Phys.* **6**, 1425 (1965).

30. A. J. Coleman, in *Quantum Statistics and the Many-Body Problem* (S. B. Trickey, W. P. Kirk, and J. W. Dufty, Eds.), Plenum Press, New York, 1975.

31. A. J. Coleman, *Phys. Rev. Lett.* **13**, 406 (1964).

32. J. Bardeen, L. N. Cooper, and J. R. Schrieffer, *Phys. Rev.* **108**, 1175 (1957).

33. P. Glansdorff and I. Prigogine, *Thermodynamic Theory of Structure, Stability, and Fluctuations*, Wiley, New York, 1971.

34. G. Nicolis and I. Prigogine, *Self-Organization in Nonequilibrium System*, Wiley, New York, 1977.

35. C. A. Chatzidimitriou-Dreismann and E. J. Brändas, *Int. J. Quantum Chem.* **37**, 155 (1990).

36. C. A. Chatzidimitriou-Dreismann and E. J, Brändas, *Ber. Bunsenges. Phys. Chem.* **93**, 1065 (1989).

37. C. A. Chatzidimitriou-Dreismann, *Int. J. Quantum Chem. Symp.* **23**, 153 (1989).

38. C. A. Chatzidimitriou-Dreismann, *Ber. Bunsenges. Phys. Chem.* **94**, 234 (1990).

39. H. Weingärtner and C. A. Chatzidimitriou-Dreismann, *Nature* **346**, 548 (1990).

40. C. A. Chatzidimitriou-Dreismann, E. J. Brändas, and E. Karlsson, *Phys. Rev.* **B**: *Rapid Commun.* **42**, 2704 (1990).

41. E. J. Brändas and C. A. Chatzidimitriou-Dreismann, *Ber. Bunsenges. Phys. Chem.*, submitted (1990).

42. R. Balescu, *Equilibrium and Nonequilibrium Statistical Mechanics*, Wiley, New York, 1975.

43. C. A. Chatzidimitriou-Dreismann, *J. Math. Phys.* **27**, 2770 (1986).

44. A. Einstein, B. Podolsky, and N. Rosen, *Phys. Rev.* **47**, 777 (1935).

45. A. Aspect, J. Dalibard, and G. Roger, *Phys. Rev. Lett.* **49**, 1804 (1982).

46. E. Lippert, C. A. Chatzidimitriou-Dreismann, and K.-H. Naumann, *Adv. Chem. Phys.* **57**, 311 (1984); see chapter 3.

47. E. Lippert, W. Rettig, V. Bonačić-Koutecký, F. Heisel, and J. A. Miehé, *Adv. Chem. Phys.* **68**, 1 (1987).

48. C. A. Chatzidimitriou-Dreismann, *Physica A* **159**, 109 (1989).

49. S. S. Schweber, *An Introduction to Relativistic Quantum Field Theory*, Harper & Row, New York, 1964.

50. H. D. Zeh, *The Physical Basis of the Direction of Time*, Springer, Berlin, 1989.

51. K. G. Denbigh, *Three Concepts of Time*, Springer, Berlin, 1981.

52. J. von Neumann, *Mathematische Grundlagen der Quantenmechanik*, Springer, Berlin, 1932.

53. B. Nagel, in Ref. 18, pp. 1–10.

54. L. D. Landau and E. M. Lifshitz, *Statistical Physics—Part 1*, 3rd ed., Pergamon, Oxford, 1980.

55. G. Gamow, *Z. Phys.* **51**, 204 (1928).

56. A. J. F. Siegert, *Phys. Rev.* **56**, 750 (1939).

57. J. Kumičák and E. Brändas, *Int. J. Quant. Chem.* **32**, 669 (1987).

58. E. Nelson, *Ann. Math.* **70**, 572 (1959).

59. This example is due to B. Nagel; see Ref. 20.
60. D. A. Micha and E. Brändas, *J. Chem. Phys.* **55**, 4792 (1971).
61. E. Brändas and P. Froelich, *Phys. Rev. A* **16**, 2207 (1977).
62. E. Brändas, *J. Mol. Spectr.* **27**, 236 (1968).
63. E. Engdahl, E. Brändas, M. Rittby, and N. Elander, *J. Math. Phys.* **27**, 2629 (1986).
64. C. E. Reid and E. J. Brändas, in Ref. 18, pp. 475–483.
65. F. R. Gantmacher, *The Theory of Matrices*, Chelsea, New York, 1959, Vol. II.
66. P.-O. Löwdin, *Int. J. Quantum Chem. Symp.* **16**, 485 (1982).
67. P.-O. Löwdin, *Int. J. Quantum Chem.* **12** (Suppl. 1), 197 (1978); *Int. J. Quantum Chem.* **21**, 275 (1982).
68. Ch. Obcemea, P. Froelich, and E. J. Brändas, *Int. J. Quantum Chem.* **S15**, 695 (1981).
69. E. Brändas and P. Froelich, *Int. J. Quantum Chem.* **13**, 619 (1978).
70. N. Moiseyev, P. R. Certain, and F. Weinhold, *Mol. Phys.* **36**, 1631 (1978).
71. N. Moiseyev and S. Friedland, *Phys. Rev. A* **22**, 618 (1980).
72. N. Moiseyev, in *Lecture Notes in Physics*, Springer, Berlin, 1985, Vol. 256, p. 122.
73. A. J. Engelmann, M. A. Natiello, M. Höghede, E. Engdahl, and E. Brändas, *Int. J. Quantum Chem.* **31**, 841 (1987).
74. For a novel approach to the concept of quantum chaos, see (a) Th. Zimmermann, L. S. Cederbaum, H.-D. Meyer, and H. Köppel, *J. Phys. Chem.* **91**, 4446 (1987); (b) Th. Zimmermann, H. Köppel, L. S. Cederbaum, G. Persch, and W. Demtröder, *Phys. Rev. Lett.* **61**, 3 (1988).
75. E. G. Larson, *Int. J. Quantum Chem. Symp.* **20**, 95 (1986).
76. P.-O. Löwdin, *Phys. Rev.* **97**, 1474 (1955).
77. C. N. Yang, *Rev. Mod. Phys.* **34**, 694 (1962).
78. J. Linderberg and Y. Öhrn, *Int. J. Quantum Chem.* **12**, 161 (1977).
79. B. Weiner and O. Goscinski, *Phys. Rev. A* **22**, 2374 (1980).
80. F. Sasaki, *Phys. Rev.* **138**, B1338 (1965).
81. M. D. Girardeau, in *Proceedings of Conference on Reduced Density Matrices with Applications to Physical and Chemical Systems*, Queens Papers on Pure and Applied Mathematics, No. 11, p. 111 (1968).
82. H. Primas, *Chemistry, Quantum Mechanics, and Reductionism*, Springer, Berlin, 1983.
83. B. d'Espagnat, *Conceptual Foundations of Quantum Mechanics*, 2nd ed., Benjamin, London, 1976.
84. J. A. Wheeler and W. H. Zurek (Eds.), *Quantum Theory and Measurement*, Princeton Univ. Press, Princeton, NJ 1983.
85. B. d'Espagnat, *Physics Reports* **110**, 201 (1984).
86. B. d'Espagnat, *In Search of Reality*, Springer, New York, 1983.
87. B. d'Espagnat, *Reality and the Physicist—Knowledge, Duration and the Quantum World*, Cambridge Univ. Press, Cambridge, 1989.
88. T. K. Lim, *Chem. Phys. Letters* **4**, 521 (1970).
89. P.-O. Löwdin, Set theory and linear algebra—some mathematical tools to be used in quantum theory. Part II: Binary product spaces and their operators, *Uppsala Technical Note* **509**, (1977).
90. J. M. Blatt, *Prog. Theoret. Phys.* (*Kyoto*) **23**, 447 (1960).

91. M. Eigen and L. de Maeyer, *Proc. Roy. Soc. (London)* **A247**, 505 (1958).

92. S. Meiboom, *J. Chem. Phys.* **34**, 375 (1961).

93. M. Eigen, *Angew. Chem.* **75**, 489 (1963); *Angew. Chem. Internat. Edit.* **3**, 1 (1964).

94. H. G. Hertz, *Chem. Scr.* **27**, 479 (1987).

95. H. G. Hertz, B. M. Braun, K. J. Müller, and R. Maurer, *J. Chem. Educ.* **64**, 777 (1987) and references cited therein.

96. J. P. Hansen and I. R. McDonald, *Theory of Simple Liquids*, Academic, London, 1976.

97. D. A. McQuarrie, *Statistical Mechanics*, Harper & Row, New York, 1976.

98. C. A. Chatzidimitriou-Dreismann, *J. Mol. Liq.* **39**, 53 (1988).

99. H. G. Hertz, M. Holz, B. M. Braun, T. Frech, R. Maurer, and K. J. Müller, *Z. Phys. Chem. NF* **141**, 133 (1984).

100. Z. Luz and S. Meiboom, *J. Am. Chem. Soc.* **86**, 4768 (1964).

101. E. J. Brändas and C. A. Chatzidimitriou-Dreismann, *Int. J. Quantum Chem. Symp.* **23**, 147 (1989).

102. A. Loewenstein and A. Szöke, *J. Am. Chem. Soc.* **84**, 1151 (1962).

103. R. E. Glick and K. C. Tewari, *J. Chem. Phys.* **44**, 546 (1966).

104. J. A. Glasel, in *Water—A Comprehensive Treatise*, F. Franks (Ed.), Plenum, New York, 1972. Vol. 1, pp. 215–254.

105. R. A. Robinson and R. H. Stokes, *Electrolyte Solutions*, Butterworths, London, 1970.

106. S. W. Rabideau and H. G. Hecht, *J. Chem. Phys.* **47**, 544 (1967).

107. H. G. Hertz, H. Versmold, and C. Yoon, *Ber. Bunsenges. Phys. Chem.* **87**, 577 (1983).

108. B. Halle and G. Karlström, *J. Chem. Soc. Faraday Trans. 2*, **79**, 1031 (1983).

109. A. Gierer and K. Wirtz, *Ann. Physik Lpz.* (6) **6**, 257 (1949).

110. D. L. Turner, *Mol. Phys.* **40**, 949 (1980).

111. R. Pfeifer and H. G. Hertz, *Ber. Bunsenges. Phys. Chem.* **94**, (November, 1990), in press.

112. G. Zunder and M. Eckert, *J. Mol. Struc. (Theochem)* **200**, 73 (1989).

113. R. H. Stokes, *J. Phys. Chem.* **65**, 1242 (1961).

114. R. Haase, P.-F. Sauermann, and K. H. Dücker, *Z. Phys. Chem. NF* **47**, 224 (1965).

115. G. S. Kell, *J. Chem. Ref. Data* **6**, 1109 (1977).

116. L. G. Longsworth, *J. Phys. Chem.* **64**, 1914 (1960).

117. O. E. Frivold, O. Hassel, and E. Hetland, *Avhandl. Norske Videnskaps Akad. Oslo I Mat. Naturv. Kl.*, No. 9, 1 (1942); cited after Ref. 118.

118. *Gmelin's Handbuch der Anorganischen Chemie*, 8th ed., *Chlor*, Suppl. Issue Part **B1**, p. 1, Verlag Chemie, Weinheim, 1969.

119. L. G. Longsworth and D. A. McInnes, *J. Am. Chem. Soc.* **59**, 1666 (1937).

120. D. G. Swain and D. F. Evans, *J. Am. Chem. Soc.* **88**, 383 (1966).

121. G. Jones and H. Fornwald, *J. Chem. Phys.* **4**, 30 (1936).

122. H. Weingärtner, *Ber. Bunsenges. Phys. Chem.* **88**, 47 (1984).

123. K. G. Weil, "Structure-transport relations in molten salts," in *Physics and Chemistry of Electrons and Ions in Condensed Matter*, J. V. Acrivos et al. (Eds.), Reidel, New York, 1984, pp. 255–271.

124. G. G. W. Greening and K. G. Weil, *Z. Naturforsch.* **39a**, 765 (1984).

125. H.-H. Emons, G. Bräutigam, and H. Vogt, *Z. Chem.* **10**, 344 (1970).

126. E. R. van Artsdalen and I. S. Yaffe, *J. Phys. Chem.* **59**, 118 (1955).

127. G. J. Janz, *Molten Salts Handbook*, Academic, New York, 1967.

128. H. Reiss, S. W. Mayer, and J. L. Katz, *J. Chem. Phys.* **35**, 820 (1961).

129. J. W. Cable and R. M. Nicklow, *Phys. Rev. B* **39**, 11732 (1989).

130. O. Hartmann, E. Karlsson, R. Wäppling, J. Chappert, A. Yaouanc, L. Asch and G. M. Kalvius, *J. Phys. F: Met. Phys.* **16**, 1593 (1986).

131. J. W. Cable, R. M. Nicklow, and N. Wakabayashi, *Phys. Rev. B* **32**, 1710 (1985).

132. E. Wäckelgård, O. Hartmann, E. Karlsson, R. Wäppling, L. Asch, G. M. Kalvius, J. Chappert, and A. Yaouanc, *Hyperfine Interactions* **31**, 325 (1986).

133. J. Muller and J. L. Olsen (Eds.), Proceedings of the International Conference on High Temperature Superconductors and Materials and Mechanism of Superconductivity, Interlaken, Switzerland, 28 February–4 March 1988, *Physica* **C153–155**, (1988).

134. J. W. Halley (Ed.), *Theories of High Temperature Superconductivity*, Addison-Wesley, Redwood City, CA, 1988.

135. R. T. Collins, Z. Schlesinger, F. Holtzberg, and C. Feild, *Phys. Rev. Lett.* **63**, 422 (1989).

136. J.-M. Imer, F. Patthey, B. Dardel, W.-D. Schneider, Y. Baer, Y. Petroff, and A. Zettl. *Phys. Rev. Lett.* **62**, 336 (1989).

137. C. G. Olson, R. Liu, A.-B. Yang, D. W. Lynch, A. J. Arko, R. S. List, B. W. Veal, Y. C. Chang, P. Z. Jiang, and A. P. Paulikas, *Science* **245**, 731 (1989).

138. J. E. Demuth, B. N. J. Persson, F. Holtzberg, and C. V. Chandrasekhar, *Phys. Rev. Lett.* **64**, 603 (1990).

139. Y. J. Uemura, G. M. Luke, B. J. Sternlieb, J. H. Brewer, J. F. Carolan, W. N. Hardy, R. Kadono, J. R. Kempton, R. F. Kiefl, S. R. Kreitzman, P. Muhlhern, T. M. Riseman, D. Ll, Williams, B. X. Yang, S. Uchida, H. Takagi, J. Gopalakrishnan, A. W. Sleight, M. A. Subramanian, C. L. Chien, M. Z. Cieplak, Gang Xiao, V. Y. Lee, B. W. Statt, C. E. Stronach, W. J. Kossler, and X. H. Yu, *Phys. Rev. Lett.* **62**, 2317 (1989).

140. O. Penrose and L. Onsager, *Phys. Rev.* **104**, 576 (1956).

141. H. Seidel, F. Hentsch, M. Mehring, J. G. Bendnorz, and K. A. Müller, *Europhys. Lett.* **5**, 647 (1988).

142. A. C. Aitken, *Determinants and Matrices*, Interscience, New York, 1951.

143. E. Karlsson, C. A. Chatzidimitriou-Dreismann, and E. J. Brändas, *Hyperfine Interactions*, in the press (1990/91).

144. E. Karlsson, E. J. Brändas, and C. A. Chatzidimitriou-Dreismann, *Phys. Scr.*, in the press (1991).

CHEMICAL KINETICS OF FLUE GAS CLEANING BY IRRADIATION WITH ELECTRONS

H. MÄTZING

Kernforschungszentrum Karlsruhe GmbH, Laboratorium für Aerosolphysik und Filtertechnik I, Karlsruhe, FRG

CONTENTS

Advances in Chemical Physics Volume LXXX, Edited by I. Prigogine and Stuart A. Rice
ISBN 0-471-53281-9 © 1991 John Wiley & Sons, Inc.

PART II: THE AGATE-CODE

PART I: CHEMICAL KINETICS OF FLUE GAS CLEANING BY ELECTRON BEAM

I. INTRODUCTION

NO_x and SO_2 emissions from fossil fuel burning power plants have been recognized as sources of atmospheric and biospheric hazards. Their avoidance is being forced presently. Existent pollutant control technologies solve this task mostly in two stages that reduce NO_x and SO_2 emissions separately, yielding economically unimportant products like N_2 and $CaSO_4$. The EBDS (Electron Beam Dry Scrubbing) process offers an economic alternative: the irradiation of the flue gas with fast (300–800 keV) electrons initiates the build-up of radical concentrations that are high enough to oxidize NO_x and SO_2 traces simultaneously. Only a minor portion of the nitrogen oxides are transformed to molecular nitrogen and dinitrogen oxide. The oxidation products are nitric and sulfuric acids. Ammonia addition induces the formation of a mixed ammonium nitrate/sulfate aerosol, which can be collected as solid and sold as fertilizer.

Originally developed in Japan (Tokunaga et al., 1978, 1984; Suzuki and Tokunaga, 1981), the EBDS process has been promoted in the United States and West Germany (Frank et al., 1985, 1988; Jordan, 1988) and has gained international recognition (Markovic, 1987). Detailed model studies have provided much insight into the chemical kinetics of the process (Nishimura et al., 1979, 1981; Person et al., 1985, 1988; Busi et al., 1987, 1988; Gentry et al., 1988), although some questions are unresolved.

The EBDS process involves very different physicochemical steps, like energy absorption, reactions in homogeneous gas phase, and heterogeneous aerosol particle and mass growth. Energy absorption produces chemically active species at concentration levels that represent a highly unstable state compared to thermal equilibrium. In this sense irradiation by e-beam causes a sudden deviation from thermodynamic equilibrium in the waste gas. Subsequent relaxation establishes a new equilibrium state that is characterized by lower NO_x/SO_2 concentrations and aerosol formation. A theoretical description of this relaxation process is hardly possible by simple thermodynamics, but requires the use of appropriate kinetic models. The AGATE-code has been developed for this purpose and the present study is widely based on its results as well as their analysis.

The goal of this study is to show how microscopic molecular interactions work together and determine the characteristics, performance and thereby the economics of the EBDS process. After a short description of the primary radiolytic events, the chemistry of the primary active species is considered. The reactions of positive ions are shown to constitute the major source of neutral radicals. These radicals are needed to convert NO_x to nitric acid and SO_2 to sulfuric acid. The OH radical turns out to be the most important radical for the formation of these acids and hence the final nitrate/sulfate aerosol. In addition, nitric acid is also produced directly from some ion–molecule reactions that work most efficiently at high concentrations of water vapor.

The oxidation of NO_x by radicals is not a simple, straightforward reaction sequence, however. Part of the intermediate NO_2 is reduced back to NO by oxygen atoms. Furthermore, intermediate HNO_2 is likely to decompose at surfaces, which acts as an OH sink. Such "back-reactions" determine the dose dependence of NO_x removal and thereby the economics of the EBDS process. Other reductive pathways yield N_2O as a gaseous by-product and also yield molecular nitrogen. The nitrogen formation is not easy to measure and therefore, the N balance is difficult to investigate experimentally.

The added ammonia partly enters into the radiation-induced radical chemistry. It favors the reduction of NO to N_2 and NO_2 to N_2O. The reactions behind are similar to those of the thermal $DeNO_x$ process.

Most of the ammonia, of course, is consumed by ammonium salt formation. Ammonia thus provides a link between gas-phase chemistry and particulate formation. The properties of the developing aerosol are investigated and heterogeneous reactions at the aerosol surface are discussed.

All these physicochemical mechanisms work together simultaneously. Kinetic models allow one to quantify the net effects of single mechanisms or reactions separately and to assess their contributions and importance for the entire process. This reveals the molecular interactions that are responsible for the measurable performance characteristics of the EBDS process, like dose dependence of removal yields and product formation or relative humidity effects, for instance.

Many results described in Part I of this chapter have been obtained from the AGATE-code. A complete listing of the reaction set is given in Part II.

II. RADIOLYSIS

A. The Fate of Fast Electrons

The interaction of electrons with matter depends both on the electron energy and on certain target properties. In the EBDS process, the energy of incident electrons is in the range of 300–800 keV typically. This energy is too low to permit close electron–nuclei interactions, the origin of bremsstrahlung.

Rather, the incident electrons transfer part of their energy to the electron shells of molecules by inelastic collisions. These collisions are also associated with momentum transfer and the electrons are readily scattered throughout the irradiated medium. A full description of the energy release by accelerated electrons is fairly complicated (Meissner, 1964) and the history of an individual electron is only accessible by numerical methods. Therefore, integral values have gained practical importance. One example is the so-called linear energy transfer (LET). It is defined as the energy of the incident electron divided by its total path length in the traversed medium (Henglein et al., 1969). The LET depends mainly on the electron energy and the density of the medium, but also on its chemical composition (Lohrmann, 1983). For 300-keV electrons in liquid water, the LET is of the order of 0.5 eV/nm; in air at ambient conditions it is roughly a factor of 1000 lower. The LET gives an idea of the electron range R, although in a rigorous sense, these values are not related. The electron range may be defined as the linear distance from the source at which the average electron energy has decayed to 1–5% of the start value. The accuracy of this definition is subject to practical requirements (Henglein et al., 1969). For 300-keV electrons in air at NTP, R is approximately 0.5 m. Also, it is reasonable to assume $R \sim \rho \cdot E^{1.35}$, where ρ is the material density and E is the incident energy between 300 and 800 keV (Landolt-Börnstein, 1952; Lohrmann, 1983).

The energy loss in single collisions varies statistically between a few eV ("distant collisions") and some tens of keV ("close collisions"). Both of these extremes are comparatively scarce and leave the contact molecules in excited states or as (excited) ions, respectively. In the latter case, secondary electrons with a kinetic energy of many keV may be produced, which may cause further ionization themseleves. In this way, tertiary and higher-order electrons result from ionization processes, which all contribute to the spatial energy distribution initiated by the primary electrons. The overall gain of excited-state molecules, direct dissociation into neutral radicals and dissociation into ion pairs is described by G-values (Willis and Boyd, 1976). These G-values are an average over the combined effects of all orders of electrons. The ionization gain is about three ion pairs per 100 eV absorbed energy in air. It is fairly independent of primary electron energy, but may depend on the dose rate (Willis and Boyd, 1976; Armstrong, 1987).

The EBDS processing of waste gases typically involves dose rates below about 1000 kGy/s, which is a much lower dose rate than may be obtained in Febetron studies. Among others, this comparison has given rise to the question of whether the performance of the EBDS process depends on the dose rate. Many different aspects are related to this subject and will be discussed throughout this chapter. At this stage, it is only considered in terms of energy deposition.

The microscopic time scale of the energy transfer can be estimated from the classical mechanics of elastic collisions neglecting the kinetic energy loss by electron–molecule interaction. As an upper limit, collision times of the order of 10^{-16} s are obtained for an energy transfer of some 10 eV to a single molecule by an electron of several hundred keV incident energy (Armstrong, 1987). This is by far the shortest time scale of any physicochemical process involved in EBDS and it is definitely beyond the scope of technical process control means. Interestingly, this corresponds to a microscopic dose rate around 3×10^{21} kGy/s for molecular weights ≈ 30 g/mol. Obviously, such processes are well beyond the scope of process control.

B. The Fate of Primary Species

Returning to macroscopic considerations, molecular excitation, homolytic dissociation, and ionization are counteracted by quenching, radical recombination, and associative ion–electron recombination, respectively. The first two "deactivation" processes are not directly related to the energy absorption and will be discussed subsequently. Ion–electron recombination can occur only when the electrons have "cooled" down to thermal energy ($kT \approx 0.01$ eV at 273 K). Thermalization takes about 1 ns in air at 1 atm (Armstrong, 1987). During this time, the primary ions may already undergo charge transfer reactions or attach to neutral molecules and form ionic clusters. Owing to Brownian motion, the positive charge (that is, a single or clustered ion) diffuses a linear distance of about 0.1 μm at NTP in the absence of external force fields. This range may be imagined as a spherical ion core, which develops around the ionization point prior to charge neutralization.

Both charge transfer and dissociative neutralization reactions produce radicals. As will be shown below, the lifetime of radicals is at least 10 ns and the quenching of excited transients takes 200 ns on the average. The diffusive motion of these species constitutes a chemical core about the point of electron impact, which is in the micrometer range. According to common terminology, this is called a spur. Along the path of energetic electrons, spurs are created "like beads on a string" (Chatterjee, 1987). The entity of spurs represents the electron track as observed in cloud chambers.

An overlap of spurs (and hence tracks) generated by different electrons can be expected to favor the recombination of active species by a local increase of their concentrations above the normal level of independent energy transfer events. Also, the chemical mechanism may change in this way, for example, through preference of alternative reaction branches. This effect has been accepted to explain the dose-rate-dependent ozone formation in the radiolysis of pure oxygen (Sauer, 1976). The dose rate, at which spur overlap occurs, can be estimated from the G-value for ionization, the spur dimension, and its evolution time. Using the figures given previously, one obtains

$$\dot{D} \approx \frac{(100/3)\,\mathrm{eV}}{(3\,\mu\mathrm{m})^3 \times 10^{-7}\,\mathrm{s}} \approx 10^{19}\,\frac{\mathrm{eV}}{\mathrm{cm}^3\,\mathrm{s}} = 2.5 \times 10^3\,\frac{\mathrm{kGy}}{\mathrm{s}}$$

for air or waste gas conditions. This rough estimate for the onset of dose rate effects is in accordance with recent model calculations on the EBDS process (Gentry et al., 1988). Also, it is well above the dose rates obtained with presently available electron accelerators in waste gas treatment. A necessary minimum does rate for the onset of measurable radiation effects will be discussed in Section III D.

III. GAS-PHASE CHEMISTRY: EXCITED SPECIES, PRIMARY RADICALS, AND IONS

Since its discovery in the early 1970s, the characteristics of the EBDS process have been discussed in terms of the chemical reactions in homogeneous gas phase, which precede and induce particulate formation (Nishimura et al., 1979, 1981). During the past decade, the discussion has been stimulated through the development of extensive computer codes on this part of the process (Busi et al., 1988; Person et al., 1988; this chapter). The results of these modeling studies provide an understanding of most experimental findings. A summary is given here and an interpretation of the various reaction types is offered.

A. Modeling Active Species Generation

A microscopic modeling of energy absorption and active species generation, for example, by Monte Carlo methods, has not been attempted in EBDS models. Rather, integral descriptions of the primary processes are in use, which relate active species formation directly to the dose rate experienced by flue gas:

$$\frac{dn}{dt} = G_n \dot{D} x_i \rho$$

In this basic equation, n is the number concentration of species n, generated from species i with mole fraction x_i in the flue gas. G_n is the corresponding gain [molecules/100 eV], as discussed previously [for details see Klassen (1987); Armstrong (1987)]. $\dot{D}\rho$ is the dose rate times the average density in units of $100 \, \text{eV/cm}^3$ s.

Two basic assumptions are inherent in this equation:

1. energy absorption can be treated as a quasicontinuous process, and
2. the probability of electron impact is proportional to the (mass) concentration of the parent species.

The first assumption is applicable, because only low LET electrons are considered, and is supported by the dose rate consideration in the preceding section. The second assumption considers the collisional cross section for electron–molecule interaction as independent of electron energy and molecule nature. This is valid for electron energies down to about 30 keV (Cole, 1969) and hence over at least 90% of the electron range.

Assumption 2 also suggests that one neglect radiolytic degradation of trace constituents in the flue gas and regard only the major components in energy absorption. From the G-values reported by Willis and Boyd (1976), the relevant stoichiometric equations read:

$$4.43\text{N}_2 \xrightarrow{100\,\text{eV}} 0.29\text{N}_2^* + 0.885\text{N}(^2D) + 0.295\text{N}(^2P)$$
$$+ 1.87\text{N}(^4S) + 2.27\text{N}_2^+ + 0.69\text{N}^+ + 2.96e^-$$

$$5.377\text{O}_2 \xrightarrow{100\,\text{eV}} 0.077\text{O}_2^* + 2.25\text{O}(^1D) + 2.8\text{O}(^3P)$$
$$+ 0.18\text{O}^* + 2.07\text{O}_2^+ + 1.23\text{O}^+ + 3.3e^-$$

$$7.33H_2O \xrightarrow{100\,eV} 0.51H_2 + 0.46O(^3P) + 4.25OH + 4.15H + 1.99H_2O^+$$
$$+ 0.01H_2^+ + 0.57OH^+ + 0.67H^+ + 0.06O^+ + 3.3e^-$$

$$7.54CO_2 \xrightarrow{100\,eV} 4.72CO + 5.16O(^3P) + 2.24CO_2^+$$
$$+ 0.51CO^+ + 0.07C^+ + 0.21O^+ + 3.03e^-$$

This representation implies some simplifications concerning the nature of electronically excited nitrogen and oxygen molecules. Dissociative states have been treated as forming atoms directly. Therefore, N_2^* and O_2^* represent the sum of all not-dissociating excited–state molecules described by Willis and Boyd (1976). In the present context, it is reasonable to treat these as $N_2(A)$ and $O_2(^1\Delta_g)$: The numerical results do not change upon variation of the corresponding G-values by a factor of two (Mätzing, 1987a). O^* denotes a highly excited O atom above the $O(^1S)$ level.

B.   Reactions of Primary Excited Species

Electronically excited-state species arise only from nitrogen and oxygen radiolysis. Those considered here have radiative lifetimes above 1 μs. The rate constants for collisional quenching are less than 10^{-14} cm^3 s^{-1} (Baulch et al., 1980, 1982, 1984; Atkinson et al., 1989), which also gives lifetimes above 1 μs at 1 atm, $T \approx 350$ K. Compared to quenching rates, chemical reaction rates are 50 times faster typically. The reactions of excited-state species with the main constituents of the flue gas therefore take place at a time scale of some hundred nanoseconds, which exceeds electron thermalization times by two orders of magnitudes. So within 200 ns they can diffuse over a linear distance up to 3 μm from their point of origin, before they are consumed (diffusion coefficient $D \approx 0.2$ cm^2/s). This range determines the spur size within which all other reactions occur.

The total gain of excited-state species can be estimated from the data in Section III A and is around 2 per 100 eV. The products of their reactions with the major waste gas constituents are H, N, and O atoms and in particular, only little OH.

Excited species can thus initiate partial NO oxidation to NO_2. Thereafter, reduction reactions become important, yielding NO and N_2O from NO_2, and N_2 from NO. In this way, primary excited species lead to an oxidation–reduction cycle between NO and NO_2, which offers stable exit paths to gaseous products only. Nitric and also sulfuric acid are not formed due to the lack of sufficient OH concentrations. Particulate formation therefore cannot be expected to originate from the generation of excited species.

C. Reactions of Primary Radicals

According to the previously given radiolytic equations, the total radical gain, ΣG (ground-state radicals), is about 3 per 100 eV for a typical flue gas containing 75% N_2, 5% O_2, 10% H_2O, and 10% CO_2. This gives a total radical production around 2×10^{15} cm^{-3} s^{-1} at 10 kGy/s under typical conditions. Consider typical termolecular recombination rate constants around 5×10^{-33} cm^6 s^{-1} and an overall concentration of 2×10^{19} molecules/cm^3. The quasistationary radical concentration is then given by

$$[\text{Radical}] = \left(\frac{2 \times 10^{15}}{(5 \times 10^{-33})(2 \times 10^{19})} \right)^{1/2} \text{cm}^{-3} \approx 10^{14} \, \text{cm}^{-3} \approx 5 \, \text{ppm}$$

On the one hand, this result shows that recombination cannot compete with primary radical production for dose rates $\geq$ 10 kGy/s. However, the types of primary radicals are not much different from those originating from excited species. The primary OH formation from water vapor is not high enough to break the oxidation–reduction cycle between NO and NO_2 and to produce nitric acid effectively. The latter holds for sulfuric acid also. So for the same reasons as given previously, primary radical formation plays a minor part for NO_x/SO_2 removal in the EBDS process.

The radical levels discussed previously are some orders of magnitude higher than those observed under tropospheric or stratospheric conditions (Levine, 1985; Seinfeld, 1986). Therefore, a simple negation of their significance for NO_x/SO_2 degradation may not readily be accepted. Rather, one likes to expect the establishment of self-sustaining reaction chains that reinforce single steps. Chain lengths of the order of 10^3–10^6 are not unusual in gas-phase or liquid-phase chemistry and would be sufficient in the present context.

In the low-temperature NO_x chemistry, such radical chains do not exist, however. One reason is the nature of the envisaged product (acid) formation: Nitrous and nitric acids can only be formed via radical addition (OH, NO_3) to NO or NO_2. Those termolecular reactions can be considered as "terminating" steps, because NO and NO_2 are of radical nature themselves. A more careful investigation of bimolecular NO_x radical reactions further shows that they can at best generate OH and O in the temperature range of interest here. The OH radical can be consumed by recombination with NO_x, while O atoms effectively counteract oxidative pathways via NO_2 + O $\rightarrow$ NO + O_2 (see following discussion).

Chain reactions in the H–C–O subsystem also must be excluded in this context, mainly because some of the propagating steps have high activation energies, but also because of depletion of radical concentrations by interactions with the N–O subsystem. Furthermore, there are only very few SO_x reactions possible in homogeneous gas phase under waste gas conditions and these do not support any chain reaction.

From these arguments it is understood that the primary radicals formed by e-beam irradiation of flue gas constitute only a very limited potential for NO_x/SO_2 degradation. By analogy, the same is recalled for·excited species formation.

D. Reactions of Primary Ions

The preceding discussion leaves the key part of the EBDS chemistry to ionic processes. Among these, charge transfer reactions (whether dissociative or not) predominate: They proceed with rate constants around $5 \times 10^{-10}\,cm^3/s$ at total molecule concentrations near $2 \times 10^{19}\,cm^{-3}$, which corresponds to individual ion lifetimes of only some $10^{-10}\,s$. Positive charge transfer reactions are 10 times as fast as electron thermalization (see previous discussion) and every positive charge can be distributed freely among 10 individual molecules, before the onset of charge neutralization reactions. A positive charge, which is created in an environment of already thermalized electrons, has a 10^7 times greater chance to react with surrounding neutral species than with negatively charged species because of concentration differences (10^{19} neutrals against 10^{12} negative charges). Vice versa, the same holds for the reactions of thermalized electrons which statistically rather recognize a neutral environment than positive charges in their neighborhood. Yet, because neutralization reactions have about 10^3 times higher rate constants than ion–molecule reactions, the aforementioned probability ratio is reduced from $10^7:1$ to $10^4:1$.

These figures, of course, must be regarded as rough estimates with an accuracy around an order of magnitude. Still they clearly demonstrate the preference of ion (electron)–neutral reactions over ionic recombination. In contrast to reactions of excited species and radicals, ion (electron)–molecule reactions naturally do not consume active species, but stabilize the charge separation initiated by e-beam treatment. It is therefore plausible to regard positive and negative ion chemistry as essentially independent or decoupled reactive pathways.

A true distinction of primary ion reactions from higher-order ion reactions, however, is neither possible nor meaningful, because the concentration of any charged species must be expected both to decay and to grow simultaneously owing to charge transfer. It is obviously impossible to discriminate between primary O_2^+ ions and those who have got their charge from N_2^+, CO_2^+, or any other candidate, for example. So, the ionization yields reported previously loose their "fingerprints" in the irradiated gas upon the onset of charge transfer reactions. This also means that charge transfer reactions redistribute the incident energy consumed in primary ionization processes.

For the subsequent discussion of the ion chemistry, it is interesting to estimate the total ionization level established in irradiated waste gas. The quasistationary ion concentrations are given by

$$\frac{dn_+}{dt} = \frac{dn_-}{dt} = G_{ion}\dot{D}\rho - k_{rec}n_+^2 \approx 0$$

where $n_+ = n_-$ are the number concentrations of all positive or negative species, respectively. $G_{ion} \approx 3$ per $100\,eV$ is the average ionization gain in the gas, $\dot{D}$ is the dose rate, and $\rho \approx 1\,kg/m^3$ is the gas density. The mutual recombination rate constant is $k_{rec} \approx 10^{-7}\,cm^3/s$. For $\dot{D} = 10\,kGy/s$, this gives $n_+ = n_- \approx 10^{10}\,cm^{-3}$, which is about 10^6 times higher than typical tropospheric ionization levels (Levine, 1985; Friedlander, 1977). Since the quasistationary charge concentrations are approximately proportional to the square root of dose rate, the dose rate must be greater than about $10^{-10}\,kGy/s$ for the ionization level to exceed the natural background. This extremely small value gives the minimum dose rate required for the onset of observable radiation effects and sets some kind of a physical lower limit, at which dose rate effects can be expected. Certainly, such a figure has no technical relevance.

E.   Positive Ion Chemistry

The charge transfer processes lead the way from comparatively unstable and short-lived ions to more stable ions and ionic clusters with longer lifetimes. Among the primary molecular ions, O_2^+ and H_2O^+ must be considered more stable than N_2^+ and CO_2^+, which is demonstrated by the reaction sequence

$$N_2^+ + O_2 \rightarrow O_2^+ + N_2 \qquad\qquad k_1 = 3.9 \times 10^{-10}\exp(-T/143)\,cm^3/s \quad (1)$$

$$N_2^+ + CO_2 \rightarrow CO_2^+ + N_2 \qquad\qquad k_2 = 8.3 \times 10^{-10}\,cm^3/s \quad\qquad\qquad (2)$$

$$CO_2^+ + O_2 \rightarrow O_2^+ + CO_2 \qquad\qquad k_3 = 6.5 \times 10^{-9}T^{-0.78}\,cm^3\,s \qquad\quad (3)$$

$$N_2^+ + H_2O \rightarrow H_2O^+ + N_2 \qquad\quad k_4 = 2 \times 10^{-9}\,cm^3/s \qquad\qquad\qquad (4)$$

$$CO_2^+ + H_2O \rightarrow H_2O^+ + CO_2 \quad k_5 = 1.7 \times 10^{-9}\,cm^3/s \qquad\qquad\quad (5)$$

$$H_2O^+ + O_2 \rightarrow O_2^+ + H_2O \qquad\quad k_6 = 2.6 \times 10^{-10}\,cm^3/s \qquad\qquad (6)$$

The corresponding reverse reactions are negligible (Anicich and Huntress, 1986), which is characteristic for this reaction type (at temperatures around $100\,°C$).

Charge transfer reactions that regenerate N_2^+ or CO_2^+ do not compete with primary ionization under usual conditions. Hence, considering N_2^+ as illustrative example, its quasistationary concentration can be estimated from

$$\frac{d[N_2^+]}{dt} \approx G_{N_2^+} \dot{D}\rho x_{N_2} - k_{tr}[N_2^+]n - k_{rec}[N_2^+]n_- \approx 0$$

Here, k_{tr} is the rate constant for the charge transfer reactions, n is the concentration of reactant neutrals, and the other symbols have the same meaning as previously.

From the preceding section n_- is known to be

$$n_- \approx \sqrt{G_{ion}\dot{D}\rho/k_{rec}}$$

so that

$$[N_2^+] \approx \frac{G(N_2^+)\dot{D}\rho x_{N_2}}{k_{tr}n + \sqrt{G_{ion}\dot{D}\rho k_{rec}}} = \frac{10^{15}}{4 \times 10^9 + 10^4}\ cm^{-3}$$

The figures hold for the typical waste gas conditions mentioned previously; in particular, $x_{N_2} \approx 0.75$, $k_{tr} \approx 10^{-9}\ cm^3/s$, $k_{rec} \approx 10^{-7}\ cm^3/s$, and $\dot{D} = 10\ kGy/s$. The example is not meant to stress absolute figures, but to point out

1. rapid charge transfer makes many ion concentrations proportional to the dose rate;
2. charge transfer usually is many orders of magnitude faster than ion neutralization.

Again, this supports the previous conclusion on the negligible significance of dose rate effects. But also, reactive pathways must be supposed closely related to the dose rate, just like the ion concentrations.

The key to this relation is obtained from a consideration of dissociative reactions which conserve the total ion concentrations and produce radicals, for example,

$$O_2^+(H_2O) + H_2O \rightarrow H_3O^+ + OH + O_2 \qquad k_7 = 2 \times 10^{-10}\ cm^3/s \qquad (7)$$

Owing to the essential quasistationarity of ion concentrations, reactions like these constitute a quasicontinuous radical source. Using $[H_2O] = 10^{18}\ cm^{-3}$ ($\approx 5\ vol\%$) and $[O_2^+(H_2O)] = 10^6\ cm^{-3}$ (that is, in the previous range for $[N_2^+]$) one obtains

$$\frac{d[OH]}{dt} \approx 2 \times 10^{-10} \times 10^{24}\ cm^{-3}\ s^{-1} \approx 10\ ppm/s$$

at $\dot{D} = 10\,\text{kGy/s}$ for the OH production rate due to reaction (7). The branching reaction (7a)

$$O_2^+(H_2O) + H_2O \rightarrow H_3O^+(OH) + O_2 \qquad k_{7a} = 1.5 \times 10^{-9}\,\text{cm}^3/\text{s} \quad (7a)$$

is roughly 10 times faster than reaction (7) and is followed by the detachment reaction

$$H_3O^+(OH) + H_2O \rightarrow H_3O^+ + OH + H_2O \qquad k_8 = 1.4 \times 10^{-9}\,\text{cm}^3/\text{s} \quad (8)$$

These sequences lead to OH production rates of the order of 100 ppm/s at $\dot{D} = 10\,\text{kGy/s}$. This obviously is the magnitude of radical production rates required for effective process performance. In analogy, other radicals are produced from

$$N^+ + H_2O, CO_2, O_2 \rightarrow N + H_2O^+, CO_2^+, O_2^+$$
$$O^+ + O_2, N_2 \rightarrow O + O_2^+, N_2^+$$
$$H^+ + O_2, H_2O, CO_2 \rightarrow H + O_2^+, H_2O^+, CO_2^+$$

The preceding example has thus revealed charge transfer processes as the major radical source. Like the parent ion concentration, radical production rates are proportional to the dose rate. Consequences thereof will be discussed subsequently.

E.   Negative Ion Chemistry

It has been pointed out that the negative ion chemistry occurs somewhat delayed compared to positive ion chemistry, because the time required for electron thermalization ($\approx 1\,\text{ns}$) is about 10 times longer than the characteristic time scale for (positive) ion–molecule reactions. The principal reactions of thermalized electrons are those of termolecular attachment, leading to O_2^-, NO_x^-, and other negative ions as well as to the corresponding ionic clusters. In analogy to the preceding chapter, one might expect another radical source to originate from negative ion chemistry. This is not the case, however: Negative ion–molecule reactions do not release radicals as "by-products." This particular property makes negative ion chemistry a fairly unimportant sideway in the EBDS process. Arguments in favor of relatively stable negative ions like NO_2^- and NO_3^- do not hold, because these are not at all liable to enter into the particulate phase; rather, they regenerate NO_x by charge neutralization. The author has tested the effect of negative ion chemistry by tentative omission from his computer model: Compared to the full mechanism, the calculated final removal efficiencies for NO_x and SO_2

changed by less than 1 ppm, which is below any realistic detection limit (Mätzing, 1987a).

F.   Ion Mutual Neutralization

Ionic recombination is of course the final fate of the charge separation initiated by impact of high-energy electrons with molecules. It is therefore a necessary condition for the degradation of the absorbed energy, which leads to the final new equilibrium state of the cleaned waste gas. Yet, for the degree of contaminant removal, ionic recombination is totally unimportant, as has been argued throughout this discussion and especially in the preceding section.

H.   Overview on the Chemistry of Primary Species

The irradiation by electron beam results in the direct formation of the following primary reactive species:

1. Excited molecules
2. Radicals in ground and excited states
3. Positive ions

Negative ions are formed from attachment of thermalized electrons to neutral species.

The primary neutral reactants (1) and (2) do not bring about substantial NO_x/SO_2 oxidation, mainly because of their inability to initiate significant OH formation. The OH radical is needed for nitric and sulfuric acid formation.

The positive ions undergo fast charge transfer reactions in which radicals are formed as "by-products." This constitutes the major radical source and in particular the only significant OH source in the EBDS process. In contrast, the chemistry of negative ions does not contribute to radical formation nor to NO_x/SO_2 degradation.

IV.   GAS-PHASE CHEMISTRY: POSITIVE IONS AND RADICALS

So far, the absorption of high-energy electrons in waste gases has been discussed, along, with the formation of primary species and their relevance for contaminant removal. The major radical source has been shown to originate from positive ion chemistry in the gas phase. Still it would be too simplified to regard positive ion–molecule reactions solely under the aspect of radical formation. In fact, they must also be considered as oxidizing reactions. Therefore, their role in contaminant removal cannot strictly be separated from the role of radicals; hence, these mechanisms are discussed together under the same heading.

A. NO_x Oxidation by Positive Ions

From the generalized theory of redox processes it is well known that electron uptake constitutes the transition to a lower oxidation state. Hence, acquirement of a positive charge (that is, release of an electron) is synonymous with oxidation.

Primary ionization can be interpreted in this way. Subsequent charge transfer processes can also be regarded as redox processes. While in the preceding sections only the major constituents of the waste gas have been considered, the focus is now on the fate of trace contaminants. Charge transfer to trace species proceeds at a 1000 times longer time scale ($\approx 10^{-7}$ s) than charge transfer to major components, simply because of the difference in concentration. The most important waste gas contaminants are NO and SO_2, which can readily be oxidized to NO^+ and SO_2^+. Of course, these ions again are liable to lose their charge to neighboring neutrals and this is the simple fate of SO_2^+ (Anicich and Huntress, 1986). But the chemistry of NO^+ offers an important alternative: NO^+ stabilizes through the attachment of one, two, or three water molecules (Fehsenfeld et al., 1971; Sutherland and Zinn, 1975). As the $NO^+(H_2O)$ associate can be imagined as a mesomeric form of protonated nitrous acid, it appears very natural that NO^+-H_2O clusters can release nitrous acid. It was Fehsenfeld et al. (1971) who pointed out this analogy between gas-phase and aqueous-phase ion chemistry. Hence, the oxidation of NO to NO^+ eventually becomes manifest through

$$NO^+(H_2O)_3 + H_2O \rightarrow HNO_2 + H_3O^+(H_2O)_2$$
$$k_9 = 2 \times 10^{-6} \exp(-3000/T) \, cm^3/s \tag{9}$$

which is only slightly opposed by the reverse reaction, $k_{-9} = 1.1 \times 10^{-8}$ $(300/T)^{2.6} \, cm^3/s$ (Fehsenfeld et al., 1971; Sutherland and Zinn, 1975). This already provides an indication that nitrous acid must be expected to form from gas-phase reactions. Nitrous acid is kinetically stable in the gas phase (Vosper, 1976; Kaiser and Wu, 1977b; Jenkin and Cox, 1987; Levine, 1985), which has particular consequences for the EBDS process to be discussed.

B. Radical Reactions

Positive charge transfer processes have been shown to produce radicals at a rate of the order of 100 ppm/s $\approx 2 \times 10^{15} \, cm^{-3} s^{-1}$ at $\dot{D} = 10 \, kGy/s$ ($T \approx 350$ K, $P \approx 1$ bar). The radical production rate is essentially proportional to the dose rate. Bimolecular radical–radical reactions may both reduce the total radical concentration, for example,

$$H + H_2O \rightarrow H_2 + O_2$$
$$H + HO_2 \rightarrow H_2O + O$$

or keep it unchanged through formation of a new radical pair, for example,

$$H + HO_2 \rightarrow 2OH$$
$$NH_2 + N \rightarrow N_2 + 2H$$

Termolecular radical recombination always depletes the available radical reservoir, the rate constants are of the order of $k_{ter} \approx 5 \times 10^{-33}$ cm^6 s^{-1}, so that $k_{ter}[M] \approx 10^{-13}$ cm^3 s^{-1}. As a careful overall estimate, radical recombination will be treated using a bimolecular rate constant of 5×10^{-12} cm^3 s^{-1}. For comparison, fast radical–molecule reactions proceed with equally high rate constants. Then, quasistationary radical concentrations $[R]$ can be estimated from

$$\frac{d[R]}{dt} \approx 2 \times 10^{15} \text{ cm}^{-3} \text{ s}^{-1}$$
$$- 5 \times 10^{-12} \text{ cm}^3/\text{s} \, [R] \, n - 5 \times 10^{-12} \text{ cm}^3/\text{s} \, [R]^2 \approx 0$$

which gives radical levels in the ppb range for neutral concentrations $n \approx 10^{16}$–10^{19} cm^{-3} at $\dot{D} \approx 10$ kGy/s. The already overestimated quadratic term can be neglected ($n \gg [R]$). This means

1. radical concentrations are proportional to the dose rate;
2. radical recombination becomes important only at high dose rates, definitely above 1000 kGy/s.

These rough estimates have been confirmed by detailed modeling studies (Gentry et al., 1988) and again exclude any dose rate effect from the more chemical side of the process. This is also in agreement with the previously discussed upper physical limit for the occurrence of dose rate effects.

Concerning the fate of radicals, two termolecular reactions must be considered:

$$O + O_2 + M \rightarrow O_3 + M$$
$$H + O_2 + M \rightarrow HO_2 + M$$

These reactions proceed with rate constants $k[M] \approx 10^{-14}$ and 3×10^{-12} cm^3/s, respectively, and thus make the hydroperoxide radical and ozone substantial oxidizers for NO. Thereby, NO_2 production is started. This results in a competition of NO, NO_2 and SO_2 for OH (Baulch et al., 1980, 1982, 1984; Atkinson et al., 1989)

$$NO + OH + M \rightarrow HNO_2 + M \qquad k_{10}[M] \approx 4 \times 10^{-12}\,cm^3/s \qquad (10)$$

$$NO_2 + OH + M \rightarrow HNO_3 + M \qquad k_{11}[M] \approx 9 \times 10^{-12}\,cm^3/s \qquad (11)$$

$$SO_2 + OH + M \rightarrow HSO_3 + M \qquad k_{12}[M] \approx 7 \times 10^{-13}\,cm^3/s \qquad (12)$$

The crucial importance of this competitive set of termolecular reactions for the EBDS process arises from the following arguments:

1. reaction (12) is practically the only important SO_2 sink in homogeneous gas phase;
2. reaction (11) is the only important source of nitric acid from neutral reactants in homogeneous gas phase;
3. reaction (10) is a very effective NO sink, but leads only to gaseous nitrous acid, which does not form an ammonium salt upon ammonia addition (Kaiser and Japar, 1978);
4. reaction (12) is followed by the fast reaction $HSO_3 + O_2 \rightarrow SO_3 + HO_2$ (Gleason et al., 1987), which immediately induces sulfuric acid formation and nucleation and simultaneously releases HO_2; its competition with reaction (10) is therefore desirable in that it both inhibits HNO_2 formation and supports the sequence

$$NO + HO_2 \rightarrow NO_2 + OH \xrightarrow{M} HNO_3$$

The last argument clearly stresses the option of simultaneous NO_x/SO_2 removal by EBDS and explains the increase of NO removal with increasing SO_2 concentration, which has been observed by experiment (Paur et al., 1989).

Despite their basic importance, these considerations do not constitute the whole story: According to the above arguments, a kind of turnover would be expected at very high SO_2 concentrations in that they would promote NO_2 formation but also inhibit nitric acid formation by consumption of OH. In this case, NO_x removal would decrease with increasing sulfate formation. Such a turnover has never been reported from experimental investigations.

One explanation for this experimental gap certainly is the ionic pathway, which also contributes to nitric acid formation from NO_2. This path is in perfect analogy to the ionic NO oxidation described above and the key reaction is

$$NO_2^+(H_2O)_2 + H_2O \rightarrow HNO_3 + H_3O^+(H_2O)$$

(Fehsenfeld et al., 1975). Obviously and with appreciation, this ionic pathway prohibits the observation of the turnover suggested above, especially because the destruction of HNO_3 by thermal electrons, albeit fast, is of negligible importance in the present context (Fehsenfeld et al., 1975).

A second and supplementary explanation stems from the observation that NO_2 (and NO) does not only enter into oxidation reactions but also into reduction reactions which are discussed below.

C. Oxidation versus Reduction

It has been pointed out that during irradiation of waste gas with energetic electrons NO is oxidized to NO_2 by O_3 and HO_2. Termolecular oxidation of NO by O atoms and oxidation by molecular oxygen are too slow for a process that must be finished at a time scale of a few seconds or less. NO oxidation and nitric acid formation have also been shown to be supported by sulfur dioxide. In this section, the focus is on reductive pathways that may both favor and oppose NO_x degradation and certainly do not support nitrate formation.

Neglecting negatively charged species, H and N atoms are favorite candidates to invoke reductive pathways. The fastest radical reaction is

$$N + NO \rightarrow N_2 + O \qquad k_{13} = 3.25 \times 10^{-11} \, cm^3/s \qquad (13)$$

(Baulch et al., 1982; Brown and Winkler, 1979). In this reaction, nitric oxide is reduced to molecular nitrogen, which is a welcome product. Under typical conditions, roughly 10% of the NO are removed in this way (Baumann et al., 1987). The molecular nitrogen can be detected only by sophisticated analytical methods that are not commonly in use. Most often, the N balance is based on NO_x input and output measurements including product nitrate. The thus observed deficiency in the N balance has first been explained by model calculations in terms of undetected molecular nitrogen. Only recently, Namba et al. (1988) have reported a direct measurement of the nitrogen formation achieved by use of isotope labeled [15]NO and mass spectrometric analysis. It will be shown in Section V that additional molecular nitrogen is formed from ammonia oxidation and NO reduction. Note that in reaction (13) an oxygen atom is released, which is a really unfavorable intermediate, as shown below.

The reduction of NO_2 by N atoms has been discussed in terms of the following reaction branches:

$$NO_2 + N \rightarrow 2NO$$
$$\rightarrow O_2 + N_2$$
$$\rightarrow N_2O + O$$

It is well established that the reaction takes the last branch exclusively (Baulch et al., 1984) thus producing N_2O as stable product and also an intermediate O atom. Up to a dose around 10 kGy, the N_2O production from this reaction is only a few ppm, since the N atoms are consumed preferentially by NO.

Fortunately, the reaction $N + O_2 \rightarrow NO + O$ has too high an activation energy to be important in the temperature range around 100 °C (Baulch et al., 1980).

The oxygen atoms attach to molecular oxygen only comparatively slowly (see previous section). Instead, they effectively reduce NO_2 to NO:

$$NO_2 + O \rightarrow NO + O_2 \qquad k_{14} = 5.2 \times 10^{-12} \exp(+200/T) \, cm^3/s \quad (14)$$

(Geers-Müller and Stuhl, 1987). This unfortunate reaction opposes NO oxidation extensively. Reaction (14) has been shown to account for the nonlinear NO removal as function of dose (Mätzing, 1989).

Also, it can be used to estimate intermediate O atom concentrations from measured NO vs dose curves. One goal for optimum process performance therefore should be to suppress reaction (14) by offering alternative reaction paths to NO_2 or O or both.

The H atoms mentioned previously preferably attach to molecular oxygen thereby forming HO_2 (see preceding discussion) which is needed for NO oxidation. Part of the HO_2 (and also of OH) recombines under formation of H_2O_2 and this recombination is favored by high concentrations of water vapor. H_2O_2 is comparatively stable under typical EBDS conditions and has a vapor pressure low enough to suggest its condensation at the particulate surface. Its calculated final concentration is around 20 ppb (Mätzing, 1987a).

To a minor extent, H atoms also reduce NO_x:

$$NO_2 + H \rightarrow NO + OH \qquad k_{15} = 5.8 \times 10^{-10} \exp(-740/T) \, cm^3/s \quad (15)$$
$$NO + H + M \rightarrow HNO + M \qquad k_{16}[M] \approx 10^{-13} \exp(+300/T) \, cm^3/s \quad (16)$$

(Baulch et al., 1983). While reaction (15) may be interpreted to favor nitrous acid formation, reaction (16) rather favors nitric acid formation via

$$HNO + O_2 \rightarrow HO_2 + NO \qquad k_{17} = 3.3 \times 10^{-14} \, cm^3/s \qquad (17)$$

$$\rightarrow OH + NO_2 \qquad k_{18} = 1.7 \times 10^{-15} \, cm^3/s \qquad (18)$$

(Hack et al., 1985a,b). But altogether, the interaction of H atoms with NO_x is of minor importance.

It has now become clear that NO_x oxidation is partly complemented by NO_x reduction through N_2 and N_2O formation. However, reductive pathways also oppose oxidative reactions in a way to decrease the removal efficiency with rising dose. Thus, NO_x removal is a nonlinear function of dose and eventually attains a saturation with increasing dose.

Therefore, it has been attempted to substitute a single, high-dose irradiation step by successive low-dose irradiation steps in order to save energy and increase efficiency. These results show that multiple irradiation may in fact increase the NO_x removal efficiency (Baumann et al., 1987).

The preceding discussion suggests multiple irradiation to be most effective, if NO_2 is removed between successive irradiation steps thus preventing its late reduction to NO.

The nonlinear dose dependence of NO_x removal has also given rise to the study of dose rate effects in the EBDS process (Wittig et al., 1988a, 1988b). In fact, both ion and radical formation rates are proportional to the dose rate, as shown previously. Since mutual recombination of ions and radicals are comparatively slow at $\dot{D} \leq 1000 \, kGy/s$, the active species concentrations are proportional to the dose rate. Hence, the overall chemical reaction proceeds the faster, the higher the dose rate. This holds for both the oxidative and reductive pathways described previously. Therefore, rising dose rate accelerates every single reaction step involved in the decay of the irradiated gas mixture from its nonequilibrium state to its new equilibrium state. The low density and viscosity of the waste gas under consideration also prohibit local overheating by high dose rates and thus prevent changes of the chemical mechanism. So the final gas composition depends only on the initial deviation from equilibrium, that is, on the absorbed dose, and not on "pulse intensity," that is, dose rate. This has been shown to be valid for dose rates less than about 1000 kGy/s (Gentry et al., 1988).

It must be added that this is often interpreted from the point of constant acceleration voltage, which for present economic reasons is in the range 300–800 kV. Commercial accelerators presently preclude dose rates above about 100 kGy/s owing to limitations in achievable current density. Still, it should be noted that the dose rate also increases with decreasing electron energy and hence along the electron path through the irradiated medium. High dose rates therefore can always be expected in a region where the electron energy has decayed to a few percent of its initial value. The design of irradiation chambers, of course, should not exceed the electron range and hence these dose rate effects become indistinguishable from wall effects.

D.   A Simplified Reaction Scheme

It is now possible to derive a simple overview on the kinetics of the EBDS process. According to the preceding discussion, positive ions are the most important irradiation products. They support NO_x oxidation both directly and via radical formation. SO_2 oxidation is initiated only by OH radicals:

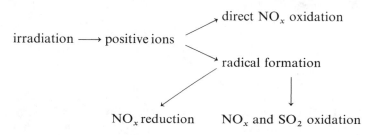

However, radical reactions do not support NO_x oxidation straightforwardly. There also exist two types of reduction reactions:

1. those yielding molecular nitrogen and N_2O as stable products; they complement oxidative NO_x removal, but do not support particle formation;
2. those reducing intermediate NO_2 back to NO; they counteract oxidative NO_x removal.

The latter mechanism is due to oxygen atoms that effectively reduce NO_2 [reaction (14)]. In consequence, any input NO molecule is trapped in the oxidation–reduction cycle between NO and NO_2, before it ends up in HNO_3, N_2, or N_2O. This corresponds to an inefficient utilization of the available radical source and also of the input energy: NO_x removal therefore becomes a nonlinear function of dose and eventually attains a saturation with increasing dose.

V.   THE CHEMISTRY OF AMMONIA

A common measure of the amount of ammonia added to the waste gas is the stoichiometry ratio, defined by

$$s = \frac{[NH_3]}{[NO_x] + 2[SO_2]}$$

For the typical s values between 0.5 and 1, the ammonia concentration is comparable or higher than those of NO_x or SO_2. Ammonia can be added at different stages of the process: Upstream, in, or downstream of the irradiation chamber; a mixture of these possibilities is also of interest. The experiments have shown that ammonia should be added to the waste gas before irradiation (Baumann et al., 1987; Frank, 1988), this giving best results for both NO_x and SO_2 removal. This finding and the almost linear correlation between N_2O formation and stoichiometry ratio clearly show that ammonia does not only act as to the neutralize the acids, but also takes an important part in the gas-phase chemistry. In this way, ammonia provides a connecting link between the gas-phase chemistry and the particle formation in the EBDS process.

A. Ammonia and Radicals

The radical chemistry of ammonia is rather complicated and has thoroughly been investigated and discussed during recent years (Dean et al., 1982; Miller et al., 1983; Lesclaux, 1984; Dransfeld et al., 1984; Cohen, 1987). Also, the modeling approaches forwarded by Miller et al. (1983) and Cohen (1987) are found to disagree somewhat. The author has made an effort to put things together and part of this is described below.

The reactions of ammonia with H and O atoms

$$NH_3 + H \rightarrow NH_2 + H_2 \qquad\qquad (19a)$$

$$NH_3 + O \rightarrow NH_2 + OH \qquad\qquad (19b)$$

proceed fairly slowly under EBDS conditions; k_{19a} has been measured by Hack et al. (1986): $k_{19a} = 9.1 \times 10^{-15} T^{1.3} \exp(-6570/T)\, cm^3/s$ and k_{19b} has been reviewed and proposed by Cohen (1987): $k_{19b} = 1.83 \times 10^{-18} T^{2.1} \exp(-2620/T)\, cm^3/s$. The major attack on NH_3 is by OH

$$NH_3 + OH \rightarrow NH_2 + H_2O \qquad\qquad (20)$$

Cohen (1987) uses $k_{20} = 8.32 \times 10^{-17} T^{1.6} \exp(-480/T)\, cm^3/s$ which for $T = 300\text{--}500\,K$ is in good agreement with the measurements by Jeffries and Smith (1986) and previous data (Miller et al., 1983; Levine, 1985).

Ammonia has also been reported to promote the recombination of HO_2 (Lii et al., 1980), but this appears negligible in the concentration regime discussed here.

The chemistry of the moderately reactive amidogen radical (NH_2), produced by reaction (20) mainly, is fairly complex and only the most obvious reaction paths are considered here. The most prominent characteristic of

NH_2 is its instability against oxidation, or, vice versa, its reducing property. Thus, it readily reduces NO and NO_2:

$$NH_2 + NO_2 \rightarrow N_2O + H_2O \qquad k_{21} = 2.2 \times 10^{-12} \exp(+650/T) \, cm^3/s$$

$$(21)$$

(DeMore et al., 1987). This reaction is the major source of N_2O in the EBDS process and accounts for the final N_2O concentration to be roughly proportional to the dose and to the stoichiometry ratio (Baumann et al., 1987).

The $NH_2 + NO$ reaction is a key step in the thermal $DeNO_x$ process. Vibrationally excited H_2O and ground-state OH radicals have been identified among the reaction products (Dreier and Wolfrum, 1984 and references therein). These findings are attributed to the reaction branches

$$NH_2 + NO \rightarrow N_2 + H_2O \qquad (22a)$$

$$\rightarrow N_2H + OH \qquad (22b)$$

(Miller et al., 1983; Cohen, 1987). Reaction (22) is known to proceed faster with decreasing temperature; from the measurements by Andresen et al. (1982) and by Silver and Kolb (1982), the approximation

$$k_{22} = k_{22a} + k_{22b} = 1.15 \times 10^{-7} T^{-1.54} \, cm^3/s$$

can be derived for $T = 300–1000$ K. This expression, taken from Cohen (1987), compares well to that used by Miller et al. (1983). The OH formation by reaction (22) presumably increases with temperature, which may be interpreted in terms of a temperature-dependent branching ratio, $k_{22a}/k_{22b} = f(T)$, although this is not a necessary conclusion (Dreier and Wolfrum, 1984; Harrison et al., 1987). It seems to be well established that $k_{22a}:k_{22b} \approx 083:0.17$ around room temperature (Dolson, 1986; Hall et al., 1986; Silver and Kolb, 1987). Around 1000 K, $k_{22a}:k_{22b} \approx 1:1$ appears plausible (Kimball-Linne and Hanson, 1986). In combination with k_{22}, these branching ratios can be interpreted by the following rate expressions, the first of which is close to earlier data reported by Hack et al., (1979):

$$k_{22a} = 1 \times 10^{-6} T^{-1.96} \, cm^3/s$$

$$k_{22b} = 1 \times 10^{-12} \exp(+330/T) \, cm^3/s$$

This derivation of the rate constants k_{22a} and k_{22b} has found good support through two very recent measurements (Atakan et al., 1989; Bulatov et al., 1989) and is believed valid at $T = 300–1000$ K. Figure 1 gives a short overview

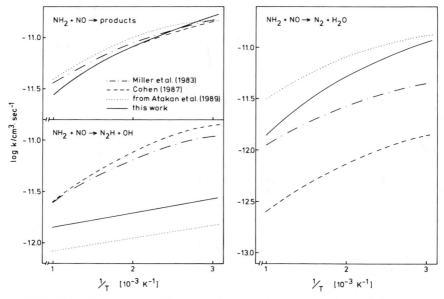

FIGURE 1. Comparison of literature data on the rate constants of the reaction $NH_2 + NO \rightarrow$ prod and of the reaction branches (22a) and (22b) used in this work.

on the literature data quoted above and compares them to the present proposal.

Reactions (21) and (22) have been discussed here in some detail, because they represent the most important NH_2 sinks in the system under consideration and provide the key to the understanding of NO_x removal by ammonia: Since NH_2 does not react significantly with the major components of the waste gas, it directly supports NO_x removal. In addition, this also holds for the side reaction

$$NH_2 + O \rightarrow HNO + H \qquad k_{23a} = 1.1 \times 10^{-9} T^{-0.5} \text{ cm}^3/\text{s} \qquad (23a)$$

$$\rightarrow NH + OH \qquad k_{23b} = 1.2 \times 10^{-11} \text{ cm}^3/\text{s} \qquad (23b)$$

(Dransfeld et al., 1984; Miller et al., 1983). Like reaction (21), the reaction (23) opposes the undesired $NO_2 + O$ reaction discussed in Section IV C. Furthermore, reaction (23) supports the oxidative pathway through regeneration of OH via reaction (23b) and indirectly, since reaction (23a) is followed by

$$HNO + O_2 \rightarrow NO + HO_2 \qquad (17)$$

$$\rightarrow NO_2 + OH \qquad (18)$$

and by

$$H + O_2 + M \rightarrow HO_2 + M$$

Especially because of its potential ability to counteract the oxidation–reduction cycle between NO and NO_2, reaction (23) is thought to be the sole radical–radical reaction of importance in the EBDS process (Mätzing, 1989). In this context it is worth noting that the preceding reaction sequence may not only be interpreted in terms of NO_x reduction, but also from the viewpoint of NH_3 oxidation.

Lesclaux (1984) has mentioned that it is desirable to study the interaction between NH_2 and sulfur dioxide, since this might become important in polluted atmospheres under certain conditions. In the EBDS, such an interaction would certainly be of interest in order to get a more complete understanding of the reaction mechanism. Recently, Ioffe et al. (1989) have reported that the reaction is termolecular and yields the adduct, NH_2SO_2. The reported rate constant is high enough to permit a possible competition between NO_x and SO_2 for NH_2 under EBDS conditions. Possibly, the amidogen–sulfur dioxide adduct might open an alternative path to SO_2 oxidation, which could partly explain the promoting role of ammonia for the removal of SO_2 (see subsequent discussion). But since there exists no clear picture about the corresponding mechanism, this reaction will not be considered here.

Besides the previously discussed favorable role of NH_2 in the EBDS process, it must not be forgotten that its formation involves the consumption of OH: $NH_3 + OH \rightarrow NH_2 + H_2O$ (20). In essence, reaction (20) works as to replace an oxidizing species (OH) by a reducing species (NH_2). The reaction branch (22b) is not fast enough to regenerate OH effectively, otherwise it would act as a catalytic cycle. Therefore, the OH consumption by ammonia could be expected to cause a decrease of HNO_3 and H_2SO_4 formation and hence a decrease of particulate formation. In fact, the concentration of ammonium nitrate has been observed to decrease somewhat with increasing ammonia concentration (Paur and Jordan, 1988b; 1989). At high relative humidity, this effect is less pronounced, since then the ionic NO_x oxidation (Sections IV B and V B) complements the oxidation by radicals. The sulfate formation, however, is greatly enhanced by the presence of ammonia (Paur and Jordan, 1988b; 1989). Presently, many authors agree that it is not feasible to interpret this experimental finding in terms of genuine radiation chemical effects (Busi et al., 1987; Frank, 1988; Jordan, 1988): Rather, it is interpreted to provide evidence for additional purely thermal or heterogeneous sulfate formation steps or both which are not easy to isolate from radiation induced reactions experimentally. These will be discussed.

B.   The Formation of Ammonium Salts

Two stable acids are formed by the gas-phase chemistry of the EBDS process, as described previously: HNO_3 and H_2SO_4. They have different physical properties and those of interest here are their vapor pressures that differ by many orders of magnitude. The vapor pressure of sulfuric acid, in particular, is so small at $T = 273–373\,K$ (Roedel, 1979; Perry et al., 1984) that the existence of gaseous sulfuric acid even becomes questionable in this temperature range. It is therefore reasonable to assume that sulfuric acid nucleates prior to reaction with ammonia. Ammonium sulfate formation thus is probably not a gas–solid transition, but rather a heterogeneous reaction determined by the rate of incorporation of ammonia into sulfuric acid droplets. When both the nucleation of sulfuric acid and the transport of ammonia to the surface of the nucleating droplets are assumed to be fast processes, it appears a good approximation to substitute the real process by the fast dummy reactions

$$NH_3 + H_2SO_4 \rightarrow NH_4HSO_4$$

$$NH_3 + NH_4HSO_4 \rightarrow (NH_4)_2SO_4$$

(Rolle and Renner, 1984; Person et al., 1985). Another argument for this abbreviated reaction modeling is the thermal stability of ammonium sulfate, which decomposes only above about 150 °C.

This treatment results in the prediction that H_2SO_4 be converted to $(NH_4)_2SO_4$ completely in the presence of sufficient ammonia. The chemical composition of the EBDS aerosol, however, shows something different: While in case of low relative humidity, the chemical composition indeed corresponds to a mixture of NH_4NO_3 and $(NH_4)_2SO_4$ (Paur et al., 1988a), some NH_4^+ deficiency shows up in case of high relative humidity (Paur and Jordan, 1988b). The NH_4^+ deficiency can be interpreted such that at high relative humidity the $NH_3–H_2SO_4$ reaction is not complete at the time scale available, that is, at the point of aerosol sampling. The formation of $(NH_4)_2SO_4$ is complete at later stages, perhaps only in the product filter under some conditions. The heterogeneous reaction between sulfuric acid droplets and gaseous ammonia can be modeled according to the concept described in Section VI.

Unlike sulfuric acid, nitric acid cannot be expected to nucleate under typical EBDS conditions and this view is strongly supported by the observation that in the absence of ammonia, no nitrate can be detected in the aerosol (Paur and Jordan, 1989). Note that this experimental fact also is an argument against ion-assisted nucleation of nitric acid. Therefore, it appears probable that ammonium sulfate and nitrate formation are independent processes that are determined only by the difference of their vapor pressures. A mixed nitrate- and sulfate-containing aerosol can thus only result from the coagulation of the two chemically different and independently developing aerosols. The condensation of NH_4NO_3 at the surface of incompletely neutralized H_2SO_4 droplets cannot definitely be excluded, but may be speculated to invoke the decomposition

$$NH_4NO_3 \xrightarrow{H_2SO_4} N_2O + 2H_2O$$

In comparison with ammonium sulfate, the ammonium nitrate has a much higher vapor pressure or, equivalently, a much lower thermal stability that is determined by the heterogeneous equilibrium

$$NH_4NO_3(s) \rightleftharpoons NH_3(g) + HNO_3(g)$$

It is not easy to extract a conclusive equilibrium constant for this process from the literature, because this equilibrium is not a unique function of temperature, but also a function of the ambient water vapor pressure (relative humidity), since ammonium nitrate is a hygroscopic substance. Equation (4) given by Stelson and Seinfeld (1982) can be interpreted to imply an average free energy of sublimation of dry NH_4NO_3 around $86.65 \, kJ/mol$ over the temperature range 293–400 K. On the other hand, Heicklen and Luria (1975) found

$$[NH_3] \cdot [HNO_3] = 5.8 \times 10^{27} \, molec^2/cm^6 \approx 10 \, ppm^2$$

as the necessary condition for the onset of NH_4NO_3 nucleation at $T = 298$ K. If this value is interpreted in terms of thermodynamic equilibrium, $\Delta G° = 62.79 \, kJ/mol$ is obtained, which is markedly smaller than that by Stelson and Seinfeld (1982). Unfortunately, the data by Heicklen and Luria (1975) hold at $T = 298$ K only. Owing to lack of further information, the "free enthalpy" derived from their data is considered independent of temperature tentatively. With this restriction one arrives at the equilibrium vapor pressures of NH_4NO_3 in Table I. They differ by about two and three orders of magnitude at $T = 298$ K and $T = 370$ K, respectively. *A priori*, this

TABLE I
Estimated Equilibrium Vapor Pressures above Dry NH_4NO_3

	$\Delta G°$ [kJ/mol]			
	86.65 (Stelson et al., 1982) $P_{NH_3} = P_{HNO_3}$		62.79 (Heicklen et al., 1975) $P_{NH_3} = P_{HNO_3}$	
	[atm]	[ppb]	[atm]	[ppb]
$T = 298$ K	2.9×10^{-8}	29	3.4×10^{-6}	3,400
$T = 370$ K	8.4×10^{-7}	840	4×10^{-5}	40,000

difference is not a contradiction, since according to the different approaches, which they are derived from, they have different meanings: The kinetic study by Heicklen and Luria (1975) refers to the maximum possible concentrations of gaseous NH_3 and HNO_3 *prior* to nucleation, while the thermodynamic study by Stelson and Seinfeld (1982) gives the corresponding minimum concentration which can be expected in the gas phase above the (flat) surface of particulate NH_4NO_3.

The preceding discussion can be summarized to interpret the apparent difference of the two data sets in terms of a supersaturation ratio S. S is given by

$$S = \exp(+2850/T)$$

using the previous assumptions. Interestingly, a decrease of supersaturation ratio with rising temperature is obtained here for a gas–solid transition with chemical reaction. If this is true, the reaction

$$NH_3 + HNO_3 \rightarrow \text{solid } NH_4NO_3$$

would have to have a higher activation energy than the reverse evaporation reaction. Certainly, more experimental data are needed on this interesting topic, since the above discussion is based on only one kinetic measurement at $T = 298$ K and this also means that the previously derived temperature dependence of S should be regarded rather as a question than as conclusive result.

The nucleation rate of NH_4NO_3 has been evaluated by Heicklen and Luria (1975) in terms of an apparent eight-order process:

$$\frac{d[NH_4NO_3]}{dt} = (1.26 \times 10^{-28}[NH_3][NHO_3])^8 \text{ cm}^{-3} \text{ min}^{-1}$$

and the second-order condensation rate constant is given by the same authors to be 2×10^{-8} cm^3/s at $T = 298$ K. These rates compare to those of the very fast ion recombination rates (Section III D). Their energies of activation would have to be as high as 25 kJ/mol in order to decrease them to the maximum possible for bimolecular reactions of neutrals. Applied to EBDS conditions, this infers nitric acid formation to be the rate-determining step for NH_4NO_3 formation, which is plausible. The author has found this concept useful to model the particulate nitrate concentrations established by EBDS processing.

It remains to be noted that the low equilibrium vapor pressures over dry NH_4NO_3 (Stelson and Seinfeld, 1982) already correspond to an almost complete consumption of gaseous substrate. With increase of the relative humidity in the gas phase, these equilibrium vapor pressures are expected to decrease. This means that only ppb amounts of substrate could be converted to NH_4NO_3 on account of increasing relative humidity, which is beyond the scope of measurable mass concentrations. Therefore, the experimentally established increase of nitrate formation with increasing relative humidity (Paur and Jordan, 1989) is not due to a simple shift of the thermodynamic equilibrium: Nitric acid formation is enhanced by increasing relative humidity via the radiation induced ionic pathway mentioned in Section IV B:

$$NO_2^+(H_2O)_2 + H_2O \rightarrow HNO_3 + H_3O^+(H_2O)$$

This pathway has recently been found adequate to model the dependence of nitrate mass concentration on relative humidity (Mätzing et al., 1988).

C.   Ammonia and Sulfur Dioxide

The direct interaction of gaseous ammonia and sulfur dioxide is known to yield ammonium sulfate under atmospheric conditions, that is, at comparatively low temperatures in the presence of oxygen and water vapor. In the context of atmospheric pollutants, this SO_2/NH_3 reaction is fairly well understood and is known to occur at a time scale of hours (Friedlander, 1977; Levine, 1985; Seinfeld, 1986). In the range of waste gas conditions, on the contrary, fairly little is known on the reaction between gaseous ammonia and sulfur dioxide, although corresponding experiments have been conducted and proposed (Hartley and Matteson, 1975; Landreth et al., 1975, and references therein). From these and similar studies it is known that a solid product can form from a mixture of gaseous ammonia, sulfur dioxide, and water in the temperature range $T = 290$–320 K.

The solid product has no uniform composition, but consists of a mixture of ammonium salts like sulfite, sulfate, pyrosulfite, and others. The N/S ratio of the solid and its water content depend in a rather complex way on the initial gas-phase composition and on temperature. Landreth et al. (1975) derived an equilibrium constant

$$\ln K_p [\mathrm{atm}^{-4}] \approx -56.5 + 26500/T[\mathrm{K}]$$

for the gas–solid equilibrium

$$2NH_3 + SO_2 + H_2O \rightleftharpoons (NH_4)_2SO_3$$

Their derivation is based on vapor pressure measurements and the authors have identified ammonium sulfite to be the *major* reaction product (Landreth et al.,1975). The above equilibrium must be regarded as an approximation for two reasons:

1. there are still doubts concerning the effective stoichiometry of the reaction;
2. the preceding equilibrium suggests a quatermolecular reaction that is very unlikely to occur in the gas phase as such, a kinetic description would certainly involve a mechanism of two or more steps.

The study by Hartley and Matteson (1975) gives some hints on the kinetics of the ammonia–sulfur dioxide reaction. From a reevaluation of the data, the author has found the following rate law for the formation of the solid product:

$$\frac{dc}{dt} = k(c_{max} - c)$$

in which c and c_{max} are the intermediate and final product concentrations, respectively, k is the apparent first-order rate constant (see Fig. 2). A rate law of this kind is well known from other studies of nucleation and condensation phenomena (Elias, 1981; Stumm and Morgan 1981). A number of experiments described by Hartley and Matteson (1975) have been evaluated in this way and k was found to be in the range $7–17 \, \mathrm{s}^{-1}$. The tentative plot in Fig. 3 may indicate a dependence of k on the stoichiometry ratio, but care must be taken not to overinterpret the data.

These examples may suffice to point out the complexity of ammonia–sulfur dioxide interaction. The problem becomes even worse, when temperatures of the order of $350 \, \mathrm{K}$ apply, that is, under EBDS conditions, because this is above or amidst the thermal decomposition range of simple ammonium sulfites.

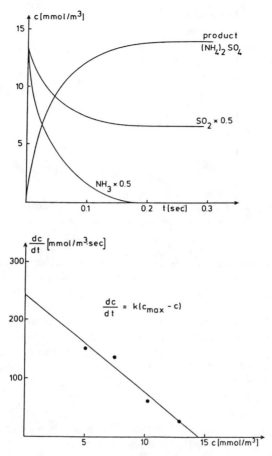

FIGURE 2. Plots of measured concentrations versus time and determination of rate constant k for product formation in experiment series D from Hartley and Matterson (1975).

From many EBDS research studies it is well known that ammonia and sulfur dioxide are degraded solely upon mixing, that is, in the absence of electron-beam irradiation. From these studies, it could be established that the so-called "thermal" NH_3–SO_2 reaction is favored by high ammonia concentration and by high relative humidity (Baumann et al., 1987; Paur and Jordan, 1989; Frank et al., 1988; Wittig et al., 1988b). Recent investigations preclude the thermal NH_3–SO_2 reaction to be an artifact invoked by the measuring technique or to generate a submicron aerosol which escapes the filter (Paur et al., 1988a). Up to date, there exists no conclusive idea on the

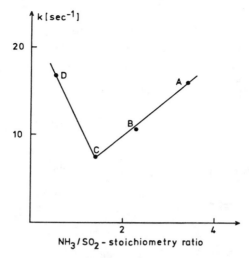

FIGURE 3. Rate constant for product formation as function of ammonia stoichiometry, evaluated for experiment series A, B, C, D reported by Hartley and Matteson (1975).

nature of this thermal NH_3–SO_2 reaction. A first, tentative proposal has been published by Jordan (1988) and Jordan et al. (1988). It involves the formation of a gaseous adduct (charge transfer complex) from the Lewis base NH_3 and the Lewis acid SO_2. This gaseous complex is thought to form ammonium sulfate upon irradiation. Such a Lewis acid–base reaction between NH_3 and SO_2 has been studied by Landreth et al. (1974) under anhydrous conditions. In the temperature range $T = 278$–$318\,K$, these authors even observed particle formation from dry NH_3 and SO_2 and give a reaction enthalpy $\Delta H = 40\,kJ/mol$ for the process

$$NH_3 \cdot SO_2\,(s) \rightleftharpoons NH_3 + SO_2$$

This ΔH value is in excellent agreement with the calculated energy for the N–S bond in that complex (Lucchese et al., 1976; Douglas and Kollmann, 1978). A kinetic study of this subject would certainly be of interest both from the modeling and application point of view.

D.   Effects of Ammonia Addition

In the radiation-induced gas-phase chemistry of the EBDS process, the OH radical has been shown to play the key part for nitric and sulfuric acid formation; NO, NO_2, and SO_2 compete for OH. The added ammonia enters into this competition: About 5% of the ammonia (that is, roughly 50 ppm usually) react with OH. Thereby, the NH_2 radical is formed which reduces NO and NO_2 to yield molecular nitrogen and N_2O as stable gaseous products. The amounts of N_2 and N_2O formed in this way are approximately equal and are around $1-2\,ppm/kGy$ typically. Up to a dose around $10\,kGy$, this is the only significant N_2O source in the EBDS process. As molecular nitrogen source, this mechanism is about half as effective as the $NO + N$ reaction (Section IV C) and readily increases the total nitrogen formation to about 10–15% of the input NO.

The major part of the ammonia remains available for ammonium salt formation. As the parent acids differ remarkably in their vapor pressures, the particulate formation is likely to involve different mechanisms:

1. ammonium nitrate is formed from gaseous nitric acid and ammonia directly;
2. ammonium sulfate is formed from a heterogeneous reaction between gaseous ammonia and nucleating sulfuric acid droplets.

High relative humidity has been reported to increase the aerosol mass. For ammonium nitrate, this can be explained by a corresponding increase of nitric acid formation. In the literature, only limited data are available on the direct interaction of ammonia and sulfur dioxide which in EBDS studies has been found to be of "thermal," that is, not radiation-induced, origin.

VI. HETEROGENEOUS CHEMISTRY

In the preceding sections, the kinetics of the EBDS process has been discussed in terms of homogeneous gas-phase reactions yielding gaseous and solid products. The importance of heterogeneous processes that occur at the surface of the evolving aerosol has implicitly been mentioned in the context of ammonium nitrate formation (Section V 8). This importance of heterogeneous chemical reactions is emphasized by experimental findings which state

1. particulate formation to depend on the relative humidity
2. removal yields to be increased by filtration

(Paur et al., 1988a; Paur and Jordan, 1989; Frank et al., 1988). Another hint to this point comes from the theoretical treatment: According to models on the pure gas–phase chemistry of the EBDS process, nitrous acid would be a major product species due to its unexpected kinetic stability discussed in Sections IV A and IV B. This is further demonstrated in the results by Busi et al. (1988) and Person et al. (1988). Contrary to its kinetic stability in the gas phase, HNO_2 is highly unstable against heterogeneous decomposition, a process that also depends on the kind of surface involved (Kaiser and Wu,

1977a). Altogether, the high HNO_2 concentrations predicted by gas-phase chemical EBDS models should either be detectable as particulate NO_2^- or be destroyed at the particulate surface. Since definitely no NO_2^- is found in the particulate samples (Jordan et al., 1986; Paur et al., 1986), the hetero-geneous decomposition of HNO_2 must play a crucial part of the NO_x chemistry in the EBDS process.

In the context of EBDS processing, the term "heterogeneous chemistry" has most often been used in a way to summarize those experimental findings that are not readily understood by genuine gas phase reactions. Recently, Busi et al. (1988) have emphasized the need for a more precise terminology. Reactions associated with the filtration process, for instance, would probably be a mixture of reactions occurring at gas–solid and gas–liquid interfaces as well as pure liquid–phase reactions. Trace gas removal across the filtration unit is easy to measure compared to processes involving particulate dispersions in the gas phase. Only the latter are addressed here in view of their relation to the gas composition and particulate mass loading transferred to the filter. In order to get an idea about the particle reactivity, one likes to have a reliable estimate at least on the physical properties (surface) of the aerosol arising from electron-beam irradiation of waste gas. Such an estimate has often been ignored in previous publications and it will be the first subject of the subsequent discussion.

A.   The Aerosol Surface

The specific surface A_s of an aerosol consisting of spherical particles with diameter d is given by

$$A_s = 6/\rho d$$

where ρ is the particulate density. The effective surface of that aerosol, $A_e \, [\mathrm{m^2/m^3}]$, is A_s times the mass concentration $c \, [\mathrm{g/m^3}]$. Therefore, $A_e \sim c/d$ is determined by the rate of particulate mass formation and particle growth in a kind of an antagonistic way:

1. nucleation increases c
2. condensation increases both c and d
3. coagulation increases d.

The isolated effect of coagulation on particle size can be estimated from the well-known theory of coagulation of monodisperse hard spheres (Friedlander, 1977; Hidy and Brock, 1970). This has been done for typical mass concen-trations and particle diameters in the range $d = 0.01–1 \, \mu m$. Under EBDS conditions, this size range spans the whole transition regime between continuous and molecular flow conditions; in the transition regime, the

TABLE II
Coagulation Time Scales ($T = 343$ K, $\rho = 1.5 \text{g/cm}^3$)

$c \, [\text{mg/Nm}^3]$	200	1000	200	1000	200	1000
$d \, [\mu\text{m}]$	0.01	0.01	0.1	0.1	1	1
$\tau_K \, [\text{s}]$	0.01	0.002	3	0.7	1000	200

coagulation rate must be expected to be up to a factor of about 5 higher than in the free molecule regime (Hidy and Brock, 1970). As for a rough estimate, the author has adopted the free molecule approximation to calculate the coagulation rate over the size range mentioned. Table II gives the characteristic time scale τ_K at which the particle number density reduces to $1/e$ of its initial valuel.

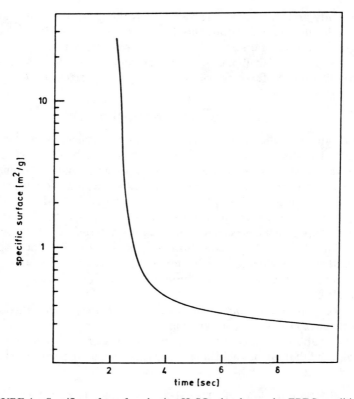

FIGURE 4.   Specific surface of nucleating H_2SO_4 droplets under EBDS conditions.

The estimate states that particle coagulation cannot yield particles with diameters much larger than about 0.1 μm, since this would require coagulation times much longer than a second, that is, a much longer time than available under EBDS conditions. For a particulate density around 1.5 g/cm^3 this gives a specific surface $A_s > 10 \, \text{m}^2/\text{g}$ of the nucleating aerosol. This result is in excellent agreement with a rigorous treatment of the nucleation and growth of sulfuric acid droplets under EBDS conditions (Paur et al., 1988 a; Bunz and Dlugi, 1990). According to this study, A_s is 30 m^2/g at the incidence of H$_2$SO$_4$ nucleation and decreases to 5 m^2/g within less than 2 s (Fig. 4).

These theoretical considerations are in good agreement with experimental size determinations of the ammonium salt particulate (Jordan et al., 1986) and permit an estimate of the rate constant for gas to particle transfer, k_{het}. From the free molecule theory, a lower limit

$$k_{het} = 0.25 \alpha \bar{v} A_e$$

is given (Hidy and Brock, 1970), in which $\bar{v}$ is the mean thermal velocity of the gas molecule and α is a dimensionless factor between 0 and 1 that gives a measure of the reaction probability. For $\bar{v} \approx 400$ m/s and $A_e \approx 5 \, \text{m}^2/\text{m}^3$, this gives $k_{het} \approx 500 \, \text{s}^{-1}$, if α is 1. This corresponds to a first-order halflife time around 1 ms. Note, of course, that gas–particle reactions need not be transport limited.

B. Heterogeneous Reactions

When introducing heterogeneous reactions into a chemical kinetics code, the problem of the stoichiometry of these reactions must be dealt with: As no detailed kinetic studies on single heterogeneous gas–solid reactions at the surface of airborne particles exist, these must either be handled with plausible assumptions or only the loss of reactants can be calculated assuming the reaction products to be unimportant or inert.

The latter approach has been used in a study on the effect of heterogeneous ion sinks in the EBDS process (Mätzing et al., 1987b). The effect has been found to be negligible, which simply reflects the fact that ion–particle transfer proceeds much slower (Section VI A) than ion–molecule reactions (Section III D).

The preceding simple approach is no longer feasible, if one considers the heterogeneous HNO$_2$ decomposition mentioned earlier: In this case, assumptions about the reaction products have to be made. Busi et al. (1988) have proposed that nitrous acid enters into the particulate phase and becomes mainly reduced to N$_2$O:

$$2HNO_2 + 2SO_2 \cdot H_2O \rightarrow 2HSO_4^- + N_2O + H_2O + 2H^+$$

In the author's opinion, this reaction should be negligibly slow, because HNO_2 in solution is likely to decompose upon warm-up and can be reduced effectively to N_2O by strongly reducing reactants like sodium amalgam only (Holleman-Wiberg, 1976). Recent environmental studies have shown that the aqueous phase oxidation of SO_2 by HNO_2 and other atmospheric constituents proceeds at time scales above half an hour (Chang et al., 1981; Seinfeld, 1986). Furthermore, the quantitative transformation of the calculated intermediate HNO_2 to H_2O would yield N_2O levels in the 50 ppm range at $D = 10$ kGy (Busi et al., 1988), which appears too high (Baumann et al., 1987).

So at present, there is no evidence for HNO_2 to promote the oxidation of SO_2 substantially under EBDS conditions. Two major heterogeneous decomposition reactions have been adoped in the AGATE-code:

$$HNO_2 \rightarrow \tfrac{1}{3}HNO_3 + \tfrac{2}{3}NO + \tfrac{1}{3}H_2O$$

$$\rightarrow 0.5NO_2 + 0.5NO + 0.5H_2O$$

The stoichiometry has been chosen by analogy with aqueous-phase chemistry and the α values have been fitted to obtain agreement with experimental results. According to this proposal, less than one-third of the HNO_2 are converted to nitric acid directly and the major decomposition product is NO. The mechanism therefore acts as an OH sink and slows down the rate of NO removal. The adjustment of the α values does not critically influence the curvature of the calculated NO versus dose line. So the proposed HNO_2 decomposition does not affect the importance of reaction (14), which still remains as major factor to oppose NO removal (Section IV C).

Nitric acid has also been considered as possible candidate for the heterogeneous oxidation of SO_2 assuming NO or NO_2 among the reaction products, for example,

$$2HNO_3 + SO_2 \rightarrow 2NO + H_2SO_4 + O_2$$

Reactions of this kind critically affect the calculated NO_x removal and nitrate concentration, because the regeneration of HNO_3 is slow. Correspondingly, only α values below 0.1 are compatible with experimental NO_x data. Hence, this pathway is of minor importance for SO_2 oxidation.

The necessity to consider heterogeneous reactions of nitrous acid arises from model calculations and is not obvious from measured NO_x and SO_2 data. In contrast, heterogeneous SO_2 removal mechanisms have been well established in EBDS research and development work (Frank, 1988; Jordan, 1988, 1989). Large amounts of SO_2 have been found to form sulfate at the filter surface (Jordan and Paur, 1988) and similar reactions have been suggested to occur at the surface of the aerosol during and after irradiation.

The increase of measured sulfate concentrations with rising relative humidity has been taken as a major argument for the importance of heterogeneous SO_2 oxidation. Apart from nitrous acid the most likely heterogeneous oxidizers are H_2O_2, O_3, OH, and HO_2, which may considerably promote the oxidation of sulfur dioxide in airborne particles or cloud droplets (Calvert et al., 1985). According to the AGATE-code, the calculated intermediate concentrations of these species do not show any pronounced dependence on the relative humidity and no such dependence has been reported from comparable models. Moreover, only ppb amounts (H_2O_2) or less (O_3, OH, HO_2) of these species can be transferred to the particulate surface at the time scale available and a very effective catalysis would be required to generate measurable amounts of sulfate thereby. A similar argument holds for the case of molecular oxygen, which supports sulfate formation in droplets only through catalysis by metal ions (Filby, 1988; Seinfeld, 1986). The occurrence of effective catalysts cannot definitely be excluded in systems under electron irradiation and should be studied separately with more simple gas mixtures containing only SO_2 and H_2O or $SO_2/O_2/H_2O$.

There are two alternative hypotheses to explain the relation between sulfate formation and relative humidity that do not employ the assumption of heterogeneous catalysis and are based on homogeneous gas-phase mechanisms. One of these speculates about a ionic mechanism that promotes sulfuric acid formation via clustering of SO_2 ions and water in a way comparable to the chemistry of NO_x ions described in previous sections. There are reports in the literature that relate SO_2 oxidation to negative ion chemistry (Brasseur and Chatel, 1983), but some parts of the mechanism still remain to be validated in independent experiments. The same holds for the chemistry of positive SO_2 ions, which apparently has not been studied as intensively as the chemistry of positive NO_x ions.

The last hypothesis is related to the radical chemistry and claims that the termolecular $SO_2 + OH$ reaction is very sensitive to water vapor as third body (Nielson et al., 1987; Person et al., 1985). The calculated intermediate OH concentration in fact is high enough to permit a more extensive SO_2 oxidation than derived from literature data on k_{12} (Section IV B) and therefore, this assumption has been investigated quantitatively with the AGATE-code (Mätzing et al., 1988). It suggests the reaction

$$SO_2 + OH + H_2O \rightarrow products \quad k_{24} = 4.4 \times 10^{-34} \exp(+2400\,K/T)\,cm^6/s$$

$$(24)$$

as major sulfate formation step at relative humidities above 20%. As shown in Fig. 5, close agreement between measured and calculated sulfate concentra-

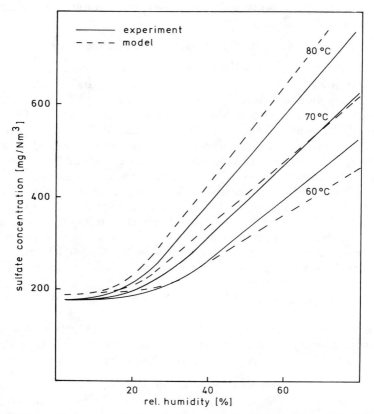

FIGURE 5. Comparison of measured and calculated sulfate concentrations as functions of relative humidity and temperature; $[SO_2] = 500$ ppm, no ammonia, $D = 10$ kGy. Experimental data (aerosol samples) from Baumann et al. (1987) and from Paur and Jordan (1988b).

tions can be achieved in this way. Reaction (12) contributes only 200 mg SO_4^{2-}/Nm^3, independent of relative humidity and temperature.

The magnitude of k_{24} corresponds to a collisional efficiency of water which at 300 K is about 75 times that of dry air. A small preexponential factor and a strongly negative formal activation energy must be chosen for reaction (24) in order to obtain a valuable model response. So the suggested reaction (24) either represents the composite of a multistep mechanism and/or involves a strongly bonded transition state, for example, one that is associated with the nucleation of sulfuric acid. If HSO_3 radicals or H atoms are taken to be direct products of reaction (24), then it is found to contribute substantially to NO oxidation via HO_2 formation. The calculated NO_x removal thus becomes a linear function of the SO_2 inlet concentration, as observed in experiments (Paur et al., 1989).

C. Effects of Gas–Particle Reactions

The size distribution of the aerosol produced by e-beam processing of flue gases is expected to have a maximum at $d \approx 0.1\,\mu\text{m}$ and a specific surface that drops from about $30\,\text{m}^2/\text{g}$ to about $5\,\text{m}^2/\text{g}$ during or shortly after the irradiation. According to model calculations, nitrous acid is formed as intermediate species that does not have substantial sinks in homogeneous reactions. Its most likely fate appears to be heterogeneous decomposition into NO_x and HNO_3. Large amounts of SO_2 are oxidized heterogeneously to sulfate at the filter surface. A similar heterogeneous oxidation has been proposed to occur at the aerosol surface. This proposal would require an effective heterogeneous catalysis in order to explain the rapid sulfate formation derived from aerosol measurements. Alternative proposals are based on ionic or radical mechanisms in homogeneous gas phase. If water is taken to be a very effective third body in the $SO_2 + OH$ reaction, the agreement between model calculation and experiment becomes quantitative.

VII. SUMMARY

Detailed theoretical and model results of the EBDS process have lead to a profound understanding of its physical and chemical fundamentals. The following results are directly related to industrial application:

1. dose rate does not affect removal efficiencies,
2. the measured NO_x removal results from both oxidative and reductive reaction pathways,
3. multistep irradiation is most effective upon intermittent NO_2 removal,
4. The EBDS process works most efficiently in case of simultaneous NO_x and SO_2 removal,
5. ammonia promotes both NO_x removal and particulate formation,
6. relative humidity crucially affects removal efficiencies,
7. the resulting aerosol must be expected to consist of particles in the submicron size range.

For theoretical and practical purposes more detailed informations are desirable on the following subjects:

1. measurements of intermediate HNO_2 concentrations,
2. NH_2 formation and consumption,
3. thermal NH_3/SO_2 reaction,
4. heterogeneous gas–particle reactions.

References

Andresen, P., Jacobs, A. Kleinermanns, C., and J. Wolfrum (1982), "Direct investigations of the NH_2 + NO reaction by laser photolysis at different temperatures," *Nineteeth Symposium (International) on Combustion*, Haifa, The Combustion Institute, Pittsburgh, PA, pp. 11–22.

Anicich, V. G. and W. T. Huntress, Jr. (1986), "A survey of bimolecular ion-molecule reactions for use in modeling the chemistry of planetary atmospheres, cometary comae and interstellar clouds," *Astrophys. J. Suppl. Ser.*, **62**, 553–672.

Armstrong, D. A. (1987), "The radiation chemistry of gases," in *Radiation Chemistry*, Farhataziz and M. A. J. Rodgers (eds.), Verlag Chemie, Weinheim, pp. 263–319.

Atakan, B., Jacobs, A., Wahl, M., Weller, R., and J. Wolfrum (1989), "Kinetic measurements and product branching ratio for the reaction NH_2 + NO at 294–1027 K," *Chem. Phys. Lett.* **155**, 609–614.

Atkinson, R., Baulch, D. L., Cox, R. A., Hampson, R. F. Jr., Kerr, J. A., and J. Troe (1989), "Evaluated kinetic and photochemical data for atmospheric chemistry," *J. Phys. Chem. Ref. Data* **18**, 881–1095.

Baulch, D. L., Cox, R. A., Hampson, R. F. Jr., Kerr, J. A., Troe, J., and R. T. Watson (1980, 1982, 1984), "Evaluated kinetic and photochemical data for atmospheric chemistry," *J. Phys. Chem. Ref. Data* **9**, 295–471 (1980); *J. Phys. Chem. Ref. Data* **11**, 327–497 (1982); *J. Phys. Chem. Ref. Data* **13**, 1259–1381 (1984).

Baulch, D. L., Drysdale, D. D., and D. G. Horne (1983), *Evaluated Kinetic Data for High Temperature Reactions*, Butterworths, London, Vol. 2.

Baumann, W., Jordan, S., Mätzing, H., Paur, H.-R., Schikarski, W., and H. Wiens (1987), "Simultane Rauchgasreinigung durch Elektronenstrahl," *Kernforschungszentrum Karlsruhe, KfK-PEF* **17**.

Brasseur, G. and A. Chatel (1983), "Modelling of stratospheric ions: a first attempt," *Annal. Geophys.* **1**, 173–185.

Brown, R. G. and C. A. Winkler (1979), "Active nitrogen," in *Dinitrogen Fixation*, F. Bottomley, and R. C. Burns (eds.), Wiley, New York, pp. 249–290.

Bulatov, V. P., Ioffe, A. A., Lozovsky, V. A., and O. M. Sarkisov (1989), "On the reaction of the NH_2 radical with NO at 295–620 K," *Chem. Phys. Lett.* **161**, 141–146.

Bunz, H. and R. Dlugi (1990), "Numerical studies of the behavior of aerosols in smog chambers," (unpublished).

Busi, F., D'Angelantonio, M., Mulazzani, Q. G., Raffaelli, V., and O. Tubertini (1987), "A kinetic model for radiation treatment of combustion gases," *Science of the Total Environment* **64**, 231–238.

Busi, F., D'Angelantonio, M., Mulazzani, Q. G., and O. Tubertini (1988), "Radiation induced NO_x/SO_2 emission control for industrial and power plants flue gas," *Radiat. Phys. Chem.* **31**, 101–108.

Calvert, J. G., Lazrus, A., Kok, G. L., Heikes, B. G., Walega, J. G., Lind, J., and C. A. Cantrell (1985), "Chemical mechanisms of acid generation in the troposphere," *Nature* **317**, 27–35.

Chang, S. G., Toossi, R., and T. Novakov (1981), "The importance of soot particles and nitrous acid in oxidizing SO_2 in atmospheric aqueous droplets," *Atoms. Environ.* **15**, 1287–1292.

Chatterjee, A. (1987), "Interaction of ionizing radiation with matter," in *Radiation Chemistry*, Farhataziz and M. A. J. Rodgers (eds.), Verlag Chemie, Weinheim, pp. 1–28.

Cohen, N. (1987), "The O + NH_3 reaction: a review," *Int. J. Chem. Kinet.* **19**, 319–362.

Cole, A. (1969), "Absorption of 20 eV to 50,000 eV electron beams in air and plastic," *Rad. Research* **38**, 7–33.

Dean, A. M., Hardy, J. E., and R. K. Lyon (1982), "Kinetics and mechanism of NH_3 oxidation," *Nineteenth Symposium (International) on Combustion*, Haifa, The Combustion Institute, Pittsburgh, PA, pp. 97–105.

DeMore, W. B., Molina, M. J., Sander, S. P., Golden, D. M., Hampson, R. F., Kurylo, M. J., Howard, C. J., and A. R. Ravishankara (1987), "Chemical kinetics and photochemical data for use in stratospheric modeling," Evaluation No. 8, JPL Publication No. 87–41.

Dolson, D. A. (1986), "Experimental determination of the OH product yield from $NH_2 + NO$ at 300 K," *J. Phys. Chem.* **90**, 6714–6718.

Douglas, J. E. and P. A. Kollmann (1978), "Trimethylamine-SO_2, the prototype strong charge-transfer complex," *J. Am. Chem. Soc.* **100**, 5226–5227.

Dransfeld, P., Hack, W., Kurzke, H., Temps, F., and H. Gg. Wagner (1984), "Direct studies of elementary reactions of NH_2-radicals in the gas phase," *Twentieth Symposium (International) on Combustion*, Ann Arbor, The Combustion Institute, Pittsburgh, PA, pp. 655–664.

Dreier, T. and J. Wolfrum (1984), "Determination of the vibrational energy distribution in N_2 and H_2O formed in the $NH_2 + NO$ reaction by CARS and IR-spectroscopy," *Twentieth Symposium (International) on Combustion*, Ann Arbor, The Combustion Institute, Pittsburgh, PA, pp. 695–702.

Elias, H. G. (1981), *Makromoleküle*, 4th ed., Basel.

Fehsenfeld, F. C., Mosesman, M., and E. E. Ferguson (1971), "Ion-molecule reactions in $NO^+ - H_2O$ system," *J. Chem. Phys.* **55**, 2120–2125.

Fehsenfeld, F. C., Howard, C. J., and A. L. Schmeltekopf (1975), "Gas phase ion chemistry of HNO_3," *J. Chem. Phys.* **63**, 2835–2841.

Filby, W. G. (1988), "Participation of free radicals in atmospheric chemistry," in *Chemical Kinetics of Small Organic Radicals*, Z. B. Alfassi (ed.), CRC Press, Boca Raton, FL, Vol. IV, pp. 31–57.

Frank, N., Kawamura, K., and G. Miller (1985), "Electron beam treatment of stack gases," *Radiat. Phys. Chem.* **25**, 35–45.

Frank, N., Hirano, Sh., and K. Kawamura (1988), "EBARA electron beam process for flue gas clean-up," *Radiat. Phys. Chem.* **31**, 57–82.

Friedlander, S. K. (1977), *Smoke, Dust and Haze*, Wiley, New York.

Geers-Müller, R. and F. Stuhl (1987), "On the kinetics of the reactions of oxygen atoms with NO_2, N_2O_4 and N_2O_3 at low temperatures," *Chem. Phys. Lett.* **135**, 263–268.

Gentry, J. W., Paur H.-R., and H. Mätzing (1988), "A modelling study on the dose rate effect on the efficiency of the EBDS-process," *Radiat. Phys. Chem.* **31**, 95–100.

Gleason, J. F., Sinha, A., and C. J. Howard (1987), "Kinetics of the gas-phase reaction $HOSO_2 + O_2 \rightarrow HO_2 + SO_3$," *J. Phys. Chem.* **91**, 719–724.

Hack, W., Schacke, H., Schröter, M., and H. Gg. Wagner (1979), "Reaction rates of NH_2-radicals with NO, NO_2, C_2H_2, C_2H_4 and other hydrocarbons," *Seventeenth Symposium (International) on Combustion*, Leeds, The Combustion Institute, Pittsburgh, PA, pp. 505–513.

Hack, W. and H. Kurzke (1985a), "The reaction of NH_2-radicals with electronically excited molecular oxygen $O_2(^1\Delta_g)$", *Ber. Bunsenges. Phys. Chem.* **89**, 86–93.

Hack, W., Kurzke, H., and H. Gg. Wagner (1985b), "Reaction of NH $(X^3\Sigma^-)$ and $O_2(^1\Delta_g)$ in the gas phase," *J. Chem. Soc., Farad. Trans. 2*, **81**, 949–961.

Hack, W., Rouveirolles, P., and H. Gg. Wagner (1986), "Direct measurements of the reactions $NH_2 + H_2 \rightleftharpoons NH_3 + H$ at temperatures from 670 to 1000 K," *J. Phys. Chem.* **90**, 2505–2511.

Hall, J. L., Zeitz, D., Stephens, J. W., Kasper, J. V. V., Glass, G. P., Curl, R. F., and F. K. Tittel

(1986), "Studies of the NH_2 + NO reaction by infrared kinetic spectroscopy," *J. Phys. Chem.* **90**, 2501–2505.

Harrison, J. A., Maclagan, R. G. A. R., and A. R. Whyte (1987), "Structures, energies and vibrational frequencies of intermediates and transition states in the reaction of NH_2 and NO," *J. Phys. Chem.* **91**, 6683–6686.

Hartley, E. M., Jr. and M. J. Matteson (1975), "Sulfur dioxide reactions with ammonia in humid air," *Ind. Eng. Chem. Fundam.* **14**, 67–72.

Heicklen, J. and M. Luria (1975), "Kinetics of homogeneous particle nucleation and growth," *Int. J. Chem. Kinet. Symp.* No. 1, p. 567–580.

Henglein, A., Schnabel, W., and J. Wendenburg (1969), *Einführung in die Strahlenchemie*, Verlag Chemie, Weinheim.

Hidy, G. M. and J. R. Brock (1970), *The Dynamics of Aerocolloidal Systems*, Pergamon, Oxford.

Holleman-Wiberg (1976), *Lehrbuch der anorganischen Chemie*, de Gruyter, Berlin.

Ioffe, A. A., Bulatov, V. P., Lozovsky, V. A., Goldenberg, M. Ya., Sarkisov, O. M., and S. Ya. Umansky (1989), "On the reaction of the NH_2 radical with SO_2 at 298–363 K," *Chem. Phys. Lett.* **156**, 425–432.

Jeffries, J. B. and G. P. Smith (1986), "Kinetics of the reaction OH + NH_3," *J. Phys. Chem.* **90**, 487–491.

Jenkin, M. E. and R. A. Cox (1987), "Kinetics of the gas phase reaction of OH with nitrous acid," *Chem. Phys. Lett.* **137**, 548–552.

Jordan, S., Paur, H.-R., Cherdron, W., and W. Lindner (1986), "Physical and chemical properties of the aerosol produced by the ES-Verfahren," *J. Aerosol Sci.* **17**, 669–675.

Jordan, S. (1988), "Progress in the electron beam treatment of stack gases," *Radiat. Phys. Chem.* **31**, 21–28.

Jordan, S., Paur, H.-R., and W. Schikarski (1988), "Simultane Entschwefelung und Entstickung mit dem Elektronenstrahlverfahren," *Physik in unserer Zeit* **19**, 8–15.

Jordan, S. and H.-R. Paur (1988), "Recent progress in flue gas treatment by electron irradiation," IAEA workshop, Madrid, 2.–4. Oct.

Kaiser, E. W. and C. H. Wu (1977a), "Measurement of the rate constant of the reaction of nitric acid with nitrous acid," *J. Phys. Chem.* **81**, 187–190.

Kaiser, E. W. and C. H. Wu (1977b), "A kinetic study of the gas phase formation and decomposition reactions of nitrous acid," *J. Phys. Chem.* **81**, 1701–1706.

Kaiser, E. W. and S. M. Japar (1978), "Upper limits to the gas phase reaction rates of HONO with NH_3 and $O(^3P)$ atoms," *J. Phys. Chem.* **82**, 2753–2754.

Kimball-Linne, M. A. and R. K. Hanson (1986), "Combustion-driven flow reactor studies of thermal $DeNO_x$ reaction kinetics," *Comb. & Flame* **64**, 337–351.

Klassen, N. V. (1987), "Primary products in radiation chemistry," in *Radiation Chemistry*, Farhataziz and M. A. J. Rodgers (eds.), Varlag Chemie, Weinheim, pp. 29–64.

Landolt-Börnstein (1952), *Zahlenwerte und Funtionen aus Physik, Chemie, Astronomie, Geophysik und Technik*, A. Eucken (ed.), Springer-Verlag, Berlin, Bd. I, 5.

Landreth, R., de Pena, R. G., and J. Heicklen (1974), "Thermodynamics of the reactions $(NH_3)_n \cdot SO_2(s) \rightleftharpoons n\ NH_3(g) + SO_2(g)$," *J. Phys. Chem.* **78**, 1378–1380.

Landreth, R., de Pena, R. G., and J. Heicklen (1975), "Thermodynamics of the reaction of ammonia and sulfur dioxide in the presence of water vapour," *J. Phys. Chem.* **79**, 1975–1978.

Lesclaux, R. (1984), "Reactivity and kinetic properties of the NH_2 radical in the gas phase," *Rev. Chem. Intermediat.* **5**, 347–392.

Levine, J. S. (1985), *The Photochemistry of Atmospheres*, Academic Press, New York.

Lii, R.-R., Gorse, R. A., Jr., Sauer, M. C., Jr., and S. Gordon (1980), "Temperature dependence of the gas-phase self-reaction of HO_2," *J. Phys. Chem.* **84**, 813–817.

Lohrmann, E. (1983), *Einführung in die Elementarteilchenphysik*, Teubner Studienbücher, Stuttgart.

Lucchese, R. R., Haber, K., and H. F. Schaefer (1976), "Charge-transfer complexes," *J. Am. Chem. Soc.* **98**, 7617–7620.

Mätzing, H. (1987a) (unpublished results).

Mätzing, H., Paur, H.-R., and J. W. Gentry (1987b), "Zur Bedeutung heterogener Ionensenken beim Elektronenstrahlverfahren," *J. Aerosol Sci.* **18**, 773–776.

Mätzing, H., Paur, H.-R., and H. Bunz (1988), "Dynamics of particulate formation in the electron beam dry scrubbing process," *J. Aerosol Sci.* **19**, 883–885.

Mätzing, H. (1989), "On the kinetics of O atoms in the radiation treatment of waste gases," *Radiat. Phys. Chem.* **33**, 81–84.

Markovic, V. (1987), "Introduction to 'Electron Beam Processing of Combustion Flue Gases," IAEA meeting held in Karlsruhe, FRG, Oct. 27–29 (1986), IAEA-TECDOC-428, Vienna.

Meissner, G. (1964), "Berechnung des Durchgangs schneller Elektronen durch Materie durch eine Kombination von analytischen und stochastischen Methoden," *Z. Naturforsch.* **19a**, 269–283.

Miller, J. A., Smooke, M. D., Green, R. M., and R. J. Kee (1983), "Kinetic modeling of the oxidation of ammonia in flames," *Comb. Sci. Technol.* **34**, 149–176.

Namba, H., Aoki, Y., Tokunaga, O., Suzuki, R., and S. Aoki (1988), "Experimental evidence of N_2 formation from NO in simulated coal-fired flue gas by electron beam irradiation," *Chem. Lett.*, 1465–1468.

Nielsen, O. J., Pagsberg, P., and A. Sillesen (1987), "Absolute rate constant for the reaction of OH with SO_2 in the presence of water at atmospheric pressure," Comm. Eur. Communities, EUR-10832, Phys. Chem. Behav. Atmosph. Pollutants, pp. 338–344.

Nishimura, K., Tokunaga, O., Washino, M. and N. Suzuki (1979), "Radiation treatment of exhaust gases, (IX) analysis of NO and NO_2 decomposition in $NO-N_2$ and NO_2-N_2 mixtures by computer simulation," *J. Nucl. Sci. Tech.* **16**, 596–604.

Nishimura, K. and N. Suzuki (1981), "Radiation treatment of exhaust gases, (XIV) analysis of NO oxidation and decomposition in dry and moist $NO-O_2-N_2$ mixtures by computer simulation," *J. Nucl. Sci. Tech.* **18**, 878–886.

Paur, H.-R., Jordan, S., Baumann, W., Cherdron, W., Lindner, W., and H. Wiens (1986), "The influence of flue gas humidity and reaction time on the aerosol formation process in the Electron Beam Dry Scrubbing Process (ES-Verfahren)," AEROSOLS: Formation and Reactivity, Proc. 2nd Int. Aerosol Conf., 22–26 Sept. 1986, Berlin, Pergamon, pp. 1024–1027.

Paur, H.-R., Jordan, S., Baumann, W., and H. Mätzing (1988a), "Aerosolbildung und heterogene Reaktionen bei der simultanen Rauchgasreinigung durch Elektronenstrahl," KfK-PEF 35, Kernforschungszentrum Karlsruhe, FRG, pp. 739–754.

Paur, H.-R. and S. Jordan (1988b), "Aerosol formation in the electron beam dry scrubbing process (ES-Verfahren)," *Radiat. Phys. Chem.* **31**, 9–14.

Paur, H.-R. and S. Jordan (1989), "The influence of SO_2 and NH_3 concentrations on the aerosol formation in the electron beam dry scrubbing process," *J. Aerosol Sci.* **20**, 7–11.

Paur, H.-R., Jordan, S., Baumann, W., and H. Mätzing (1989), "Flue gas treatment by electron beam irradiation," Radtech Europe, Florence, 9.–11. Oct.

Perry, R. H., Green, D. W., and J. O. Maloney (1984), *"Perry's Chemical Engineers' Handbook"*, 6th ed., McGraw Hill, New York.

Person, J. C., Ham, D. O., and A. A. Boni (1985), "Unified Projection of the Performance and Economics of Radiation-Initiated NO_x/SO_x Emission Control Technologies, Final Report," Report 1985, PSI-TR-259/542.

Person, J. C., and D. O. Ham (1988), "Removal of SO_2 and NO_x from stack gases by electron beam irradiation," *Radiat. Phys. Chem.* **31**, 1–8.

Roedel, W. (1979), "Measurement of sulfuric acid saturation vapour pressure," *J. Aerosol Sci.* **10**, 375–386.

Rolle, W. and E. Renner (1984), "Modellberechnungen des Transports, der chemischen Umwandlung und der Abscheidung von SO_2 und NO_x unter variablen atmosphärischen Bedingungen," *Staub-Reinhalt. Luft* **44**, 480–487.

Sauer, M. C., Jr. (1976), "Pulse studies of gases," in *Advances in Radiation Chemistry*, M. Burton and J. L. Magee (eds.), Wiley, New York, Vol. 5, pp. 97–184.

Seinfeld, J. H. (1986), *Atmospheric chemistry and physics of air pollution*, Wiley, New York.

Silver, J. A. and C. E. Kolb (1982), "Kinetic measurements for the reaction of $NH_2 + NO$ over the temperature range 294–1215 K," *J. Phys. Chem.* **86**, 3240–3246.

Silver, J. A. and C. E. Kolb (1987), "A reevaluation of the branching ratio for the reaction of NH_2 with NO," *J. Phys. Chem.* **91**, 3713–3714.

Stelson, A. W. and J. H. Seinfeld (1982), "Relative humidity and temperature dependence of the ammonium nitrate dissociation constant," *Atoms. Environ.* **16**, 983–992.

Stumm, W. and J. J. Morgan (1981), *Aquatic Chemistry*, 2nd ed., Wiley, New York.

Sutherland, C. D. and J. Zinn (1975), "Chemistry computations for irradiated hot air," Los Alamos Scientific Laboratory Informal Report LA-6055-MS, Los Alamos, NM.

Suzuki, N. and O. Tokunaga (1981), "Radiation chemical studies on electron beam treatment of exhaust gases in basic research on electron beam desulfurization and denitrification process," Final report of research project.

Tokunaga, O., Nishimura, N. Suzuki, N. and M. Washino (1978), "Radiation treatment of exhaust gases IV. Oxidation of NO in the moist mixture of O_2 and N_2," *Radiat. Phys. Chem.* **11**, 117–122.

Tokunaga, O. and N. Suzuki (1984), "Radiation chemical reaction in NO_x and SO_2 removals from flue gas," *Radiat. Phys. Chem.* **24**, 145–165.

Vosper, A. J. (1976), "Dinitrogen trioxide XIII. Hydration of gaseous dinitrogen trioxide and a reinvestigation of its dissociation," *J. Chem. Soc., Dalton Trans., Part I*, pp. 135–138.

Willis, C. and A. W. Boyd (1976), "Excitation in the radiation chemistry of inorganic gases," *Radiat. Phys. Chem.* **8**, 71–111.

Wittig, S., Spiegel, G., Platzer, K.-H., and U. Willibald (1988a), "The performance characteristics of the electron-beam-technique: detailed studies at the (ITS) flue gas facility," *Radiat. Phys. Chem.* **31**, 83–93.

Wittig, S., Spiegel, G., Platzer, K.-H., and U. Willibald (1988b), "Simultane Rauchgasreinigung (Entschwefelung, Denitrierung) durch Elektronenstrahl," KfK-PEF 35, Kernforschungszentrum Karlsruhe, FRG, pp. 755–767.

PART II: THE AGATE-CODE

I.   INTRODUCTION

Part II of this report gives a complete listing of the AGATE-code, which has been developed for a quantitative description and investigation of the chemical kinetics pertaining to the EBDS process. The code is continuously being updated, for example, concerning new data on temperature-dependent reactions or specific mechanisms. The present listing gives the code status in Spring 1990.

II.   THE STRUCTURE OF THE AGATE-CODE

Input values to the AGATE-code are waste gas composition in concentrations by volume, temperature, pressure, the irradiation dose D_0 in J/g units and a parameter σ [s] which determines the width of the Gaussian dose rate profile:

$$\dot{D} = \frac{D_0}{\sigma\sqrt{2\pi}} \exp\left[-\frac{1}{2}\left(\frac{t-\tau}{\sigma}\right)^2 \right]$$

From the input data, τ is calculated by giving $\dot{D}$ a small positive start value at $t = 0$. The second time derivative of the dose, $\ddot{D} = \dot{D}(\tau - t)/\sigma^2$, is used to calculate the actual dose rate at time t by integration. The active species concentrations are calculated from

$$\frac{dn}{dt} = G_n \dot{D} x_i \rho$$

where G_n is the G-value for species n obtained by irradiation of component i with mole fraction x_i. ρ is the average gas density under input conditions. By choosing G_n [molecules/100 eV], $\dot{D}$ [100 eV/g s], and ρ [g/cm^3], dn/dt results in [cm^{-3} s]. The G-values are listed in Table I.

The reactions initiated by active species generation are listed in Tables II–VIII. The net reaction rates for each reaction (or irradiation step) are calculated between successive integration steps from the stored rate constants and the actual reactant concentrations. These net rates are used to calculate the time derivative of each single species according to the stoichiometry of each reaction. The integration is performed with a commercial subroutine (Subroutine Library Specification, Harwell 1979) which employs a Gear algorithm for the integration of stiff ordinary differential equations. The computational accuracy, based on atomic mass balance, is $10^{-11}\%$ or better.

III. DATA REPRESENTATION

The data representation in Tables I–VIII is in principle self-explaining and various notes are included in the tables. Some additional remarks are summarized below.

Table I. The irradiation G-values are tabulated according to a convenience widely used in the literature. For inclusion into the numerical calculation, these steps have been converted to stoichiometric equations. The electrons supplied by the accelerator do not appear explicitly in the code. Those electrons which originate from ionization processes are arbitrarily termed "hot electrons" in the tabulation. They are allowed to thermalize before entering into reactive steps (Table III). This means that all thermal electrons are treated as having equal properties determined by the low-energy tail of the true thermal energy distribution.

Tables II–VI give the reactions implemented in the AGATE-code. All rate constants are in cm^3, s units. Some reactions (for example, mutual ion recombination) have rate constants which are a sum of bimolecular and termolecular contributions. In this context, M denotes the overall molecule concentration $[cm^{-3}]$; M has been set to 99% of the value obtained from the ideal gas law.

Where available, the temperature dependence of rate constants is included using $T[K]$.

Some rate constants depend explicitly on the concentration of water vapor; in this case, $[H_2O]$ is in $[cm^{-3}]$.

Table VII compiles almost all third-order reactions of neutrals. Where indicated, k_0 is used with the assumption that the low-pressure limit applies. In the other cases, k_0, k_∞, and F_C are given, from which

$$k = \frac{k_1}{1 + k_1/k_\infty} F_C^{\{1 + [(\log k_1/k_\infty)/(0.75 - \log F_C)]^2\}^{-1}}$$

is obtained with $k_1 = k_0[M]$. For reactions in the H_2–O_2–CO system, the third body efficiencies given by Warnatz (10) are used to calculate k_1:

H_2	O_2	N_2	H_2O	CO	CO_2	Ar	He
1.0	0.4	0.4	6.5	0.75	1.5	0.35	0.35

Table VIII shows some heterogeneous reactions that have been considered so far. According to Section VI of Part I rate constants of the form $k_{het} = 0.25\alpha$

$\bar{v}A_e\,[\mathrm{s}^{-1}]$ are used, where $\bar{v}$ is the thermal velocity of the gaseous species considered and A_e is the effective (actual) particulate surface. In case that two different species are transfered to the particulate surface, the minimum of their different velocities is used. Table VIII gives α values which hold over a wide range of conditions.

Ammonium nitrate formation is modeled according to the eighth-order expression given by Heicklen et al. (73) using

$$[HNO_3]\cdot[NH_3] \geqq M^2 \exp(-7500/T)$$

as the necessary condition for the onset of nucleation.

Ammonium sulfate formation has so far been modeled by two fast dummy reactions forming NH_4HSO_4 and $(NH_4)_2SO_4$, respectively.

TABLE I
Radiolytic Processes[a]

	G (molecules/100 eV)
$N_2 \rightarrow N_2^*$	0.29
$N_2 \rightarrow N(^2D)$	0.885
$N_2 \rightarrow N(^2P)$	0.295
$N_2 \rightarrow N(^4S)$	1.18
$N_2 \rightarrow N_2^+ + e^-$	2.27
$N_2 \rightarrow N^+ + N + e^-$	0.69
$O_2 \rightarrow O_2^*$	0.077
$O_2 \rightarrow O(^3P) + O(^1D)$	1.82
$O_2 \rightarrow O^3(P) + O^*$ (above ^1S level)	0.18
$O_2 \rightarrow O(^3P) + O^+ + e^-$	0.80
$O_2 \rightarrow O(^1D) + O^+ + e^-$	0.43
$O_2 \rightarrow O_2^+ + e^-$	2.07
$H_2O \rightarrow H_2 + O$	0.45
$H_2O \rightarrow OH + H$	3.58
$H_2O \rightarrow H_2O^+ + e^-$	1.99
$H_2O \rightarrow H_2^+ + O + e^-$	0.01
$H_2O \rightarrow H^+ + OH + e^-$	0.67
$H_2O \rightarrow OH^+ + H + e^-$	0.57
$H_2O \rightarrow O^+ + H_2 + e^-$	0.06
$CO_2 \rightarrow CO + O$	4.51
$CO_2 \rightarrow CO_2^+ + e^-$	2.24
$CO_2 \rightarrow CO^+ + O + e^-$	0.51
$CO_2 \rightarrow C^+ + 2O + e^-$	0.07
$CO_2 \rightarrow O^+ + CO + e^-$	0.21

[a] O_2^* and N_2^* are treated like $O_2(^1\Delta_g)$ and $N_2(A)$ in the model. Data are from Refs. 3 and 60. Electrons arising from radiolysis are allowed to thermalize prior to reactions (see Table III).

TABLE II
Positive Ion–Neutral Reactions

1	$N_2^+ + O_2$	$\to O_2^+ + N_2$	$k = 3.9 \times 10^{-10} \exp(-T/143)$	(27)
2	$N_2^+ + O_2$	$\to NO^+ + NO$	$k = 1.0 \times 10^{-17}$	(16)
3	$N_2^+ + H_2O$	$\to H_2O^+ + N_2$	$k = 2.0 \times 10^{-9}$	(26)
4	$N_2^+ + CO_2$	$\to CO_2^+ + N_2$	$k = 8.3 \times 10^{-10}$	(16, 59)
5	$N_2^+ + NO$	$\to NO^+ + N_2$	$k = 3.3 \times 10^{-10}$	(26)
6	$N_2^+ + NO_2$	$\to NO_2^+ + N_2$	$k = 3.0 \times 10^{-10}$	(26)
7	$N_2^+ + SO_2$	$\to SO_2^+ + N_2$	$k = 5.0 \times 10^{-10}$	(26)
8	$N_2^+ + NH_3$	$\to NH_3^+ + N_2$	$k = 1.9 \times 10^{-9}$	(26)
9	$N_2^+ + O_3$	$\to O_2^+ + O + N_2$	$k = 1.0 \times 10^{-10}$	(16)
10	$N_2^+ + N_2O$	$\to N_2O^+ + N_2$	$k = 3.0 \times 10^{-10}$	(16)
11	$N_2^+ + H_2$	$\to H_2^+ + N_2$	$k = 4.0 \times 10^{-10}$	(16)
12	$N_2^+ + CO$	$\to CO^+ + N_2$	$k = 7.0 \times 10^{-11}$	(16)
13	$N_2^+ + OH$	$\to OH^+ + N_2$	$k = 4.6 \times 10^{-10}$	(26)
14	$N_2^+ + H$	$\to H^+ + N_2$	$k = 2.5 \times 10^{-10}$	(16, 59)
15	$N_2^+ + O$	$\to NO^+ + N$	$k = 1.4 \times 10^{-10}$	(16, 59)
16	$N_2^+ + O$	$\to NO^+ + N(^2D)$	$k = 1.8 \times 10^{-10}(300/T)$	(4, 16, 78)
17	$N_2^+ + O$	$\to O^+ + N_2$	$k = 6.0 \times 10^{-12}$	(16, 55, 75)
18	$N_2^+ + N$	$\to N^+ + N_2$	$k = 1.0 \times 10^{-11}$	(55)
19	$N_2^+ + N + M$	$\to N_3^+ + M$	$k = 1.0 \times 10^{-29}(300/T)$	(55)
20	$N_2^+ + N_2 + M$	$\to N_4^+ + M$	$k = 5.0 \times 10^{-29}(300/T)$	(16)
21	$N^+ + O_2$	$\to NO^+ + O$	$k = 2.6 \times 10^{-10}$	(16)
22	$N^+ + O_2$	$\to N + O_2^+$	$k = 3.0 \times 10^{-10}$	(2, 26)
23	$N^+ + O_2$	$\to O^+ + NO$	$k = 3.6 \times 10^{-11}$	(26)
24	$N^+ + H_2O$	$\to H_2O^+ + N$	$k = 2.4 \times 10^{-9}$	(26)
25	$N^+ + H_2O$	$\to NO^+ + H_2$	$k = 2.4 \times 10^{-10}$	(26)
26	$N^+ + CO_2$	$\to CO_2^+ + N$	$k = 1.3 \times 10^{-9}$	(26)
27	$N^+ + CO_2$	$\to CO^+ + NO$	$k = 2.5 \times 10^{-10}$	(26, 59)
28	$N^+ + NO$	$\to NO^+ + N$	$k = 4.1 \times 10^{-10}$	(26)
29	$N^+ + NO$	$\to N_2^+ + O$	$k = 5.0 \times 10^{-11}$	(26)

(Continued)

TABLE II. (Continued)

30	$N^+ + NO_2$	$\rightarrow NO^+ + NO$	$k = 5.0 \times 10^{-10}$	(16)
31	$N^+ + NO_2$	$\rightarrow NO_2^+ + N$	$k = 3.0 \times 10^{-10}$	(16)
32	$N^+ + NH_3$	$\rightarrow NH_3^+ + N$	$k = 1.7 \times 10^{-9}$	(8, 26)
33	$N^+ + O_3$	$\rightarrow NO^+ + O_2$	$k = 5.0 \times 10^{-10}$	(16)
34	$N^+ + N_2O$	$\rightarrow NO^+ + N_2$	$k = 5.5 \times 10^{-10}$	(26)
35	$N^+ + H_2$	$\rightarrow H_2^+ + N$	$k = 2.0 \times 10^{-10}$	(26)
36	$N^+ + CO$	$\rightarrow CO^+ + N$	$k = 4.9 \times 10^{-10}$	(26)
37	$N^+ + OH$	$\rightarrow OH^+ + N$	$k = 3.4 \times 10^{-10}$	(16, 59)
38	$N^+ + OH$	$\rightarrow NO^+ + H$	$k = 3.4 \times 10^{-10}$	(59)
39	$N^+ + H$	$\rightarrow H^+ + N$	$k = 3.0 \times 10^{-10}$	(16)
40	$N^+ + O$	$\rightarrow O^+ + N$	$k = 1.0 \times 10^{-12}$	(16)
41	$N^+ + O + M$	$\rightarrow NO^+ + M$	$k = 1.0 \times 10^{-29} (300/T)$	(16)
42	$N^+ + N + M$	$\rightarrow N_2^+ + M$	$k = 1.0 \times 10^{-29} (300/T)$	(16)
43	$N^+ + N_2 + M$	$\rightarrow N_3^+ + M$	$k = 1.8 \times 10^{-29} (300/T)$	(16)
44	$N_3^+ + O_2$	$\rightarrow NO^+ + O + N_2$	$k = 4.0 \times 10^{-7} (300/T)^{0.5}$ $+ 3.0 \times 10^{-25} (300/T)^{2.5} M$	(16)
45	$N_3^+ + NO$	$\rightarrow NO^+ + N + N_2$	$k = 4.0 \times 10^{-7} (300/T)^{0.5}$ $+ 3.0 \times 10^{-25} (300/T)^{2.5} M$	(16)
46	$N_3^+ + NO_2$	$\rightarrow NO_2^+ + N + N_2$	$k = 4.0 \times 10^{-7} (300/T)^{0.5}$ $3.0 \times 10^{-25} (300/T)^{2.5} M$	(16)
47	$N_3^+ + N$	$\rightarrow N_2^+ + N_2$	$k = 4.0 \times 10^{-7} (300/T)^{0.5}$ $+ 3.0 \times 10^{-25} (300/T)^{2.5} M$	(16)
48	$N_4^+ + H_2O$	$\rightarrow H_2O^+ + 2N_2$	$k = 3.0 \times 10^{-10}$	(16)
49	$N_4^+ + CO_2$	$\rightarrow CO_2^+ + 2N_2$	$k = 3.0 \times 10^{-10}$	(16)
50	$N_4^+ + N_2O$	$\rightarrow N_2O^+ + 2N_2$	$k = 3.0 \times 10^{-10}$	(16)
51	$N_4^+ + OH$	$\rightarrow OH^+ + 2N_2$	$k = 3.0 \times 10^{-10}$	(16)
52	$O_2^+ + N_2$	$\rightarrow NO^+ + NO$	$k = 1.0 \times 10^{-16}$	(16)
53	$O_2^+ + NO$	$\rightarrow NO^+ + O_2$	$k = 3.5 \times 10^{-10}$	(26, 75)
54	$O_2^+ + NO_2$	$\rightarrow NO_2^+ + O_2$	$k = 6.0 \times 10^{-10}$	(16)
55	$O_2^+ + NH_3$	$\rightarrow NH_3^+ + O_2$	$k = 2.4 \times 10^{-9}$	(26)

No.	Reaction		k	Ref.
56	$O_2^+ + N_2O_5$	$\rightarrow NO_2^+ + NO_3 + O_2$	$k = 8.8 \times 10^{-10}$	(68)
57	$O_2^+ + N$	$\rightarrow NO^+ + O$	$k = 1.8 \times 10^{-10}$	(4, 55)
58	$O_2^+ + 2O_2$	$\rightarrow O_4^+ + O_2$	$k = 2.8 \times 10^{-30}(300/T)$	(16)
59	$O_2^+ + H_2O + M$	$\rightarrow O_2^+(H_2O) + M$	$k = 2.8 \times 10^{-28}(300/T)$	(16, 65)
60	$O^+ + N_2$	$\rightarrow N_2^+ + O$	$k = 9.0 \times 10^{-11} T^{-0.7}$	(2, 59)
61	$O^+ + N_2$	$\rightarrow NO^+ + N$	$k = 1.2 \times 10^{-12} + 5.4 \times 10^{-29}(300/T)M$	(16, 26)
62	$O^+ + O_2$	$\rightarrow O_2^+ + O$	$k = 6.6 \times 10^{-10} T^{-0.55}$	(26, 59)
63	$O^+ + H_2O$	$\rightarrow H_2O^+ + O$	$k = 2.7 \times 10^{-9}$	(26, 59)
64	$O^+ + CO_2$	$\rightarrow O_2^+ + CO$	$k = 1.0 \times 10^{-9}$	(7, 59)
65	$O^+ + NO$	$\rightarrow NO^+ + O$	$k = 1.0 \times 10^{-12}$	(26)
66	$O^+ + NO_2$	$\rightarrow NO_2^+ + O$	$k = 1.6 \times 10^{-9}$	(16)
67	$O^+ + NO_2$	$\rightarrow NO^+ + O_2$	$k = 5.0 \times 10^{-10}$	(16)
68	$O^+ + SO_2$	$\rightarrow O_2^+ + SO$	$k = 8.0 \times 10^{-10}$	(26)
69	$O^+ + NH_3$	$\rightarrow NH_3^+ + O$	$k = 1.2 \times 10^{-9}$	(59)
70	$O^+ + O_3$	$\rightarrow O_2^+ + O_2$	$k = 1.1 \times 10^{-10}$	(16)
71	$O^+ + N_2O$	$\rightarrow N_2O^+ + O$	$k = 5.0 \times 10^{-10}$	(16, 26)
72	$O^+ + H_2$	$\rightarrow OH^+ + H$	$k = 1.8 \times 10^{-9}$	(16, 59)
73	$O^+ + OH$	$\rightarrow OH^+ + O$	$k = 3.3 \times 10^{-11}$	(16, 59)
74	$O^+ + OH$	$\rightarrow H^+ + O_2$	$k = 2.7 \times 10^{-11}(T/300)^{0.13}$	(16)
75	$O^+ + OH$	$\rightarrow O_2^+ + H$	$k = 3.6 \times 10^{-10}$	(59)
76	$O^+ + H$	$\rightarrow H^+ + O$	$k = 6.8 \times 10^{-11}$	(16, 59)
77	$O_4^+ + H_2O$	$\rightarrow O_2^+(H_2O) + O_2$	$k = 1.8 \times 10^{-9}$	(2, 65)
78	$O_4^+ + NO$	$\rightarrow NO^+ + 2O_2$	$k = 5.0 \times 10^{-10}$	(16)
79	$O_4^+ + NO_2$	$\rightarrow NO_2^+ + 2O_2$	$k = 3.0 \times 10^{-10}$	(16)
80	$O_4^+ + O$	$\rightarrow O_2^+ + O_3$	$k = 3.0 \times 10^{-10}$	(16)
81	$O_2^+(H_2O) + H_2O$	$\rightarrow H_3O^+ + OH + O_2$	$k = 2.0 \times 10^{-10}$	(2, 16, 65)
82	$O_2^+(H_2O) + H_2O$	$\rightarrow H_3O^+(OH) + O_2$	$k = 1.5 \times 10^{-9}$	(2, 16, 65)
83	$O_2^+(H_2O) + NO_2$	$\rightarrow NO_2^+ + H_2O + O_2$	$k = 3.0 \times 10^{-10}$	(16)
84	$O_2^+(H_2O) + NO$	$\rightarrow NO^+ + H_2O + O_2$	$k = 1.0 \times 10^{-10}$	(16)
85	$H_2O^+ + O_2$	$\rightarrow O_2^+ + H_2O$	$k = 2.6 \times 10^{-10}$	(26, 59)
86	$H_2O^+ + H_2O$	$\rightarrow H_3O^+ + OH$	$k = 1.7 \times 10^{-9}$	(3, 26, 59)
87	$H_2O^+ + NO$	$\rightarrow NO^+ + H_2O$	$k = 6.0 \times 10^{-10}$	(26, 59)

(Continued)

TABLE II. (Continued)

#	Reactants	Products	k	Ref.
88	$H_2O^+ + NO_2$	$\rightarrow NO_2^+ + H_2O$	$k = 3.0 \times 10^{-10}$	(16)
89	$H_2O^+ + NH_3$	$\rightarrow NH_3^+ + H_2O$	$k = 2.2 \times 10^{-9}$	(26)
90	$H_2O^+ + NH_3$	$\rightarrow NH_4^+ + OH$	$k = 9.0 \times 10^{-10}$	(26)
91	$H_2O^+ + H_2$	$\rightarrow H_3O^+ + H$	$k = 8.7 \times 10^{-10}$	(16, 26, 59)
92	$H_2O^+ + OH$	$\rightarrow H_3O^+ + O$	$k = 6.9 \times 10^{-10}$	(26)
93	$H_2O^+ + O$	$\rightarrow O_2^+ + H_2$	$k = 5.5 \times 10^{-11}$	(26, 59)
94	$H_2O^+ + N$	$\rightarrow NO^+ + H_2$	$k = 9.0 \times 10^{-11}$	(26)
95	$H_3O^+ + N_2O_5$	$\rightarrow NO_2^+(H_2O) + HNO_3$	$k = 5.5 \times 10^{-10}$	(68, 78)
96	$H_3O^+ + N_2O_5$	$\rightarrow NO_2^+ + H_2O + HNO_3$	$k = 5.5 \times 10^{-10}$	(68, 78)
97	$H_3O^+ + NH_3$	$\rightarrow NH_4^+ + H_2O$	$k = 2.2 \times 10^{-9}$	(26)
98	$H_3O^+ + HNO_3$	$\rightarrow NO_2^+(H_2O) + H_2O$	$k = 1.6 \times 10^{-9}$	(62)
99	$H_3O^+ + OH + O_2$	$\rightarrow O_2^+(H_2O) + H_2O$	$k = 1.6 \times 10^{-27}(T/300)^{0.55}$	(16)
100	$H_3O^+ + OH + M$	$\rightarrow H_3O^+(OH) + M$	$k = 5.4 \times 10^{-25}(300/T)^{0.7}$	(16)
101	$H_3O^+ + H_2O + M$	$\rightarrow H_3O^+(H_2O) + M$	$k = 3.7 \times 10^{-27}(300/T)$	(16)
102	$H_3O^+(OH) + H_2O$	$\rightarrow H_3O^+ + H_2O + OH$	$k = 1.4 \times 10^{-9}$	(2)
103	$H_3O^+(OH) + H_2O$	$\rightarrow H_3O^+(H_2O) + OH$	$k = 3.0 \times 10^{-9}$	(65)
104	$H_3O^+(OH) + NO$	$\rightarrow NO^+ + 2H_2O$	$k = 3.0 \times 10^{-10}$	(16)
105	$H_3O^+(OH) + NO_2$	$\rightarrow NO_2^+ + 2H_2O$	$k = 3.0 \times 10^{-10}$	(16)
106	$H_3O^+(H_2O) + H_2O + M$	$\rightarrow H_3O^+(H_2O)_2 + M$	$k = 3.0 \times 10^{-27}(300/T)$	(16, 60)
107	$H_3O^+(H_2O) + OH$	$\rightarrow H_3O^+(OH) + H_2O$	$k = 2.1 \times 10^{-11}(300/T)^{1.46}\exp(-4503/T)$	(16)
108	$H_3O^+(H_2O)_2 + M$	$\rightarrow H_3O^+(H_2O) + H_2O + M$	$k = 5.9 \times 10^{-1}(T/300)^{0.54}\exp(-11220/T)$	(16)
109	$H_3O^+(H_2O)_2 + HNO_2$	$\rightarrow NO^+(H_2O)_3 + H_2O$	$k = 1.1 \times 10^{-8}(300/T)^{2.6}$	(16)
110	$CO_2^+ + O_2$	$\rightarrow O_2^+ + CO_2$	$k = 6.5 \times 10^{-9}\,T^{-0.78}$	(59)
111	$CO_2^+ + H_2O$	$\rightarrow H_2O^+ + CO_2$	$k = 1.7 \times 10^{-9}$	(59)
112	$CO_2^+ + NO$	$\rightarrow NO^+ + CO_2$	$k = 1.2 \times 10^{-10}$	(26)
113	$CO_2^+ + NO_2$	$\rightarrow NO_2^+ + CO_2$	$k = 3.0 \times 10^{-10}$	(16)
114	$CO_2^+ + SO_2$	$\rightarrow SO_2^+ + CO_2$	$k = 1.5 \times 10^{-9}$	(26)
115	$CO_2^+ + NH_3$	$\rightarrow NH_3^+ + CO_2$	$k = 1.9 \times 10^{-9}$	(26)
116	$CO_2^+ + N_2O$	$\rightarrow N_2O^+ + CO_2$	$k = 3.0 \times 10^{-10}$	(16)
117	$CO_2^+ + OH$	$\rightarrow OH^+ + CO_2$	$k = 3.0 \times 10^{-10}$	(16)

No.	Reactants	Products	Rate constant	Ref.
118	$CO_2^+ + O$	$\rightarrow CO + O_2^+$	$k = 1.3 \times 10^{-10}$	(7, 59)
119	$CO_2^+ + O$	$\rightarrow O^+ + CO_2$	$k = 1.3 \times 10^{-10}$	(16, 59)
120	$CO_2^+ + H$	$\rightarrow H^+ + CO_2$	$k = 1.9 \times 10^{-10}$	(16, 59)
121	$CO^+ + O_2$	$\rightarrow O_2^+ + CO$	$k = 1.0 \times 10^{-10}$	(39)
122	$CO^+ + H_2O$	$\rightarrow H_2O^+ + CO$	$k = 1.3 \times 10^{-10}$	(16, 59)
123	$CO^+ + CO_2$	$\rightarrow CO_2^+ + CO$	$k = 8.5 \times 10^{-10}$	(3, 16, 59)
124	$CO^+ + NO$	$\rightarrow NO^+ + CO$	$k = 3.3 \times 10^{-10}$	(7)
125	$CO^+ + NO_2$	$\rightarrow NO_2^+ + CO$	$k = 3.0 \times 10^{-10}$	(16)
126	$CO^+ + SO_2$	$\rightarrow SO_2^+ + CO$	$k = 2.0 \times 10^{-10}$	(26)
127	$CO^+ + NH_3$	$\rightarrow NH_3^+ + CO$	$k = 2.0 \times 10^{-9}$	(26)
128	$CO^+ + N_2O$	$\rightarrow N_2O^+ + CO$	$k = 3.0 \times 10^{-10}$	(16)
129	$CO^+ + OH$	$\rightarrow OH^+ + CO$	$k = 2.5 \times 10^{-10}$	(16, 59)
130	$CO^+ + OH$	$\rightarrow CO_2^+ + H$	$k = 2.1 \times 10^{-10}$	(59)
131	$CO^+ + H$	$\rightarrow O^+ + CO$	$k = 1.4 \times 10^{-10}$	(7)
132	$CO^+ + H$	$\rightarrow H^+ + CO$	$k = 3.0 \times 10^{-10}$	(16)
133	$C^+ + O_2$	$\rightarrow CO^+ + O$	$k = 3.0 \times 10^{-10}$	(39)
134	$C^+ + O_2$	$\rightarrow O^+ + CO$	$k = 6.0 \times 10^{-10}$	(39)
135	$C^+ + CO_2$	$\rightarrow CO^+ + CO$	$k = 1.5 \times 10^{-9}$	(3, 26)
136	$C^+ + NO$	$\rightarrow N^+ + CO$	$k = 7.0 \times 10^{-10}$	(26)
137	$C^+ + OH$	$\rightarrow CO^+ + H$	$k = 7.7 \times 10^{-10}$	(59)
138	$H^+ + O_2$	$\rightarrow O_2^+ + H$	$k = 1.2 \times 10^{-9}$	(26)
139	$H^+ + H_2O$	$\rightarrow H_2O^+ + H$	$k = 8.2 \times 10^{-9}$	(26)
140	$H^+ + H_2O + M$	$\rightarrow H_3O^+ + M$	$k = 2.0 \times 10^{-27}$	(3, 16)
141	$H^+ + CO_2$	$\rightarrow CO_2^+ + H$	$k = 1.2 \times 10^{-9} \exp(-1942/T)$	(16)
142	$H^+ + NO$	$\rightarrow NO^+ + H$	$k = 7.0 \times 10^{-10}$	(26, 59)
143	$H^+ + NO_2$	$\rightarrow NO_2^+ + H$	$k = 3.0 \times 10^{-10}$	(16)
144	$H^+ + NH_3$	$\rightarrow NH_3^+ + H$	$k = 5.2 \times 10^{-9}$	(26)
145	$H^+ + N_2O$	$\rightarrow N_2O^+ + H$	$k = 3.0 \times 10^{-10}$	(16)
146	$H^+ + OH$	$\rightarrow OH^+ + H$	$k = 2.0 \times 10^{-9}$	(16, 59)
147	$H^+ + O$	$\rightarrow O^+ + H$	$k = 3.8 \times 10^{-10}$	(26)
148	$H_2^+ + O_2$	$\rightarrow O_2^+ + H_2$	$k = 8.0 \times 10^{-10}$	(26)

(Continued)

TABLE II. (Continued)

149	$H_2^+ + O_2$	$\rightarrow H_2O^+ + O$	$k = 1.9 \times 10^{-9}$	(26)
150	$H_2^+ + H_2O$	$\rightarrow H_2O^+ + H_2$	$k = 3.9 \times 10^{-9}$	(26)
151	$H_2^+ + H_2O$	$\rightarrow H_3O^+ + H$	$k = 3.4 \times 10^{-9}$	(26)
152	$H_2^+ + CO_2$	$\rightarrow CO_2^+ + H_2$	$k = 1.4 \times 10^{-9}$	(26)
153	$H_2^+ + CO_2$	$\rightarrow CO^+ + H_2O$	$k = 1.4 \times 10^{-9}$	(26)
154	$H_2^+ + CO$	$\rightarrow CO^+ + H_2$	$k = 6.44 \times 10^{-10}$	(26)
155	$H_2^+ + NO$	$\rightarrow NO^+ + H_2$	$k = 1.1 \times 10^{-9}$	(26)
156	$H_2^+ + NH_3$	$\rightarrow NH_3^+ + H_2$	$k = 5.7 \times 10^{-9}$	(26)
157	$H_2^+ + OH$	$\rightarrow H_2O^+ + H$	$k = 7.5 \times 10^{-10}$	(59)
158	$H_2^+ + OH$	$\rightarrow OH^+ + H_2$	$k = 7.5 \times 10^{-10}$	(59)
159	$H_2^+ + H$	$\rightarrow H^+ + H_2$	$k = 6.4 \times 10^{-10}$	(26)
160	$OH^+ + O_2$	$\rightarrow O_2^+ + OH$	$k = 2.0 \times 10^{-10}$	(26)
161	$OH^+ + H_2O$	$\rightarrow H_3O^+ + O$	$k = 1.5 \times 10^{-9}$	(3, 16)
162	$OH^+ + H_2O$	$\rightarrow H_2O^+ + OH$	$k = 1.5 \times 10^{-9}$	(26)
163	$OH^+ + NO$	$\rightarrow NO^+ + OH$	$k = 4.6 \times 10^{-10}$	(26)
164	$OH^+ + NO_2$	$\rightarrow NO_2^+ + OH$	$k = 3.0 \times 10^{-10}$	(16)
165	$OH^+ + NH_3$	$\rightarrow NH_3^+ + OH$	$k = 1.2 \times 10^{-9}$	(26)
166	$OH^+ + NH_3$	$\rightarrow NH_4^+ + O$	$k = 1.2 \times 10^{-9}$	(26)
167	$OH^+ + N_2O$	$\rightarrow N_2O^+ + OH$	$k = 5.0 \times 10^{-10}$	(16, 26)
168	$OH^+ + N_2O$	$\rightarrow NO^+ + HNO$	$k = 1.7 \times 10^{-10}$	(16, 26)
169	$OH^+ + H_2$	$\rightarrow H_2O^+ + H$	$k = 1.3 \times 10^{-9}$	(16, 59)
170	$OH^+ + OH$	$\rightarrow H_2O^+ + O$	$k = 7.0 \times 10^{-10}$	(59)
171	$NO^+ + H_2O + M$	$\rightarrow NO^+(H_2O) + M$	$k = 1.5 \times 10^{-28} (300/T)$	(16, 65)
172	$NO^+ + O_3$	$\rightarrow NO_2^+ + O_2$	$k = 1.0 \times 10^{-14}$	(16), but not in (26)
173	$NO^+ + N_2O_5$	$\rightarrow NO_2^+ + 2NO_2$	$k = 5.9 \times 10^{-10}$	(68)
174	$NO^+ + N + M$	$\rightarrow N_2O^+ + M$	$k = 1.0 \times 10^{-29} (300/T)$	(16)
175	$NO^+(H_2O) + H_2O + M$	$\rightarrow NO^+(H_2O)_2 + M$	$k = 1.1 \times 10^{-27} (300/T)$	(16)
176	$NO^+(H_2O) + NO_2$	$\rightarrow NO_2(H_2O) + NO$	$k = 2.0 \times 10^{-15}$	(62)
177	$NO^+(H_2O) + NH_3$	$\rightarrow NH_4^+ + HNO_2$	$k = 1.0 \times 10^{-9}$	(66)
178	$NO^+(H_2O)_2 + M$	$\rightarrow NO^+(H_2O) + H_2O + M$	$k = 1.0 \times 10^{-14}$	(65)

	Reaction	Rate constant	Ref.
179	$NO^+(H_2O)_2 + H_2O + M \rightarrow NO^+(H_2O)_3 + M$	$k = 2.0 \times 10^{-27}(300/T)$	(16, 65)
180	$NO^+(H_2O)_2 + NH_3 \rightarrow NH_4^+ + HNO_2 + H_2O$	$k = 1.0 \times 10^{-9}$ NH_4^+/H_2O clusters neglected	(66)
181	$NO^+(H_2O)_3 + M \rightarrow NO^+(H_2O)_2 + H_2O + M$	$k = 1.5 \times 10^{-12}$	(16, 65)
182	$NO^+(H_2O)_3 + H_2O \rightarrow H_3O^+(H_2O)_2 + HNO_2$	$k = 2.0 \times 10^{-6} \exp(-3000/T)$	(16, 65)
183	$NO^+(H_2O)_3 + NH_3 \rightarrow NH_4^+ + HNO_2 + 2H_2O$	$k = 1.0 \times 10^{-9}$ NH_4^+/H_2O clusters neglected	(66)
184	$NO_2^+ + NO \rightarrow NO^+ + NO_2$	$k = 2.9 \times 10^{-10}$	(16)
185	$NO_2^+ + H_2O + N_2 \rightarrow NO_2^+(H_2O) + N_2$	$k = 5.0 \times 10^{-28}$	(62)
186	$NO_2^+(H_2O) + NH_3 \rightarrow NH_4^+ + HNO_3$	$k = 6.4 \times 10^{-10}$	(62)
187	$NO_2^+(H_2O) + N_2 \rightarrow NO_2^+ + H_2O + N_2$	$k = 5.0 \times 10^{-15}$	(62)
188	$NO_2^+(H_2O) + NO \rightarrow NO^+(H_2O) + NO_2$	$k = 3.1 \times 10^{-11}$	(62)
189	$NO_2^+(H_2O) + H_2O + N_2 \rightarrow NO_2^+(H_2O)_2 + N_2$	$k = 2.0 \times 10^{-27}$	(62)
190	$NO_2^+(H_2O)_2 + H_2O \rightarrow H_3O^+(H_2O) + HNO_3$	$k = 2.0 \times 10^{-10}$	(62)
191	$N_2O^+ + O_2 \rightarrow O_2^+ + N_2O$	$k = 2.3 \times 10^{-10}$	(16, 26)
192	$N_2O^+ + O_2 \rightarrow NO^+ + NO_2$	$k = 4.0 \times 10^{-11}$	(26)
193	$N_2O^+ + H_2O \rightarrow H_2O^+ + N_2O$	$k = 3.0 \times 10^{-10}$	(16)
194	$N_2O^+ + NO \rightarrow NO^+ + N_2O$	$k = 2.0 \times 10^{-10}$	(16, 26)
195	$N_2O^+ + NO_2 \rightarrow NO_2^+ + N_2O$	$k = 2.2 \times 10^{-10}$	(16, 26)
196	$N_2O^+ + NO_2 \rightarrow NO^+ + N_2 + O_2$	$k = 3.3 \times 10^{-10}$	(26)
197	$N_2O^+ + N_2O \rightarrow NO^+ + NO + N_2$	$k = 1.2 \times 10^{-11}$	(26)
198	$N_2O^+ + CO \rightarrow CO_2^+ + N_2$	$k = 1.0 \times 10^{-10}$	(26)
199	$N_2O^+ + OH \rightarrow OH^+ + N_2O$	$k = 9.2 \times 10^{-11}(300/T)^{0.46}\exp(-3326/T)$	(16)
200	$SO_2^+ + O_2 \rightarrow O_2^+ + SO_2$	$k = 2.8 \times 10^{-10}$	(26)
201	$SO_2^+ + NO \rightarrow NO^+ + SO_2$	$k = 7.0 \times 10^{-11}$	(26)
202	$NH_3^+ + H_2O \rightarrow NH_4^+ + OH$	$k = 6.0 \times 10^{-11}$	(26, 59)
203	$NH_3^+ + NO \rightarrow NO^+ + NH_3$	$k = 7.3 \times 10^{-10}$	(26)
204	$NH_3^+ + NH_3 \rightarrow NH_4^+ + NH_2$	$k = 2.0 \times 10^{-9}$	(26)
205	$NH_3^+ + OH \rightarrow NH_4^+ + O$	$k = 7.0 \times 10^{-10}$	(3, 26)
206	$NH_3^+ + NH_2 \rightarrow NH_4^+ + NH$	$k = 1.0 \times 10^{-11}$	(59)

TABLE III

Negative Ion–Neutral Reactions

				acc. to
207	$e^- $ (hot)	$\rightarrow e^-$ (thermal)	$k = 1.0 \times 10^{+9}$	(60)
208	$e^- + 2O_2$	$\rightarrow O_2^- + O_2^*$	$k = 1.2 \times 10^{-29}(300/T)^{1.38} \times \exp(-6602/T)$	(16)
209	$e^- + O_2 + M$	$\rightarrow O_2^- + M$	$k = 3.0 \times 10^{-31}\ M$ is not O_2	(5, 16)
210	$e^- + NO + M$	$\rightarrow NO^- + M$	$k = 8.0 \times 10^{-31}$	(16)
211	$e^- + NO_2 + M$	$\rightarrow NO_2^- + M$	$k = 1.5 \times 10^{-30}$	(16)
212	$e^- + HNO_3$	$\rightarrow NO_2^- + OH$	$k = 5.0 \times 10^{-8}$	(62)
213	$e^- + O_3$	$\rightarrow O^- + O_2$	$k = 9.0 \times 10^{-12}(300/T)^{1.5}$	(5, 16)
214	$e^- + O_3 + M$	$\rightarrow O_3^- + M$	$k = 1.0 \times 10^{-30}$	(16)
215	$e^- + NO_3 + M$	$\rightarrow NO_3^- + M$	$k = 1.0 \times 10^{-30}$	(16)
216	$O^- + O_2 + M$	$\rightarrow O_3^- + M$	$k = 1.1 \times 10^{-30}(300/T)$	(16)
217	$O^- + O_2^*$	$\rightarrow e^- + O_3$	$k = 3.0 \times 10^{-10}$	(16)
218	$O^- + CO_2 + M$	$\rightarrow CO_3^- + M$	$k = 8.0 \times 10^{-29}(300/T)$	(16)
219	$O^- + NO$	$\rightarrow e^- + NO_2$	$k = 3.1 \times 10^{-10}(300/T)^{0.83}$	(16, 69)
220	$O^- + NO + M$	$\rightarrow NO_2^- + M$	$k = 1.0 \times 10^{-29}(300/T)$	(16)
221	$O^- + NO_2$	$\rightarrow NO_2^- + O$	$k = 1.2 \times 10^{-9}$	(16)
222	$O^- + N_2O$	$\rightarrow NO^- + NO$	$k = 2.3 \times 10^{-10}$	(16)
223	$O^- + O_3$	$\rightarrow O_3^- + O$	$k = 6.5 \times 10^{-10}$	(5, 16)
224	$O^- + CO$	$\rightarrow e^- + CO_2$	$k = 6.0 \times 10^{-10}(300/T)^{0.32}$	(16)
225	$O^- + H_2$	$\rightarrow e^- + H_2O$	$k = 6.5 \times 10^{-10}(300/T)^{0.19}$	(16)
226	$O^- + O$	$\rightarrow e + O_2$	$k = 1.9 \times 10^{-10}$	(5, 16)
227	$O^- + N$	$\rightarrow e^- + NO$	$k = 2.0 \times 10^{-10}$	(16)
228	$O^- + NO_3$	$\rightarrow NO_3^- + O$	$k = 3.0 \times 10^{-10}$	(16)
229	$O_2^- + O_2 + M$	$\rightarrow O_4^- + M$	$k = 3.5 \times 10^{-31}(300/T)$	(16)

230 $O_2^- + O_2^*$ $\rightarrow e^- + 2O_2$	$k = 2.0 \times 10^{-10}$	(16)
231 $O_2^- + H_2O + M$ $\rightarrow O_2^-(H_2O) + M$	$k = 3.0 \times 10^{-28}(300/T)$	(16)
232 $O_2^- + NO_2$ $\rightarrow NO_2^- + O_2$	$k = 8.0 \times 10^{-10}$	(16)
233 $O_2^- + HNO_3$ $\rightarrow NO_3^- + HO_2$	$k = 2.8 \times 10^{-10}$	(62)
234 $O_2^- + O_3$ $\rightarrow O_3^- + O_2$	$k = 5.0 \times 10^{-10}$	(5, 16, 78)
235 $O_2^- + H_2$ $\rightarrow e^- + H_2O_2$	$k = 1.0 \times 10^{-9}$	(16)
236 $O_2^- + O$ $\rightarrow O^- + O_2$	$k = 1.5 \times 10^{-10}$	(5, 78)
237 $O_2^- + O$ $\rightarrow e^- + O_3$	$k = 1.5 \times 10^{-10}$	(5, 78)
238 $O_2^- + H$ $\rightarrow e^- + HO_2$	$k = 1.0 \times 10^{-9}$	(16)
239 $O_2^- + N$ $\rightarrow e^- + NO_2$	$k = 5.0 \times 10^{-10}$	(16)
240 $O_2^- + NO_3$ $\rightarrow NO_3^- + O_2$	$k = 5.0 \times 10^{-10}$	(16)
241 $O_2^-(H_2O) + NO$ $\rightarrow NO_3^- + H_2O$	$k = 3.0 \times 10^{-10}$	(16)
242 $O_2^-(H_2O) + NO_2$ $\rightarrow NO_2^- + O_2 + H_2O$	$k = 3.0 \times 10^{-10}$	(16)
243 $O_2^-(H_2O) + O_3$ $\rightarrow O_3^- + O_2 + H_2O$	$k = 8.0 \times 10^{-10}$	(16)
244 $O_2^-(H_2O) + O$ $\rightarrow O^- + O_2 + H_2O$	$k = 3.0 \times 10^{-10}$	(67)
245 $O_2^-(H_2O) + NO_3$ $\rightarrow NO_3^- + O_2 + H_2O$	$k = 3.0 \times 10^{-10}$	(16)
246 $O_2^-(H_2O) + N$ $\rightarrow e^- + NO_2 + H_2O$	$k = 1.0 \times 10^{-10}$	(16)
247 $O_3^- + CO_2$ $\rightarrow CO_3^- + O_2$	$k = 4.8 \times 10^{-10}$	(5, 16, 78)
248 $O_3^- + NO$ $\rightarrow NO_2^- + O_2$	$k = 1.0 \times 10^{-11}$	(16)
249 $O_3^- + NO_2$ $\rightarrow NO_2^- + O_3$	$k = 2.8 \times 10^{-10}$	(16)
250 $O_3^- + NO_2$ $\rightarrow NO_3^- + O_2$	$k = 2.0 \times 10^{-11}$	(16)
251 $O_3^- + SO_2$ $\rightarrow SO_3^- + O_2$	$k = 1.7 \times 10^{-9}$	(81)
252 $O_3^- + O$ $\rightarrow O_2^- + O_2$	$k = 2.5 \times 10^{-10}$	(78)
253 $O_3^- + O$ $\rightarrow e^- + 2O_2$	$k = 1.0 \times 10^{-11}$	(16)
254 $O_3^- + NO_3$ $\rightarrow NO_3^- + O_3$	$k = 5.0 \times 10^{-10}$	(16)
255 $NO^- + M$ $\rightarrow e^- + NO + M$	$k = 2.1 \times 10^{-11}(300/T)^{1.54}$ $\times \exp(-278/T)$	(16)

(Continued)

TABLE III. (Continued)

	Reaction	Rate constant	Ref.
256	$NO^- + O_2 \rightarrow O_2^- + NO$	$k = 5.0 \times 10^{-10}$	(16)
257	$NO^- + NO_2 \rightarrow NO_2^- + NO$	$k = 3.0 \times 10^{-10}$	(16)
258	$NO^- + O_3 \rightarrow O_3^- + NO$	$k = 3.0 \times 10^{-10}$	(16)
259	$NO^- + O \rightarrow O^- + NO$	$k = 3.0 \times 10^{-10}$	(16)
260	$NO^- + NO_3 \rightarrow NO_3^- + NO$	$k = 3.0 \times 10^{-10}$	(16)
261	$NO_2^- + NO_2 \rightarrow NO_3^- + NO$	$k = 2.0 \times 10^{-13}$	(5, 16, 62)
262	$NO_2^- + HNO_3 \rightarrow NO_3^- + HNO_2$	$k = 1.6 \times 10^{-9}$	(8, 62)
263	$NO_2^- + N_2O_5 \rightarrow NO_3^- + 2NO_2$	$k = 6.5 \times 10^{-10}$	(67)
264	$NO_2^- + O_3 \rightarrow NO_3^- + O_2$	$k = 5.0 \times 10^{-10}$	(5, 16, 78)
265	$NO_2^- + NO_3 \rightarrow NO_3^- + NO_2$	$k = 5.0 \times 10^{-11}$	(16)
266	$NO_3^- + NO \rightarrow NO_2^- + NO_2$	$k = 4.3 \times 10^{-11} (300/T)^{0.35}$ $\times \exp(-3788/T)$	(16)
267	$NO_3^- + O + CO_2 \rightarrow CO_3^- + NO_3$	$k = 8.2 \times 10^{-34} (300/T)^{1.16}$ $\times \exp(-2619/T)$	(16)
268	$CO_3^- + NO \rightarrow NO_2^- + CO_2$	$k = 1.0 \times 10^{-11}$	(5, 16, 78)
269	$CO_3^- + NO_2 \rightarrow NO_3^- + CO_2$	$k = 2.0 \times 10^{-10}$	(5, 78)
270	$CO_3^- + HNO_3 \rightarrow NO_3^- + OH + CO_2$	$k = 8.0 \times 10^{-10}$	(62)
271	$CO_3^- + N_2O_5 \rightarrow NO_3^- + NO_3 + CO_2$	$k = 2.8 \times 10^{-10}$	(67)
272	$CO_3^- + SO_2 \rightarrow SO_3^- + CO_2$	$k = 2.3 \times 10^{-10}$	(81)
273	$CO_3^- + O \rightarrow O_2^- + CO_2$	$k = 1.1 \times 10^{-10}$	(78)
274	$CO_3^- + NO_3 \rightarrow NO_3^- + CO_2 + O$	$k = 3.0 \times 10^{-10}$	(16)
275	$SO_3^- + H_2O \rightarrow e^- + H_2SO_4$	$k = 2.0 \times 10^{-11}$	(81)

TABLE IV

Ionic Recombination Reactions in the Gas Phase

No.	Reaction		Rate constant	Ref.
276	$N_2^+ + e^-$	$\rightarrow N + N(^2D)$	$k = 2.2 \times 10^{-7}(300/T)^{0.39}$ $+ 6.0 \times 10^{-27}(300/T)^{2.5}\,M$	(2, 16)
277	$N_2^+ + e^-$	$\rightarrow N_2$	$k = 4.0 \times 10^{-12}(300/T)^{0.7}$ $+ 6.0 \times 10^{-27}(300/T)^{2.5}\,M$	(16)
278	$N_2^+ + O_2^-$	$\rightarrow N_2 + O_2$	$k = 1.6 \times 10^{-7}(300/T)^{0.5}$ $+ 3.0 \times 10^{-25}(300/T)^{2.5}\,M$	(16)
279	$N_2^+ + O^-$	$\rightarrow N_2 + O$	$k = 4.0 \times 10^{-7}(300/T)^{0.5}$ $+ 3.0 \times 10^{-25}(300/T)^{2.5}\,M$	(16)
280	$N_2^+ + O_3^-$	$\rightarrow N_2O + O_2$	$k = 4.0 \times 10^{-7}(300/T)^{0.5}$ $+ 3.0 \times 10^{-25}(300/T)^{2.5}\,M$	(16)
281	$N_2^+ + NO^-$	$\rightarrow N_2 + NO$	$k = 4.0 \times 10^{-7}(300/T)^{0.5}$ $+ 3.0 \times 10^{-25}(300/T)^{2.5}\,M$	(16)
282	$N_2^+ + NO_2^-$	$\rightarrow NO_2 + N_2$	$k = 4.0 \times 10^{-7}(300/T)^{0.5}$ $+ 3.0 \times 10^{-25}(300/T)^{2.5}\,M$	(16)
283	$N_2^+ + NO_3^-$	$\rightarrow NO_2 + N_2O$	$k = 4.0 \times 10^{-7}(300/T)^{0.5}$ $+ 3.0 \times 10^{-25}(300/T)^{2.5}\,M$	(16)
284	$N_2^+ + CO_3^-$	$\rightarrow CO_2 + N_2 + O$	$k = 4.0 \times 10^{-7}(300/T)^{0.5}$ $+ 3.0 \times 10^{-25}(300/T)^{2.5}\,M$	(16)
285	$N_2^+ + O_4^-$	$\rightarrow N_2 + 2O_2$	$k = 4.0 \times 10^{-7}(300/T)^{0.5}$ $+ 3.0 \times 10^{-25}(300/T)^{2.5}\,M$	(16)
286	$N_2^+ + O_2^-(H_2O)$	$\rightarrow N_2 + O_2 + H_2O$	$k = 5.0 \times 10^{-7}(300/T)^{0.5}$ $+ 3.0 \times 10^{-25}(300/T)^{2.5}\,M$	(16)
287	$N^+ + e^-$	$\rightarrow N$	$k = 3.5 \times 10^{-12}(300/T)^{0.7}$ $+ 6.0 \times 10^{-27}(300/T)^{2.5}\,M$	(4, 16)
288	$N^+ + O^-$	$\rightarrow N + O$	$k = 2.6 \times 10^{-7}(300/T)^{0.5}$ $+ 3.0 \times 10^{-25}(300/T)^{2.5}\,M$	(16)
289	$N^+ + O_2^-$	$\rightarrow N + O_2$	$k = 4.0 \times 10^{-7}(300/T)^{0.5}$ $+ 3.0 \times 10^{-25}(300/T)^{2.5}\,M$	(16)

(Continued)

373

TABLE IV. (Continued)

290	$N^+ + O_3^-$	$\rightarrow NO + O_2$	$k = 4.0 \times 10^{-7} (300/T)^{0.5}$ $+ 3.0 \times 10^{-25} (300/T)^{2.5} M$	(16)
291	$N^+ + NO^-$	$\rightarrow O + N_2$	$k = 4.0 \times 10^{-7} (300/T)^{0.5}$ $+ 3.0 \times 10^{-25} (300/T)^{2.5} M$	(16)
292	$N^+ + NO_2^-$	$\rightarrow 2NO$	$k = 4.0 \times 10^{-7} (300/T)^{0.5}$ $+ 3.0 \times 10^{-25} (300/T)^{2.5} M$	(16)
293	$N^+ + NO_3^-$	$\rightarrow NO + NO_2$	$k = 4.0 \times 10^{-7} (300/T)^{0.5}$ $+ 3.0 \times 10^{-25} (300/T)^{2.5} M$	(16)
294	$N^+ + CO_3^-$	$\rightarrow NO + CO_2$	$k = 4.0 \times 10^{-7} (300/T)^{0.5}$ $+ 3.0 \times 10^{-25} (300/T)^{2.5} M$	(16)
295	$N^+ + O_2^-(H_2O)$	$\rightarrow NO_2 + H_2O$	$k = 5.0 \times 10^{-7} (300/T)^{0.5}$ $+ 3.0 \times 10^{-25} (300/T)^{2.5} M$	(16)
296	$N_3^+ + e^-$	$\rightarrow N_2 + N$	$k = 6.0 \times 10^{-27} (300/T)^{2.5} M$ $+ 7.0 \times 10^{-7} (300/T)$	(16)
297	$N_3^+ + O^-$	$\rightarrow NO + N_2$	$k = 4.0 \times 10^{-7} (300/T)^{0.5}$ $+ 3.0 \times 10^{-25} (300/T)^{2.5} M$	(16)
298	$N_3^+ + O_2^-$	$\rightarrow NO_2 + N_2$	$k = 4.0 \times 10^{-7} (300/T)^{0.5}$ $+ 3.0 \times 10^{-25} (300/T)^{2.5} M$	(16)
299	$N_3^+ + O_3^-$	$\rightarrow NO_3 + N_2$	$k = 4.0 \times 10^{-7} (300/T)^{0.5}$ $+ 3.0 \times 10^{-25} (300/T)^{2.5} M$	(16)
300	$N_3^+ + NO^-$	$\rightarrow N_2 + N_2O$	$k = 4.0 \times 10^{-7} (300/T)^{0.5}$ $+ 3.0 \times 10^{-25} (300/T)^{2.5} M$	(16)
301	$N_3^+ + NO_2^-$	$\rightarrow O_2 + 2N_2$	$k = 4.0 \times 10^{-7} (300/T)^{0.5}$ $+ 3.0 \times 10^{-25} (300/T)^{2.5} M$	(16)
302	$N_3^+ + NO_3^-$	$\rightarrow NO + NO_2 + N_2$	$k = 4.0 \times 10^{-7} (300/T)^{0.5}$ $+ 3.0 \times 10^{-25} (300/T)^{2.5} M$	(16)
303	$N_3^+ + CO_3^-$	$\rightarrow NO + CO_2 + N_2$	$k = 4.0 \times 10^{-7} (300/T)^{0.5}$ $+ 3.0 \times 10^{-25} (300/T)^{2.5} M$	(16)
304	$N_3^+ + O_2^-(H_2O)$	$\rightarrow N_2 + NO_2 + H_2O$	$k = 5.0 \times 10^{-7} (300/T)^{0.5}$ $+ 3.0 \times 10^{-25} (300/T)^{2.5} M$	(16)

305	$N_3^+ + O_4^-$	$\rightarrow N_2 + NO_2 + O_2$	$k = 4.0 \times 10^{-7}(300/T)^{0.5}$ $+ 3.0 \times 10^{-25}(300/T)^{2.5} M$	(16)
306	$N_4^+ + e^-$	$\rightarrow 2N_2$	$k = 2.0 \times 10^{-6}(300/T)$ $+ 6.0 \times 10^{-27}(300/T)^{2.5} M$	(16)
307	$N_4^+ + O_2^-$	$\rightarrow 2N_2 + O_2$	$k = 4.0 \times 10^{-7}(300/T)^{0.5}$ $+ 3.0 \times 10^{-25}(300/T)^{2.5} M$	(16)
308	$N_4^+ + NO^-$	$\rightarrow 2N_2 + NO$	$k = 4.0 \times 10^{-7}(300/T)^{0.5}$ $+ 3.0 \times 10^{-25}(300/T)^{2.5} M$	(16)
309	$N_4^+ + NO_2^-$	$\rightarrow 2N_2 + NO_2$	$k = 4.0 \times 10^{-7}(300/T)^{0.5}$ $+ 3.0 \times 10^{-25}(300/T)^{2.5} M$	(16)
310	$N_4^+ + NO_3^-$	$\rightarrow N_2 + NO_2 + N_2O$	$k = 4.0 \times 10^{-7}(300/T)^{0.5}$ $+ 3.0 \times 10^{-25}(300/T)^{2.5} M$	(16)
311	$N_4^+ + O_2^-(H_2O)$	$\rightarrow 2N_2 + H_2O + O_2$	$k = 5.0 \times 10^{-7}(300/T)^{0.5}$ $+ 3.0 \times 10^{-25}(300/T)^{2.5} M$	(16)
312	$N_4^+ + O_4^-$	$\rightarrow 2N_2 + 2O_2$	$k = 4.0 \times 10^{-7}(300/T)^{0.5}$ $+ 3.0 \times 10^{-25}(300/T)^{2.5} M$	(16)
313	$O_2^+ + e^-$	$\rightarrow O + O(^1D)$	$k = 2.1 \times 10^{-7}(300/T)^{0.55}$	(16, 75)
314	$O_2^+ + e^-$	$\rightarrow O_2$	$k = 4.0 \times 10^{-12}(300/T)^{0.7}$ $+ 6.0 \times 10^{-27}(300/T)^{2.5} M$	(4, 16)
315	$O_2^+ + O^-$	$\rightarrow O_2 + O$	$k = 9.6 \times 10^{-8}(300/T)^{0.5}$	(16)
316	$O_2^+ + O_2^-$	$\rightarrow 2O_2$	$k = 4.2 \times 10^{-7}(300/T)^{0.5}$ $+ 3.0 \times 10^{-25}(300/T)^{2.5} M$	(16)
317	$O_2^+ + O_3^-$	$\rightarrow 2O_2 + O$	$k = 4.0 \times 10^{-7}(300/T)^{0.5}$ $+ 3.0 \times 10^{-25}(300/T)^{2.5} M$	(16)
318	$O_2^+ + NO^-$	$\rightarrow NO + O_2$	$k = 4.0 \times 10^{-7}(300/T)^{0.5}$ $+ 3.0 \times 10^{-25}(300/T)^{2.5} M$	(16)
319	$O_2^+ + NO_2^-$	$\rightarrow NO_2 + O_2$	$k = 4.1 \times 10^{-7}(300/T)^{0.5}$ $+ 3.0 \times 10^{-25}(300/T)^{2.5} M$	(16)
320	$O_2^+ + NO_3^-$	$\rightarrow NO_2 + O_2 + O$	$k = 1.3 \times 10^{-7}(300/T)^{0.5}$ $+ 3.0 \times 10^{-25}(300/T)^{2.5} M$	(16)

(Continued)

375

TABLE IV. (Continued)

321	$O_2^+ + CO_3^-$	$\rightarrow CO_2 + O_2$	$k = 4.0 \times 10^{-7}(300/T)^{0.5}$ $+ 3.0 \times 10^{-25}(300/T)^{2.5} M$	(16)
322	$O_2^+ + O_2^-(H_2O)$	$\rightarrow 2O_2 + H_2O$	$k = 5.0 \times 10^{-7}(300/T)^{0.5}$ $+ 3.0 \times 10^{-25}(300/T)^{2.5} M$	(16)
323	$O^+ + e^-$	$\rightarrow O$	$k = 4.0 \times 10^{-12}(T/300)^{0.7}$	(2, 16)
324	$O^+ + O^-$	$\rightarrow 2O$	$k = 2.7 \times 10^{-7}(300/T)^{0.5}$ $+ 3.0 \times 10^{-25}(300/T)^{2.5} M$	(16)
325	$O^+ + O_2^-$	$\rightarrow O + O_2$	$k = 4.0 \times 10^{-7}(300/T)^{0.5}$ $+ 3.0 \times 10^{-25}(300/T)^{2.5} M$	(16)
326	$O^+ + O_3^-$	$\rightarrow 2O_2$	$k = 4.0 \times 10^{-7}(300/T)^{0.5}$ $+ 3.0 \times 10^{-25}(300/T)^{2.5} M$	(16)
327	$O^+ + NO^-$	$\rightarrow NO + O$	$k = 4.0 \times 10^{-7}(300/T)^{0.5}$ $+ 3.0 \times 10^{-25}(300/T)^{2.5} M$	(16)
328	$O^+ + NO_2^-$	$\rightarrow NO + O_2$	$k = 4.0 \times 10^{-7}(300/T)^{0.5}$ $+ 3.0 \times 10^{-25}(300/T)^{2.5} M$	(16)
329	$O^+ + NO_3^-$	$\rightarrow O_2 + NO_2$	$k = 4.0 \times 10^{-7}(300/T)^{0.5}$ $+ 3.0 \times 10^{-25}(300/T)^{2.5} M$	(16)
330	$O^+ + CO_3^-$	$\rightarrow O_2 + CO_2$	$k = 4.0 \times 10^{-7}(300/T)^{0.5}$ $+ 3.0 \times 10^{-25}(300/T)^{2.5} M$	(16)
331	$O^+ + O_2^-(H_2O)$	$\rightarrow O_3 + H_2O$	$k = 5.0 \times 10^{-7}(300/T)^{0.5}$ $+ 3.0 \times 10^{-25}(300/T)^{2.5} M$	(16)
332	$O_4^+ + e^-$	$\rightarrow 2O_2$	$k = 2.0 \times 10^{-6}(300/T)$ $+ 6.0 \times 10^{-27}(300/T)^{2.5} M$	(16)
333	$O_4^+ + O^-$	$\rightarrow O_2 + O_3$	$k = 4.0 \times 10^{-7}(300/T)^{0.5}$ $+ 3.0 \times 10^{-25}(300/T)^{2.5} M$	(16)
334	$O_4^+ + O_2^-$	$\rightarrow 2O_2 + 2O$	$k = 2.0 \times 10^{-6}$	(2)
335	$O_4^+ + O_2^-$	$\rightarrow 3O_2$	$k = 4.0 \times 10^{-7}(300/T)^{0.5}$ $+ 3.0 \times 10^{-25}(300/T)^{2.5} M$	(16)
336	$O_4^+ + O_3^-$	$\rightarrow 2O_2 + O_3$	$k = 4.0 \times 10^{-7}(300/T)^{0.5}$ $+ 3.0 \times 10^{-25}(300/T)^{2.5} M$	(16)

No.	Reaction		Rate constant	Ref.
337	$O_4^+ + NO^-$	$\rightarrow O_2 + NO_3$	$k = 4.0 \times 10^{-7}(300/T)^{0.5} + 3.0 \times 10^{-25}(300/T)^{2.5} M$	(16)
338	$O_4^+ + NO_2^-$	$\rightarrow 2O_2 + NO_2$	$k = 4.0 \times 10^{-7}(300/T)^{0.5} + 3.0 \times 10^{-25}(300/T)^{2.5} M$	(16)
339	$O_4^+ + NO_3^-$	$\rightarrow O_2 + NO_2 + O_3$	$k = 4.0 \times 10^{-7}(300/T)^{0.5} + 3.0 \times 10^{-25}(300/T)^{2.5} M$	(16)
340	$O_4^+ + CO_3^-$	$\rightarrow O_2 + CO_2 + O_3$	$k = 4.0 \times 10^{-7}(300/T)^{0.5} + 3.0 \times 10^{-25}(300/T)^{2.5} M$	(16)
341	$O_4^+ + O_2^-(H_2O)$	$\rightarrow 3O_2 + H_2O$	$k = 5.0 \times 10^{-7}(300/T)^{0.5} + 3.0 \times 10^{-25}(300/T)^{2.5} M$	(16)
342	$O_2^+(H_2O) + e^-$	$\rightarrow H_2O + O_2$	$k = 1.5 \times 10^{-6}(300/T)^{0.2}$	(16)
343	$O_2^+(H_2O) + O^-$	$\rightarrow H_2O + O_3$	$k = 5.0 \times 10^{-7}(300/T)^{0.5} + 5.0 \times 10^{-26}(300/T)^{2.5} M$	(16)
344	$O_2^+(H_2O) + O_2^-$	$\rightarrow H_2O + 2O_2$	$k = 5.0 \times 10^{-7}(300/T)^{0.5} + 3.0 \times 10^{-25}(300/T)^{2.5} M$	(16)
345	$O_2^+(H_2O) + O_3^-$	$\rightarrow H_2O + O_2 + O_3$	$k = 5.0 \times 10^{-7}(300/T)^{0.5} + 3.0 \times 10^{-25}(300/T)^{2.5} M$	(16)
346	$O_2^+(H_2O) + O_4^-$	$\rightarrow H_2O + 3O_2$	$k = 4.0 \times 10^{-7}(300/T)^{0.5} + 3.0 \times 10^{-25}(300/T)^{2.5} M$	(16)
347	$O_2^+(H_2O) + NO^-$	$\rightarrow H_2O + NO_3$	$k = 5.0 \times 10^{-7}(300/T)^{0.5} + 3.0 \times 10^{-25}(300/T)^{2.5} M$	(16)
348	$O_2^+(H_2O) + NO_2^-$	$\rightarrow H_2O + O_2 + NO_2$	$k = 5.0 \times 10^{-7}(300/T)^{0.5} + 3.0 \times 10^{-25}(300/T)^{2.5} M$	(16)
349	$O_2^+(H_2O) + NO_3^-$	$\rightarrow H_2O + O_3 + NO_2$	$k = 5.0 \times 10^{-7}(300/T)^{0.5} + 3.0 \times 10^{-25}(300/T)^{2.5} M$	(16)
350	$O_2^+(H_2O) + CO_3^-$	$\rightarrow H_2O + O_2 + CO_2 + O$	$k = 5.0 \times 10^{-7}(300/T)^{0.5} + 3.0 \times 10^{-25}(300/T)^{2.5} M$	(16)
351	$O_2^+(H_2O) + O_2^-(H_2O)$	$\rightarrow 2O_2 + 2H_2O$	$k = 5.0 \times 10^{-7}(300/T)^{0.5} + 3.0 \times 10^{-25}(300/T)^{2.5} M$	(16)
352	$H_2O^+ + e^-$	$\rightarrow OH + H$	$k = 6.6 \times 10^{-6} T^{-0.5}$	(36)
353	$H_2O^+ + e^-$	$\rightarrow H_2 + O$	$k = 2.4 \times 10^{-6} T^{-0.5}$	(36)

(Continued)

377

TABLE IV. (Continued)

	Reaction		Rate constant	Ref.
354	$H_2O^+ + e^-$	$\rightarrow 2H + O$	$k = 3.0 \times 10^{-6} T^{-0.5}$	(36)
355	$H_2O^+ + e^- + M$	$\rightarrow H_2O + M$	$k = 6.0 \times 10^{-27}(300/T)^{2.5}$	(16)
356	$H_2O^+ + O^-$	$\rightarrow H_2O + O$	$k = 4.0 \times 10^{-7}(300/T)^{0.5}$ $+ 3.0 \times 10^{-25}(300/T)^{2.5} M$	(16)
357	$H_2O^+ + O_2^-$	$\rightarrow H_2O + O_2$	$k = 4.0 \times 10^{-7}(300/T)^{0.5}$ $+ 3.0 \times 10^{-25}(300/T)^{2.5} M$	(16)
358	$H_2O^+ + O_3^-$	$\rightarrow H_2O + O_3$	$k = 4.0 \times 10^{-7}(300/T)^{0.5}$ $+ 3.0 \times 10^{-25}(300/T)^{2.5} M$	(16)
359	$H_2O^+ + O_4^-$	$\rightarrow 2O_2 + H_2O$	$k = 4.0 \times 10^{-7}(300/T)^{0.5}$ $+ 3.0 \times 10^{-25}(300/T)^{2.5} M$	(16)
360	$H_2O^+ + NO^-$	$\rightarrow NO + H_2O$	$k = 4.0 \times 10^{-7}(300/T)^{0.5}$ $+ 3.0 \times 10^{-25}(300/T)^{2.5} M$	(16)
361	$H_2O^+ + NO_2^-$	$\rightarrow NO_2 + H_2O$	$k = 4.0 \times 10^{-7}(300/T)^{0.5}$ $+ 3.0 \times 10^{-25}(300/T)^{2.5} M$	(16)
362	$H_2O^+ + NO_3^-$	$\rightarrow NO_3 + H_2O$	$k = 4.0 \times 10^{-7}(300/T)^{0.5}$ $+ 3.0 \times 10^{-25}(300/T)^{2.5} M$	(16)
363	$H_2O^+ + CO_3^-$	$\rightarrow H_2O + CO_2 + O$	$k = 4.0 \times 10^{-7}(300/T)^{0.5}$ $+ 3.0 \times 10^{-25}(300/T)^{2.5} M$	(16)
364	$H_2O^+ + O_2^-(H_2O)$	$\rightarrow O_2 + 2H_2O$	$k = 5.0 \times 10^{-7}(300/T)^{0.5}$ $+ 3.0 \times 10^{-25}(300/T)^{2.5} M$	(16)
365	$H_3O^+ + e^-$	$\rightarrow H_2O + H$	$k = 6.5 \times 10^{-7}(300/T)^{0.5}$ $+ 6.0 \times 10^{-27}(300/T)^{2.5} M$	(16, 26)
366	$H_3O^+ + e^-$	$\rightarrow OH + 2H$	$k = 6.5 \times 10^{-7}(300/T)^{0.5}$	(26)
367	$H_3O^+ + O^-$	$\rightarrow H_2O + OH$	$k = 4.0 \times 10^{-7}(300/T)^{0.5}$ $+ 3.0 \times 10^{-25}(300/T)^{2.5} M$	(16)
368	$H_3O^+ + O_2^-$	$\rightarrow H_2O + O_2 + H$	$k = 2.0 \times 10^{-6}$	est., (60)
369	$H_3O^+ + O_2^-$	$\rightarrow H_2O + HO_2$	$k = 4.0 \times 10^{-7}(300/T)^{0.5}$	(16)
370	$H_3O^+ + O_3^-$	$\rightarrow H_2O + OH + O_2$	$k = 4.0 \times 10^{-7}(300/T)^{0.5}$ $+ 3.0 \times 10^{-25}(300/T)^{2.5} M$	(16)

371	$H_3O^+ + NO^-$	$\rightarrow H_2O + H + NO$	$k = 4.0 \times 10^{-7} (300/T)^{0.5}$ $+ 3.0 \times 10^{-25} (300/T)^{2.5} M$	(16)
372	$H_3O^+ + NO_2^-$	$\rightarrow H_2O + H + NO_2$	$k = 4.0 \times 10^{-7} (300/T)^{0.5}$ $+ 3.0 \times 10^{-25} (300/T)^{2.5} M$	(16)
373	$H_3O^+ + NO_3^-$	$\rightarrow H_2O + OH + NO_2$	$k = 4.0 \times 10^{-7} (300/T)^{0.5}$ $+ 3.0 \times 10^{-25} (300/T)^{2.5} M$	(16)
374	$H_3O^+ + CO_3^-$	$\rightarrow H_2O + CO_2 + OH$	$k = 4.0 \times 10^{-7} (300/T)^{0.5}$ $+ 3.0 \times 10^{-25} (300/T)^{2.5} M$	(16)
375	$H_3O^+ + O_2^-(H_2O)$	$\rightarrow 2H_2O + HO_2$	$k = 5.0 \times 10^{-7} (300/T)^{0.5}$ $+ 3.0 \times 10^{-25} (300/T)^{2.5} M$	(16)
376	$H_3O^+(OH) + e^-$	$\rightarrow H_2O + H + OH$	$k = 3.0 \times 10^{-6} (300/T)$	(16)
377	$H_3O^+(OH) + e^- + M$	$\rightarrow 2H_2O + M$	$k = 5.0 \times 10^{-26} (300/T)^{0.5}$	(16)
378	$H_3O^+(OH) + O^-$	$\rightarrow H_2O + H_2O_2$	$k = 5.0 \times 10^{-7} (300/T)^{0.5}$ $+ 3.0 \times 10^{-25} (300/T)^{2.5} M$	(16)
379	$H_3O^+(OH) + O_2^-$	$\rightarrow 2H_2O + O_2$	$k = 5.0 \times 10^{-7} (300/T)^{0.5}$ $+ 3.0 \times 10^{-25} (300/T)^{2.5} M$	(16)
380	$H_3O^+(OH) + O_3^-$	$\rightarrow H_2O + H_2O_2 + O_2$	$k = 5.0 \times 10^{-7} (300/T)^{0.5}$ $+ 3.0 \times 10^{-25} (300/T)^{2.5} M$	(16)
381	$H_3O^+(OH) + NO^-$	$\rightarrow 2H_2O + NO$	$k = 5.0 \times 10^{-7} (300/T)^{0.5}$ $+ 3.0 \times 10^{-25} (300/T)^{2.5} M$	(16)
382	$H_3O^+(OH) + NO_2^-$	$\rightarrow 2H_2O + NO_2$	$k = 5.0 \times 10^{-7} (300/T)^{0.5}$ $+ 3.0 \times 10^{-25} (300/T)^{2.5} M$	(16)
383	$H_3O^+(OH) + NO_3^-$	$\rightarrow H_2O + NO_2 + H_2O_2$	$k = 5.0 \times 10^{-7} (300/T)^{0.5}$ $+ 3.0 \times 10^{-25} (300/T)^{2.5} M$	(16)
384	$H_3O^+(OH) + CO_3^-$	$\rightarrow H_2O + CO_2 + H_2O_2$	$k = 5.0 \times 10^{-7} (300/T)^{0.5}$ $+ 3.0 \times 10^{-25} (300/T)^{2.5} M$	(16)
385	$H_3O^+(OH) + O_2^-(H_2O)$	$\rightarrow 3H_2O + O_2$	$k = 5.0 \times 10^{-7} (300/T)^{0.5}$ $+ 3.0 \times 10^{-25} (300/T)^{2.5} M$	(16)
386	$H_3O^+(H_2O) + e^-$	$\rightarrow 2H_2O + H$	$k = 2.8 \times 10^{-6} (300/T)^{0.2}$ $+ 5.0 \times 10^{-26} (300/T)^{2.5} M$	(16)
387	$H_3O^+(H_2O) + O^-$	$\rightarrow 2H_2O + OH$	$k = 5.0 \times 10^{-7} (300/T)^{0.5}$ $+ 3.0 \times 10^{-25} (300/T)^{2.5} M$	(16)

(Continued)

379

TABLE IV. (*Continued*)

388	$H_3O^+(H_2O) + O_2^-$	$\rightarrow 2H_2O + HO_2$	$\begin{aligned} k = &\ 5.0 \times 10^{-7}(300/T)^{0.5} \\ &+ 3.0 \times 10^{-25}(300/T)^{2.5}\,M \end{aligned}$	(16)
389	$H_3O^+(H_2O) + O_3^-$	$\rightarrow 2H_2O + O_2 + OH$	$\begin{aligned} k = &\ 5.0 \times 10^{-7}(300/T)^{0.5} \\ &+ 3.0 \times 10^{-25}(300/T)^{2.5}\,M \end{aligned}$	(16)
390	$H_3O^+(H_2O) + NO^-$	$\rightarrow 2H_2O + H + NO$	$\begin{aligned} k = &\ 5.0 \times 10^{-7}(300/T)^{0.5} \\ &+ 3.0 \times 10^{-25}(300/T)^{2.5}\,M \end{aligned}$	(16)
391	$H_3O^+(H_2O) + NO_2^-$	$\rightarrow 2H_2O + NO_2 + H$	$\begin{aligned} k = &\ 5.0 \times 10^{-7}(300/T)^{0.5} \\ &+ 3.0 \times 10^{-25}(300/T)^{2.5}\,M \end{aligned}$	(16)
392	$H_3O^+(H_2O) + NO_3^-$	$\rightarrow 2H_2O + NO_2 + OH$	$\begin{aligned} k = &\ 5.0 \times 10^{-7}(300/T)^{0.5} \\ &+ 3.0 \times 10^{-25}(300/T)^{2.5}\,M \end{aligned}$	(16)
393	$H_3O^+(H_2O) + CO_3^-$	$\rightarrow 2H_2O + CO_2 + OH$	$\begin{aligned} k = &\ 5.0 \times 10^{-7}(300/T)^{0.5} \\ &+ 3.0 \times 10^{-25}(300/T)^{2.5}\,M \end{aligned}$	(16)
394	$H_3O^+(H_2O) + O_2^-(H_2O)$	$\rightarrow HO_2 + 3H_2O$	$\begin{aligned} k = &\ 5.0 \times 10^{-7}(300/T)^{0.5} \\ &+ 3.0 \times 10^{-25}(300/T)^{2.5}\,M \end{aligned}$	(16)
395	$H_3O^+(H_2O)_2 + e^-$	$\rightarrow 3H_2O + H$	$\begin{aligned} k = &\ 5.1 \times 10^{-6}(300/T)^{0.2} \\ &+ 5.0 \times 10^{-26}(300/T)^{2.5}\,M \end{aligned}$	(16)
396	$H_3O^+(H_2O)_2 + O^-$	$\rightarrow 3H_2O + OH$	$\begin{aligned} k = &\ 5.0 \times 10^{-7}(300/T)^{0.5} \\ &+ 3.0 \times 10^{-25}(300/T)^{2.5}\,M \end{aligned}$	(16)
397	$H_3O^+(H_2O)_2 + O_2^-$	$\rightarrow 3H_2O + HO_2$	$\begin{aligned} k = &\ 5.0 \times 10^{-7}(300/T)^{0.5} \\ &+ 3.0 \times 10^{-25}(300/T)^{2.5}\,M \end{aligned}$	(16)
398	$H_3O^+(H_2O)_2 + O_3^-$	$\rightarrow 3H_2O + O_2 + OH$	$\begin{aligned} k = &\ 5.0 \times 10^{-7}(300/T)^{0.5} \\ &+ 3.0 \times 10^{-25}(300/T)^{2.5}\,M \end{aligned}$	(16)
399	$H_3O^+(H_2O)_2 + O_4^-$	$\rightarrow 3H_2O + O_2 + HO_2$	$\begin{aligned} k = &\ 5.0 \times 10^{-7}(300/T)^{0.5} \\ &+ 3.0 \times 10^{-25}(300/T)^{2.5}\,M \end{aligned}$	(16)
400	$H_3O^+(H_2O)_2 + NO^-$	$\rightarrow 3H_2O + H + NO$	$\begin{aligned} k = &\ 5.0 \times 10^{-7}(300/T)^{0.5} \\ &+ 3.0 \times 10^{-25}(300/T)^{2.5}\,M \end{aligned}$	(16)
401	$H_3O^+(H_2O)_2 + NO_2^-$	$\rightarrow 3H_2O + NO_2 + H$	$\begin{aligned} k = &\ 5.0 \times 10^{-7}(300/T)^{0.5} \\ &+ 3.0 \times 10^{-25}(300/T)^{2.5}\,M \end{aligned}$	(16)
402	$H_3O^+(H_2O)_2 + NO_3^-$	$\rightarrow 3H_2O + NO_2 + OH$	$\begin{aligned} k = &\ 5.0 \times 10^{-7}(300/T)^{0.5} \\ &+ 3.0 \times 10^{-25}(300/T)^{2.5}\,M \end{aligned}$	(16)

No.	Reaction		Rate constant	Ref.
403	$H_3O^+(H_2O)_2 + CO_3^-$	$\rightarrow 3H_2O + CO_2 + OH$	$k = 5.0 \times 10^{-7}(300/T)^{0.5}$ $+3.0 \times 10^{-25}(300/T)^{2.5} M$	(16)
404	$H_3O^+(H_2O)_2 + O_2^-(H_2O)$	$\rightarrow 4H_2O + HO_2$	$k = 5.0 \times 10^{-7}(300/T)^{0.5}$ $+3.0 \times 10^{-25}(300/T)^{2.5} M$	(16)
405	$CO_2^+ + e^-$	$\rightarrow CO + O$	$k = 4.0 \times 10^{-7}(300/T)^{0.5}$	(26)
406	$CO_2^+ + e^- + M$	$\rightarrow CO_2 + M$	$k = 6.0 \times 10^{-27}(300/T)^{0.5}$	(26)
407	$CO_2^+ + O^-$	$\rightarrow CO_2 + O$	$k = 4.0 \times 10^{-7}(300/T)^{0.5}$ $+3.0 \times 10^{-25}(300/T)^{2.5} M$	(16)
408	$CO_2^+ + O_2^-$	$\rightarrow CO_2 + O_2$	$k = 4.0 \times 10^{-7}(300/T)^{0.5}$ $+3.0 \times 10^{-25}(300/T)^{2.5} M$	(16)
409	$CO_2^+ + O_3^-$	$\rightarrow CO_2 + O_3$	$k = 4.0 \times 10^{-7}(300/T)^{0.5}$ $+3.0 \times 10^{-25}(300/T)^{2.5} M$	(16)
410	$CO_2^+ + O_4^-$	$\rightarrow CO_2 + 2O_2$	$k = 4.0 \times 10^{-7}(300/T)^{0.5}$ $+3.0 \times 10^{-25}(300/T)^{2.5} M$	(16)
411	$CO_2^+ + NO^-$	$\rightarrow CO_2 + NO$	$k = 4.0 \times 10^{-7}(300/T)^{0.5}$ $+3.0 \times 10^{-25}(300/T)^{2.5} M$	(16)
412	$CO_2^+ + NO_2^-$	$\rightarrow CO_2 + NO_2$	$k = 4.0 \times 10^{-7}(300/T)^{0.5}$ $+3.0 \times 10^{-25}(300/T)^{2.5} M$	(16)
413	$CO_2^+ + NO_3^-$	$\rightarrow CO_2 + NO_3$	$k = 4.0 \times 10^{-7}(300/T)^{0.5}$ $+3.0 \times 10^{-25}(300/T)^{2.5} M$	(16)
414	$CO_2^+ + CO_3^-$	$\rightarrow 2CO_2 + O$	$k = 4.0 \times 10^{-7}(300/T)^{0.5}$ $+3.0 \times 10^{-25}(300/T)^{2.5} M$	(16)
415	$CO_2^+ + O_2^-(H_2O)$	$\rightarrow CO_2 + O_2 + H_2O$	$k = 4.0 \times 10^{-7}(300/T)^{0.5}$ $+3.0 \times 10^{-25}(300/T)^{2.5} M$	(16)
416	$CO^+ + e^- + M$	$\rightarrow CO + M$	$k = 5.0 \times 10^{-7}(300/T)^{0.5}$ $+3.0 \times 10^{-25}(300/T)^{2.5} M$	(16)
417	$CO^+ + O^-$	$\rightarrow CO + O$	$k = 6.0 \times 10^{-27}(300/T)^{0.5}$	(16)
418	$CO^+ + O_2^-$	$\rightarrow CO_2 + O$	$k = 4.0 \times 10^{-7}(300/T)^{0.5}$ $+3.0 \times 10^{-25}(300/T)^{2.5} M$	(16)
419	$CO^+ + O_3^-$	$\rightarrow CO_2 + O_2$	$k = 4.0 \times 10^{-7}(300/T)^{0.5}$ $+3.0 \times 10^{-25}(300/T)^{2.5} M$	(16)

(Continued)

381

TABLE IV. (Continued)

420	$CO^+ + O_4^-$	$\rightarrow CO_2 + O_2 + O$	$k = 4.0 \times 10^{-7} (300/T)^{0.5}$ $\quad + 3.0 \times 10^{-25} (300/T)^{2.5} M$	(16)
421	$CO^+ + NO^-$	$\rightarrow CO + NO$	$k = 4.0 \times 10^{-7} (300/T)^{0.5}$ $\quad + 3.0 \times 10^{-25} (300/T)^{2.5} M$	(16)
422	$CO^+ + NO_2^-$	$\rightarrow CO_2 + NO$	$k = 4.0 \times 10^{-7} (300/T)^{0.5}$ $\quad + 3.0 \times 10^{-25} (300/T)^{2.5} M$	(16)
423	$CO^+ + NO_3^-$	$\rightarrow CO_2 + NO_2$	$k = 4.0 \times 10^{-7} (300/T)^{0.5}$ $\quad + 3.0 \times 10^{-25} (300/T)^{2.5} M$	(16)
424	$CO^+ + CO_3^-$	$\rightarrow 2CO_2$	$k = 4.0 \times 10^{-7} (300/T)^{0.5}$ $\quad + 3.0 \times 10^{-25} (300/T)^{2.5} M$	(16)
425	$CO^+ + O_2^-(H_2O)$	$\rightarrow CO_2 + O + H_2O$	$k = 5.0 \times 10^{-7} (300/T)^{0.5}$ $\quad + 3.0 \times 10^{-25} (300/T)^{2.5} M$	(16)
426	$H^+ + e^-$	$\rightarrow H$	$k = 3.5 \times 10^{-12} (300/T)^{0.7}$ $\quad + 6.0 \times 10^{-27} (300/T)^{2.5} M$	(16)
427	$H^+ + O^-$	$\rightarrow H + O$	$k = 4.0 \times 10^{-7} (300/T)^{0.5}$ $\quad + 3.0 \times 10^{-25} (300/T)^{2.5} M$	(16)
428	$H^+ + O_2^-$	$\rightarrow H + O_2$	$k = 4.0 \times 10^{-7} (300/T)^{0.5}$ $\quad + 3.0 \times 10^{-25} (300/T)^{2.5} M$	(16)
429	$H^+ + O_3^-$	$\rightarrow OH + O_2$	$k = 4.0 \times 10^{-7} (300/T)^{0.5}$ $\quad + 3.0 \times 10^{-25} (300/T)^{2.5} M$	(16)
430	$H^+ + NO^-$	$\rightarrow NO + H$	$k = 4.0 \times 10^{-7} (300/T)^{0.5}$ $\quad + 3.0 \times 10^{-25} (300/T)^{2.5} M$	(16)
431	$H^+ + NO_2^-$	$\rightarrow NO + OH$	$k = 4.0 \times 10^{-7} (300/T)^{0.5}$ $\quad + 3.0 \times 10^{-25} (300/T)^{2.5} M$	(16)
432	$H^+ + NO_3^-$	$\rightarrow NO_2 + OH$	$k = 4.0 \times 10^{-7} (300/T)^{0.5}$ $\quad + 3.0 \times 10^{-25} (300/T)^{2.5} M$	(16)
433	$H^+ + CO_3^-$	$\rightarrow CO_2 + OH$	$k = 4.0 \times 10^{-7} (300/T)^{0.5}$ $\quad + 3.0 \times 10^{-25} (300/T)^{2.5} M$	(16)
434	$H^+ + O_2^-(H_2O)$	$\rightarrow H_2O + HO_2$	$k = 5.0 \times 10^{-7} (300/T)^{0.5}$ $\quad + 3.0 \times 10^{-25} (300/T)^{2.5} M$	(16)
435	$H_2^+ + e^-$	$\rightarrow 2H$	$k = 2.3 \times 10^{-8} (100/T)^{0.29}$	(16)

436	$OH^+ + e^-$	$\rightarrow H + O$	$k = 2.0 \times 10^{-7}$	(16)
437	$OH^+ + e^- + M$	$\rightarrow OH + M$	$k = 6.0 \times 10^{-27}(300/T)^{2.5}$	(16)
438	$OH^+ + O^-$	$\rightarrow H + O_2$	$k = 4.0 \times 10^{-7}(300/T)^{0.5}$ $+ 3.0 \times 10^{-25}(300/T)^{2.5} M$	(16)
439	$OH^+ + O_2^-$	$\rightarrow OH + O_2$	$k = 4.0 \times 10^{-7}(300/T)^{0.5}$ $+ 3.0 \times 10^{-25}(300/T)^{2.5} M$	(16)
440	$OH^+ + O_3^-$	$\rightarrow HO_2 + O_2$	$k = 4.0 \times 10^{-7}(300/T)^{0.5}$ $+ 3.0 \times 10^{-25}(300/T)^{2.5} M$	(16)
441	$OH^+ + NO^-$	$\rightarrow OH + NO$	$k = 4.0 \times 10^{-7}(300/T)^{0.5}$ $+ 3.0 \times 10^{-25}(300/T)^{2.5} M$	(16)
442	$OH^+ + NO_2^-$	$\rightarrow OH + NO_2$	$k = 4.0 \times 10^{-7}(300/T)^{0.5}$ $+ 3.0 \times 10^{-25}(300/T)^{2.5} M$	(16)
443	$OH^+ + NO_3^-$	$\rightarrow HNO_2 + O_2$	$k = 4.0 \times 10^{-7}(300/T)^{0.5}$ $+ 3.0 \times 10^{-25}(300/T)^{2.5} M$	(16)
444	$OH^+ + CO_3^-$	$\rightarrow H + O_2 + CO_2$	$k = 4.0 \times 10^{-7}(300/T)^{0.5}$ $+ 3.0 \times 10^{-25}(300/T)^{2.5} M$	(16)
445	$OH^+ + O_2^-(H_2O)$	$\rightarrow OH + O_2 + H_2O$	$k = 5.0 \times 10^{-7}(300/T)^{0.5}$ $+ 3.0 \times 10^{-25}(300/T)^{2.5} M$	(16)
446	$NO^+ + e^-$	$\rightarrow NO$	$k = 4.0 \times 10^{-12}(300/T)^{0.7}$ $+ 6.0 \times 10^{-27}(300/T)^{2.5} M$	(4, 16)
447	$NO^+ + e^- + M$	$\rightarrow N + O + M$	$k = 1.0 \times 10^{-27}$	(4)
448	$NO^+ + e^-$	$\rightarrow O + N(^2D)$	$k = 4.3 \times 10^{-7}(300/T)^{0.8}$	(16, 55, 75)
449	$NO^+ + O^-$	$\rightarrow O + NO$	$k = 4.9 \times 10^{-7}(300/T)^{0.5}$ $+ 3.0 \times 10^{-25}(300/T)^{2.5} M$	(16)
450	$NO^+ + O_2^-$	$\rightarrow NO + O_2$	$k = 4.0 \times 10^{-7}(300/T)^{0.5}$ $+ 3.0 \times 10^{-25}(300/T)^{2.5} M$	(16)
451	$NO^+ + O_3^-$	$\rightarrow NO + O + O_2$	$k = 4.0 \times 10^{-7}(300/T)^{0.5}$ $+ 3.0 \times 10^{-25}(300/T)^{2.5} M$	(16)
452	$NO^+ + NO^-$	$\rightarrow O_2 + N_2$	$k = 4.0 \times 10^{-7}(300/T)^{0.5}$ $+ 3.0 \times 10^{-25}(300/T)^{2.5} M$	(16)
453	$NO^+ + NO_2^-$	$\rightarrow NO_2 + N + O$	$k = 1.0 \times 10^{-7}$	(4)

(Continued)

TABLE IV. (Continued)

454	$NO^+ + NO_2^- \rightarrow NO + NO_2$	$k = 3.5 \times 10^{-7}(300/T)^{0.5}$ $+ 3.0 \times 10^{-25}(300/T)^{2.5}\,M$	(16)
455	$NO^+ + NO_3^- \rightarrow NO_3 + N + O$	$k = 1.0 \times 10^{-7}$	(4)
456	$NO^+ + NO_3^- \rightarrow 2NO_2$	$k = 4.0 \times 10^{-7}(300/T)^{0.5}$ $+ 3.0 \times 10^{-25}(300/T)^{2.5}\,M$	(16)
457	$NO^+ + CO_3^- \rightarrow NO_2 + CO_2$	$k = 4.0 \times 10^{-7}(300/T)^{0.5}$ $+ 3.0 \times 10^{-25}(300/T)^{2.5}\,M$	(16)
458	$NO^+ + O_4^- \rightarrow 2O_2 + NO$	$k = 4.0 \times 10^{-7}(300/T)^{0.5}$ $+ 3.0 \times 10^{-25}(300/T)^{2.5}\,M$	(16)
459	$NO^+ + O_2^-(H_2O) \rightarrow NO_3 + H_2O$	$k = 5.0 \times 10^{-7}(300/T)^{0.5}$ $+ 3.0 \times 10^{-25}(300/T)^{2.5}\,M$	(16)
460	$NO^+(H_2O) + e^- + M \rightarrow H_2O + NO + M$	$k = 1.0 \times 10^{-6}(300/T)^{0.2}$ $+ 5.0 \times 10^{-26}(300/T)^{2.5}\,M$	(16)
461	$NO^+(H_2O) + O^- \rightarrow H_2O + NO_2$	$k = 5.0 \times 10^{-7}(300/T)^{0.5}$ $+ 3.0 \times 10^{-25}(300/T)^{2.5}\,M$	(16)
462	$NO^+(H_2O) + O_2^- \rightarrow H_2O + O_2 + NO$	$k = 5.0 \times 10^{-7}(300/T)^{0.5}$ $+ 3.0 \times 10^{-25}(300/T)^{2.5}\,M$	(16)
463	$NO^+(H_2O) + O_3^- \rightarrow H_2O + O_2 + NO_2$	$k = 5.0 \times 10^{-7}(300/T)^{0.5}$ $+ 3.0 \times 10^{-25}(300/T)^{2.5}\,M$	(16)
464	$NO^+(H_2O) + NO^- \rightarrow H_2O + O_2 + N_2$	$k = 5.0 \times 10^{-7}(300/T)^{0.5}$ $+ 3.0 \times 10^{-25}(300/T)^{2.5}\,M$	(16)
465	$NO^+(H_2O) + NO_2^- \rightarrow H_2O + NO_2 + NO$	$k = 5.0 \times 10^{-7}(300/T)^{0.5}$ $+ 3.0 \times 10^{-25}(300/T)^{2.5}\,M$	(16)
466	$NO^+(H_2O) + NO_3^- \rightarrow H_2O + 2NO_2$	$k = 5.0 \times 10^{-7}(300/T)^{0.5}$ $+ 3.0 \times 10^{-25}(300/T)^{2.5}\,M$	(16)
467	$NO^+(H_2O) + CO_3^- \rightarrow H_2O + CO_2 + NO + O$	$k = 5.0 \times 10^{-7}(300/T)^{0.5}$ $+ 3.0 \times 10^{-25}(300/T)^{2.5}\,M$	(16)
468	$NO^+(H_2O) + O_2^-(H_2O) \rightarrow NO_3 + 2H_2O$	$k = 5.0 \times 10^{-7}(300/T)^{0.5}$ $+ 3.0 \times 10^{-25}(300/T)^{2.5}\,M$	(16)
469	$NO^+(H_2O)_2 + e^- + M \rightarrow 2H_2O + NO + M$	$k = 2.0 \times 10^{-6}(300/T)^{0.2}$ $+ 5.0 \times 10^{-26}(300/T)^{2.5}\,M$	(16)

384

470	$NO^+(H_2O)_2 + O^-$	$\rightarrow 2H_2O + NO_2$	$k = 5.0 \times 10^{-7}(300/T)^{0.5}$ $+ 3.0 \times 10^{-25}(300/T)^{2.5} M$	(16)
471	$NO^+(H_2O)_2 + O_2^-$	$\rightarrow 2H_2O + O_2 + NO$	$k = 5.0 \times 10^{-7}(300/T)^{0.5}$ $+ 3.0 \times 10^{-25}(300/T)^{2.5} M$	(16)
472	$NO^+(H_2O)_2 + O_3^-$	$\rightarrow 2H_2O + O_2 + NO_2$	$k = 5.0 \times 10^{-7}(300/T)^{0.5}$ $+ 3.0 \times 10^{-25}(300/T)^{2.5} M$	(16)
473	$NO^+(H_2O)_2 + NO^-$	$\rightarrow 2H_2O + O_2 + N_2$	$k = 5.0 \times 10^{-7}(300/T)^{0.5}$ $+ 3.0 \times 10^{-25}(300/T)^{2.5} M$	(16)
474	$NO^+(H_2O)_2 + NO_2^-$	$\rightarrow 2H_2O + NO + NO_2$	$k = 5.0 \times 10^{-7}(300/T)^{0.5}$ $+ 3.0 \times 10^{-25}(300/T)^{2.5} M$	(16)
475	$NO^+(H_2O)_2 + NO_3^-$	$\rightarrow 2H_2O + 2NO_2$	$k = 5.0 \times 10^{-7}(300/T)^{0.5}$ $+ 3.0 \times 10^{-25}(300/T)^{2.5} M$	(16)
476	$NO^+(H_2O)_2 + CO_3^-$	$\rightarrow 2H_2O + CO_2 + NO + O$	$k = 5.0 \times 10^{-7}(300/T)^{0.5}$ $+ 3.0 \times 10^{-25}(300/T)^{2.5} M$	(16)
477	$NO^+(H_2O)_2 + O_2^-(H_2O)$	$\rightarrow NO_3 + 3H_2O$	$k = 5.0 \times 10^{-7}(300/T)^{0.5}$ $+ 3.0 \times 10^{-25}(300/T)^{2.5} M$	(16)
478	$NO^+(H_2O)_3 + e^- + M$	$\rightarrow 3H_2O + NO + M$	$k = 3.0 \times 10^{-6}(300/T)^{0.2}$ $+ 5.0 \times 10^{-25}(300/T)^{2.6} M$	(16)
479	$NO^+(H_2O)_3 + O^-$	$\rightarrow 3H_2O + NO_2$	$k = 5.0 \times 10^{-7}(300/T)^{0.5}$ $+ 3.0 \times 10^{-25}(300/T)^{2.5} M$	(16)
480	$NO^+(H_2O)_3 + O_2^-$	$\rightarrow 3H_2O + O_2 + NO$	$k = 5.0 \times 10^{-7}(300/T)^{0.5}$ $+ 3.0 \times 10^{-25}(300/T)^{2.5} M$	(16)
481	$NO^+(H_2O)_3 + O_3^-$	$\rightarrow 3H_2O + O_2 + NO_2$	$k = 5.0 \times 10^{-7}(300/T)^{0.5}$ $+ 3.0 \times 10^{-25}(300/T)^{2.5} M$	(16)
482	$NO^+(H_2O)_3 + NO^-$	$\rightarrow 3H_2O + O_2 + N_2$	$k = 5.0 \times 10^{-7}(300/T)^{0.5}$ $+ 3.0 \times 10^{-25}(300/T)^{2.5} M$	(16)
483	$NO^+(H_2O)_3 + NO_2^-$	$\rightarrow 3H_2O + NO_2 + NO$	$k = 5.0 \times 10^{-7}(300/T)^{0.5}$ $+ 3.0 \times 10^{-25}(300/T)^{2.5} M$	(16)
484	$NO^+(H_2O)_3 + NO_3^-$	$\rightarrow 3H_2O + 2NO_2$	$k = 5.0 \times 10^{-7}(300/T)^{0.5}$ $+ 3.0 \times 10^{-25}(300/T)^{2.5} M$	(16)
485	$NO^+(H_2O)_3 + CO_3^-$	$\rightarrow 3H_2O + CO_2 + NO + O$	$k = 5.0 \times 10^{-7}(300/T)^{0.5}$ $+ 3.0 \times 10^{-25}(300/T)^{2.5} M$	(16)

(Continued)

385

TABLE IV. (Continued)

	Reaction		k	Ref.
486	$NO^+(H_2O)_3 + O_4^-$	$\rightarrow 3H_2O + NO_3 + O_2$	$k = 5.0 \times 10^{-7}(300/T)^{0.5}$ $+ 3.0 \times 10^{-25}(300/T)^{2.5}\,M$	(16)
487	$NO^+(H_2O)_3 + O_2^-(H_2O)$	$\rightarrow 4H_2O + NO_3$	$k = 5.0 \times 10^{-7}(300/T)^{0.5}$ $+ 3.0 \times 10^{-25}(300/T)^{2.5}\,M$	(16)
488	$NO_2^+ + e^-$	$\rightarrow NO + O$	$k = 3.0 \times 10^{-7}(300/T)^{0.5}$	(16)
489	$NO_2^+ + e^- + M$	$\rightarrow NO_2 + M$	$k = 6.0 \times 10^{-27}(300/T)^{0.5}$	(16)
490	$NO_2^+ + O^-$	$\rightarrow NO + O_2$	$k = 4.0 \times 10^{-7}(300/T)^{0.5}$ $+ 3.0 \times 10^{-25}(300/T)^{2.5}\,M$	(16)
491	$NO_2^+ + O_2^-$	$\rightarrow NO_2 + O_2$	$k = 4.0 \times 10^{-7}(300/T)^{0.5}$ $+ 3.0 \times 10^{-25}(300/T)^{2.5}\,M$	(16)
492	$NO_2^+ + O_3^-$	$\rightarrow NO_3 + O_2$	$k = 4.0 \times 10^{-7}(300/T)^{0.5}$ $+ 3.0 \times 10^{-25}(300/T)^{2.5}\,M$	(16)
493	$NO_2^+ + NO^-$	$\rightarrow N_2O + O_2$	$k = 4.0 \times 10^{-7}(300/T)^{0.5}$ $+ 3.0 \times 10^{-25}(300/T)^{2.5}\,M$	(16)
494	$NO_2^+ + NO_2^-$	$\rightarrow 2O_2 + N_2$	$k = 4.0 \times 10^{-7}(300/T)^{0.5}$ $+ 3.0 \times 10^{-25}(300/T)^{2.5}\,M$	(16)
495	$NO_2^+ + NO_3^-$	$\rightarrow NO_2 + NO + O_2$	$k = 4.0 \times 10^{-7}(300/T)^{0.5}$ $+ 3.0 \times 10^{-25}(300/T)^{2.5}\,M$	(16)
496	$NO_2^+ + CO_3^-$	$\rightarrow NO_3 + CO_2$	$k = 4.0 \times 10^{-7}(300/T)^{0.5}$ $+ 3.0 \times 10^{-25}(300/T)^{2.5}\,M$	(16)
497	$NO_2^+ + O_4^-$	$\rightarrow 2O_2 + NO_2$	$k = 4.0 \times 10^{-7}(300/T)^{0.5}$ $+ 3.0 \times 10^{-25}(300/T)^{2.5}\,M$	(16)
498	$NO_2^+ + O_2^-(H_2O)$	$\rightarrow NO_2 + O_2 + H_2O$	$k = 5.0 \times 10^{-7}(300/T)^{0.5}$ $+ 3.0 \times 10^{-25}(300/T)^{2.5}\,M$	(16)
499	$N_2O^+ + e^-$	$\rightarrow N_2 + O$	$k = 2.0 \times 10^{-7}$	(16)
500	$N_2O^+ + e^- + M$	$\rightarrow N_2O + M$	$k = 6.0 \times 10^{-27}(300/T)^{2.5}$	(16)
501	$N_2O^+ + O^-$	$\rightarrow N_2 + O_2$	$k = 4.0 \times 10^{-7}(300/T)^{0.5}$	(16)
502	$N_2O^+ + O_2^-$	$\rightarrow N_2O + O_2$	$k = 4.0 \times 10^{-7}(300/T)^{0.5}$ $+ 3.0 \times 10^{-25}(300/T)^{2.5}\,M$	(16)

	Reaction	Rate constant	Ref.
503	$N_2O^+ + O_3^-$ → $N_2 + 2O_2$	$k = 4.0 \times 10^{-7}(300/T)^{0.5}$ $+ 3.0 \times 10^{-25}(300/T)^{2.5}M$	(16)
504	$N_2O^+ + NO^-$ → $NO_2 + N_2$	$k = 4.0 \times 10^{-7}(300/T)^{0.5}$ $+ 3.0 \times 10^{-25}(300/T)^{2.5}M$	(16)
505	$N_2O^+ + NO_2^-$ → $N_2 + NO_3$	$k = 4.0 \times 10^{-7}(300/T)^{0.5}$ $+ 3.0 \times 10^{-25}(300/T)^{2.5}M$	(16)
506	$N_2O^+ + NO_3^-$ → $2NO + NO_2$	$k = 4.0 \times 10^{-7}(300/T)^{0.5}$ $+ 3.0 \times 10^{-25}(300/T)^{2.5}M$	(16)
507	$N_2O^+ + CO_3^-$ → $N_2 + CO_2 + O_2$	$k = 4.0 \times 10^{-7}(300/T)^{0.5}$ $+ 3.0 \times 10^{-25}(300/T)^{2.5}M$	(16)
508	$N_2O^+ + O_2^-(H_2O)$ → $N_2O + O_2 + H_2O$	$k = 5.0 \times 10^{-7}(300/T)^{0.5}$ $+ 3.0 \times 10^{-25}(300/T)^{2.5}M$	(16)
509	$NH_3^+ + e^-$ → $NH_2 + H$	$k = 3.0 \times 10^{-7}(300/T)^{0.5}$	(26)
510	$NH_3^+ + e^-$ → $NH + 2H$	$k = 3.0 \times 10^{-7}(300/T)^{0.5}$	(26)
511	$NH_4^+ + e^-$ → $NH_3 + H$	$k = 3.0 \times 10^{-7}(300/T)^{0.5}$	(26)
512	$NH_4^+ + e^-$ → $NH_2 + 2H$	$k = 3.0 \times 10^{-7}(300/T)^{0.5}$	(26)
513	$NH_4^+ + NO_2^-$ → $NH_3 + NO_2 + H$	$k = 5.0 \times 10^{-7}(300/T)^{0.5}$ $+ 3.0 \times 10^{-25}(300/T)^{2.5}M$	est., (16)
514	$NH_4^+ + NO_3^-$ → $NH_3 + NO_2 + OH$	$k = 4.0 \times 10^{-7}(300/T)^{0.5}$ $+ 3.0 \times 10^{-25}(300/T)^{2.5}M$	est., (16)
515	$NH_4^+ + O_4^-$ → $2O_2 + NH_3 + H$	$k = 4.0 \times 10^{-7}(300/T)^{0.5}$ $+ 3.0 \times 10^{-25}(300/T)^{2.5}M$	est., (16)

TABLE V
Reactions of Excited Species

516	$N_2^* + N_2$	$\rightarrow 2N_2$	$k = 2.7 \times 10^{-11}$	(41, 46)
517	$N_2^* + O_2$	$\rightarrow N_2 + O_2^*$	$k = 1.0 \times 10^{-12}$	(41, 46, 48)
518	$N_2^* + O_2$	$\rightarrow N_2 + 2O$	$k = 2.0 \times 10^{-12}$	(48)
519	$N_2^* + O_2$	$\rightarrow N_2O + O$	$k = 3.0 \times 10^{-14}$	(48)
520	$N_2^* + O_2$	$\rightarrow N_2O + O(^1D)$	$k = 3.0 \times 10^{-14}$	(48)
521	$N_2^* + H_2O$	$\rightarrow OH + H + N_2$	$k = 4.2 \times 10^{-11}$	(47)
522	$N_2^* + CO_2$	$\rightarrow CO + O + N_2$	$k = 1.5 \times 10^{-10}$	(46)
523	$N_2^* + NO$	$\rightarrow NO + N_2$	$k = 1.5 \times 10^{-10}$	(87)
524	$N_2^* + NO_2$	$\rightarrow NO + O + N_2$	$k = 1.0 \times 10^{-12}$	(est.)
525	$N_2^* + SO_2$	$\rightarrow SO + O + N_2$	$k = 5.0 \times 10^{-11}$	(8, 42)
526	$N_2^* + NH_3$	$\rightarrow NH_2 + H + N_2$	$k = 4.6 \times 10^{-11}$	(28)
527	$N_2^* + NH_3$	$\rightarrow NH + H_2 + N_2$	$k = 9.0 \times 10^{-11}$	(28)
528	$N_2^* + N_2O$	$\rightarrow 2N_2 + O$	$k = 8.0 \times 10^{-11}$	(46)
529	$N_2^* + N_2O$	$\rightarrow N_2 + N + NO$	$k = 8.0 \times 10^{-11}$	(46)
530	$N_2^* + H_2O_2$	$\rightarrow N_2 + 2OH$	$k = 2.0 \times 10^{-11}$	(25)
531	$N_2^* + H_2$	$\rightarrow N_2 + 2H$	$k = 3.8 \times 10^{-15}$	(28)
532	$N_2^* + N$	$\rightarrow N_2 + N$	$k = 5.0 \times 10^{-11}$	(42)
533	$N_2^* + O$	$\rightarrow N_2 + O^*$	$k = 2.3 \times 10^{-11}$	(31, 41)
534	$N_2^* + NH_2$	$\rightarrow N_2 + NH + H$	$k = 1.66 \times 10^{-11}$	(28)
535	$N(^2D) + N_2$	$\rightarrow N + N_2$	$k = 9.4 \times 10^{-14} \exp(-510/T)$	(30, 44)
536	$N(^2D) + O_2$	$\rightarrow NO + O$	$k = 3.5 \times 10^{-13} T^{0.5}$	(30, 31, 44)
536	$N(^2D) + CO_2$	$\rightarrow NO + CO$	$k = 4.0 \times 10^{-13}$	(30, 31)
538	$N(^2D) + NO$	$\rightarrow N_2 + O$	$k = 7.0 \times 10^{-11}$	(30)
539	$N(^2D) + NO_2$	$\rightarrow N_2O + O$	$k = 1.5 \times 10^{-13}$	(19)
540	$N(^2D) + NO_2$	$\rightarrow 2NO$	$k = 1.1 \times 10^{-13}$	(19)
541	$N(^2D) + N_2O$	$\rightarrow N_2 + NO$	$k = 1.2 \times 10^{-11} \exp(-570/T)$	(30, 31, 44)
542	$N(^2D) + NH_3$	$\rightarrow NH + NH_2$	$k = 7.0 \times 10^{-11}$	(40)
543	$N(^2D) + O$	$\rightarrow N + O$	$k = 7.0 \times 10^{-13}$	(75)
544	$N(^2P) + N_2$	$\rightarrow N_2 + N$	$k = 2.0 \times 10^{-18}$	(44)
545	$N(^2P) + O_2$	$\rightarrow NO + O$	$k = 2.0 \times 10^{-12}$	(44, 75)
546	$N(^2P) + NO_2$	$\rightarrow N_2O + O$	$k = 1.5 \times 10^{-13}$	(19)
547	$N(^2P) + NO_2$	$\rightarrow 2NO$	$k = 1.1 \times 10^{-13}$	(19)

	Reaction		Rate constant		Ref.
548	N(^2P) + NH$_3$	→ NH + NH$_2$	$k = 7.0 \times 10^{-11}$		(40)
549	N(^2P) + O	→ N(^2D) + O	$k = 1.0 \times 10^{-11}$		(75)
550	O$_2^*$ + M	→ O$_2$ + M	$k = 5.0 \times 10^{-19}$		mean value (1)
551	O$_2^*$ + O$_3$	→ 2O$_2$ + O	$k = 5.2 \times 10^{-11} \exp(-2840/T)$		(1)
552	O$_2^*$ + O	→ O$_2$ + O*	$k = 1.7 \times 10^{-10}$		(2)
553	O$_2^*$ + H	→ OH + O	$k = 1.83 \times 10^{-13} \exp(-1550/T)$		(58, 84)
554	O$_2^*$ + HO$_2$	→ O$_2$ + HO$_2$	$k = 1.66 \times 10^{-12}$		(88)
555	O$_2^*$ + HO$_2$	→ OH + O + O$_2$	$k = 1.66 \times 10^{-10}$		(28)
556	O$_2^*$ + NH$_2$	→ HNO + OH	$k = 1.0 \times 10^{-14}$		(14, 15)
557	O$_2^*$ + NH	→ NO + OH	$k = 1.0 \times 10^{-14}$		(14, 15)
558	O$_2^*$ + O*	→ O$_2$ + O	$k = 1.7 \times 10^{-10}$		(2, 75)
559	O(^1D)	→ O + $h\nu$	$k = 6.3 \times 10^{-3}$		(3)
560	O(^1D) + M	→ O + M	$k = 1.8 \times 10^{-11} \exp(+110/T)$	(M = N$_2$, CO$_2$)	(1, 32)
561	O(^1D) + N$_2$ + M	→ N$_2$O + M	$k = 3.5 \times 10^{-37} (300/T)^{0.6}$		(32)
562	O(^1D) + O$_2$	→ O + O$_2^*$	$k = 2.7 \times 10^{-11} \exp(+67/T)$		(1)
563	O(^1D) + O$_2$	→ O + O$_2$	$k = 5.0 \times 10^{-12} \exp(+67/T)$		(1)
564	O(^1D) + H$_2$O	→ H$_2$ + O$_2$	$k = 2.3 \times 10^{-12}$		(1)
565	O(^1D) + H$_2$O	→ 2OH	$k = 2.2 \times 10^{-10}$		(1, 2)
566	O(^1D) + H$_2$O	→ H$_2$O + O	$k = 1.2 \times 10^{-11}$		(1)
567	O(^1D) + NH$_3$	→ NH$_2$ + OH	$k = 2.9 \times 10^{-10}$		(32, 38)
568	O(^1D) + NH$_3$	→ NH + H$_2$O	$k = 2.9 \times 10^{-11}$		(89)
569	O(^1D) + O$_3$	→ 2O + O$_2$	$k = 1.2 \times 10^{-10}$		(1)
570	O(^1D) + O$_3$	→ 2O$_2$	$k = 1.2 \times 10^{-10}$		(1, 60)
571	O(^1D) + H$_2$	→ H + OH	$k = 1.1 \times 10^{-10}$		(1, 2)
572	O(^1D) + N$_2$O	→ 2NO	$k = 6.7 \times 10^{-11}$		(1, 2, 32)
573	O(^1D) + N$_2$O	→ N$_2$ + O$_2$	$k = 4.9 \times 10^{-11}$		(1, 2, 32)
574	O(^1D) + NO$_2$	→ NO + O$_2$	$k = 1.4 \times 10^{-10}$		(44, 90)
575	O(^1D) + H$_2$O$_2$	→ OH + HO$_2$	$k = 5.2 \times 10^{-10}$		(91)
576	O*	→ O$^+$ + e^-	$k = 2.0 \times 10^{-7}$		(2)
577	O* + O$_2$	→ O$_2^*$ + O	$k = 4.9 \times 10^{-12} \exp(-850/T)$		(2, 63)
578	O$^+$ + H$_2$O	→ H$_2$O + O	$k = 7.0 \times 10^{-11}$		(64)
579	O$^+$ + CO$_2$	→ CO$_2$ + O	$k = 3.0 \times 10^{-11} \exp(-1315/T)$		(63)
580	O$^+$ + NH$_3$	→ NH$_3$ + O	$k = 5.0 \times 10^{-10}$		(64)
581	O$^+$ + O$_3$	→ 2O$_2$	$k = 1.2 \times 10^{-10}$		est. (1, 60)

TABLE VI
Bimolecular Reactions among (Ground-State) Neutrals

No.	Reaction	Rate constant	Ref.
582	$OH + H \rightarrow H_2 + O$	$k = 1.38 \times 10^{-14}\, T \exp(-3500/T)$	(24)
583	$OH + OH \rightarrow H_2O + O$	$k = 1.0 \times 10^{-11} \exp(-500/T)$	(1)
584	$H_2O + O \rightarrow 2OH$	$k = 2.5 \times 10^{-14}\, T^{1.14} \exp(-8624/T)$	(10)
585	$H_2O + H \rightarrow OH + H_2$	$k = 7.6 \times 10^{-16}\, T^{1.6} \exp(-9281/T)$	(10)
586	$H_2 + OH \rightarrow H_2O + H$	$k = 1.66 \times 10^{-16}\, T^{1.6} \exp(-1578/T)$	(10, 32)
587	$H_2 + HO_2 \rightarrow H_2O_2 + H$	$k = 1.2 \times 10^{-12} \exp(-9329/T)$	(10)
588	$H_2O_2 + H \rightarrow HO_2 + H_2$	$k = 2.8 \times 10^{-12} \exp(-1875/T)$	(10)
589	$H_2O_2 + O \rightarrow OH + HO_2$	$k = 1.4 \times 10^{-12} \exp(-2000/T)$	(1, 32)
590	$H_2O_2 + OH \rightarrow H_2O + HO_2$	$k = 1.2 \times 10^{-11} \exp(-720/T)$	(10)
591	$H_2O_2 + H \rightarrow H_2O + OH$	$k = 2.8 \times 10^{-12} \exp(-1875/T)$	(10, 20)
592	$O_3 + M \rightarrow O_2 + O + M$	$k = 4.13 \times 10^{-30}\, T^{-1.25} \exp(-11500/T)$	(9, 10)
593	$O_3 + O \rightarrow 2O_2$	$k = 8.0 \times 10^{-12} \exp(-2060/T)$	(32)
594	$O_3 + H \rightarrow OH + O_2$	$k = 1.4 \times 10^{-10} \exp(-470/T)$	(1, 32)
595	$O_3 + HO_2 \rightarrow OH + 2O_2$	$k = 1.8 \times 10^{-14} \exp(-680/T)$	(2, 49, 43)
596	$O_3 + OH \rightarrow HO_2 + O_2$	$k = 1.6 \times 10^{-12} \exp(-1000/T)$	(1, 32)
597	$H + HO_2 \rightarrow H_2 + O_2$	$k = 3.2 \times 10^{-11} \exp(-350/T)$	(1, 10, 20, 32)
598	$H + HO_2 \rightarrow 2OH$	$k = 3.0 \times 10^{-10} \exp(-500/T)$	(1, 10, 20, 32)
599	$H + HO_2 \rightarrow H_2O + O$	$k = 9.0 \times 10^{-12}$	(1, 10, 32)
600	$O + OH \rightarrow H + O_2$	$k = 2.3 \times 10^{-11} \exp(+110/T)$	(1)
601	$O + HO_2 \rightarrow OH + O_2$	$k = 2.9 \times 10^{-11} \exp(+200/T)$	(1)
602	$O_2 + N \rightarrow NO + O$	$k = 4.4 \times 10^{-12} \exp(-3220/T)$	(1, 23)
603	$O_2 + HNO \rightarrow NO_2 + OH$	$k = 1.66 \times 10^{-15}$	(14, 15)
604	$O_2 + HNO \rightarrow NO + HO_2$	$k = 3.32 \times 10^{-14}$	(14, 15)
605	$O_2 + HSO_3 \rightarrow SO_3 + HO_2$	$k = 1.34 \times 10^{-12} \exp(-330/T)$	(22, 29, 32)
606	$O_2 + SO \rightarrow SO_2 + O$	$k = 2.4 \times 10^{-13} \exp(-2370/T)$	(32)
607	$O_2 + NH_2 \rightarrow HNO + OH$	$k = 3.0 \times 10^{-16}$	(54, 88)
608	$O_2 + NH \rightarrow NO + OH$	$k = 1.26 \times 10^{-13} \exp(-764/T)$	(14, 15, 20)
609	$CO_2 + N \rightarrow NO + CO$	$k = 3.2 \times 10^{-13} \exp(-1711/T)$	(16)
610	$NO + O_3 \rightarrow NO_2 + O_2$	$k = 1.8 \times 10^{-12} \exp(-1370/T)$	(1)

611	$NO + N$	$\rightarrow N_2 + O$	$k = 3.25 \times 10^{-11}$	(1, 32, 61)
612	$NO + NO_3$	$\rightarrow 2NO_2$	$k = 3.0 \times 10^{-11}$	(1, 32, 77)
613	$NO + NO_3$	$\rightarrow 2NO + O_2$	$k = 7.3 \times 10^{-12}(300/T)^{0.23}\exp(-947/T)$	(16)
614	$NO + HO_2$	$\rightarrow NO_2 + OH$	$k = 3.7 \times 10^{-12}\exp(+240/T)$	(1, 32)
615	$NO + NH_2$	$\rightarrow N_2 + H_2O$	$k = 1.0 \times 10^{-6}\,T^{-1.96}$	(this evaluation)
616	$NO + NH_2$	$\rightarrow N_2H + OH$	$k = 1.0 \times 10^{-12}\exp(+330/T)$	(this evaluation)
617	$NO + NH$	$\rightarrow N_2 + OH$	$k = 5.15 \times 10^{-11}$	(14, 15, 53)
618	$NO + N_2H$	$\rightarrow N_2 + HNO$	$k = 8.3 \times 10^{-11}$	(20)
619	$NO + N_2H_2$	$\rightarrow N_2O + NH_2$	$k = 5.0 \times 10^{-12}$	(20)
620	$NO_2 + N$	$\rightarrow N_2O + O$	$k = 3.0 \times 10^{-12}$	(1)
621	$NO_2 + O_3$	$\rightarrow NO_3 + O_2$	$k = 1.2 \times 10^{-13}\exp(-2450/T)$	(1, 32, 77)
622	$NO_2 + NO_3$	$\rightarrow NO_2 + NO + O_2$	$k = 2.3 \times 10^{-13}\exp(-1600/T)$	(34)
623	$NO_2 + O$	$\rightarrow NO + O_2$	$k = 5.21 \times 10^{-12}\exp(+202/T)$	(1, 50)
624	$NO_2 + H$	$\rightarrow NO + OH$	$k = 5.8 \times 10^{-10}\exp(-740/T)$	(17, 34)
625	$NO_2 + OH$	$\rightarrow NO + HO_2$	$k = 3.03 \times 10^{-11}\exp(-3400/T)$	(80)
626	$NO_2 + SO$	$\rightarrow NO + SO_2$	$k = 1.4 \times 10^{-11}$	(1, 32)
627	$NO_2 + SO_2$	$\rightarrow NO + SO_3$	$k = 8.8 \times 10^{-30}$	(6)
628	$NO_2 + NH_2$	$\rightarrow N_2O + H_2O$	$k = 2.2 \times 10^{-12}\exp(+650/T)$	(32, 92)
629	$NO_2 + NH$	$\rightarrow N_2O + OH$	$k = 1.61 \times 10^{-11}$	(53)
630	$N_2O_4 + M$	$\rightarrow 2NO_2 + M$	$k = 4.2 \times 10^{-7}\exp(-5550/T)$	(34)
631	$N_2O_4 + H_2O$	$\rightarrow HNO_3 + HNO_2$	$k = 1.97 \times 10^{-12}\exp(-3185/T)$	(evaluated from 85)
632	$N_2O_4 + NH_3$	$\rightarrow HNO_3 + N_2 + H_2O$	$k = 5.54 \times 10^{-14}\exp(-1410/T)$	(evaluated from 85)
633	$2HNO_2$	$\rightarrow NO + NO_2 + H_2O$	$k = 1.0 \times 10^{-20}$	(16, 70)
634	$HNO_2 + O$	$\rightarrow NO_2 + OH$	$k = 3.0 \times 10^{-15}$	(72)
635	$HNO_2 + OH$	$\rightarrow NO_2 + H_2O$	$k = 1.8 \times 10^{-11}\exp(-390/T)$	(35)
636	$HNO_3 + HNO_2$	$\rightarrow 2NO_2 + H_2O$	$k = 1.6 \times 10^{-17}$	(71)
637	$HNO_3 + OH$	$\rightarrow H_2O + NO_3$	$k = 4.63 \times 10^{-18}\exp(-715/T)M + 1$ $k = 1.9 \times 10^{-33}\exp(+725/T)M/k$ $k = k + 7.2 \times 10^{-15}\exp(+785/T)$	(32, 58)
638	$HNO_3 + O$	$\rightarrow NO_3 + OH$	$k = 3.0 \times 10^{-15}$	(72)
639	$HNO_3 + H$	$\rightarrow HNO_2 + OH$	$k = 1.0 \times 10^{-13}$	(34, 56)

(Continued)

TABLE VI. (Continued)

	Reaction	Rate constant	Ref.
640	$HO_2NO_2 + O \rightarrow OH + NO_2 + O_2$	$k = 7.0 \times 10^{-11} \exp(-3370/T)$	(2, 32)
641	$HO_2NO_2 + OH \rightarrow NO_2 + H_2O + O_2$	$k = 1.3 \times 10^{-12} \exp(+380/T)$	(2)
642	$N_2O + NH \rightarrow N_2 + HNO$	$k = 1.66 \times 10^{-13} T^{-0.5} \exp(-1500/T)$	(20, 24)
643	$N_2O_5 + H_2O \rightarrow 2HNO_3$	$k = 3.1 \times 10^{-21}$	(2)
644	$NO_3 + NO_3 \rightarrow 2NO_2 + O_2$	$k = 7.5 \times 10^{-12} \exp(-3000/T)$	(16, 34)
645	$NO_3 + O \rightarrow NO_2 + O_2$	$k = 1.7 \times 10^{-11}$	(32, 77, 94)
646	$NO_3 + H \rightarrow NO_2 + OH$	$k = 1.1 \times 10^{-10}$	(93)
647	$NO_3 + OH \rightarrow NO_2 + HO_2$	$k = 2.3 \times 10^{-11}$	(86, 93)
648	$NO_3 + HO_2 \rightarrow HNO_3 + O_2$	$k = 2.0 \times 10^{-13} \exp(+550/T)$	(37, 86)
649	$NO_3 + HO_2 \rightarrow NO_2 + OH + O_2$	$k = 8.0 \times 10^{-13} \exp(+550/T)$	(37, 86)
650	$NO_3 + CO \rightarrow NO_2 + CO_2$	$k = 1.6 \times 10^{-11} \exp(-3250/T)$	(16)
651	$SO_2 + HO_2 \rightarrow SO_3 + OH$	$k = 2.25 \times 10^{-16} \exp(+300/T)$	(32)
652	$SO_3 + H_2O \rightarrow H_2SO_4$	$k = 6.0 \times 10^{-15}$	(32)
653	$SO_3 + N \rightarrow SO_2 + NO$	$k = 5.3 \times 10^{-16}$	(97)
654	$SO + O_3 \rightarrow SO_2 + O_2$	$k = 3.6 \times 10^{-12} \exp(-1100/T)$	(42)
655	$SO + HO_2 \rightarrow SO_2 + OH$	$k = 9.0 \times 10^{-19}$	(32)
656	$SO + OH \rightarrow SO_2 + H$	$k = 9.0 \times 10^{-10} (T/300)^{0.5}$	(2, 32)
657	$CO + OH \rightarrow CO_2 + H$	$k = 1.5 \times 10^{-13} (1 + 0.6P)$ P [atm]	(26, 32)
658	$O_3 + N \rightarrow NO + O_2$	$k = 1.0 \times 10^{-16}$	(32)
659	$O_3 + NH_2 \rightarrow NH_2O + O_2$	$k = 2.0 \times 10^{-12} \exp(-700/T)$	(1, 23)
660	$NH_3 + OH \rightarrow NH_2 + H_2O$	$k = 8.32 \times 10^{-17} T^{1.6} \exp(-480/T)$	(74)
661	$NH_3 + O \rightarrow NH_2 + OH$	$k = 1.83 \times 10^{-18} T^{2.1} \exp(-2620/T)$	(76, 79)
662	$NH_3 + H \rightarrow NH_2 + H_2$	$k = 9.1 \times 10^{-15} T^{1.3} \exp(-6570/T)$	(34, 79)
663	$NH_2 + H_2O \rightarrow NH_3 + OH$	$k = 1.56 \times 10^{-11} \exp(-7350/T)$	(83)
664	$NH_2 + H_2 \rightarrow NH_3 + H$	$k = 4.15 \times 10^{-12} \exp(-4420/T)$	(95)
665	$NH_2 + H_2O_2 \rightarrow NH_3 + HO_2$	$k = 5.00 \times 10^{-13}$	(82, 83)
666	$NH_2 + NH_2O \rightarrow NH_3 + HNO$	$k = 8.3 \times 10^{-12}$	(91)
667	$NH_2 + OH \rightarrow NH + H_2O$	$k = 7.47 \times 10^{-12} \exp(-1100/T)$	(74)
668	$NH_2 + H \rightarrow NH + H_2$	$k = 1.15 \times 10^{-10} \exp(-1825/T)$	(20)
669	$NH_2 + HO_2 \rightarrow NH_3 + O_2$	$k = 1.00 \times 10^{-13}$	(est., 28, 82)

No.	Reaction	Rate constant	References
670	$NH_2 + HO_2 \rightarrow HNO + H_2O$	$k = 1.00 \times 10^{-13}$	(est., 82)
671	$NH_2 + HO_2 \rightarrow NH_2O + OH$	$k = 2.00 \times 10^{-11}$	(est., 82, 96)
672	$NH_2 + HNO \rightarrow NH_3 + NO$	$k = 2.0 \times 10^{-10} \exp(-500/T)$	(20, 21, 51, 79)
673	$NH_2 + O \rightarrow HNO + H$	$k = 1.1 \times 10^{-9} T^{-0.5}$	(13, 20)
674	$NH_2 + O \rightarrow NH + OH$	$k = 1.2 \times 10^{-11}$	(13)
675	$NH_2 + N \rightarrow N_2 + 2H$	$k = 1.2 \times 10^{-10}$	(45, 57)
676	$2NH_2 + M \rightarrow N_2H_4 + M$	$k = 2.5 \times 10^{-11}$ high-pressure limit for $P > 250$ torr	(20, 54)
677	$2NH_2 \rightarrow N_2H_2 + H_2$	$k = 1.23 \times 10^{-12} \exp(-1250/T)$	(51, 54, 79)
678	$NH_2 + N_2H \rightarrow N_2 + NH_3$	$k = 1.66 \times 10^{-11}$	(20)
679	$NH_2 + NH \rightarrow N_2H_3$	$k = 7.2 \times 10^{-11}$	(54)
680	$NH_2 + NH \rightarrow N_2H_2 + H$	$k = 6.8 \times 10^{-11}$	(20, 79)
681	$NH + OH \rightarrow HNO + H$	$k = 3.3 \times 10^{-11}$	(20)
682	$NH + OH \rightarrow N + H_2O$	$k = 8.3 \times 10^{-13} T^{0.5} \exp(-1000/T)$	(20)
683	$NH + N \rightarrow N_2 + H$	$k = 1.1 \times 10^{-11} T^{0.5}$	(2, 45)
684	$NH + O \rightarrow NO + H$	$k = 3.3 \times 10^{-11}$	(20)
485	$NH + O \rightarrow N + OH$	$k = 1.66 \times 10^{-12} T^{0.5} \exp(-50/T)$	(20)
686	$2NH \rightarrow N_2 + H_2$	$k = 1.2 \times 10^{-10}$	(54)
687	$NH + H \rightarrow N + H_2$	$k = 5.0 \times 10^{-11}$	(20, 28)
688	$NH + HO_2 \rightarrow NH_2 + O_2$	$k = 7.7 \times 10^{-11}$	(14, 15, 54)
689	$NH + N_2H_3 \rightarrow N_2H_2 + NH_2$	$k = 3.32 \times 10^{-11}$	(20)
690	$N_2H_4 + H \rightarrow N_2H_3 + H_2$	$k = 2.2 \times 10^{-11} \exp(-1250/T)$	(2, 20, 54)
691	$N_2H_4 + O \rightarrow N_2H_3 + OH$	$k = 5.0 \times 10^{-12}$	(52)
692	$N_2H_4 + O \rightarrow N_2H_2 + H_2O$	$k = 1.41 \times 10^{-10} \exp(-600/T)$	(20)
693	$N_2H_4 + NH_2 \rightarrow N_2H_3 + NH_3$	$k = 6.5 \times 10^{-12} \exp(-750/T)$	(20, 54)
694	$N_2H_4 + OH \rightarrow N_2H_3 + H_2O$	$k = 8.3 \times 10^{-12} \exp(-500/T)$	(20)
695	$N_2H_3 + H \rightarrow N_2 + 2H_2$	$k = 1.0 \times 10^{-11}$	(2)
696	$N_2H_3 + H \rightarrow 2NH_2$	$k = 2.7 \times 10^{-12}$	(2, 20, 54)
697	$2N_2H_3 \rightarrow N_2H_4 + N_2 + H_2$	$k = 6.0 \times 10^{-11}$	(2)
698	$2N_2H_3 \rightarrow 2NH_3 + N_2$	$k = 1.7 \times 10^{-10}$	(2, 54)

(Continued)

TABLE VI. (*Continued*)

699	$N_2H_3 + O$	$\rightarrow NH_2 + HNO$	$k = 1.66 \times 10^{-11}$	(20)
700	$N_2H_3 + O$	$\rightarrow N_2H_2 + OH$	$k = 8.3 \times 10^{-12} \exp(-2500/T)$	(20)
701	$N_2H_3 + OH$	$\rightarrow N_2H_2 + H_2O$	$k = 1.66 \times 10^{-12} \exp(-500/T)$	(20)
702	$N_2H_2 + O$	$\rightarrow N_2H + OH$	$k = 3.32 \times 10^{-11} \exp(-500/T)$	(20)
703	$N_2H_2 + O$	$\rightarrow NO + NH_2$	$k = 1.66 \times 10^{-11}$	(20)
704	$N_2H_2 + OH$	$\rightarrow N_2H + H_2O$	$k = 1.66 \times 10^{-11} \exp(-500/T)$	(20)
705	$N_2H_2 + H$	$\rightarrow N_2H + H_2$	$k = 8.30 \times 10^{-11} \exp(-500/T)$	(20)
706	$N_2H_2 + NH$	$\rightarrow N_2H + NH_2$	$k = 1.66 \times 10^{-11} \exp(-500/T)$	(20)
707	$N_2H_2 + NH_2$	$\rightarrow N_2H + NH_3$	$k = 1.66 \times 10^{-11} \exp(-500/T)$	(20)
708	$N_2H + H$	$\rightarrow N_2 + H_2$	$k = 6.14 \times 10^{-11} \exp(-1500/T)$	(20)
709	$N_2H + OH$	$\rightarrow N_2 + H_2O$	$k = 5.0 \times 10^{-11}$	(20)
710	$N_2H + O$	$\rightarrow N_2 + OH$	$k = 1.66 \times 10^{-11} \exp(-2500/T)$	(20)
711	$N_2H + O$	$\rightarrow N_2O + H$	$k = 1.66 \times 10^{-11} \exp(-1500/T)$	(20)
712	$HNO + H$	$\rightarrow H_2 + NO$	$k = 9.0 \times 10^{-12}$	(8, 20)
713	$HNO + H$	$\rightarrow NH + OH$	$k = 3.0 \times 10^{-20}$	(14, 15)
714	$HNO + OH$	$\rightarrow H_2O + NO$	$k = 6.0 \times 10^{-11}$	(14, 15, 20)
715	$HNO + O_3$	$\rightarrow NO + OH + O_2$	$k = 3.3 \times 10^{-14}$	(74)
716	$HNO + O$	$\rightarrow NO + OH$	$k = 1.66 \times 10^{-13}$	(20)
717	$2HNO$	$\rightarrow H_2O + N_2O$	$k = 6.56 \times 10^{-12} \exp(-2500/T)$	(20)
718	$NH_2O + O_3$	$\rightarrow NH_2 + 2O_2$	$k = 1.5 \times 10^{-14}$	(74)
719	$NH_2O + O_2$	$\rightarrow HNO + HO_2$	$k = 2.7 \times 10^{-18}$	(2)
720	$NH_2O + O$	$\rightarrow NH_2 + O_2$	$k = 6.64 \times 10^{-11}$	(91, 96)
721	$N + OH$	$\rightarrow NO + H$	$k = 5.8 \times 10^{-11}$	(1, 34)
722	$N + HO_2$	$\rightarrow NO + OH$	$k = 2.2 \times 10^{-11}$	(33)

TABLE VII
Termolecular Reactions

No.	Reaction	Rate expression	Ref.
723	$O_2 + O + M \rightarrow O_3 + M$	$k_0 = 6.0 \times 10^{-34} (300/T)^{2.3}$ $k_\infty = 2.8 \times 10^{-12}, F_c = \exp(-T/696)$	(1, 11, 12)
724	$O_2 + H + M \rightarrow HO_2 + M$	$k_0 = 5.9 \times 10^{-32} (300/T)$	(1)
725	$NO + O + M \rightarrow NO_2 + M$	$k_0 = 7.5 \times 10^{-11} (T/300)^{0.5}, F_c = \exp(-T/498)$	(1)
726	$NO + OH + M \rightarrow HNO_2 + M$	$k_0 = 1.0 \times 10^{-31} (300/T)^{1.6}$ $k_\infty = 3.0 \times 10^{-11} (T/300)^{0.3}, F_c = \exp(-T/1850)$	(1)
727	$NO + H + M \rightarrow HNO + M$	$k_\infty = 7.4 \times 10^{-31} (300/T)^{2.6}$ $k_\infty = 1.0 \times 10^{-11}, F_c = \exp(-T/1300)$ $k = 1.0 \times 10^{-32} \exp(+300/T)$	(8, 34)
728	$2NO_2 + M \rightarrow N_2O_4 + M$	$k = 4.7 \times 10^{-35} \exp(+860/T)$ corrected for K_p	(34)
729	$NO_2 + NO_3 + M \rightarrow N_2O_5 + M$	$k_0 = 2.7 \times 10^{-30} (300/T)^{3.4}$ $k_\infty = 2.0 \times 10^{-12} (T/300)^{0.2}, F_c = 0.34$	(1)
730	$NO_2 + O + M \rightarrow NO_3 + M$	$k_0 = 9.0 \times 10^{-32} (300/T)^{2.0}$ $k_\infty = 2.2 \times 10^{-11}, F_c = \exp(-T/1300)$	(16, 32)
731	$NO_2 + OH + M \rightarrow HNO_3 + M$	$k_0 = 2.6 \times 10^{-30} (300/T)^{2.9}$ $k_\infty = 5.2 \times 10^{-11}, F_c = \exp(-T/353)$	(1)
732	$HO_2 + NO_2 + M \rightarrow HO_2NO_2 + M$	$k_0 = 1.7 \times 10^{-31} (300/T)^{3.2}$ $k_\infty = 4.7 \times 10^{-12}, F_c = 0.6$	(1)
733	$N_2O_5 + M \rightarrow NO_2 + NO_3$	$k_0 = 2.2 \times 10^{-3} (300/T)^{4.4} \exp(-11080/T)$ $k_\infty = 9.7 \times 10^{+14} (T/300)^{0.1} \exp(-11080/T), F_c = 0.34$	(1)
734	$HO_2NO_2 + M \rightarrow HO_2 + NO_2 + M$	$k_0 = 4.5 \times 10^{-6} \exp(-10000/T)$ $k_\infty = 3.4 \times 10^{+14} \exp(-10420/T), F_c = 0.6$	(1)
735	$SO_2 + OH + M \rightarrow HSO_3 + M$	$k_0 = 5.0 \times 10^{-31} (300/T)^{3.3}$ $k_\infty = 2.0 \times 10^{-12}, F_c = \exp(-T/380)$	(1, 32)
736	$SO_2 + O + M \rightarrow SO_3 + M$	$k = 4.0 \times 10^{-32} \exp(-1000/T)$	(1)
737	$SO_3 + O + M \rightarrow SO_2 + O_2 + M$	$k = 8.1 \times 10^{-30}$	(1)

(Continued)

TABLE VII. (Continued)

738	$CO + O + M$	$\rightarrow CO_2 + M$	$k = 6.6 \times 10^{-33} \exp(-2173/T)$	(10)
739	$O + O + M$	$\rightarrow O_2 + M$	$k = 2.76 \times 10^{-31}/T$ k_0 is used	(10)
740	$H + H + M$	$\rightarrow H_2 + M$	$k = 2.67 \times 10^{-31} T^{0.6}$ k_0 is used	(10)
741	$OH + H + M$	$\rightarrow H_2O + M$	$k = 1.1 \times 10^{-23} T^{-2.6}$	(10)
742	$2OH + M$	$\rightarrow H_2O_2 + M$	$k_0 = 6.9 \times 10^{-31} (300/T)^{0.8} (1 + 5[H_2O]/M)$ $k_\infty = 2.0 \times 10^{-11}, F_c = \exp(-150/T)$	(1, 98)
743	$OH + HO_2(+M)$	$\rightarrow H_2O + O_2 (+M)$	$k = 4.8 \times 10^{-11} \exp(+250/T)$	(1, 32)
744	$2HO_2 (+M)$	$\rightarrow H_2O_2 + O_2 (+M)$	$k = 2.2 \times 10^{-13} \exp(+600/T)$ $+1.9 \times 10^{-33} (M - [H_2O]) \exp(+980/T)$ $k = k(1 + 1.4 \times 10^{-21} \exp(+2200/T)[H_2O])$	(1, 32)
745	$N + N + M$	$\rightarrow N_2 + M$	$k = 8.3 \times 10^{-34} \exp(+500/T) T^{-0.5}$	(34)
746	$N + O + M$	$\rightarrow NO + M$	$k = 1.8 \times 10^{-31} T^{-0.5}$ k_0 is used	(34)
747	$NH_2 + H + M$	$\rightarrow NH_3 + M$	$k = 6.0 \times 10^{-30}$	(2, 84)
748	$NH + H + M$	$\rightarrow NH_2 + M$	$k = 5.5 \times 10^{-32} T^{-0.5}$	(20)

TABLE VIII
Particle Formation and Heterogeneous Reactions

$NH_3 + H_2SO_4$	$\rightarrow NH_4HSO_4$	(see text)
$NH_3 + NH_4HSO_4$	$\rightarrow (NH_4)_2SO_4$	(see text)
$NH_3 + HNO_3$	$\rightarrow NH_4NO_3$	(see text)
HNO_2	$\rightarrow \frac{1}{3}HNO_3 + \frac{2}{3}NO + \frac{1}{3}H_2O$	$\alpha = 0.1$
HNO_2	$\rightarrow 0.5NO + 0.5NO_2 + 0.5H_2O$	$\alpha = 0.01$
HNO_2	$\rightarrow NO + 0.5H_2O + 0.25O_2$	$\alpha = 1.0 \times 10^{-3}$
$2HNO_2 + SO_2$	$\rightarrow 2NO + H_2SO_4$	$\alpha = 1.0 \times 10^{-2a}$
$HNO_2 + HNO_3$	$\rightarrow 2NO_2 + H_2O$	$\alpha = 5.0 \times 10^{-3}$
HNO_3	$\rightarrow NO + 0.5H_2O + 0.75O_2$	$\alpha = 2.0 \times 10^{-4}$
HNO_3	$\rightarrow NO_2 + 0.5H_2O + 0.25O_2$	$\alpha = 5.0 \times 10^{-3}$
HNO_3	$\rightarrow 0.5NO_2 + 0.5NO_3 + 0.5H_2O$	$\alpha = 3.0 \times 10^{-5}$
$2HNO_3 + SO_2$	$\rightarrow 2NO + H_2SO_4 + O_2$	$\alpha = 1.0 \times 10^{-2a}$
$2HNO_3 + SO_2$	$\rightarrow 2NO_2 + H_2SO_4$	$\alpha = 1.0 \times 10^{-5a}$

[a] At relative humidity $> 20\%$.

Acknowledgments

The author is grateful to Prof. Dr. W. Schikarski and Dr. S. Jordan for their guidance and support of this work. Helpful discussions with Dr. H.-R. Paur are gratefully acknowledged. My particular thank is given to Mrs. B. Mathes for preparing the manuscript. Part of this work was funded by PEF (Projekt Europäisches Forschungszentrum für Massnahmen zur Luftreinhaltung) under contract no. 86/006/03.

References

1. Baulch, D. L., Cox, R. A., Hampson, R. F. Jr., Kerr, J. A., Troe, J. and R. T. Watson, "Evaluated kinetic and photochemical data for atmospheric chemistry," *J. Phys. Chem. Ref. Data* **9**, 295–471 (1980); *J. Phys. Chem. Ref. Data* **11**, 327–497 (1982); *J. Phys. Chem. Ref. Data* **13**, 1259–1381 (1984); *J. Phys. Chem. Ref. Data* **18**, 881–1095 (1989) (Atkinson, R. et al.).

2. Levine, J. S. (1985), *The Photochemistry of Atmospheres*, Academic Press.

3. Willis, C. and A. W. Boyd (1976), "Excitation in the radiation chemistry of inorganic gases," *Radiat. Phys. Chem.* **8**, 71–111.

4. Niles, F. E. (1970), "Airlike discharges with CO_2, NO, NO_2 and N_2O as impurities," *J. Chem. Phys.* **52**, 408–424.

5. Ferguson, E. E., Fehsenfeld, F. C. and D. L. Albritton (1979), "Ion Chemistry of the Earth's Atmosphere," in *Gas Phase Ion Chemistry*, M. T. Bowers (ed.), Academic Press, New York, pp. 45–82.

6. Daubendiek, R. L. and J. G. Calvert (1975), "A study of the N_2O_5-SO_2-O_3 reaction system," *Environ. Lett.* **8**, 103–116.

7. Bortner, M. H. and R. Baurer (1978), "Defense Nuclear Agency Reaction Rate Handbook," 2nd ed., Revision Number 7, NASA STAR Technical Report Issue 11.

8. Person, J. C., Ham, D. O. and A. A. Boni (1985), "Unified Projection of the Performance

and Economics of Radiation-Initiated NO_x/SO_x Emission Control Technologies, Final Report," Report 1985, PSI–TR–259/542.

9. Benson, S. W. and A. E. Axworthy, Jr. (1975), "Mechanism of the gas phase, thermal decomposition of ozone," *J. Chem. Phys.* **26**, 1718–1726.

10. Warnatz, J. (1984), "Rate coefficients in the C/H/O system," in *Combustion Chemistry*, W. C. Gardiner, Jr. (ed.), Springer, New York, pp. 197–360.

11. Lin, C. L. and M. T. Leu (1982), "Temperature and third-body dependence of the rate constant for the reaction $O + O_2 + M \rightarrow O_3 + M$," *Int. J. Chem. Kinet.* **14**, 417–434.

12. Croce de Cobos, A. E. and J. Troe (1984), "High pressure range of the recombination $O + O_2 \rightarrow O_3$," *Int. J. Chem. Kinet.* **16**, 1519–1530.

13. Dransfeld, P., Hack, W., Kurzke, H., Temps, F. and H. Gg. Wagner (1984), "Direct studies of elementary reactions of NH_2-radicals in the gas phase," *20th Symposium (International) on Combustion,* Ann Arbor. The Combustion Institute, Pittsburgh, PA, pp. 655–663.

14. Hack, W. and H. Kurzke (1985), "The reaction of NH_2-radicals with electronically excited molecular oxygen O_2 $(^1\Delta_g)$," *Ber. Bunsenges. Phys. Chem.* **89**, 86–93.

15. Hack, W., Kurzke, H., and H. Gg. Wagner (1985), "Reaction of NH $(X^3\Sigma^-)$ and $O_2(^1\Delta_g)$ in the gas phase," *J. Chem. Soc., Farad. Trans. 2,* **81**, 949–961.

16. Sutherland, C. D. and J. Zinn (1975), "Chemistry computations for irradiated hot air," Los Alamos Scientific Laboratory Informal Report LA-6055-MS, Los Alamos, NM.

17. Gehring, M., Hoyermann, K., Schacke, H., and J. Wolfrum (1972), "Direct studies of some elementary steps for the formation and destruction of nitric oxide in the H-N-O-system," *14th Symposium (International) on Combustion,* Pennsylvania State University, The Combustion Institute, Pittsburgh, PA, pp. 99–105.

18. Nakashima, K., Takagi, H., and H. Nakamura (1986), "Dissociative recombination of H_2^+, HD^+ and D_2^+ by collisions with slow electrons," *J. Chem. Phys.* **86**, 726–737.

19. Iwata, R., Ferrieri, R. A., and A. P. Wolf (1986), "Rate constant determination of the reaction of metastable $N(^2D, {}^2P)$ with NO_2 using moderated nuclear recoil atoms," *J. Phys. Chem.* **90**, 6722–6726.

20. Miller, J. A., Smooke, M. D., Green, R. M., and R. J. Kee (1983), "Kinetic modeling of the oxidation of ammonia in flames," *Comb. Sci. Technol.* **34**, 149–176.

21. Chen, A. T., Malte, Ph.C. and M. M. Thornton (1984), "Sulfur-nitrogen interaction in stirred flames," *20th Symposium (International) on Combustion,* Ann Arbor. The Combustion Institute, Pittsburgh, PA, pp. 769–777.

22. Martin, D., Jourdain, L., and G. LeBras (1986), "Discharge flow measurements of the rate constants for the reactions $OH + SO_2 + He$ and $HOSO_2 + O_2$ in relation with the atmospheric oxidation of SO_2," *J. Phys. Chem.* **90**, 4143–4147.

23. Barnett, A. J., Marston, G. and R. P. Wayne (1987), "Kinetics and chemiluminescence in the reaction of N-atoms with O_2 and O_3," *J. Chem. Soc. Farad. Trans. 2,* **83**, 1453–1463.

24. Loirat, H., Caralp, F., Destriau, F., and R. Lesclaux (1987), "Oxidation of CO by N_2O between 1076 and 1228 K: Determination of the rate constant of the exchange reaction, "*J. Phys. Chem.* **91**, 6538–6542.

25. Tao, W., Golde, M. F., Ho, G. H., and A. M. Moyle (1987), "Energy transfer from metastable electronically exicted N_2, Ar, Kr, and Xe to CH_3OH, H_2O_2, CH_3NH_2, and N_2H_4," *J. Chem. Phys.* **87**, 1045–1053.

26. Prasad, S. S. and W. T. Huntress, Jr. (1980), "A model for gas phase chemistry in interstellar clouds: I. The basic model, library of chemical reactions and chemistry among C, N and O compounds," *Astrophys. J. Suppl. Ser.* **43**, 1–35.

Anicich, V. G. and W. T. Huntress, Jr. (1986), "A survery of bimolecular ion–molecule reactions for use in modeling the chemistry of planetary atmospheres, cometary comae and interstellar clouds," *Astrophys. J. Suppl. Ser.* **62**, 553–672.

27. Gaucherel, P., Marquette, J. B., Rebrion, C., Poissant, G., Dupeyrat, G. and B. R. Rowe (1986), "Temperature dependence of slow charge-exchange reactions: $N_2^+ + O_2$ from 8 to 163 K," *Chem. Phys. Lett.* **132**, 63–66.

28. Kurzke, H. (1985), "Untersuchungen der Reaktionen von elektronisch angeregten Molekülen mit Atomen und Radikalen in der Gasphase," Dissertation, Universität Göttingen; Max-Planck-Institut für Strömungsforschung, Bericht 2/1985.

29. Gleason, J. F. and C. J. Howard (1988), "Temperature dependence of the gas-phase reaction $HOSO_2 + O_2 \rightarrow HO_2 + SO_3$," *J. Phys. Chem.* **92**, 3414–3417.

30. Lin, Ch. and F. Kaufman (1971), "Reactions of metastable nitrogen atoms," *J. Chem. Phys.* **55**, 3760–3770.

31. Piper, L. G. (1982), "The excitation of O (^1S) in the reaction between $N_2(A^3\sum_u^+)$ and O (^3P)," *J. Phys. Chem.* **77**, 2373–2377.

    Piper, L. G., Donahue, M. E. and W. T. Rawlins (1987), "Rate coefficients for $N(^2D)$ reactions," *J. Phys. Chem.* **91**, 3883–3888.

32. DeMore, W. B., Molina, M. J., Sander, S. P., Golden, D. M., Hampson, R. F., Kurylo, M. J., Howard, C. J., and A. R. Ravishankara (1987), "Chemical Kinetics and Photochemical Data for Use in Stratospheric Modeling," Evaluation No. 8, JPL Publication No. 87–41.

33. Brune, Wm, H., Schwab, J. J., and J. G. Anderson (1983), "Laser magnetic resonance, resonance fluorescence and resonance absorption studies of the reaction kinetics of $O + OH \rightarrow H + O_2, O + HO_2 \rightarrow OH + O_2$, $N + OH \rightarrow H + NO$ and $N + HO_2 \rightarrow$ products at 300 K between 1 and 5 torr," *J. Phys. Chem.* **87**, 4503–4514.

34. Baulch, D. L., Drysdale, D. D., and D. G. Horne (1983), *Evaluated Kinetic Data for High Temperature Reactions*, Butterworths, London, Vol. 2.

35. Jenkin, M. E. and R. A. Cox (1987), "Kinetics of the gas phase reaction of OH with nitrous acid," *J. Phys. Lett.* **137**, 548–552.

36. Rowe, B. R., Vallée, F., Queffelec, J. L., Gomet, J. C, and M. Morlais (1988), "The yield of oxygen and hydrogen atoms through dissociative recombination of H_2O^+ ions with electrons," *J. Chem. Phys.* **88**, 845–850.

37. Hoffmann, A. and R. Zellner (1988), paper presented at the 10th Int. Symposium on Gas Kinetics, Swansey.

38. Cheskis, S. G., Iogansen, A. A., Sarkisov, O. M., and A. A. Titov (1985), "Laser photolysis of ozone in the presence of ammonia. Formation and decay of vibrationally excited NH_2 radicals," *Chem. Phys. Lett.* **120**, 45–49.

    Cheskis, S. G., Iogansen, A. A., Kulakov, P. V., Sarkisov, O. M., and A. A. Titov (1988), "Laser photolysis in the presence of ammonia. Vibrationally excited OH radicals," *Chem. Phys. Lett.* **143**, 348–352.

39. Miller, T. M., Wetterskog, R. E., and J. F. Paulson (1984), "Temperature dependence of the ion-molecule reactions $N^+ + CO$, $C^+ + NO$, and C^+, CO^+, $CO_2^+ + O_2$ from 90–450 K," *J. Chem. Phys.* **80**, 4922–4925.

40. Ghosh, K. K. (1984), "Spectro-photometric studies of the reactions between ammonia, hydrogen sulfide and formaldehyde and excited nitrogen atoms," *Physica* **124C**, 395–398.

41. Thomas, J. M. and F. Kaufman (1985), "Rate constants of the reactions of metastable $N_2(A^3\sum_u^+)$ in $v = 0, 1, 2$ and 3 with ground state O_2 and O," *J. Chem. Phys.* **83**, 2900–2903.

    Thomas, J. M., F. Kaufman, and M. F. Golde (1987), "Rate constants for electronic quenching of $N_2(A^3\sum_u^+$, $v = 0-6)$ by O_2, NO, CO, N_2O and C_2H_4," *J. Chem. Phys.* **86**, 6885–6892.

42. Brown, R. G. and C. A. Winkler (1979), "Active nitrogen" in *Dinitrogen Fixation*, Bottomley, F. and R. C. Burns (eds.), Wiley, New York, pp. 249–290.

43. Wang, X., Suto, M., and L. C. Lee (1988), "Reaction rate constants of $HO_2 + O_3$ in the temperature range 233–400 K," *J. Chem. Phys.* **88**, 896–899.

44. Schofield, K. (1979), "Critically evaluated rate constants for gaseous reactions of several electronically excited species," *J. Phys. Chem. Ref. Data*, **8**, 723–763.

45. Whyte, A. R. and L. F. Phillips (1984), "Products of reaction of nitrogen atoms with NH_2," *J. Phys. Chem.* **88**, 5670–5673.

46. Young, R. A., Black, G. and T. G. Slanger (1969), "Vacuum-ultraviolet photolysis of N_2O.II. Deactivation of $N_2(A^3\sum_u^+)$ and $N_2(B^3\Pi_g)$," *J. Chem. Phys.* **50**, 303–308.

47. Slanger, T. G. (1983), "Reactions of electronically excited diatomic molecules," in *Reactions of Small Transient Species*, A. Fontijn and M. A. A. Clyne (eds.), Academic Press, New York, pp. 231–310.

48. Iannuzzi, M. P., Jeffries, J. B., and F. Kaufman (1982), "Product channels of the $N_2(A^3\sum_u^+) + O_2$ interaction," *Chem. Phys. Lett.* **87**, 570–574.

49. Sinha, A., Lovejoy, E. R., and C. J. Howard (1987), "Kinetic study of the reaction of HO_2 with ozone," *J. Chem. Phys.* **87**, 2122–2127.

50. Geers-Müller, R. and F. Stuhl (1987), "On the kinetics of the reactions of oxygen atoms with NO_2, N_2O_4 and N_2O_3 at low temperatures," *Chem. Phys. Lett.* **135**, 263–268.

51. Dean, A. M., Hard, J. E., and R. K. Lyon (1982), "Kinetics and mechanism of NH_3 oxidation," *Nineteenth Symposium (International) on Combustion*, Haifa, The Combustion Institute, Pittsburgh, PA, pp. 97–105.

52. Chobanyan, S. A. and T. G. Mkryan (1985), "Kinetics of the gas-phase reaction of atomic oxygen with hydrazine," *Khim. Fiz.* **4**, 1577–1578.

53. Harrison, J. A., Whyte, A. R., and L. F. Phillips (1986), "Kinetics of reactions of NH with NO and NO_2," *Chem. Phys. Lett.* **129**, 346–352.

54. Pagsberg, P. B., Eriksen, J., and H. C. Christensen (1979), "Pulse radiolysis of gaseous ammonia-oxygen mixtures," *J. Phys. Chem.* **83**, 582–590.

55. O'Neil, R. R., Lee, E. T. P., and E. R. Huppi (1979), "Auroral O (^1S) production and loss processes: Ground based measurements of the artificial auroral experiment precede," *J. Geophys. Res.* **84**, 823–833.

56. Bérces, T., Förgeteg, S., and F. Márta (1970), "Kinetics of photolysis of nitric acid vapour," *Trans. Farad. Soc.* **66**, 648–655.

57. Dransfeld, P. and H. Gg. Wagner (1987), "Investigation of the gas phase reaction $N + NH_2 \rightarrow N_2 + 2H$ at room temperature," *J. Phys. Chem. NF* **153**, 89–97.

58. Zellner, R. (1986), "Chemistry of free radicals in the atmosphere," *Oxid. Commun.* **9**, 255–300.

59. Viala, Y. P. (1986), "Chemical equilibrium from diffuse to dense interstellar clouds. I. Galactic molecular clouds," *Astron. Astrophys. Suppl. Ser.* **64**, 391–437.

60. Armstrong, D. A. (1987), "The radiation chemistry of gases," in *Radiation Chemistry*, Farhataziz and M. A. J. Rodgers, (eds.), Verlag Chemie, Weinheim, pp. 263–319.

61. Morgan, J. E., Phillips, L. F., and H. J. Schiff (1962), "Studies of vibrationally excited nitrogen using mass spectrometric and calorimeter-probe techniques," *Disc. Farad. Soc.* **33**, 118–127.

62. Fehsenfeld, F. C., Howard, C. J., and A. L. Schmeltekopf (1975), "Gas phase ion chemistry of HNO_3," *J. Chem. Phys.* **63**, 2835–2841.

63. Filseth, S. V. Stuhl, F. and K. H. Welge (1970), "Collisional deactivation of $O(^1S)$," *J. Chem. Phys.* **52**, 239–243.

64. Atkinson, R. and K. H. Welge (1972), "Temperature dependence of $O(^1S)$ deactivation by CO_2, O_2, N_2 and Ar," *J. Chem. Phys.* **57**, 3689–3693.

65. Fehsenfeld, F. C., Mosesman, M., and E. E. Ferguson (1971), "Ion-molecule reactions in an O_2^+-H_2O system," *J. Chem. Phys.* **55**, 2115–2120.

66. Fehsenfeld, F. C. and E. E. Ferguson (1971), "Fast reactions $NO^+(H_2O)_n + NH_3 \rightarrow NH_4^+(H_2O)_{n-1} + HNO_2$, $n = 1$–3, and $NO^+ \cdot NH_3 \rightarrow NH_4^+ + ONNH_2$," *J. Chem. Phys.* **54**, 439–440.

67. Viggiano, A. A. and J. F. Paulson (1984), "Reactions of negative ions," AFGL-TR-84-0176, Air Force Geophysics Larboratory, Hanscom, Mass.

68. Davidson, J. A., Viggiano, A. A., Howard, C. J., Dotan, I., Fehsenfeld, F. C., Albritton, D. L. and E. E. Ferguson (1978), "Rate constant for the reactions of O_2^+, NO_2^+, NO^+, H_2O^+, CO_3, NO_2 and halide ions with N_2O_5 at 300 K," *J. Chem. Phys.* **68**, 2085–2087.

69. Viggiano, A. A. and J. F. Paulson (1983), "Temperature dependence of associative detachment reactions," *J. Chem. Phys.* **79**, 2241–2245.

70. Kaiser, E. W. and C. H. Wu (1977), "A kinetic study of the gas phase formation and decomposition reactions of nitrous acid," *J. Phys. Chem.* **81**, 1701–1706.

71. Kaiser, E. W. and C. H. Wu (1977), "Measurement of the rate constant of the reaction of nitric acid with nitrous acid," *J. Phys. Chem.* **81**, 187–190.

72. Kaiser, E. W. and S. M. Japar (1978), "Upper limits to the gas phase reaction rates of HONO with NH_3 and O (^3P) atoms," *J. Phys. Chem.* **82**, 2753–2754.

73. Heicklen, J. and M. Luria (1975), "Kinetics of homogeneous particle nucleation and growth," *Int. J. Chem. Kinet. Symp. No. 1*, 567–580.

74. Hack, W., Horie, O., and H. Gg. Wagner (1981), "The rate of the reaction of NH_2 with O_3," *Ber. Bunsenges. Phys. Chem.* **85**, 72–78.

75. Torr, M. R. and D. G. Torr (1982), "The role of metastable species in the thermosphere," *Rev. Geophys. Space Phys.* **20**, 91–144.

76. Jeffries, J. B. and G. P. Smith (1986), "Kinetics of the reaction $OH + NH_3$," *J. Phys. Chem.* **90**, 487–491.

77. Graham, R. A. and H. S. Johnston (1978), "The photochemistry of NO_3 and the kinetics of the N_2O_5-O_3 system," *J. Phys. Chem.* **82**, 254–268.

78. E. E. Ferguson (1979), "Ion-molecule reactions in the atmosphere," in *Kinetics of Ion-Molecule Reactions*, P. Ausloos (ed), Plenum Press, New York, pp. 377–403.

79. Cohen, N. (1987), "The $O + NH_3$ reaction: a review," *Int. J. Chem. Kinet.* **19**, 319–362.

80. Howard, C. J. (1980), "Kinetic study of the equilibrium $HO_2 + NO \rightleftharpoons OH + NO_2$ and the thermochemistry of HO_2," *J. Am. Chem. Soc.* **102**, 6937–6941.

81. Fehsenfeld, F. C. and E. E. Ferguson (1974), "Laboratory studies of negative ion reactions with atmospheric trace constituents," *J. Chem. Phys.* **61**, 3181–3193.

82. Lesclaux, R. (1984), "Reactivity and kinetic properties of the NH_2 radical in the gas phase," *Rev. Chem. Intermediat.* **5**, 347–392.

83. Hack, W. Rouveirolles, P., and H. Gg. Wagner (1986), "Direct measurements of the reactions $NH_2 + H_2 \rightleftharpoons NH_3 + H$ at temperatures from 670 to 1000 K," *J. Phys. Chem.* **90**, 2505–2511.

84. Hack, W. and H. Kurzke (1986), "Kinetic study of the elementary chemical reaction $H(^2S_{1/2}) + O_2(^1\Delta_g) \rightarrow OH(^2\pi) + O(^3P)$ in the gas phase," *J. Phys. Chem.* **90**, 1900–1906.

85. Mearns, A. M. and K. Ofosu-Asiedu (1984), "Kinetics of reaction of low concentration mixtures of oxides of nitrogen, ammonia and water vapour," *J. Chem. Tech. Biotechnol.* **34A**, 341–349.

Mearns, A. M. and K. Ofosu-Asiedu (1984), "Ammonium nitrate formation in low concentration mixtures of oxides of nitrogen and ammonia," *J. Chem. Tech. Biotechnol.* **34A**, 350–354.

86. Mellouki, A., Le Bras, G., and G. Poulet (1988), "Kinetics of the reactions of NO_3 with OH and HO_2," *J. Phys. Chem.* **92**, 2229–2234.

87. Meyer, J. A., Setser, D. W., and W. G. Clark (1972), "Rate constants for quenching of $N_2(A^3\sum_u^+)$ in active nitrogen," *J. Phys. Chem.* **76**, 1–9.

88. Hack, W. (1986), "Bimolekulare Elementarprozesse elektronisch angeregter Teilchen in der Gasphase," Habilitationsschrift, Universität Göttingen; Max-Planck-Institut für Strömungsforschung, Bericht 19/1986.

89. Breckenridge, W. H. (1983), "Reactions of electronically excited atoms," in *Reactions of small transient species*, A. Fontijn and M. A. A. Clyne (eds.), Academic Press, New York, pp. 157–230.

90. Shibuya, K., Nagai, H., Imajo, T., Obi, K., and I. Tanaka (1986), "Formation mechanism of vibrationally excited O_2 molecules in the multiphoton absorption of NO_2," *J. Chem. Phys.* **85**, 5061–5067.

91. Dransfeld, P. (1986), "Untersuchung des Mechanismusses spezieller Radikalreaktionen im H-N-C-O-system," Dissertation, Universität Göttingen; Max-Planck-Institut für Strömungsforschung, Bericht 17/1986.

92. Bulatov, V. P., Ioffe, A. A., Lozovsky, V. A. and O. M. Sarkisov (1989), "On the reaction of the NH_2 radical with NO_2 at 295–620 K," *Chem. Phys. Lett.* **159**, 171–174.

93. Boodaghians, R. B., Canosa-Mas, C. E., Carpenter, P. J., and R. P. Wayne (1988), "The reactions of NO_3 with OH and H," *J. Chem. Phys., Farad. Trans. 2* **84**, 931–948.

94. Canosa-Mas, C. E., Carpenter, P. J., and R. P. Wayne (1989), "The reaction of NO_3 with atomic oxygen," *J. Chem. Soc. Farad. Trans. 2* **85**, 697–707.

95. Dransfeld, P., Hack, W., Jost, W., Rouveirolles, P., and H. Gg, Wagner (1987), "Amidogen + water → hydroxyl + ammonia reaction in a flow discharge reactor," *Khim. Fiz.* **6**, 1668–1676.

96. Bozzelli, J. W. and A. M. Dean (1989), "Energized complex quantum Rice–Ramsperger–Kassel analysis on reactions of amidogen with hydroperoxyl, oxygen and oxygen atoms," *J. Phys. Chem.* **93**, 1058–1065.

97. Wang, X., Jin, Y. G., Suto, M., and L. C. Lee (1988), "Rate constant of the gas phase reaction of SO_3 with H_2O," *J. Chem. Phys.* **89**, 4853–4860.

98. Zellner, R., Ewig, F., Paschke, R., and G. Wagner (1988), "Pressure and temperature dependence of the gas-phase recombination of hydroxyl radicals," *J. Phys. Chem.* **92**, 4184–4190.

A THEORETICAL STUDY OF ORIGINS OF RESONANCE RAMAN AND RESONANCE FLUORESCENCE USING A SPLIT-UP OF THE EMISSION CORRELATION FUNCTION

H. KONO

Department of Basic Technology, Faculty of Engineering, Yamagata University, Yonezawa, Japan

Y. NOMURA and Y. FUJIMURA

Department of Chemistry, Faculty of Science, Tohoku University, Sendai, Japan

CONTENTS

Advances in Chemical Physics Volume LXXX, Edited by I. Prigogine and Stuart A. Rice ISBN 0-471-53281-9 © 1991 John Wiley & Sons, Inc.

I. INTRODUCTION

Resonance secondary emission (RSE) has long been a controversial subject in molecular or solid-state science [1–4]. In spite of the great effort that went into the research on the light-induced spontaneous emission process, a number of issues remain unsolved. From a theoretical point of view, satisfactory consensus has not yet been reached even for rather simplified models such as the three-level molecular system whose interaction with the surrounding heat bath can be treated as an instantaneous event [the Markovian (impact) limit]. Here the heat bath means the phonon field in a crystal lattice or the ensemble of colliding perturbers. (We select one of the two according to circumstances.) A three-level system is pictured in Fig. 1 where molecular states g, m, and f are taken to be the initial, intermediate (excited), and final states in RSE, respectively. The light impinging on the molecular system, E_I, is quasiresonant with the $g \to m$ transition, while the black-body radiation field (emitted field), E_S, induces the $m \to f$ transition. Among the most fundamental issues mutually related is whether the two steps of excitation leading to emission are sequential or simultaneous, that is, whether an emission step $(m \to f)$ sets in after the completion of the absorption step $(g \to m)$ or these two steps simultaneously take place.

Experimental results imply that in general both types of processes coexist 5–11. The observed spectra include the component that is assigned to fluorescence from intermediate states [the fluorescencelike (FL) component] and the one that exhibits energetic and temporal correlation with the incident light [the Ramanlike (RL) component]. The former is called resonance fluorescence and the latter is called resonance Raman. A variety of theoretical studies have also supported the existence of these two components [2, 12–15]. The view that resonance fluorescence is a sequential process while resonance Raman is a simultaneous one has been widely accepted but not yet grounded upon convincing argument.

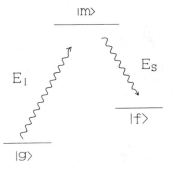

FIGURE 1.   Energy scheme of a three-level system interacting with an incident field E_I and an emitted field E_S. The levels g, m, and f are the initial, intermediate, and final states in RSE, respectively.

A proposition to establish the preceding view was based on the dual Feynman diagrams [16] as shown in Fig. 2. These diagrams represent how the molecular density matrix $\hat{\sigma}$ develops with time in the RSE process. The two horizontal lines in each diagram designate the axes along which time passes from left to right. The upper line indicates the time evolution of the bra vector and the lower line indicates that of the ket vector. The vertices specify the times at which the bra and ket states encounter interaction with the incident or emitted field. For instance, the upper line in the diagram A shows that the initial state $\langle g|$ is perturbed at time t_2 by interaction with the incident light to be transferred to the intermediate state $\langle m|$ and then this bra state is perturbed at t_1 by interaction with the emitted field to end up in the final state $\langle f|$.

In the diagram A, the absorption and emission steps are inseparable. It is also seen that the intermediate state population vanishes, that is, $\langle m|\hat{\sigma}|m\rangle = 0$, while the off-diagonal density matrix element $\langle g|\hat{\sigma}|f\rangle$ (abbreviated hereafter as σ_{gf}) is nonzero. On the other hand, in the diagrams B and C, the absorption and emission steps are separable and the intermediate state is populated, that is, $\sigma_{mm} \neq 0$. Owing to these facts, the idea had prevailed that resonance Raman is the process characterized by the diagram A and

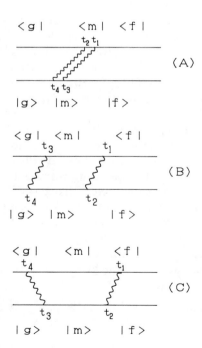

FIGURE 2. Typical dual Feynman diagrams that symbolize time evolution of the molecular density matrix in the RSE process. The two horizontal lines in each diagram denote the axes along which time passes from left to right. The upper line represents the time evolution of the bra vector; the lower line represents that of the ket vector. The intersecting points with the left wavy line indicate interaction with the incident field; the intersecting points with the right one indicate interaction with the emitted field.

resonance fluorescence is the one characterized by the diagrams B and C [2, 17].

Careful examinations [4, 18–21] have however proved that the diagram A generates not only a RL line shape but also a negative-going FL one and that these two are exactly offset each other in integrated intensity or in time resolution measurement (with no frequency resolution). It is also found that the time-resolves spectrum for the diagrams B and C contains, in addition to a FL component (or FL ones), the RL one of which intensity conforms to the temporal change in the incident pulse profile. The interpretation based on the diagrams is puzzling. The diagram that represents a time evolution of the bra and ket states does not necessarily correspond to a physical process. Hence, we conclude that the diagrammatic representation is not an effective method of painting an unpuzzling and vivid picture of RSE, although it is useful in deriving analytical expressions in a systematic way.

Another proposition has been made by Huber [22]. Noting that the transition probability (the line shape function) of RSE can be formulated by Fourier transforming the two-time correlation function of a molecular polarizability operator $\hat{\alpha}(t)$, he has broken up the operator into the thermally averaged term $\langle \hat{\alpha}(t) \rangle$ and the fluctuating term $\delta\hat{\alpha}(t)$. In the Markovian limit, the resulting thermally averaged part of the transition probability function, that is, the Fourier transform of $|\langle \hat{\alpha}(t) \rangle|^2$, exhibits only a RL line shape if the incident light has no spectral broadening (if it is stationary and monochromatic) and the fluctuating part, that is, the Fourier transform of $\langle \delta\hat{\alpha}^\dagger \delta\hat{\alpha}(t) \rangle$, exhibits only a FL one. The latter part is equivalent to what is known as the FL component due to molecular pure dephasing (elastic interaction between the molecule and the heat bath) [12–15]. His approach thus provides important views of RSE, especially in the absence of spectral broadening.

There are an infinite number of ways of writing the transition probability (or the emission spectrum intensity) as a sum of terms. We would like to divide the total expression so that each divided one may be labeled either "fluorescence" or "Raman." However, if the spectral broadening is present, there would exist terms that fall into neither category (interferencelike terms). Although the terminology, "fluorescence" and "Raman," is useful as has been shown in experimental works, one must recognize that these labels are of limited utility in general. We want to clarify in what situations and to what extent classification into "fluorescence" and "Raman" works well.

The points upon which we shall concentrate our thought are listed as follows:

(i) What definitions of "fluorescence" and "Raman" should be adopted in dividing the total expression of the emission spectrum (or alternatives to the emission spectrum)? Are the absorption and emission steps described by the "fluorescence" term(s) separable? Are those described by the "Raman" term inseparable?

(ii) From what kind of molecular states are the FL and RL components arising?

(iii) How is the excited-state population related with the FL and RL components?

(iv) What role do the remaining terms (the interferencelike ones) play?

(v)   In what way does molecular pure dephasing (MPD) induce FL emission?

We use here a novel method of viewing RSE to discuss the preceding points, not clinging to the diagrammatic representation; we divide the correlation function of the emitted field component into several parts so that each divided one generates either a RL or a FL component (or an interferencelike one). Although individual components are not observable, this approach enables us to know what physical factors trigger what spectrum components and to see in perspective the relation among the incident light, the scattered light, and the time evolution of the molecule. It will be shown that the way the total emission correlation function is partitioned is the same whether the incident light is stationary or not. For example, in any excitation condition, a certain divided correlation function characterizes RL emission; for that correlation function, the frequency-resolved spectrum obtained in stationary excitation exhibits only a RL line shape and the time-resolved spectrum (photon counting rate) in pulsed excitation also exhibits only a RL temporal profile. We do not need to alter the partitioning according to whether the incident light is stationary or not.

Particular emphasis should be placed on the point that the form of each correlation function itself reveals whether the corresponding RSE process is sequential or simultaneous. By tracing the time evolution of divided correlation functions, the mechanisms of resonance fluorescence and resonance Raman can be argued more concretely and the validity and limitation of classification into the two categories can be fruitfully discussed. We believe that a full grasp of RSE in the Markovian limit serves as a competent guide in comprehending various aspects of RSE.

When the interaction between the molecule and the heat bath does not belong to the Markovian case, the creation and annihilation of phonons lasts a long time extending from absorption to emission, which smears the separation of resonance Raman and resonance fluorescence. Such cases have been extensively studied by the stochastic [23, 24] or dynamical model [25–27]. (In the latter, the dynamics of the whole system is evaluated without introducing phenomenological parameters such as random frequencies in the stochastic model.) Toyozawa et al. [26c] have, on the basis of the dynamical model, presented the two complementary standpoints ("duality") in understanding the diversity of RSE; that is, RSE can be viewed as a superposition of Raman processes *accompanied by the creation and annihilation of phonons*, and, on the other hand, it can be viewed as an emission process *correlated with absorption*. Investigation of non-Markovian effects is essential to comprehending RSE. We here apply the previously mentioned partition technique to the stochastic model and extract components that do not arise in the Markov approximation. The mechanisms of resonance Raman and resonance fluorescence and the applicability of the two categories are discussed in non-Markovian cases.

This chapter is organized as follows. Sections II–VI are devoted to the analysis of RSE in the Markovian limit. In Section II we briefly review the density matrix of the emitted field for the conventional three-level system, from which the emission correlation function is derived [28, 29]. In Section III we split the emission correlation function into the three types (FL, RL, and interferencelike terms) using a realistic molecular relaxation scheme and a model of the incident light correlation function. In Section IV we analytically or numerically examine what frequency or time resolution spectra or both each correlation function presents. The meanings of the three types of divided correlation functions are discussed in Section V; the origin of FL emission due to MPD is discussed in Section VI. In Section VII, RSE in non-Markovian cases is discussed using the stochastic model. Finally, conclusions are summarized in Section VIII. It is hoped that this article enables one to be free from conceptual confusion pertaining to RSE and to gain common ground for accumulating advanced knowledge of RSE.

II.   THEORETICAL BACKGROUND

We would like to treat resonance secondary emission (RSE) in a molecular system coupled to a heat bath. We assume that the molecule contains three levels as shown in Fig. 1. The energy of each level is given by $\hbar\omega_j$ ($j = g, m$, or f). In the following, $\hbar\omega_{ij}$ denotes the energy difference:

$$\hbar\omega_{ij} = \hbar(\omega_i - \omega_j) \qquad (2.1)$$

It is also assumed that the interaction between the molecule and the heat bath happens in a moment [the Markov (impact) approximation] [30–33]. For example, suppose that the duration of the molecule–perturber collision is very much shorter than the other relaxation times [11]. In this approximation the relaxation processes caused by the interaction are described by rate constants [34–36]. The discussion under the Markov approximation continues until the end of Section VI.

We shall discuss the familiar case that the level spacing between g and f is as large as a molecular vibrational frequency and the level m belongs to an electronically excited state. In this case, the line width in an optical transition is much narrower than the level spacings (the isolated line approximation is applicable) [35–37]: only two kinds of rate constants are needed, namely, the longitudinal relaxation constant Γ_j (the population decay rate of level j) and the molecular pure dephasing (MPD) constant Γ'_{ij} resulting from elastic collision (which disrupts the phase coherence between i and j but does not bring about population decay). The transverse relaxation constant (the overall coherence loss rate) is then given by [13, 14, 38]

$$\Gamma_{ij} = (\Gamma_i + \Gamma_j)/2 + \Gamma'_{ij} \qquad (2.2)$$

which is equal to the line width in the optical transition between i and j.

The electric field component of an emitted field, $E_S^{(+)}$, is expressed by a linear combination of photon creation operators a^+ (Ref. 39)

$$E_S^{(+)} = -i(\hbar/2\varepsilon_0 v)^{1/2} \sum_\lambda \omega_\lambda^{1/2} a_\lambda^+ \qquad (2.3)$$

where λ denotes the emitted mode characterized by frequency ω_λ, ε_0 is the permittivity, and v is the volume of the emitted field (the subscript S is used to refer to the emitted field). It is the emission correlation function G_S that is directly connected to the observed spectrum in RSE

$$G_S(t, t') = \langle E_S^{(+)}(t) E_S^{(-)}(t') \rangle, \qquad (2.4)$$

where $E_S^{(-)}$ is the hermitian adjoint of $E_S^{(+)}$ and $\langle \cdots \rangle$ denotes the ensemble average over all modes of the emitted field. The goal in this section is to derive the expression of G_S for the present model system.

It has been established that the correlation function $G_S(t, t')$ is linked with the spectral distribution of the emitted radiation energy per unit area until time t, $g(\omega, t)$ [29]:

$$G_S(t, t') = \frac{1}{c\varepsilon_0} \int_{-\infty}^{\infty} d\omega \exp\left[i\omega(t - t')\right] \frac{\partial}{\partial t} g(\omega, t) \qquad \text{for } t \geq t' \qquad (2.5)$$

From the definition of $g(\omega, t)$, one can readily note the relation

$$g(\omega, t) = \frac{vh\omega^3}{(2\pi c)^3} K' \langle \omega | \rho_S(t) | \omega \rangle \qquad (2.6)$$

where $\rho_S(t)$ is the density matrix of the emitted field and the diagonal component $\langle\omega|\rho_S(t)|\omega\rangle$ is equivalent to the number of emitted photons with frequency ω. The factor K' is to be determined by the relative configuration of the photodetector and the molecular ensemble.

The density matrix $\rho_S(t)$ is obtained by tracing the solution of the Liouville equation for the total system (molecular system + heat bath + incident and emitted fields) over the variables of the molecular system, heat bath, and incident field. The reduced density matrix $\rho_S(t)$ can be written in terms of a perturbation series of the molecule-radiation field interactions. The fourth-order solution in the perturbative density matrix treatment suffices for the case of weak incident lights and has been derived by a number of researchers [12–15, 18–27]. (It is beyond the scope of this paper to consider strong incident lights [40–42].)

We shall begin with the final result of the fourth-order solution $\rho_S^{(4)}(t)$ [18]

$$\langle\omega|\rho_S^{(4)}(t)|\omega\rangle = \frac{\omega|\mu_{gm}\mu_{mf}|^2}{4\varepsilon_0 v\hbar^3} \int_{-\infty}^{t} dt_1 \int_{-\infty}^{t_1} dt_2 \int_{-\infty}^{t_2} dt_3 \int_{-\infty}^{t_3} dt_4$$

$$\times \{e^{-i\omega(t_1-t_3)}A(t_1,t_2,t_3,t_4) + e^{-i\omega(t_1-t_2)}B(t_1,t_2,t_3,t_4)$$

$$+ e^{-i\omega(t_1-t_2)}C(t_1,t_2,t_3,t_4) + \text{c.c.}\} \tag{2.7}$$

where μ_{ij} denotes the transition dipole matrix element between levels i and j and the three terms $A, B,$ and C are given by

$$A(t_1,t_2,t_3,t_4) = \exp\left[i\omega_{mf}(t_1-t_3) - i\omega_{mg}(t_2-t_4) - \Gamma_{mg}(t_3-t_4)\right.$$

$$\left. - \Gamma_{fg}(t_2-t_3) - \Gamma_{mf}(t_1-t_2)\right]G_I(t_2,t_4) \tag{2.8a}$$

$$B(t_1,t_2,t_3,t_4) = \exp\left[i\omega_{mf}(t_1-t_2) - (i\omega_{mg} - \Gamma_{mg})(t_3-t_4)\right.$$

$$\left. - \Gamma_m(t_2-t_3) - \Gamma_{mf}(t_1-t_2)\right]G_I(t_3,t_4) \tag{2.8b}$$

$$C(t_1,t_2,t_3,t_4) = \exp\left[i\omega_{mf}(t_1-t_2) + (i\omega_{mg} - \Gamma_{mg})(t_3-t_4)\right.$$

$$\left. - \Gamma_m(t_2-t_3) - \Gamma_{mf}(t_1-t_2)\right]G_I(t_4,t_3) \tag{2.8c}$$

with the incident field correlation function $G_I(t,t')$

$$G_I(t,t') = \langle E_I^{(+)}(t)E_I^{(-)}(t')\rangle \tag{2.9}$$

The first three terms in Eq. (2.7) are represented distinctively by the dual Feynman diagrams A, B, and C in Fig. 2. The electric dipole interaction Hamiltonian in the rotating wave approximation has been used in the derivation of Eq. (2.7) [41, 43]. This is justified since the incident lights we are treating are weak and near-resonant with the $g \to m$ transition.

We shall now substitute Eq. (2.7) into Eq. (2.5). The terms A, B, and C in Eq. (2.7) bring forth the corresponding correlation functions G_S^A, G_S^B, and G_S^C:

$$G_S(t, t') = G_S^A(t, t') + G_S^B(t, t') + G_S^C(t, t') \tag{2.10}$$

where these three are expressed as, for $t \geqq t'$,

$$G_S^A(t, t') = K \int_{t'}^{t} dt_1 \int_{-\infty}^{t'} dt_2 A(t, t_1, t', t_2) \tag{2.11a}$$

$$G_S^B(t, t') = K \int_{-\infty}^{t'} dt_1 \int_{-\infty}^{t_1} dt_2 B(t, t', t_1, t_2) \tag{2.11b}$$

$$G_S^C(t, t') = K \int_{-\infty}^{t'} dt_1 \int_{-\infty}^{t_1} dt_2 C(t, t', t_1, t_2) \tag{2.11c}$$

with

$$K = \frac{\omega_{mf}^4 |\mu_{gm}\mu_{mf}|^2}{32\varepsilon_0^2 \hbar^2 \pi^3 c^4} K' \tag{2.12}$$

The factor ω^4 appearing in the integrand of Eq. (2.5) has been replaced with the constant ω_{mf}^4, since the other factor has peaks solely around ω_{mf}.

The sum of G_S^B and G_S^C is written as

$$G_S^{BC}(t, t') \equiv G_S^B(t, t') + G_S^C(t, t')$$
$$= K \exp\left[(i\omega_{mf} - \Gamma_{mf})(t - t')\right] P(t'; \Gamma_m, \Gamma'_{mg} + \tfrac{1}{2}\Gamma_g, \omega_{mg}) \tag{2.13}$$

where $P(t; x, y, z)$ is defined as

$$P(t; x, y, z) \equiv 2 \operatorname{Re} \int_{-\infty}^{t} dt_1 \int_{-\infty}^{t_1} dt_2 \exp\left[- iz(t_1 - t_2)\right.$$
$$\left. - \frac{x}{2}(2t - t_1 - t_2) - y(t_1 - t_2) \right] G_I(t_1, t_2) \tag{2.14}$$

The function $P(t; \Gamma_m, \Gamma'_{mg} + \tfrac{1}{2}\Gamma_g, \omega_{mg}) |\mu_{mg}|^2 / \hbar^2$ is equal to $\sigma_{mm}(t)$, that is, nothing but the population of the intermediate state m at time t [15]. For a stationary incident light, the correlation function G_S^{BC} forms a Lorentzian emission spectrum centered at frequency ω_{mf} with width Γ_{mf}. [If the incident light is pulsed, the temporal change in the intermediate state population brings about an extra width (which is time-dependent).] Since this emission looks like the total fluorescence from the intermediate state m, one might agree to the intuitive view that the diagrams B and C represent the step-by-step process via the state m (which will prove wrong). As will be shown, this type of process however emerges also from the A term.

The A term is an enigma. A naive look at the diagram A in Fig. 2 tempts us to think that the A term produces a Ramanlike (RL) component but does not produce a fluorescencelike (FL) one. If it were the case, the RL and FL components could be uniquely assigned to the dual Feynman diagrams (the diagrammatic interpretation). However, the story does not come to an end. For example, take the fact that the time-resolved spectrum $G_S^A(t, t)$ is nil, that is, all the frequency components arising from the A term are integrated to be missing [note that $\sigma_{mm} = 0$ for the diagram A]. This cancellation connotes that the A term could be split into two or more terms.

We have to add that in the time-resolved spectrum $G_S^{BC}(t, t)$ a RL temporal change is seen as well as a FL component decaying at the longitudinal relaxation rate Γ_m. All frequency components are counted in the time-resolved intensity $G_S^{BC}(t, t)$, since $G_S^A(t, t) = 0$. This is also unfavorable to the diagrammatic interpretation.

Let us subdivide the correlation function G_S^A (in an attempt to extract physical processes involved in G_S):

$$G_S^A(t, t') = G_S^{AR}(t, t') - G_S^{AF}(t, t') \tag{2.15}$$

where

$$G_S^{AR}(t, t') \equiv K \exp\left[i\omega_{mf}(t - t')\right] \int_{-\infty}^{t} dt_1 \int_{-\infty}^{t'} dt_2 \exp\left[-i\omega_{mg}(t_1 - t_2)\right.$$

$$\left. - \Gamma_{mg}(t' - t_2) - \Gamma_{mf}(t - t_1) - \Gamma_{fg}(t_1 - t')\right] G_I(t_1, t_2) \tag{2.16}$$

$$G_S^{AF}(t, t') \equiv K \exp\left[(i\omega_{mf} - \Gamma_{mf})(t - t')\right] \int_{-\infty}^{t'} dt_1 \int_{-\infty}^{t'} dt_2 \exp\left[-i\omega_{mg}(t_1 - t_2)\right.$$

$$\left. - \Gamma_{mg}(t' - t_2) - \Gamma_{mf}(t' - t_1) - \Gamma_{fg}(t_1 - t')\right] G_I(t_1, t_2) \tag{2.17}$$

The correlation function G_S^{AF} gives rise to a FL spectrum similar to that from G_S^{BC}. In the next section we deal with the correlation functions G_S^{AF} and G_S^{BC} in a bundle and subdivide G_S^{AR} into some parts.

III. DIVIDED EMISSION CORRELATION FUNCTIONS

Before recombining or decomposing the emission correlation functions obtained, we make the model of relaxation processes as simple as possible, of course, in conformity with realistic conditions. To begin with, we note that Γ'_{fg} is much smaller than Γ'_{mf} or Γ'_{mg}, since g and f are assumed to belong to the same electronic state (vibrational dephasing is much less rapid than electronic dephasing) [9, 44]. Thus, as usual, we set

$$\Gamma'_{fg} \ll \Gamma'_{mf}; \quad \Gamma'_{fg} \ll \Gamma'_{mg}; \quad \Gamma'_{mf} = \Gamma'_{mg} = \Gamma' \tag{3.1}$$

We also assume that the initial state g is stable:

$$\Gamma_g = 0 \tag{3.2}$$

Then, the width in absorption, Γ_{mg}, is given by

$$\Gamma_{mg} = \tfrac{1}{2}\Gamma_m + \Gamma' \tag{3.3}$$

and that in fluorescence, Γ_{mf}, is given by

$$\Gamma_{mf} = \tfrac{1}{2}(\Gamma_m + \Gamma_f) + \Gamma' \tag{3.4}$$

The present Markovian model is the same as adopted in previous works (Ref. 21).

On the conditions (3.1) and (3.2), we have

$$G_S^{AR}(t, t') = K \exp\left[(i\omega_{mf} - \Gamma_{fg})(t - t')\right] \int_{-\infty}^{t} dt_1 \int_{-\infty}^{t'} dt_2$$

$$\exp\left[-i\omega_{mg}(t_1 - t_2) - \Gamma_{mg}(t + t' - t_1 - t_2)\right] G_I(t_1, t_2) \tag{3.5}$$

$$G_S^{PF}(t, t') \equiv G_S^{BS}(t, t') - G_S^{AF}(t - t')$$

$$= K \exp\left[(i\omega_{mf} - \Gamma_{mf})(t - t')\right] \{P(t'; \Gamma_m, \Gamma', \omega_{mg}) - P(t'; 2\Gamma_{mg}, 0, \omega_{mg})\} \tag{3.6}$$

of which sum is the total correlation function

$$G_S(t, t') = G_S^{AR}(t, t') + G_S^{PF}(t, t') \tag{3.7}$$

When spectral broadening of the incident light is absent, the correlation functions G_S^{AR} and G_S^{PF} are reduced, respectively, to the average and fluctuating parts in the RSE transition probability function given by Huber [22]. The newly defined correlation function G_S^{PF} is not susceptible of the conventional diagrammatic representation. The compact form of G_S^{PF} facilitates its interpretation as will be discussed in Section VI. It should be emphasized that G_S^{PF} generates always a positive FL component alone *in any excitation condition* and is hence thought to reflect a "real" physical process. This FL component is ascribed to MPD since $G_S^{P\dot{F}}$ vanishes if the MPD constant Γ' is zero.

The correlation function G_S^{AR} to be subdivided includes not only a RL component but still also includes FL ones due to spectral broadening. The origin of spectral broadening is twofold. In general, a radiation field does not possess perfect first-order coherence [45]: the phase of the light wave fluctuates because of, for example, collisions between the radiating molecules in the light source. The phase fluctuation, which is called "light-induced dephasing" in Ref. 21, causes a broadening to the spectrum of incident light. Let us call this the spectral broadening due to phase fluctuation. In pulsed excitation, the time variation in light amplitude brings about "Fourier broadening" effect. The temporal profile of a pulse amplitude is expressed by a linear combination of continuous waves; the coefficient of each frequency component is determined by Fourier transform of the pulse amplitude [as shown in the forthcoming Eq. (3.30)]. It is known that the frequency distribution spreads over about the size of the inverse pulse duration time. To incorporate these broadening mechanisms into our theory, we use the following form [28, 36, 46]

$$G_I(t, t') = \exp[i\omega_I(t - t')]\, \phi(t)\phi(t')\xi(|t - t'|) \tag{3.8}$$

where ω_I is the central frequency, $\phi(t)$ the pulse amplitude, and $\xi(|t - t'|)$ the stationary correlation function that describes phase fluctuation (it is assumed that the phase fluctuation process is stationary) [36].

Substituting Eq. (3.8) into Eq. (3.5) and dividing the area of integration therein, we have

$$G_S^{AR}(t, t') = G_S^A(t, t') + G_S^{AF}(t, t') \tag{3.9}$$

where, with the detuning frequency $\Delta_I \equiv \omega_{mg} - \omega_I$,

$$G_S^A(t, t') = K \exp[(i\omega_{mf} - \Gamma_{fg})(t - t')] \int_{t'}^{t} dt_1 \int_{-\infty}^{t'} dt_2 \exp[-i\Delta_I(t_1 - t_2)$$

$$- \Gamma_{mg}(t + t' - t_1 - t_2)]\xi(|t_1 - t_2|)\phi(t_1)\phi(t_2) \tag{3.10}$$

$$G_S^{AF}(t,t') = K \exp\left[(i\omega_{mf} - \Gamma_{mf})(t-t')\right] \int_{-\infty}^{t'} dt_1 \int_{-\infty}^{t'} dt_2 \exp\left[-i\Delta_I(t_1 - t_2)\right.$$

$$\left. - \Gamma_{mg}(2t' - t_1 - t_2)\right]\xi(|t_1 - t_2|)\phi(t_1)\phi(t_2) \tag{3.11}$$

We now want to extract from Eq. (3.11) the FL component due to phase fluctuation. For this purpose, it is helpful to specify the function ξ, say,

$$\xi(|\tau|) = \exp\left[-\Gamma_I|\tau|\right] \tag{3.12}$$

In this case, Eq. (3.11) is written as

$$G_S^{AF}(t,t') = K \exp\left[(i\omega_{mf} - \Gamma_{mf})(t-t')\right] P'(t'; 2\Gamma_{mg}, \Gamma_I, \Delta_I) \tag{3.13}$$

where

$$P'(t; x, y, z) \equiv 2\,\mathrm{Re} \int_{-\infty}^{t} dt_1 \int_{-\infty}^{t_1} dt_2 \exp\left[-iz(t_1 - t_2)\right.$$

$$\left. - \frac{x}{2}(2t - t_1 - t_2) - y(t_1 - t_2)\right]\phi(t_1)\phi(t_2) \tag{3.14}$$

Comparing Eq. (3.13) with Eq. (2.13), we find that the phase disruption rate Γ_I of the incident light has an analogy to the MPD constant Γ'. The FL component due to phase fluctuation may be disclosed by a subtraction like in Eq. (3.6). We therefore introduce a function of the form

$$G_D'(t,t') = K \exp\left[(i\omega_{mf} - \Gamma_{mf})(t-t')\right] X(t') \tag{3.15}$$

and subtract it from Eq. (3.13):

$$G_S^{EF}(t,t') \equiv G_S^{AF}(t,t') - G_D'(t,t')$$

$$= K \exp\left[(i\omega_{mf} - \Gamma_{mf})(t-t')\right]\{P'(t'; 2\Gamma_{mg}, \Gamma_I, \Delta_I) - X(t')\} \tag{3.16}$$

The as yet unknown function $X(t)$ will be determined later so that G_S^{EF} disappears when $\Gamma_I = 0$ (so that Γ_I is responsible for the appearance of G_S^{EF}).

The residual part $\tilde{G}_S$ is thus given by adding G_D' to G_S^A

$$\tilde{G}_S(t,t') \equiv G_S^A(t,t') + G_D'(t,t') \tag{3.17}$$

Upon substituting Eq. (3.12) into Eq. (3.10) and making use of integration by parts, we find

$$G_S^A(t, t') = \frac{K[(\Delta_I^2 + \Gamma_{mg}^2 - \Gamma_I^2) + 2i\Gamma_I\Delta_I]}{(\Delta_I^2 + \Gamma_{mg}^2 - \Gamma_I^2)^2 + 4\Gamma_I^2\Delta_I^2}$$
$$\times \{\phi(t)\phi(t')\exp[(i\omega_{mf} - i\Delta_I - \Gamma_{fg} - \Gamma_I)(t - t')]$$
$$- \phi(t')\phi(t')\exp[(i\omega_{mf} - \Gamma_{mf})(t - t')]\} + G_S^{AT}(t, t') \tag{3.18}$$

where

$$G_S^{AT}(t, t') = \frac{K \exp[(i\omega_{mf} - \Gamma_{mf})(t - t') - 2\Gamma_{mg}t']}{(i\Delta_I + \Gamma_{mg} + \Gamma_I)(-i\Delta_I + \Gamma_{mg} - \Gamma_I)}$$
$$\times \{[\phi(t')e^{(-i\Delta_I + \Gamma_{mg} - \Gamma_I)t'} - \phi(t)e^{(-i\Delta_I + \Gamma_{mg} - \Gamma_I)t}]\Phi_1(t', \Gamma_I)$$
$$- \phi(t')e^{(-i\Delta_I + \Gamma_{mg} - \Gamma_I)t'}\phi_2(t, t') + \Phi_1(t', \Gamma_I)\Phi_2(t, t')\} \tag{3.19}$$

with the definitions

$$\Phi_1(t, \gamma) \equiv \int_{-\infty}^{t} d\tau \frac{d\phi(\tau)}{d\tau} \exp[(i\Delta_I + \Gamma_{mg} + \gamma)\tau] \tag{3.20a}$$

$$\Phi_2(t, t') \equiv \int_{t'}^{t} d\tau \frac{d\phi(\tau)}{d\tau} \exp[(-i\Delta_I + \Gamma_{mg} - \Gamma_I)\tau] \tag{3.20b}$$

We add G_D' to Eq. (3.18) according to Eq. (3.17), then splitting up the sum into three parts

$$\tilde{G}_S(t, t') = G_S^R(t, t') + G_S^{XT}(t, t') + G_S^M(t, t') \tag{3.21}$$

where

$$G_S^R(t, t') = \frac{K[(\Delta_I^2 + \Gamma_{mg}^2 - \Gamma_I^2)\phi(t)\phi(t')}{(\Delta_I^2 + \Gamma_{mg}^2 - \Gamma_I^2)^2 + 4\Gamma_I^2\Delta_I^2} \exp[(i\omega_{mf} - i\Delta_I - \Gamma_{fg} - \Gamma_I)(t - t')] \tag{3.22}$$

$$G_S^{XT}(t, t') = K \left\{ X(t') - \frac{(\Delta_I^2 + \Gamma_{mg}^2 - \Gamma_I^2)\phi(t')\phi(t')}{(\Delta_I^2 + \Gamma_{mg}^2 - \Gamma_I^2)^2 + 4\Gamma_I^2\Delta_I^2} \right\}$$
$$\times \exp[(i\omega_{mf} - \Gamma_{mf})(t - t')] + G_S^{AT}(t, t') \tag{3.23}$$

$$G_S^M(t, t') = \frac{2K\Gamma_I\Delta_I i}{(\Delta_I^2 + \Gamma_{mg}^2 - \Gamma_I^2)^2 + 4\Gamma_I^2\Delta_I^2}$$
$$\times \{\phi(t)\phi(t')\exp[(i\omega_{mf} - i\Delta_I - \Gamma_{fg} - \Gamma_I)(t - t')]$$
$$- \phi(t')\phi(t')\exp[(i\omega_{mf} - \Gamma_{mf})(t - t')]\} \tag{3.24}$$

It will be shown that G_S^R generates a RL component and G_S^M generates a mixture of RL and FL characters. The appearance of the interferencelike term G_S^M is attributed to phase fluctuation Γ_I.

The pending function $X(t)$ is present in G_S^{EF} and in G_S^{XT}. We look for a form which satisfies the following two conditions. The first one, as mentioned before, is that G_S^{EF} should be zero when $\Gamma_I = 0$. The second one is that G_S^{XT} should vanish when the incident light is stationary; that is, $X(t)$ is to be chosen so that for a stationary incident light the FL component due to phase fluctuation arises only from G_S^{EF}. Although $X(t)$ cannot be uniquely determined by the preceding two conditions alone, we choose here the following form as $X(t)$ (by using hindsight)

$$X(t) = \frac{(\Delta_I^2 + \Gamma_{mg}^2 - \Gamma_I^2)(\Delta_I^2 + \Gamma_{mg}^2)}{(\Delta_I^2 + \Gamma_{mg}^2 - \Gamma_I^2)^2 + 4\Gamma_I^2\Delta_I^2} P'(t; 2\Gamma_{mg}, 0, \Delta_I) \tag{3.25}$$

It will be demonstrated in the next section how reasonable this choice is. Inserting Eq. (3.25) into Eq. (3.23), one can decompose G_S^{XT} as follows:

$$G_S^{XT}(t, t') = G_S^{TF}(t, t') + G_S^{T}(t, t') \tag{3.26}$$

where

$$T_S^{TF}(t, t') = \frac{K(\Delta_I^2 + \Gamma_{mg}^2 - \Gamma_I^2)}{(\Delta_I^2 + \Gamma_{mg}^2 - \Gamma_I^2)^2 + 4\Gamma_I^2\Delta_I^2}$$
$$\times \exp[(i\omega_{mf} - \Gamma_{mf})(t - t') - 2\Gamma_{mg}t']|\Phi_1(t', 0)|^2 \tag{3.27}$$

$$G_S^{T}(t, t') = \frac{K(\Delta_I^2 + \Gamma_{mg}^2 - \Gamma_I^2)}{(\Delta_I^2 + \Gamma_{mg}^2 - \Gamma_I^2)^2 + 4\Gamma_I^2\Delta_I^2} \exp[(i\omega_{mf} - \Gamma_{mf})(t - t')]$$
$$\times \{-2\phi(t')e^{-\Gamma_{mg}t'} \operatorname{Re}[\Phi_1(t', 0)e^{-i\Delta_I t'}]\} + G_S^{AT}(t, t') \tag{3.28}$$

The final form of G_S^{EF} is written as [see Eq. (3.16)]

$$G_S^{EF}(t, t') = K \exp[(i\omega_{mf} - \Gamma_{mf})(t - t')]\left\{P'(t'; 2\Gamma_{mg}, \Gamma_I, \Delta_I)\right.$$
$$\left. - \frac{(\Delta_I^2 + \Gamma_{mg}^2 - \Gamma_I^2)(\Delta_I^2 + \Gamma_{mg}^2)}{(\Delta_I^2 + \Gamma_{mg}^2 - \Gamma_I^2)^2 + 4\Gamma_I^2\Delta_I^2} P'(t'; 2\Gamma_{mg}, 0, \Delta_I)\right\} \tag{3.29}$$

The correlation function G_S^{TF} represents the FL component due to Fourier broadening. Replacing the pulse amplitude $\phi(t)$ with its Fourier transform

$$\phi(t) = \int_{-\infty}^{\infty} d\omega \, A(\omega) e^{-i\omega t} \tag{3.30}$$

we have

$$\Phi_1(t,0) = -i \int_{-\infty}^{\infty} d\omega \frac{\omega A(\omega)}{i(\Delta_I - \omega) + \Gamma_{mg}} e^{(i\Delta_I - i\omega + \Gamma_{mg})t} \tag{3.31}$$

If the pulse amplitude is a slowly varying function of time, $A(\omega)$ is peaked around $\omega = 0$ [in the stationary limit, $A(\omega) = \phi_0 \delta(\omega)$], then, $\Phi_1(t,0)$ in Eq. (3.27) is small. The value of $\Phi_1(t,0)$ is large if $A(\omega)$ is so wide-spreading that it has appreciable values around $\omega = \Delta_I$. The width of the frequency distribution function $A(\omega)$ becomes larger as the pulse amplitude changes more drastically.

In summary, we have split the emission correlation function into six parts, that is, three FL components (G_S^{PF}, G_S^{EF}, and G_S^{TF}), one RL component (G_S^R), and two interferencelike ones (G_S^M and G_S^T):

$$G_S = G_S^{PF} + G_S^{EF} + G_S^{TF} + G_S^R + G_S^M + G_S^T \tag{3.32}$$

which will undergo close examination in the following sections. Although the division is artificial (not unique) except the separation between G_S^{PF} and the others, each correlation function has distinctive characters. For example, in stationary excitation, G_S^{TF} and G_S^T are obviously zero.

Among three FL components there are the ones due to MPD (G_S^{PF}), due to phase fluctuation (G_S^{EF}), and due to Fourier broadening (G_S^{TF}). It is useful to present how their fluorescence intensities fall off after the incident light has fully decayed:

$$\frac{dG_S^{PF}(t,t)}{dt} = -\Gamma_m G_S^{PF}(t,t) + 2\Gamma'[G_S^{EF}(t,t) + G_S^{TF}(t,t)] \tag{3.33a}$$

$$\frac{dG_S^{EF}(t,t)}{dt} = -2\Gamma_{mg} G_S^{EF}(t,t) \tag{3.33b}$$

$$\frac{dG_S^{TF}(t,t)}{dt} = -2\Gamma_{mg} G_S^{TF}(t,t) \tag{3.33c}$$

The total fluorescence intensity (the sum of the three components) decays at the longitudinal relaxation rate Γ_m but each FL component does not because of MPD. A detailed discussion is presented in Section VI.

IV. ANALYTICAL AND NUMERICAL RESULTS

In this section we illustrate what spectra the six correlation functions obtained in Section III present. Both stationary and transient responses are examined; (i) the stationary excitation in which the incident light amplitude is given by a constant ϕ_0

$$G_I(t, t') = \phi_0^2 \exp[i\omega_I(t - t') - \Gamma_I|t - t'|] \tag{4.1}$$

and (ii) the pulsed excitation in which

$$G_I(t, t') = \phi(t)\phi(t') \exp[i\omega_I(t - t') - \Gamma_I|t - t'|] \tag{4.2}$$

The pulse amplitude $\phi(t)$ takes here the double exponential form [47]

$$\phi(t) = \begin{cases} \phi_0 [\exp(-\gamma_1 t) - \exp(-\gamma_2 t)], & t \geq 0 \\ 0, & t < 0 \end{cases} \tag{4.3}$$

For stationary excitation, the emission can be fully frequency-resolved, in principle. This kind of spectrum, $L(\omega)$, is usually calculated from the emission correlation function by making use of the Wiener–Khintchine relation [28, 39]. For pulsed excitation, two kinds of experiments can be carried out: perfect time-resolution experiment (that asks only for time resolution); time- and frequency-resolution experiments. The fully time-resolved spectrum $I(t)$ which is obtained by measuring the instantaneous power of emission with a "white" photodetector is nothing but the emission intensity $G_S(t, t)$.

Eberly and Wódkiewicz [28] have formulated the time-dependent "physical spectrum" on the basis of the "actual" counting rate of a photodetector. This generalized spectrum is applicable to those experiments that require both time and frequency resolution. When the tunable element, the filter, in a detector is a Fabry–Perot interferometer, it acts to average the emission signal both over its Lorentzian passband width Γ_D around the "setting" frequency ω_D and over its characteristic response time $1/\Gamma_D$ (the response function exponentially decays with time): the photon counting rate at time t, $R(t; \omega_D, \Gamma_D)$, takes the form [28]

$$R(t; \omega_D, \Gamma_D) = 2\Gamma_D \operatorname{Re} \int_{-\infty}^{t} dt_1 \int_{-\infty}^{t_1} dt_2 \exp[-(\tfrac{1}{2}\Gamma_D - i\omega_D)(t - t_1)$$
$$- (\tfrac{1}{2}\Gamma_D + i\omega_D)(t - t_2)]G_S(t_1, t_2) \tag{4.4}$$

This expression has been already applied to problems of RSE [20, 29, 48, 49] and is our mainstay in calculating time- and frequency-resolved spectra. In order to obtain a "good" time- and frequency-resolved spectrum, the spectral filter must satisfy two mutually contradictory requirements at once; a temporal response fast enough to track the emission in time domain (a large Γ_D is desired) and a frequency response sharp enough to resolve the emission in frequency domain or to resolve RL and FL components (a small Γ_D is desired). A compromise is necessary in adjusting the value of Γ_D to the purpose. It would not be amiss to point out that Eq. (4.4) is proportional to the population of an excited state with longitudinal relaxation constant Γ_D and frequency ω_D, that is, $P(t; \Gamma_D, 0, \omega_D)$ [of course, the driving field in Eq. (2.14) is replaced with the emitted field].

We calculate the above three types of spectra for each correlation function. Numerical examples are presented in Figs. 3–5. The parameters for widths are expressed in units of Γ_m. Figures 3a–3c display fully frequency-resolved spectra. The units in line shape intensity are arbitrary but the same throughout Figs. 3a–3c. The parameters used are as follows: in Fig. 3a, $\Gamma_f = 0.01$, $\Gamma' = 0.1$, $\Gamma'_{fg} = 0.001$, $\Gamma_I = 1$, and $\Delta_I = 5$; in Fig. 3b, the same as in Fig. 3a except the larger MPD constants ($\Gamma' = 1$, $\Gamma'_{fg} = 0.01$); in Fig. 3c the same as in Fig. 3a except the larger phase fluctuation ($\Gamma_I = 10$). Figures 4a–4c display

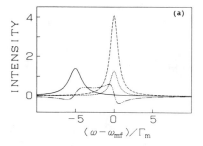

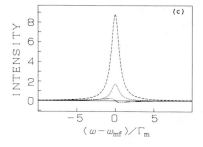

FIGURE 3. Numerical examples of the fully frequency-resolved spectrum. The solid, dotted, broken, and dashed-double dotted line designate, respectively, the G_S^R, G_S^{PF}, G_S^{EF}, and G_S^M components. The parameters used are as follows (in units of Γ_m): (a) $\Gamma_f = 0.01$, $\Gamma' = 0.1$, $\Gamma'_{fg} = 0.001$, $\Gamma_I = 1$, and $\Delta_I = 5$; (b) the same as in Fig. 3a except the larger MPD constants ($\Gamma' = 1$ and $\Gamma'_{fg} = 0.01$); (c) the same as in Fig. 3a except $\Gamma_I = 10$.

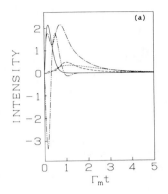

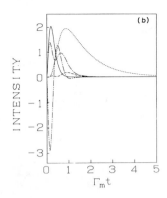

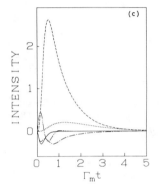

FIGURE 4. Numerical examples of the fully time-resolved spectrum for the double exponential pulse ($\gamma_1 = 5.001$, $\gamma_2 = 1.1\gamma_1$). The solid, dotted, broken, dashed-dotted, and dashed-double dotted lines designate, respectively, the G_S^R, G_S^{PF}, G_S^{EF}, G_S^{TF}, and G_S^T components. The G_S^R component changes according to the pulse profile. The parameters in Figs. 4a–4c are identical with those in Figs. 3a–3c.

fully time-resolved spectra for the double exponential pulse ($\gamma_1 = 5.001$ and $\gamma_2 = 1.1\gamma_1$). The width parameters used in Figs. 4a–4c are, respectively, identical with those in Figs. 3a–3c. The emission intensities in Figs. 4a–4c are plotted in the same units. The solid, dotted, broken, and dashed-dotted lines in Figs. 3 and 4 designate the G_S^R, G_S^{PF}, G_S^{EF}, and G_S^{TF} components, respectively (in Fig. 3, G_S^{TF} is zero); the dashed-double dotted lines in Figs. 3a–3c indicate the G_S^M component and those in Figs. 4a–4c indicate the G_S^T component.

Time- and frequency-resolved spectra for the six correlation functions are drawn in Figs. 5a–5f. The width and pulse parameters (for the double exponential pulse) are identical with those in Fig. 4a. The passband width Γ_D is taken to be $\Gamma_D = 2\Gamma_m$. The incident pulse profile appears to the right of each spectrum.

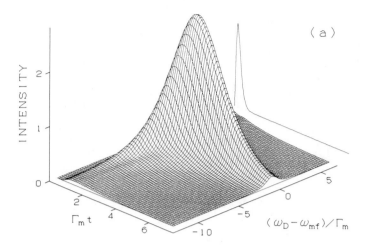

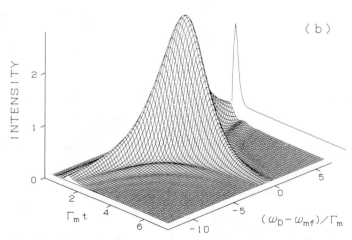

FIGURE 5. Time- and frequency-resolved spectra for the six correlation functions: (a) G_S^{PF}; (b) G_S^{EF}; (c) G_S^{TF}; (d) G_S^R; (e) G_S^M; (f) G_S^T. In all the spectra, the width and pulse parameters (for the double exponential pulse) are identical with those in Fig. 4a. The passband width Γ_D is set to be $\Gamma_D = 2\Gamma_m$. The incident pulse profile appears to the right of each spectrum.

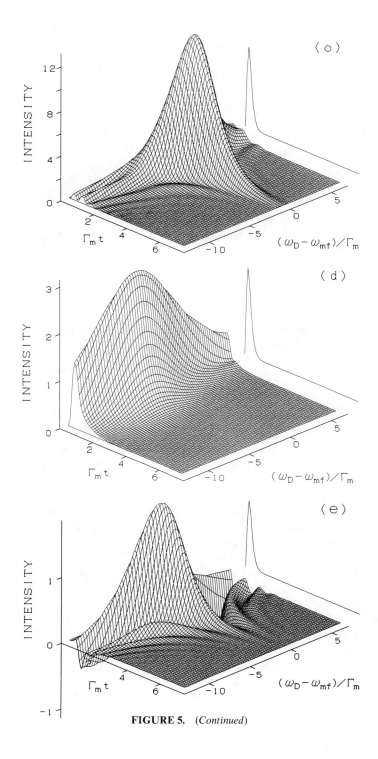

FIGURE 5. (*Continued*)

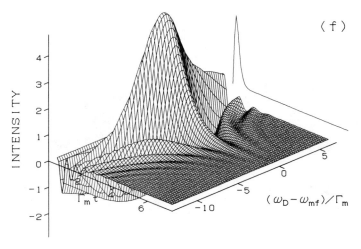

FIGURE 5. (*Continued*)

A. The Fluorescencelike Component due to Molecular Pure Dephasing: The Correlation Function G_S^{PF}

The fully frequency-resolved spectrum for G_S^{PF} becomes Lorentzian

$$L^{PF}(\omega) = \frac{\Gamma_{mf}}{(\omega - \omega_{mf})^2 + \Gamma_{mf}^2} \left\{ \frac{(\Gamma_{mg} + \Gamma_I)}{\Delta_I^2 + (\Gamma_{mg} + \Gamma_I)^2} \right\} \frac{4\Gamma'}{\Gamma_m \Gamma_{mg}} \qquad (4.5)$$

This line shape is of FL emission from the intermediate state m. The spectrum intensity of this component increases linearly with the MPD constant Γ' when Γ' is relatively small. Numerically calculated line shapes are displayed in Figs. 3a–3c (dotted line). Figure 3a compared with Fig. 3b, we note that as Γ' becomes larger this component increases relative to the other components. The variation in the MPD constant Γ' influences G_S^{PF} (and G_S^{EF}) more strongly than the other components G_S^R and G_S^M.

As shown by the dotted lines in Figs. 4a–4c, the fully time-resolved intensity for G_S^{PF}, $I^{PF}(t)$, grows monotonically until the incident light intensity fully decays (until $t \simeq 1.0/\Gamma_m$), following which it attenuates. [The pulse profile $|\phi(t)|^2$ takes the same shape as indicated by the solid line in Fig. 4a or 4b.] It should be pointed out that this component increases with the accumulated pulse energy $\int_{-\infty}^{t} |\phi(\tau)|^2 d\tau$ and decreases slower than at rate Γ_m because of MPD [see Eq. (3.33a)]. These findings suggest the idea that the component arising from G_S^{PF} is ascribed to a joint effect of MPD and incident light. The detailed discussion is presented in Section VI.

Figure 5a depicts a time- and frequency-resolved spectrum for G_S^{PF}. (The width and pulse parameters in Figs. 5a–5f are identical with those in Fig. 4a.) In the frequency domain, this spectrum (symmetric around the central frequency ω_{mf}) is broader than the Lorentzian of the stationary width Γ_{mf}, since an extra width is superimposed on the stationary width owing both to temporal change in $I^{PF}(t)$ and to the frequency average over passband width Γ_D. [Note that $G_S^{PF}(t, t') = I^{PF}(t)e^{(i\omega_{mf} - \Gamma_{mf})(t - t')}$.] At early times the spectrum shows broader line shapes, which reflects the relatively rapid change in $I^{PF}(t)$ until $t \simeq 1.0/\Gamma_m$. Swinging around to the time domain, we note that the frequency components near ω_{mf} fall off slower than the fully time-resolved spectrum $I^{PF}(t)$ on account of the finite response time in detection (cf. the dotted line in Fig. 4a). As the setting frequency is tuned far from ω_{mf}, the frequency component decays faster: the central components survive eventually. That is, as time goes by, the line shape in the frequency domain becomes narrower and approaches the stationary one.

B. The Fluorescencelike Components due to Spectral Broadening: The Correlation Functions G_S^{EF} and G_S^{TF}

We must point out here that a line cannot be drawn between the Fourier broadening effect and the broadening effect due to phase fluctuation if both coexist. As a matter of fact, the Fourier broadening effect is not completely eliminated in G_S^{EF}. These two broadening effects are not thoroughly independent. Moreover, the two correlation functions G_S^{EF} and G_S^{TF} are indistinguishable in the following senses: both have the factor $\exp[(i\omega_{mf} - \Gamma_{mf})(t - t')]$ that generates a FL component; their time-resolved intensities diminish at the identical rate Γ_{mg}, as shown in Eqs. (3.33b) and (3.33c). The idea may be proposed that these two correlation functions should be put together. We however examine each correlation function individually throughout the article, since thereby one can readily discuss the behavior of the combined correlation function.

1.   The Correlation Function G_S^{EF}

It has been known that spectral broadening due to phase fluctuation is effective in introducing FL features in the emission spectrum [18, 21]. We further analyze the role of phase fluctuation using the present classification. The fully frequency-resolved spectrum for G_S^{EF} exhibits a Lorentzian of FL emission

$$L^{EF}(\omega) = \frac{\Gamma_{mf}}{(\omega - \omega_{mf})^2 + \Gamma_{mf}^2} \left\{ \frac{2\Gamma_I}{[\Delta_I^2 + (\Gamma_{mg} + \Gamma_I)^2]\Gamma_{mg}} \right.$$

$$\left. + \frac{4\Gamma_I(\Gamma_I - \Gamma_{mg})}{(\Delta_I^2 + \Gamma_{mg}^2 - \Gamma_I^2)^2 + 4\Gamma_I^2\Delta_I^2} \right\} \tag{4.6}$$

The broken lines in Figs. 3a–3c show how L^{EF} varies with parameters. The intensity of L^{EF} becomes smaller with increasing Γ', as is demonstrated in comparison between Figs. 3a and 3b. This is interpreted as follows: the absorption line shape gets wider with increasing Γ', the spectral broadening due to phase fluctuation thus become difficult to cover the whole range of the absorption line shape. If Γ_I is larger than the other width parameters as in Fig. 3c, the line shape of the total intensity is governed by L^{EF}.

Fully time resolved spectra for G_S^{EF}, $I^{EF}(t)$, are drawn in Figs. 4a–4c (broken line). It is not at the longitudinal relaxation rate Γ_m but at the rate $2\Gamma_{mg}$ that the emission intensity I^{EF} falls off after the passage of the incident pulse [cf. Eq. (3.33b)]: the spectrum $I^{EF}(t)$ decays faster than $I^{PF}(t)$.

The time- and frequency-resolved spectrum for G_S^{EF} is drawn in Fig. 5b. This spectrum is wider than Γ_{mf} and shows at early times broad features, as in the case of G_S^{PF}. The appearance of small undulation is understandable because the time variation in $I^{EF}(t)$ is slightly faster than in $I^{PF}(t)$ as seen in Fig. 4a $[G_S^{EF}(t, t') = I^{EF}(t')e^{(i\omega_{mf} - \Gamma_{mf})(t - t')}]$.

If the spectrum intensity for each correlation function is in any case positive, the idea is acceptable that each correlation function reflects a "real" physical process. The function $L^{EF}(\omega)$ is positive unless $\Gamma_I < \Gamma_{mg}$. The problem lies in the narrow band excitation limit $(\Gamma_I \ll \Gamma_{mg})$ where the value of $L^{EF}(\omega)$ can be negative. This defect may be inevitable, since in this limit the spectral broadening due to phase fluctuation is too small to cover the absorption width Γ_{mg}.

The problem is however not serious if the line shape of the RL component, L^R, which is given later, is taken into account. In the narrow band excitation limit, the ratio in integrated intensity of L^R to L^{EF} becomes

$$L^R : L^{EF} = 1 : \frac{\Gamma_I}{\Gamma_{mg}} \left\{ 1 - \frac{2\Gamma_{mg}^2}{\Delta_I^2 + \Gamma_{mg}^2} \right\} \qquad (4.7)$$

It follows that the FL component due to phase fluctuation is, irrespective of its sign, buried in the RL component (L^{EF} does not affect the overall spectral features). This is also the case for the time-resolved or the time- and frequency-resolved spectrum for G_S^{EF}. To conclude, when the component due to phase fluctuation goes negative, it does not affect the overall spectral features.

2.   The Correlation Function F_S^{TF}

The correlation function G_S^{TF} is in general expressed as

$$G_S^{TF}(t, t') = I^{TF}(t') \exp[(i\omega_{mf} - \Gamma_{mf})(t - t')] \qquad (4.8)$$

where I^{TF} is the fully time-resolved spectrum. As is shown in Figs. 4a and 4b, the value of I^{TF} (see the dashed-dotted lines) bounces up at the times when the pulse profile changes drastically, that is, at the turn-on time $t = 0$ and at the turn-off time τ_p, dropping off at rate $2\Gamma_{mg}$ after the termination of the incident light. From comparison between Figs. 4a and 4b, it turns out that this component is weakened as the absorption line shape gets broader. In the broad-band excitation limit the quantity I^{TF} can be negative, but it is so small that it merges into the FL component due to phase fluctuation as shown in Fig. 4c. According to our classification, the Fourier broadening effect is mostly absorbed into G_S^{EF} than into G_S^{TF} when Γ_I is larger than the other widths.

The time- and frequency-resolved spectrum is shown in Fig. 5c. In a region near ω_{mf} the frequency components change much more smoothly than as anticipated from Fig. 4a where $I^{TF}(t)$ shows conspicuous undulation. That is because the temporal response of the filter is not fast enough to resolve the spectra undulation in time (in the present case, $1/\Gamma_D = 0.5/\Gamma_m$).

C.   The Ramanlike Component: The Correlation Function G_S^R

The fully frequency-resolved spectrum for G_S^R takes a Lorentzian form

$$L^R(\omega) = \frac{\Gamma_{fg} + \Gamma_I}{(\omega - \omega_{mf} + \Delta_I)^2 + (\Gamma_{fg} + \Gamma_I)^2} \frac{2(\Delta_I^2 + \Gamma_{mg}^2 - \Gamma_I^2)}{(\Delta_I^2 + \Gamma_{mg}^2 - \Gamma_I^2)^2 + 4\Gamma_I^2\Delta_I^2} \qquad (4.9)$$

This component has the central frequency of $\omega_I - \omega_{fg}$, and the width of $\Gamma_{fg} + \Gamma_I$ as shown by the solid lines in Figs. 3a and 3b, indicating energetic correlation with the incident light. The fully time-resolved spectra for G_S^R in Figs. 4a and 4b (solid line) follow the incident pulse profile, indicating perfect temporal correlation with the incident light. The component arising from G_S^R constitutes the characteristic features of resonance Raman.

The time- and frequency-resolved spectrum for G_S^R shown in Fig. 5d is broader than the width of the pulse profile (on account of the finite response time of the detector). In the frequency domain, the spectrum is centered at the "Raman" frequency $\omega_{mf} - \Delta_I$ ($= -5\Gamma_m$ for the present set of parameters) but much broader than the stationary "Raman" width $\Gamma_{fg} + \Gamma_I$. The broad line shape mainly echoes the Fourier broadening of the incident pulse.

In the broad band excitation limit where $\Gamma_I \gg \Gamma_{mg}$ and $\Gamma_I \gg \Delta_I$ (for "Δ_I" in the inequalities read "$|\Delta_I|$"), the frequency-resolved intensity L^R goes negative. Let us here remind the reader that in this limit the FL component due to phase fluctuation [$L^{EF}(\omega)$ in Eq. (4.6)] overweighs L^R

$$L^{EF}:L^R = 1: -\Gamma_{mg}/\Gamma_I \qquad (4.10)$$

Similarly, in the same limit, the fully time-resolved spectrum for G_S^R, $I^R(t)$, has a negative sign but it is much smaller than $I^{EF}(t)$. It is demonstrated in Figs. 3c and 4c, where the broad band excitation limit is nearly the case, the G_S^R contributes little to the entire spectral intensity; FL components are dominant as pointed out by Heitler (for the fully frequency-resolved spectrum) [50]. The unphysical negative sign in the broad-band excitation limit is not so obstructive of considering that G_S^R characterizes resonance Raman.

As Γ' or Γ_I becomes larger the RL component diminishes. This may be explained by the "sandwich" effect [21]. The key is to realize that the "absorption" time becomes shorter with increasing Γ' or Γ_I. [The "absorption" time may be defined as the difference between the time at which the molecule interacts with the incident light in bra space and the corresponding time in ket space (like between t_3 and t_4 in the diagram B in Fig. 2); it is inversely proportional to Γ' or Γ_I]. There is less opportunity for an emission event to occur during shorter absorption times. In other words, there is less correlation between absorption and emission, which results in the weakening of RL emission relative to FL emission.

D.   The Interferencelike Components: The Correlation Functions G_S^M and G_S^T

Of two interferencelike terms, only G_S^M survives for stationary excitation. It will be seen that the spectra for the interferencelike components take both positive and negative values (in each spectrum), unlike the four correlation functions examined so far.

1.   The Correlation Function G_S^M

The fully frequency-resolvd spectrum for G_S^M (dashed-double dotted line in Figs. 3a–3c) is

$$L^M(\omega) = \frac{4\Gamma_I \Delta_I}{(\Delta_I^2 + \Gamma_{mg}^2 - \Gamma_I^2)^2 + 4\Gamma_I^2 \Delta_I^2} \left\{ \frac{\omega - \omega_{mf} + \Delta_I}{(\omega - \omega_{mf} + \Delta_I)^2 + (\Gamma_{fg} + \Gamma_I)^2} \right.$$
$$\left. - \frac{\omega - \omega_{mf}}{(\omega - \omega_{mf})^2 + \Gamma_{mf}^2} \right\} \tag{4.11}$$

This line shape function may be considered the difference between the RL and FL *emission processes* in the real part of the complex susceptibility (the difference in dispersion). The line shape has positive and negative regions, which reflects the fact that the integrated intensity is zero

$$\int L^M(\omega)\,d\omega = 0 \tag{4.12}$$

or that $G_S^M(t,t) = 0$ (the intensity of the fully time-resolved spectrum disappears).

To estimate the magnitude of L^M we compare it with two components, L^{EF} and L^R. It can be shown that L^M is much smaller than L^R in the narrow-band excitation limit ($\Gamma_I \ll \Gamma_{mg}$) and is much smaller than L^{EF} in the broad-band excitation limit where $\Gamma_I \gg \Gamma_{mg}$ and $\Gamma_I \gg \Delta_I$ (see Fig. 3c). The L^M component is also proved to be smaller than L^R when ω_I is tuned far from resonance ($\Delta_I \gg \Gamma_{mg}$ and $\Delta_I \gg \Gamma_I$). For resonant excitation ($\Delta_I \ll \Gamma_{mg}$), we find that L^M is smaller than L^R or L^{EF} as long as $|\Gamma_{mg} - \Gamma_I| > \Delta_I$. In these limits, G_S^M has no role to play. Even in intermediate cases such as in Figs. 3a and 3b, the L^M component is not so large as to affect the overall spectral profile.

The time- and frequency-resolved spectrum for the intermediate case in Fig. 4a is shown in Fig. 5e. The spectrum indicates an interference character that in the time domain the spectral intensity attenuates slower than the RL component (Fig. 5d) but faster than the FL components. As known from the scale in the ordinate, this spectrum makes only a little contribution to the total intensity.

To conclude, the G_S^M component is relatively small in any excitation condition.

2. The Correlation Function G_S^T

The correlation function G_S^T can be expressed in the form [see Eq. (3.28)]

$$G_S^T(t,t') = I^T(t') \exp[(i\omega_{mf} - \Gamma_{mf})(t - t')] + G_S^{AT}(t,t') \qquad (4.13)$$

where $I^T(t)$ is the fully time-resolved spectrum for G_S^T [note that $G_S^{AT}(t,t) = 0$]. In the case of the rectangular pulse [that is, $\phi(t) = \phi_0$ for $0 < t < \tau_p$; otherwise, $\phi(t) = 0$], $I^T(t)$ is

$$I^T(t) = -\frac{2K(\Delta_I^2 + \Gamma_{mg}^2 - \Gamma_I^2)\phi_0^2}{(\Delta_I^2 + \Gamma_{mg}^2 - \Gamma_I^2)^2 + 4\Gamma_I^2\Delta_I^2} (\cos \Delta_I t)e^{-\Gamma_{mg}t}, \quad 0 < t < \tau_p \quad (4.14a)$$

$$= 0, \quad \text{otherwise} \qquad (4.14b)$$

For the pulse duration, $I^T(t)$ is damped at rate Γ_{mg} while modulated by a "detuning" oscillation with period Δ_I^{-1}. These features are generally observed; for the double exponential pulse, see the dashed-double dotted lines in Fig. 4a–4c. The presence of this "transient" component G_S^T causes the total emission intensity to lag behind the pulse profile. Caution must be exercised on situations in which the incident pulse of exponential form [that is, $\phi(t) = \phi_0 \exp(\gamma t)$ for $t \leq 0$] is turned on in the infinite past, $t = -\infty$ [15, 20, 21]. In those unrealistic cases the total emission intensity rises with the pulse profile.

The time- and frequency-resolved spectrum for the double exponential pulse is shown in Fig. 5f. In the time-frequency domain, the role of G_S^T does not seem as crucial as in the case where only time resolution is required [in Fig. 4a, the function $I^T(t)$ (dashed-double dotted line) has the largest absolute value among all the components]. The time- and frequency-resolved spectrum for G_S^T is so energetically broadened as to take appreciable values even outside the domain of Fig. 5f. [The correlation function G_S^R, G_S^M, and G_S^T that explicitly include the pulse amplitude function $\phi(t)$ project the Fourier broadening of the incident pulse on the time- and frequency-resolved spectra.] Such a temporal behavior as in the fully time-resolved spectrum is recovered by the integration of the spectrum intensity with respect to frequency or the use of larger Γ_D.

V.   THE RAMAN STATE AND THE PROPER EXCITED STATE

Many conceptual issues touching upon RSE may be reduced to the question as to why the total FL component is not simply proportional to the excited-state diagonal element of the molecular density matrix $\hat{\sigma}$, that is, σ_{mm}. It is the total intensity including the RL component, $G_S(t, t)$, that is in proportion to σ_{mm} [cf. Eqs. (2.11) and (2.13)] [11, 15, 18]. We shall make a scrutiny into the problem on the basis of results obtained so far.

First of all, we investigate what is the source of the RL component. Its root can be made clear by focusing on the molecular state that generates the RL component. We neglect relaxation constants for the moment, since they are expected to be dispensable in this context. [Note that the reductions in relaxation constants enhance the RL component.] Let H be the Hamiltonian for a composite system consisting of a molecule and an incident field. Provided that at time t_0 the molecule was in the initial state with the one photon pulse of central frequency ω_I, $|g, \hbar\omega_I\rangle$, the time evolution for the composite system is found to be

$$|\Psi(t)\rangle = \exp[-iH(t-t_0)/\hbar]|g, \hbar\omega_I\rangle$$

$$= e^{-i(\omega_g + \omega_I)(t-t_0)}\left[|g, \hbar\omega_I\rangle - \frac{1}{i\hbar}R(t, t_0)\right] \qquad (5.1)$$

with

$$R(t, t_0) = \sum_m \int_{t_0}^{t} d\tau\, \mu_{mg}\phi(\tau)e^{-i(\omega_{mg} - \omega_I)(t-\tau)}|m, 0\rangle \qquad (5.2)$$

where the propagator $\exp[-iH(t-t_0)/\hbar]$ has been expanded up to the first order of the molecule-incident light interaction. The symbol $|m, 0\rangle$ stands for intermediate states m dressed with no photons (we are in general dealing with a multilevel intermediate manifold). If $\phi(t)$ is constant, the wave function $R(t = \infty, t_0 = 0)$ in Eq. (5.2) becomes equivalent to the so-called Raman wavefunction in the time-dependent formulation of the Kramers–Heisenberg–Dirac sum-over-states expression (which formulation is addressed to the stationary excitation case) [51–53].

We now extract from Eq. (5.2) the component that *adiabatically* follows the temporal change in the incident light. Applying to Eq. (5.2) the method of integration by parts as used in the derivation of G_S^R, we have

$$|\Psi(t)\rangle = e^{-i(\omega_g + \omega_I)(t - t_0)}|g, \hbar\omega_I\rangle + |\Psi_R(t)\rangle + |\Psi_F(t)\rangle \qquad (5.3)$$

where the states $|\Psi_R\rangle$ and $|\psi_F\rangle$ are expressed as

$$|\Psi_R(t)\rangle = -e^{-i(\omega_g + \omega_I)(t - t_0)} \sum_m C_m(t)|m, 0\rangle \qquad (5.4)$$

$$|\Psi_F(t)\rangle = \sum_m \frac{\mu_{mg}e^{-i\omega_m(t - t_0)}}{\hbar(\omega_{mg} - \omega_I)} \int_{t_0}^t d\tau \frac{d\phi(\tau)}{d\tau} e^{i(\omega_{mg} - \omega_I)(\tau - t_0)}|m, 0\rangle \qquad (5.5)$$

with

$$C_m(t) = \frac{\mu_{mg}\phi(t)}{\hbar(\omega_{mg} - \omega_I)} \qquad (5.6)$$

It is worth notice that the state $|\Psi_R\rangle$ *adiabatically* follows the change in the incident light. Equation (5.4) indicates that the state $|\Psi_R\rangle$ is modulated at frequency $\omega_g + \omega_I$ and the coefficients C_m change according to the pulse amplitude $\phi(t)$ (i.e., this state is forced to oscillate by the incident light). [It should be noted that the frequency $\omega_g + \omega_I$ is not proper to the states $|m, 0\rangle$ involved in $|\Psi_R\rangle$. The frequency ω_m is proper to $|m, 0\rangle$, of course.] We will demonstrate below that these points characterize the RL component and that $|\Psi_R\rangle$ leads to the RL emission generating correlation function G_S^R. Let us therefore call $|\Psi_R\rangle$ the "Raman" state. It turns out that this state characterizes RL emission more generally than the Raman wavefunction (in the sense that the Raman state can describe time-dependent RL emission).

We are now interested in finding out what spectra the Raman state exhibits. We note that the correlation function of the emitted field is related with that of the $m \to f$ transition dipole operator $\hat{\mu}$ [40, 41]

$$G_S(t, t') \propto \langle \hat{\mu}(t)\hat{\mu}(t')\rangle \qquad (5.7)$$

where the quantity in $\langle \cdots \rangle$ is averaged over the composite system density matrix $\hat{\sigma}_c$

$$\langle \hat{\mu}(t)\hat{\mu}(t') \rangle = \mathrm{Tr}\hat{\sigma}_c(t_0)e^{iH(t-t_0)/\hbar}\hat{\mu}e^{-iH(t-t')/\hbar}\hat{\mu}e^{-iH(t'-t_0)/\hbar} \tag{5.8}$$

Starting from the initial condition $\hat{\sigma}_c(t_0) = |g, \hbar\omega_I \rangle \langle g, \hbar\omega_I|$, we have

$$\langle \hat{\mu}(t)\hat{\mu}(t') \rangle = \langle \Psi(t)| \hat{\mu}e^{-iH(t-t')/\hbar}\hat{\mu} | \Psi(t') \rangle \tag{5.9}$$

This is the correlation function that describes the properties of the emission from $|\Psi(t)\rangle$.

If we put the Raman state instead of $|\Psi(t)\rangle$ in Eq. (5.9), we obtain the correlation function that represents the emission component from the Raman state (assuming a weak incident light)

$$\langle \Psi_R(t)| \hat{\mu}e^{-iH(t-t')/\hbar}\hat{\mu} | \Psi_R(t') \rangle$$

$$= \left| \sum_m \frac{\mu_{gm}\mu_{mf}}{\hbar(\omega_{mg} - \omega_I)} \right|^2 \phi(t)\phi(t') \exp\left[i(\omega_I - \omega_{fg})(t - t')\right] \tag{5.10}$$

In the single intermediate level case [54], the preceding correlation function equals to G_S^R [except for the relaxation constants in Eq. (3.22)]. The pulse amplitude $\phi(t)$ in Eq. (5.10) derives from the coefficients C_m of states $|m, 0\rangle$ and the factor $\exp[i(\omega_I - \omega_{fg})(t - t')]$ proceeds from the fact that the Raman state oscillates at frequency $\omega_g + \omega_I$ [if it oscillated at ω_m, the factor should be $e^{i\omega_{mf}(t-t')}$]. These two points characterize the RL emission. It is of secondary importance that the states $|g, \hbar\omega_I\rangle$ and $|m, 0\rangle$ are linearly combined in Eq. (5.3). The phase relationship between $|g, \hbar\omega_I\rangle$ and $|m, 0\rangle$ is indifferent to the construction of the RL emission generating correlation function (as long as the emissions from molecules do not interfere with each other [43]).

Confining ourselves to the single intermediate level case, we obtain the following density matrix for the Raman state

$$\hat{\sigma}^R(t) \equiv \mathrm{Tr}_I |\Psi_R(t)\rangle \langle \Psi_R(t)|$$

$$= \sigma^R_{mm}(t)|m\rangle \langle m| \tag{5.11}$$

where $\mathrm{Tr}_I(\cdots)$ denotes the trace over the incident field. The diagonal component σ^R_{mm} is given by $|C_m|^2$, that is, the Raman state has a nonzero excited-state population. This element σ^R_{mm} must be included in the total diagonal element σ_{mm} but does not contribute to FL emission. Noting that the population is in proportion to the fully time-resolved spectrum

$[= G_S(t, t)]$, we obtain an expression of σ_{mm}^R including relaxation constants [cf. Eq. (3.22)]

$$\sigma_{mm}^R(t) = \frac{|\mu_{gm}|^2}{\hbar^2} \left\{ \frac{(\Delta_I^2 + \Gamma_{mg}^2 - \Gamma_I^2)\phi^2(t)}{(\Delta_I^2 + \Gamma_{mg}^2 - \Gamma_I^2)^2 + 4\Gamma_I^2\Delta_I^2} \right\} \qquad (5.12)$$

If the small constant Γ_{fg} is eliminated in Eq. (3.22), the correlation function $G_S^R(t, t')$ can be written in the form

$$G_S^R(t, t') = \frac{K(\Delta_I^2 + \Gamma_{mg}^2 - \Gamma_I^2)}{(\Delta_I^2 + \Gamma_{mg}^2 - \Gamma_I^2)^2 + 4\Gamma_I^2\Delta_I^2} G_I(t, t')\exp[-i\omega_{fg}(t - t')] \qquad (5.13)$$

which ascertains that the emission characterized by G_S^R perfectly correlates with the incident light (both in amplitude and in frequency). Equation (5.13) suggests that the correlation with the incident light lasts from t' until t; that is, the absorption step has not been completed by time t'. In the Raman process, the absorption and emission steps are therefore considered inseparable (both steps continue from t' until t).

It is interesting that the correlation functions G_S^{PF}, G_S^{EF}, and G_S^{TF} which generate FL components have a form in common. Each is written as the product of its fully time-resolved spectrum intensity at time t' and the FL emission generating stationary correlation function $\exp[(i\omega_{mf} - \Gamma_{mf})(t - t')]$. See, for example, Eq. (4.8). The fact that the central frequency is ω_{mf} indicates that in the spectral broadening and MPD mechanisms a photon is emitted from the "proper" excited state m (the emission step lasts from t' until t). [The "proper" state refers to the one (m) that evolves temporally at its intrinsic frequency corresponding to the state energy (ω_m), that is, at the Bohr frequency.] This indication may be paraphrased in such a way that an emission step is initiated after the absorption step is completed.

To elucidate this point, we use an expression equivalent to Eq. (5.8)

$$\langle \hat{\mu}(t)\hat{\mu}(t') \rangle = \text{Tr}\hat{\sigma}_c(t')e^{iH(t-t')/\hbar}\hat{\mu}e^{-iH(t-t')/\hbar}\hat{\mu} \qquad (5.14)$$

where

$$\hat{\sigma}_c(t) = e^{-iH(t-t_0)/\hbar} \hat{\sigma}_c(t_0) e^{iH(t-t_0)/\hbar} \qquad (5.15)$$

If an absorption step is completed by time t' for some reasons, $\hat{\sigma}_c(t')$ should take the form which expresses a statistical two-state distribution

$$\hat{\sigma}_c(t') = h(t')|m, 0\rangle\langle m, 0| + |g, \hbar\omega_I\rangle\langle g, \hbar\omega_I| \qquad (5.16)$$

where $h(t)$ is the probability of finding the molecule in $|m, 0\rangle$. Substituting Eq. (5.16) into (5.14), we have

$$\langle \hat{\mu}(t)\hat{\mu}(t')\rangle = |\mu_{mf}|^2 h(t') \exp[i\omega_{mf}(t - t')] \qquad (5.17)$$

The condition that the absorption and emission steps are separable certainly leads to the product form of the excited state population at time t' and $\exp[i\omega_{mf}(t - t')]$.

If $\hat{\sigma}_c(t')$ includes off-diagonal elements such as $|m, 0\rangle\langle g, \hbar\omega_I|$ (this is the case for the process that generates the RL component), the absorption step has not been completed yet by t'. For example, in the RL emission generating process, $\hat{\sigma}_c(t')$ must take the form of

$$\{|g, \hbar\omega_I\rangle + C_m(t')|m, 0\rangle\}\{\langle g, \hbar\omega_I| + C_m^*(t')\langle m, 0|\}$$

In this sense, the fact that the states $|g, \hbar\omega_I\rangle$ and $|m, 0\rangle$ are linearly combined is important to the RL emission generating process.

The excited-state populations for the spectra broadening and MPD mechanisms can be obtained from the corresponding emission correlation functions. The density matrix for the spectra broadening mechanisms, $\hat{\sigma}^{SB}$, is expressed as

$$\hat{\sigma}^{SB}(t) = \sigma_{mm}^{EF}(t)|m\rangle\langle m| + \sigma_{mm}^{TF}(t)|m\rangle\langle m| \qquad (5.18)$$

where σ_{mm}^{EF} is the density matrix element stemming from G_S^{EF} and σ_{mm}^{TF} is that from G_S^{TF} [see Eqs. (3.27) and (3.29)]. As will be demonstrated in Section VI, MPD collision can boost the initial-state molecule into the "proper" excited state while teaming up with an incident light (this process tends to enhance FL emission). The density matrix for the MPD mechanism can be written as

$$\hat{\sigma}^{PF}(t) = \sigma_{mm}^{PF}(t)|m\rangle\langle m| \qquad (5.19)$$

where σ_{mm}^{PF} derives from Eq. (3.6). The molecules excited through the spectral broadening and MPD mechanisms are in the "proper" excited state. To put all the processes together with the transient component σ_{mm}^T from G_S^T, we have

$$\bar{\sigma}_{mm}(t) \equiv \sigma_{mm}^{PF}(t) + \sigma_{mm}^{EF}(t) + \sigma_{mm}^{TF}(t) + \sigma_{mm}^R(t) + \sigma_{mm}^T(t) \qquad (5.20)$$

As is obvious, $\bar{\sigma}_{mm}(t)$ amounts to the diagonal matrix element $\sigma_{mm}(t)$ $[= |\mu_{gm}|^2 P'(t; \Gamma_m, \Gamma', \Delta_I)/\hbar^2]$. That is, not only those in the "proper" excited state but also those in the "Raman" state are counted among the excited state molecules.

The state $\Psi_F(t)$ is associated with the FL component due to Fourier broadening; one can easily find that $G_S^{TF}(t,t') \propto e^{i\omega_{mf}(t-t')}\langle\Psi_F(t')|\Psi_F(t')\rangle$. Let us therefore call the state $\Psi_F(t)$ the Fourier broadening state. To mention here the meaning of the "transient" component G_S^T might be in order, since the function G_S^T can be expressed in terms of $\Psi_R(t)$ and $\Psi_F(t)$. We neglect the effects of phase fluctuation: G_S^{EF} and G_S^M disappear. In this case, G_S^T is given as

$$
\begin{aligned}
G_S^T(t,t') &= G_S^{AR}(t,t') - G_S^R(t,t') - G_S^{TF}(t,t') \\
&\propto e^{-i\omega_f(t-t')}\{\langle\Psi_R(t)| + \langle\Psi_F(t)|\}\{|\Psi_R(t')\rangle + |\Psi_F(t')\rangle\} \\
&\quad - e^{-i\omega_f(t-t')}\langle\Psi_R(t)|\Psi_R(t')\rangle - e^{-i\omega_{mf}(t-t')}\langle\Psi_F(t')|\Psi_F(t')\rangle \\
&\propto e^{-i\omega_f(t-t')}\langle\Psi_F(t)|\Psi_F(t')\rangle - e^{-i\omega_{mf}(t-t')}\langle\Psi_F(t')|\Psi_F(t')\rangle \\
&\quad + e^{-i\omega_f(t-t')}\langle\Psi_R(t)|\Psi_F(t')\rangle + e^{-i\omega_f(t-t')}\langle\Psi_F(t)|\Psi_R(t')\rangle
\end{aligned}
\tag{5.21}
$$

The last two terms in the final version represent the interference in emission between the Raman and Fourier broadening states. The first term is understood as a process where the absorption and emission steps occur between t' and t; the second one is understood as a process where the emission step sets in after the absorption step completed at t'. The difference between the first and second terms corresponds to the last term $[\propto \Phi_1(t', \Gamma_I)\Phi_2(t, t')]$ in the function G_S^{AT} included in G_S^T [see Eqs. (3.19) and (3.28)]. The presence of this difference suggests that until around the time $\tau_p + 1/\Gamma_D$ the absorption and emission steps are regarded as inseparable even if the Raman state Ψ_R were not created. (τ_p is the turn-off time of the pulse and $\tau_p + 1/\Gamma_D$ is the time at which the finite response time $1/\Gamma_D$ of the detector has passed since the

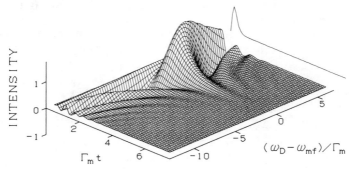

FIGURE 6.   A time- and frequency-resolved spectrum for the last term in Eq. (3.19). The parameters are the same as in Fig. 5.

incident light was turned off.) The time- and frequency-resolved spectrum generated by the last term in Eq. (3.19) is depicted in Fig. 6. From a comparison with Fig. 5f, it is evident that the interference between the Raman and Fourier broadening states is more important.

It should be pointed out that after $\tau_p + 1/\Gamma_D$ only FL components survive, that is, the absorption and emission steps are considered separable because the effects of all the interference terms and of the Raman one then vanish. [Note that the first two terms in Eq. (5.21) cancel each other or that $\Phi_2(t, t') = 0$ if t and t' are larger than τ_p.] This result is consistent with the experimental fact that after the passage of an incident pulse the fully time-resolved spectrum exhibits only FL components.

VI. THE SPECTRAL BROADENING MECHANISMS AND THE MOLECULAR PURE DEPHASING MECHANISM

There are three mechanisms that cause FL emission; namely, the spectral broadening mechanism due to phase fluctuation, the Fourier broadening mechanism, and the molecular pure dephasing (MPD) mechanism. The first two are concerned with the nature of incident light, which become therefore inoperative after the incident light pulse has gone away, and the last one is concerned with the molecule-bath interaction.

The spectral broadening mechanisms are simpler than the MPD mechanism. An incident light that is energetically broadened (because of phase fluctuation or Fourier broadening) can prepare a "proper" excited state. If the energy broadening of the incident light is allowed to cover the width of an excited state, the light excites coherently all the frequency components that the excited state is composed of [36]. The excited state prepared this way shows what the "proper" excited state is supposed to show; for example, the FL emission line shape.

The reader may however wonder why the FL components due to the spectral broadening mechanisms are damped at the rate of $2\Gamma_{mg}(=\Gamma_m + 2\Gamma')$ but not at Γ_m. This peculiarity stems from an effect of MPD. Besides this point, it is not yet well understood why MPD causes FL emission at all. The rest of this section is spent on examining what role MPD plays in RSE.

We shall rewrite Eq. (3.6) in the following form so as to clarify the role of MPD

$$G_S^{PF}(t, t') \propto \{P(t'; \Gamma_m, \Gamma', \omega_{mg}) - P(t'; 2\Gamma_{mg}, 0, \omega_{mg})\}$$

$$\propto 2\,\mathrm{Re}\int_{-\infty}^{t'} dt_1 \int_{-\infty}^{t_1} dt_2 \exp[-(i\omega_{mg} + \Gamma_{mg})(t_1 - t_2)]G_I(t_1, t_2)$$

$$\times \exp[-\Gamma_m(t' - t_1)]\{1 - \exp[-2\Gamma'(t' - t_1)]\} \qquad (6.1)$$

This form confirms that the FL component arising from G_S^{PF} is ascribed to the MPD collision which takes place after the absorption step (lasting from t_2 until t_1). In Eq. (6.1), except for the pure dephasing factor $1\text{-}\exp[-2\Gamma'(t'-t_1)]$, the integral is the population of the excited state m and the integrand is interpreted as follows: the molecule interacts with the incident light between t_2 and t_1 (absorption step), then transferring from the excited state to the other states through longitudinal relaxation processes. The factor $\exp[-\Gamma_m(t'-t_1)]$ represents the probability of remaining in the excited state after the absorption step. On the one hand, the pure dephasing factor is the probability that the molecule experiences a MPD collision after the time t_1. Equation (6.1) thus expresses, among all excited molecules, how many molecules have experienced a MPD collision. (Recall that the category of "excited" molecule contains those both in the proper excited state and in the Raman state. Not all the excited molecules produce FL emission.)

The molecules excited through the spectral broadening mechanisms (whose number is given at the outset by $\sigma_{mm}^{EF} + \sigma_{mm}^{TF}$) are subject to MPD (and, of course, subject to longitudinal relaxation). Since MPD does not bring about longitudinal relaxation, MPD collisions leave those molecules remaining in the same proper excited state. Those excited molecules which suffered from a MPD collision are however counted in Eq. (6.1), that is, among the proper excited molecules prepared through the MPD mechanism. In short, because of the presence of MPD, the values of σ_{mm}^{EF} and σ_{mm}^{TF} are damped at rate $\Gamma_m + 2\Gamma'$ and σ_{mm}^{PF} is damped slower than at Γ_m. These behaviors are in accord with what Eqs. (3.33) signify.

The "pure" MPD mechanism also exists; the proper excited state can be prepared even if the molecule is not excited through the spectral broadening mechanism. In what follows we explain how MPD collision prepares the proper excited state without the aid of the spectral broadening mechanisms.

The role of MPD can be more closely investigated by using second-order time-dependent perturbation theory. As the perturbation V, two interactions should be taken into account; namely, the molecule-bath interaction V^C and the interaction between the molecule and the incident light, $-\mu E_I$:

$$V = -\mu E_I + V^C \tag{6.2}$$

Let H_0 be the unperturbed Hamiltonian for the system (molecule + heat bath + incident field); then the total Hamiltonian H is

$$H = H_0 + V \tag{6.3}$$

What we want to calculate is the probability w_t of finding the molecule in the state m at time t provided that it was in $|g, \hbar\omega_I\rangle$ at time t_0 (the state $|g, \hbar\omega_I\rangle$ stands for the molecular state g with one photon pulse of central frequency ω_I, as in Section V)

$$w_t(m, 0, p'|g, \hbar\omega_I, p) = |\langle p'|\langle m, 0|e^{-iH(t-t_0)/\hbar}|g, \hbar\omega_I\rangle|p\rangle|^2 \qquad (6.4)$$

where p and p' denote initial and final states of the heat bath. Expanding the propagator $\exp[-iH(t-t_0)/\hbar]$ in powers of V, we have for the second-order term

$$
\begin{aligned}
&w_t(m, 0, p'|g, \hbar\omega_I, p) \\
&= \frac{1}{\hbar^4}\left| \int_{t_0}^{t} dt_I \int_{t_0}^{t_1} dt_2 \langle p'|\langle m, 0|V(t_1)V(t_2)|g, \hbar\omega_1\rangle|p\rangle \right|^2 \\
&= \frac{1}{\hbar^4}\left| \int_{t_0}^{t} dt_1 \int_{t_0}^{t_1} dt_2 \{\langle m, 0| - \mu(t_1)E_I(t_1)|g, \hbar\omega_I\rangle\langle p'|V_{gg}^C(t_2)|p\rangle \right. \\
&\qquad \left. + \langle p'|V_{mm}^C(t_1)|p\rangle\langle m, 0| - \mu(t_2)E_I(t_2)|g, \hbar\omega_I\rangle\} \right|^2 \qquad (6.5)
\end{aligned}
$$

where $V_{mm}^C \equiv \langle g|V^C|g\rangle$, $V_{mm}^C \equiv \langle m|V^C|m\rangle$, and the time-dependent operators are defined as

$$\hat{O}(t) \equiv \exp[+iH_0(t-t_0)/\hbar]\hat{O}\exp[-iH_0(t-t_0)/\hbar] \qquad (6.6)$$

Equation (6.5) indicates that the collision which does not alter the molecular-state pairs with the molecule–light interaction $-\mu E_I$ pushing the initial-state molecule into the (proper) excited state. This process functions as a "collision-induced absorption" effect, whether or not the spectral broadening mechanisms are present (see subsequent discussion).

Assuming that the incident light is a monochromatic continuous wave with frequency ω_I (that is, in the absence of spectral broadening), a routine manipulation provides the transition probability per unit time averaged over all populated heat bath states (ρ_B is the density matrix for the heat bath)

$$
\begin{aligned}
W(m, 0|g, \hbar\omega_I) &= \lim_{t\to\infty} \frac{d}{dt} \sum_{\substack{p,p' \\ (p\neq p')}} \langle p|\rho_B|p\rangle w_t(m, 0, p'|g, \hbar\omega_I, p) \\
&= \frac{2\pi(\mu_{gm}\phi_0)^2}{\hbar^3\Delta_I^2} \sum_{p,p'} \langle p|\rho_B|p\rangle|\langle p'|\Delta V^C|p\rangle|^2 \delta(E_{p'} - E_p + \hbar\Delta_I)
\end{aligned}
$$

$$(6.7)$$

with

$$\phi_0 = (\hbar\omega_I/2\varepsilon_0 v)^{1/2} \tag{6.8}$$

$$\Delta V^C = \sum_{\substack{p_1,p_2 \\ (p_1 \neq p_2)}} |p_1\rangle\langle p_1|(V_{gg}^C - V_{mm}^C)|p_2\rangle\langle p_2|, \tag{6.9}$$

where E_p is the energy of the heat bath state p and $E_{p'}$ that of p'. The δ function in Eq. (6.7) renders energy conservation between the states $|g,\hbar\omega_I\rangle|p\rangle$ and $|m,0\rangle|p'\rangle$. It should be noted that at this stage $E_p \neq E_{p'}$ unless the detuning $\Delta_I = 0$: the collision process that we have been treating in the "pure" MPD mechanism is not elastic, contrary to our expectation. This type of collision may be called "quasielastic," since the collision must feed or absorb the energy difference $\hbar\Delta_I$ [42a].

However, the Markov approximation invoked, the collision becomes elastic. This is understood through the following expression equivalent to Eq. (6.7)

$$W(m,0|g,\hbar\omega_I) = \frac{(\mu_{gm}\phi_0)^2}{\hbar^4\Delta_I^2} \int_{-\infty}^{\infty} d\tau \langle \Delta V^C(\tau)\Delta V^C\rangle_B e^{i\Delta_I \tau} \tag{6.10}$$

where $\langle \cdots \rangle_B \equiv Tr\,\rho_B \cdots$. We shall suppose that the correlation function $\langle \Delta V^C(\tau)\Delta V^C\rangle_B$ collapses faster than the oscillation period of its rear exponential function [upon letting τ_c be the correlation time, $\tau_c \ll \Delta_I^{-1}$], which is a necessary condition of the Markov approximation [30–33]. The MPD constant Γ' then comes out in Eq. (6.10) as

$$W(m,0|g,\hbar\omega_I) = \frac{2(\mu_{gm}\phi_0)^2}{\hbar^2\Delta_I^2}\Gamma' \tag{6.11}$$

where the explicit form of Γ' is

$$\Gamma' = \frac{1}{2\hbar^2} \int_{-\infty}^{\infty} d\tau \langle \Delta V^C(\tau)\Delta V^C\rangle_B \tag{6.12}$$

This expression agrees with the well-known form of the so-called MPD constant [36, 44]. Equation (6.11) multiplied by the ratio of the spontaneous emission rate to the longitudinal relaxation one Γ_m yields the integrated intensity of the MPD-induced FL component [see Eq. (4.5)].

The following δ function appears from Eq. (6.12)

$$\delta(E_p - E_{p'})$$

which means that upon introducing the Markov approximation the "quasi-elastic" collision process becomes elastic. The price to be paid is the energy conservation law in Eq. (6.7). The violation of the energy conservation law leads to a contradiction: according to Eq. (4.5) the FL component due to MPD would be observed even if the excitation wavelength is extremely off-resonant. We define $S(\Delta_I)$ as the integrated intensity of an emission line shape $L(\omega)$ [$S(\Delta_I)$ is called the excitation spectrum]

$$S(\Delta_I) = \int d\omega L(\omega) \qquad (6.13)$$

In the Markov approximation, the FL component due to MPD, $S^{PF}(\Delta_I)$, decreases at large detunings in proportion to $1/\Delta_I^2$ [see Eq. (6.11) or (4.5)] and the RL one $S^R(\Delta_I)$ decreases in the same way: the ratio of the FL component to the RL component, $S^{PF}(\Delta_I)/S^R(\Delta_I)$, becomes a constant of $2\Gamma'/\Gamma_m$ regardless of the excitation wavelength (we have assumed that no FL component due to phase fluctuation exists, that is, $\Gamma_I = 0$), which is unacceptable from an experimental point of view. We must pay heed to the fact that in the far wing regions of a line shape the Markov approximation no longer holds [30–33]; quantitative discussion based on this approximation is not applicable to the regions where $\Delta_I \gg \Gamma_{mg}$. If one use Eq. (6.10), the FL component due to MPD decreases as $1/\Delta_I^4$ or more rapidly: the ratio S^{PF}/S^R is hence proportional to $1/\Delta_I^2$, as pointed out by Ron and Ron [55] who have derived an expression similar to Eq. (6.10) in a different way. A more detailed discussion of non-Markovian effects is given in Section VII.

The "pure" MPD mechanism can be visualized by using a set of potential surfaces depicted in Fig. 7. Molecule–perturber interaction potentials are plotted against the distance between the molecule and the colliding particle (perturber). The lowest potential curve represents the interaction of the initial state molecule with the perturber (that is, V_{gg}^C). If one photon of frequency ω_I is added to this potential, we have the potential for the dressed state $|g, \hbar\omega_I\rangle$. This second potential is indicated by the solid line connecting two points i and $1'$. The third solid line stands for the interaction of the excited molecule with the perturber (V_{mm}^C). The second and third solid lines intersect at the point X but the level crossing is lifted by the molecule–light interaction. The resultant noncrossing potentials are shown by the broken lines.

Let us consider the transition probability from the asymptotic state $|g, \hbar\omega_I\rangle$ to the asymptotic state $|m, 0\rangle$. Suppose that the perturber is initially far away from (the point i corresponds to this situation) and is moving in close to the molecule. After some scrimmage among the molecule, perturber, and incident light, the composite system may fall apart into the molecule in the excited state and the perturber (the point f). Including the relative motion,

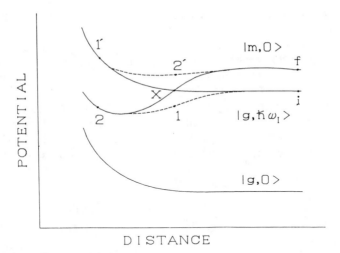

FIGURE 7. Illustration of molecule–perturber interaction potentials. The potentials are plotted against the distance between the molecule and the perturber. The lowest potential curve denotes the interaction of the initial state molecule with the perturber; the second potential indicated by the solid line connecting two points i and $1'$ denotes the one for the dressed state $|g, \hbar\omega_I\rangle$; the third solid line stands for the interaction of the excited molecule with the perturber. The potential crossing X is lifted by interaction with the incident light, which gives rise to the two noncrossing potentials (broken line). The collision process turns the initial-state molecule into the proper excited state in cooperation with the incident light. There are two main routes in going from the initial asymptotic state i to the final asymptotic state $f: i \to 1' \to 2' \to f$ (Route I); $i \to 1 \to 2 \to f$ (Route II).

we set the initial state as $|g, \hbar\omega_I\rangle|p\rangle$ and the final state as $|m, 0\rangle|p'\rangle$. The symbols p and p' may be taken as the momenta of relative motion.

There are two main routes in going from i to f:

$$i \to 1' \to 2' \to f \quad \text{(Route I)}$$

$$i \to 1 \to 2 \to f \quad \text{(Route II)}$$

We simplify the collision scheme by considering these two routes alone [56]. Our interpretation of these two routes is as follows: in Route I, starting from i the system passes the crossing point X to the point $1'$ with the change in relative motion from $|p\rangle$ to $|p'\rangle$ (a "quasielastic" collision); then it goes back to f through the point $2'$. The first process is induced by V^C and the second one is induced by the molecule–light interaction. In Route II, the system passes the point 1 (by the molecule–light interaction) and goes back to f

through the point X (by V^C). Routes I and II are associated with the first and second terms in Eq. (6.5), respectively.

In Route I or the first term in Eq. (6.5), the molecule experiences a (quasielastic) MPD collision $(i \rightarrow 1')$ *before* the onset of the molecule–light interaction $(1' \rightarrow 2' \rightarrow f)$. This seems to be inconsistent with the explanation of Eq. (6.1) that the factor $1 - \exp[-2\Gamma'(t' - t_1)]$ is ascribed to the MPD collision following *after* the absorption step. This discrepancy arises from the fact that the transition probability per unit time, dw_t/dt, has been evaluated at $t = \infty$. This procedure in reaching the transition probability W is valid when the transient effect due to turning on the interactions suddenly at $t = t_0$ is small (this is the case in the Markovian regime): the transition rate then manifests the energ conservation law

$$\lim_{t \to \infty} \frac{d}{dt} w_t(m, 0, p'|g, \hbar\omega_I, p) \propto \left| \left\langle p' \left| \frac{V_{gg}^C}{E_{p'} - E_p} + \frac{V_{mm}^C}{\hbar\Delta_I} \right| p \right\rangle \right|^2 \delta(\hbar\Delta_I + E_{p'} - E_p)$$

(6.14)

Because of the δ-function, the first term in Eq. (6.14) becomes $-V_{gg}^C/\hbar\Delta_I$, which term can be obtained also by replacing the first term in Eq. (6.5) with

$$\int_{t_0}^t dt_1 \int_{t_0}^{t_1} dt_2 \langle p'| - V_{gg}^C(t_1)|p\rangle \langle m, 0| - \mu(t_2)E_I(t_2)|g, \hbar\omega_I\rangle \qquad (6.15)$$

Apart from the signs, the process where a MPD collision takes place before photon absorption is equivalent to the one where a MPD collision takes place after photon absorption [the interpretation of Eq. (6.15)]. We must here pay attention to the negative sign of $V_{gg}^C(t_1)$ in Eq. (6.15). The negative sign ensures that the transition probability W vanishes when the energies of the two levels vary in the same way (that is, when $V_{gg}^C = V_{mm}^C$).

We hence conclude that generation of FL emission in the "pure" MPD mechanism consists of three steps: (i) ground-state molecules are initially excited by incident light to the Raman state of which energy is different from that of the proper excited state by the detuning Δ_I; (ii) some of Raman state molecules are then transferred into the proper excited state by a quasielastic (MPD) collision; (iii) finally, the proper excited molecules produce a FL component.

The discussions developed in this section confirm that the processes associated with G_S^{PF} are "real." It is thus a natural consequence that G_S^{PF} always generates a positive spectrum [as obvious from the form of Eq. (6.1)]. The division into G_S^{PF} and G_S^{AR} as shown in Eq. (3.7) is natural rather than artificial, since it properly reflects a statistical distribution caused by MPD

(G_S^{AR} also generates a positive spectrum). Some part of molecules behave according to the mechanism described by G_S^{PF}; the rest behave according to the mechanism described by G_S^{AR}.

To summarize the results obtained so far, we present Fig. 8 in which all the paths leading to emission are schematically illustrated. The proper excited states appearing in the paths are symbolized by the line shape painted in a dark tone and the Raman states are symbolized by the bold line. The molecule is initially excited into the proper excited state or the Raman state. The former is created by the spectral broadening mechanisms and the latter is created by the adiabatic following mechanism (in which the molecule is excited adiabatically following the change in the incident pulse). The ratio depends on various parameters such as Δ_I and Γ_I. In the next stage, the molecule may suffer a MPD collision (indicated by the solid arrow) or may not (open arrow). Whether the molecule in the proper excited state suffers a MPD collision (Path $F2$) or not (Path $F1$), the molecule produces a FL component. On the other hand, the created Raman state eventually produces a FL or RL component according as the molecule suffers a MPD collision (Path $F3$) or not (Path R). (In Path R, a line cannot be clearly drawn between the three stages.) In Path $F3$ (that we have called the "pure" MPD mechanism), the Raman state is converted into the proper excited state (which means that the RL component is reduced as Γ' increases).

As indicated in the last column in Fig. 8, Path $F1$ corresponds to the sum of G_S^{EF} and G_S^{TF}; the sum of Paths $F2$ and $F3$ corresponds to G_S^{PF}; Path R corresponds to G_S^R. The interference between Path $F1$ and Path R is represented by G_S^M and G_S^T. After the incident pulse has gone away, the sum of the three FL components ($F1 + F2 + F3$) decays at rate Γ_m but each FL component does not; the emission component via Path $F1$ decays faster than at Γ_m because of the population leakage to Path $F2$ by MPD and the component via Path $F2$ decays slower than at Γ_m because of the population supply from Path $F1$. The concepts of Raman state and proper excited state which are brought forth in this article are of practical use, as will be indicated also in the non-Markovian discussion given in the next section.

The existence of the MPD effect and spectral broadening ones has been pointed out in previous works. For example, Melinger and Albrecht [20, 21] have noted these effects by calculating the total spectrum. However, their treatment is basically rooted in the diagrammatic representation [where the spectra associated with the A, B, and C terms of Eqs. (2.8) become the synosure] and therefore the question as to what kinds of molecular states (Raman state or proper excited state) produce what kind of emission is not discussed. Such a succinct and useful scheme as in Fig. 8 is not reached. The mechanism of FL emission due to MPD cannot be revealed within the framework of the conventional diagrammatic representation.

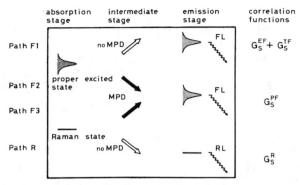

FIGURE 8. Schematic illustration of paths of absorption leading to emission. The line shapes painted in a dark tone symbolize the proper excited state and the bold lines symbolize the Raman state. The molecule is initially excited into the proper excited state or the Raman state. The former is created by the spectral broadening mechanisms and the latter is created by the adiabatic following mechanism. In the next stage, the molecule may suffer a MPD collision (indicated by the solid arrow) or may not (open arrow). Whether the molecule in the proper excited state suffers a MPD collision (Path $F2$) or not (Path $F1$), the molecule produces a FL component. The created Raman state eventually produces a FL or RL component according as the molecule suffers a MPD collision (Path $F3$) or not (Path R). In Path $F3$, the Raman state is converted into the proper excited state. The last column indicates emission correlation functions corresponding to the three components in the emission stage. Path $F1$ is represented by the sum of G_S^{EF} and G_S^{TF}; the sum of Paths $F2$ and $F3$ is represented by G_S^{PF}; Path R is represented by G_S^R.

VII. NON-MARKOVIAN EFFECT: AN ANALYSIS BASED ON THE STOCHASTIC MODEL

A. Derivation of Emission Correlation Functions

In this section we discuss effects of MPD without invoking the Markov approximation, that is, without using the MPD constants. To that end, we employ a stochastic model where the molecular energies are assumed to fluctuate owing to the random force exerted by the heat bath. In this model, the molecular energy differences are replaced with stochastic functions of time

$$\omega_{mg} \to \omega_{mg} + \delta\omega_{mg}(t) \qquad (7.1a)$$

$$\omega_{mf} \to \omega_{mf} + \delta\omega_{mf}(t) \qquad (7.1b)$$

The functions $\delta\omega_{mg}(t)$ and $\delta\omega_{mf}(t)$ are treated as classical variables [57]. We set these two random variables equal, as has been done in Eq. (3.1) [$\delta\omega_{fi}(t)$

is taken to be very small. In this section, $\delta\omega_{fi}(t) = 0$]:

$$\delta\omega_{mg}(t) = \delta\omega_{mf}(t) \equiv \delta\omega(t) \qquad (7.2)$$

The ensemble average of the random frequency $\delta\omega(t)$ can be set zero without loss of generality:

$$\langle \delta\omega(t) \rangle = 0 \qquad (7.3)$$

(Hereafter ω_{mg} and ω_{mf} are regarded as the mean energy differences.)

As usual, we assume that the modulation in the random frequency is described by a Gaussian process [30, 58]. In the Gaussian process, the $2n$th correlation function $\langle \delta\omega(t_1)\cdots\delta\omega(t_{2n}) \rangle$ can be given by the sum of all distinguishable products of second order correlation functions, that is, any order correlation function is given by using the second order correlation function. The form that is used most commonly for the second-order correlation function is [24, 30]

$$\langle \delta\omega(\tau)\delta\omega(\tau') \rangle = \Delta^2 \exp(-\Lambda|\tau - \tau'|) \qquad (7.4)$$

where Δ and Λ denote the amplitude and rate of frequency modulation, respectively.

The stochastic model can be incorporated into the theory of RSE. We start with substituting Eqs. (7.1) into Eqs. (2.8) as follows:

$$\exp[-i\omega_{mg}(t_f - t_i)] \to \exp\left\{-i\int_{t_i}^{t_f} dt[\omega_{mg} + \delta\omega(t)]\right\} \qquad (7.5a)$$

$$\exp[-i\omega_{mf}(t_f - t_i)] \to \exp\left\{-i\int_{t_i}^{t_f} dt[\omega_{mf} + \delta\omega(t)]\right\} \qquad (7.5b)$$

The ensemble averages of Eqs. (2.8) then results in [making use of the properties of the Gaussian process and Eq. (7.3)] [59]

$$A(t_1, t_2, t_3, t_4)$$

$$= \exp\left[\left(i\omega_{mf} - \frac{\Gamma_f}{2}\right)(t_1 - t_3) - i\omega_{mg}(t_2 - t_4) - \frac{\Gamma_m}{2}(t_1 - t_2 + t_3 - t_4)\right.$$

$$\left. - f(t_1, t_2, t_3, t_4) + \Delta f(t_1, t_2, t_3, t_4)\right]G_I(t_2, t_4) \qquad (7.6a)$$

$B(t_1, t_2, t_3, t_4)$

$$= \exp\left[\left(i\omega_{mf} - \frac{\Gamma_f + \Gamma_m}{2}\right)(t_1 - t_2) - i\omega_{mg}(t_3 - t_4) - \frac{\Gamma_m}{2}(2t_2 - t_3 + t_4)\right.$$

$$\left. - f(t_1, t_2, t_3, t_4) + \Delta f(t_1, t_2, t_3, t_4)\right]G_I(t_3, t_4) \qquad (7.6\text{b})$$

$C(t_1, t_2, t_3, t_4)$

$$= \exp\left[\left(i\omega_{mf} - \frac{\Gamma_f + \Gamma_m}{2}\right)(t_1 - t_2) - i\omega_{mg}(t_3 - t_4) - \frac{\Gamma_m}{2}(2t_2 + t_3 - t_4)\right.$$

$$\left. - f(t_1, t_2, t_3, t_4) - \Delta f(t_1, t_2, t_3, t_4)\right]G_I(t_4, t_3) \qquad (7.6\text{c})$$

where

$$f(t_1, t_2, t_3, t_4) = g(t_1, t_2) + g(t_3, t_4) \qquad (7.7)$$

$$\Delta f(t_1, t_2, t_3, t_4) = g(t_1, t_4) + g(t_2, t_3) - g(t_2, t_4) - g(t_1, t_3) \qquad (7.8)$$

with

$$g(t_f, t_i) = \int_{t_i}^{t_f} dt_1 \int_{t_i}^{t_1} dt_2 \langle \delta\omega(t_2)\delta\omega(t_1)\rangle \qquad (7.9)$$

The functions f and Δf determine how long the correlation between the four vertices (t_1, t_2, t_3, t_4) last. The emission correlation functions G_S^A, G_S^B, and G_S^C for the stochastic model are given by substituting Eqs. (7.6) into Eqs. (2.11) [60]. Here we have assumed that longitudinal relaxation is Markovian, since a large change in internal energy of molecule is caused by a large momentum or high velocity of the colliding particle.

Upon the substitution of Eq. (7.4) into Eq. (7.9), we get

$$g(t_f, t_i) = \frac{\Delta^2}{\Lambda^2}[\exp(-\Lambda|t_f - t_i|) - 1 + \Lambda|t_f - t_i|)$$

$$\equiv g(t_f - t_i) \qquad (7.10)$$

Since the process described by Eq. (7.4) is stationary, Eq. (7.10) depends only on the time difference $|t_f - t_i|$ [although the correlation function to be inserted into Eq. (7.9) does not need to be stationary]. In the fast modulation limit, that is, when $\Lambda \gg \Delta$, we recover the Markovian result

$$g(\tau) \simeq \frac{\Delta^2}{\Lambda} \tau \qquad (7.11)$$

The constant Δ^2/Λ corresponds to the MPD constant Γ'. On the other hand, in the slow modulation (static) limit is obtained when $\Lambda \ll \Delta$. In this case, a Taylor expansion of Eq. (7.10) yields

$$g(\tau) \simeq \Delta^2 \tau^2/2 \qquad (7.12)$$

which leads to an inhomogeneous broadening of a Gaussian line shape.

B.   The Jump Process due to Frequency Modulation

The emission correlation functions obtained from Eqs. (7.6) reflect the non-Markovian character of frequency modulation and many other effects discussed in the preceding sections as well. To make discussions as simple as possible and to concentrate upon non-Markovian effect in RSE, we utilize Fig. 8. As shown below, the schematic illustration therein serves as a guiding principle even in investigating non-Markovian effect.

We first look through the role of spectral broadening in the present model. Irrespective of the frequency modulation rate, spectral broadening can in the absorption stage create the proper excited state (that could be homogeneously or inhomogeneously broadened). The molecule in the proper excited state remains in the same kind of excited state that produces FL emission, whether the molecule undergoes a rapid frequency modulation in the intermediate stage or not (Path $F2$ or $F1$). The only difference is that the emission component in Path $F1$ decays faster than at the longitudinal relaxation constant Γ_m and the emission component in Path $F2$ decays slower than at Γ_m (after the incident pulse has fully decayed). The sum of the FL one via Path $F3$ and these two components is damped at Γ_m. The character of frequency modulation does not alter these features. The only non-Markovian effect on Paths $F1$ and $F2$ is that the absorption and emission line shapes of the proper excited state becomes Gaussian as the case approaches the static limit. We eliminate Paths $F1$ and $F2$ from subsequent discussions (that is, the excitation condition is assumed to be stationary and monochromatic).

We next pay our attention to the path where frequency modulation converts the Raman state into the proper excited state (Path $F3$). This path in the stochastic model, which is similar to the "pure" MPD mechanism presented in Section VI, is viewed as follows: the ground-state molecule is excited initially to the Raman state by the incident light; the Raman state is then transferred into the proper excited state by frequency modulation;

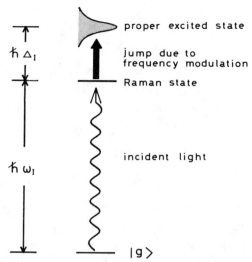

FIGURE 9. Conversion of the Raman state into the proper excited state due to frequency modulation. The ground-state molecule is excited initially to the Raman state by the incident light; the Raman state is then transferred into the proper excited state by frequency modulation; the proper excited state created produces a FL component. Since the Raman and proper excited state differ in energy by the detuning $\hbar\Delta_I$, the Raman state must exchange the amount $\hbar\Delta_I$ with the heat bath in order to be converted into the proper excited state. The mechanism in the stochastic model is devised by analogy with Path F_3 in the Markovian limit.

the proper excited state created produces a FL component. This scheme is illustrated in Fig. 9. Since the Raman and proper excited states differ in energy by the detuning $\hbar\Delta_I$, the Raman state must exchange the amount of detuning energy with the heat bath in order to be transferred into the proper excited state. This "absorption" or "feed" process of heat bath energy (for example, phonon energy) occurs if frequency modulation is adequately fast. The event that the random frequency $\delta\omega(t)$ increases by an amount $\Delta E/\hbar$ indicates that the Raman state absorbs the (phonon) energy ΔE from the heat bath. If $\delta\omega(t)$ varies before the emission step gets under way, within the life time $1/\Gamma_m$, the proper excited state would be prepared.

We try to calculate the probability of finding the molecule in the proper excited state at time t_f provided that it was in the Raman state at time t_i. The "effective" phonon field $\mathscr{E}(t)$ that causes the transition may be written as

$$\mathscr{E}(t) = \frac{1}{\sqrt{2}}[\delta\omega(t) - \delta\omega(t + t_f - t_i)] \tag{7.13}$$

This form ensures that the transition does not occur unless the random frequency varies during the period from t_i to t_f. Substituting $\mathscr{E}(t)$ into the Schrödinger equation, we have in the interaction picture

$$i\frac{\partial a_R}{\partial t} = \mathscr{E}(t)a_E e^{-i\Delta_I t} \tag{7.14a}$$

$$i\frac{\partial a_E}{\partial t} = \mathscr{E}(t)a_R e^{+i\Delta_I t} \tag{7.14b}$$

where a_R and a_E are the probability amplitudes of the Raman and proper excited states, respectively. The initial condition is that $a_R = 1$ and $a_E = 0$ at time t_i, because we have confined ourselves to Paths $F3$ and R. Since the variation in $\mathscr{E}(t)$ obeys Gaussian statistics, we get

$$|a_E(t_f)|^2 = 1 - \exp\left[-2\,\mathrm{Re}\int_{t_i}^{t_f}dt_1\int_{t_i}^{t_1}dt_2 e^{-i\Delta_I(t_1-t_2)}\langle\mathscr{E}(t_2)\mathscr{E}(t_1)\rangle\right] \tag{7.15}$$

Taking into account longitudinal relaxation of the proper excited state, we convert Eq. (7.15) into an approximate relation:

$$|\bar{a}_E(t_f)|^2 \simeq 1 - \exp[-J(t_f, t_i)] \tag{7.16}$$

where $|\bar{a}_E|^2$ is the probability of finding the molecule that *went through* the proper excited state and

$$J(t_f, t_i) = 2\,\mathrm{Re}\int_{t_i}^{t_f}dt_1\int_{t_i}^{t_1}dt_2\exp\left[-\left(\frac{\Gamma_m}{2}+i\Delta_I\right)(t_1-t_2)\right]\langle\mathscr{E}(t_2)\mathscr{E}(t_1)\rangle \tag{7.17}$$

The function J is the probability derived from first-order perturbation theory. It should be noted that the transition probability per unit time $\lim_{t\to\infty}d(J(t,0))/dt$ corresponds to the integral in Eq. (6.10). When frequency modulation is stationary, $J(t_f, t_i)$ becomes a function of time difference $\tau = t_f - t_i$

$$J(\tau) \equiv \Delta^2\,\mathrm{Re}\,\{(2-e^{-\Lambda\tau})[\tau+(e^{-\zeta_+\tau}-1)/\zeta_+]/\zeta_+$$
$$-e^{-\Lambda\tau}[\tau+(e^{-\zeta_-\tau}-1)/\zeta_-]/\zeta_-\} \tag{7.18}$$

where

$$\zeta_+ = \frac{\Gamma_m}{2}+\Lambda+i\Delta_I \tag{7.19a}$$

$$\zeta_- = \frac{\Gamma_m}{2} - \Lambda + i\Delta_I \qquad (7.19\text{b})$$

The FL component due to frequency modulation will be defined in the following.

C. Rearrangement of Emission Correlation Functions

We make preparatory arrangements for splitting the total correlation function $G_S(t, t')$, that is, the sum of $G_S^A, G_S^B,$ and G_S^C, into some components. Let us here define three terms $A_0, B_0,$ and C_0 as

$$A(t_1, t_2, t_3, t_4) \equiv A_0(t_1, t_2, t_3, t_4)e^{\Delta f(t_1, t_2, t_3, t_4)} \qquad (7.20\text{a})$$

$$B(t_1, t_2, t_3, t_4) \equiv B_0(t_1, t_2, t_3, t_4)e^{\Delta f(t_1, t_2, t_3, t_4)} \qquad (7.20\text{b})$$

$$C(t_1, t_2, t_3, t_4) \equiv C_0(t_1, t_2, t_3, t_4)e^{-\Delta f(t_1, t_2, t_3, t_4)} \qquad (7.20\text{c})$$

The total population of excited state at time t' is in proportion to

$$\int_{-\infty}^{t'} dt_1 \int_{-\infty}^{t_1} dt_2 [B_0(t', t', t_1, t_2) + C_0(t', t', t_1, t_2)] \qquad (7.21)$$

It should be noted that $A_0, B_0,$ and C_0 are factorized as a product of four factors; for instance,

$$A_0(t_1, t_2, t_3, t_4) = U_1(t_1 - t_2)U_2(t_2 - t_3)U_3(t_3 - t_4)G_I(t_2, t_4) \qquad (7.22)$$

where $U_1, U_2,$ and U_3 are propagators independently averaged for the non-overlapping time intervals $t_1 - t_2, t_2 - t_3, t_3 - t_4$. The so-called factorization approximation in RSE can therefore be obtained when setting $\Delta f(t_1, t_2, t_3, t_4) = 0$ for $t_1 \geq t_2 \geq t_3 \geq t_4$ [24, 31a]. It has been reported that the factorization approximation is valid in the range of the fast modulation limit (or in the Markovian limit) and in the off-resonance excitation limit as well (it strictly holds in the fast modulation limit) [61].

We have known from Eqs. (3.6) and (6.1) that in the Markovian limit, the FL component due to MPD, $G_S^{PF}(t, t')$, is factorized into three factors: (i) the FL emission generating stationary correlation function $\exp[(i\omega_{mf} - \Gamma_{mf})(t - t')]$; (ii) the total population of excited state at time t'; (iii) the probability that the excited molecule experiences a MPD collision between t' and t_1 (t_1 is the time at which the absorption step ends). The stochastic model version of this component is the FL component due to frequency modulation and the above three factors are then expressed by: (i)

$$\exp\left[\left(i\omega_{mf} - \frac{\Gamma_f + \Gamma_m}{2}\right)(t - t') - g(t, t')\right]$$

(ii) Eq. (7.21); (iii) Eq. (7.16). The FL component due to frequency modulation can thus be written as

$$G_S^{PF}(t, t') \equiv K \int_{-\infty}^{t'} dt_1 \int_{-\infty}^{t_1} dt_2 [B_0(t, t', t_1, t_2) + C_0(t, t', t_1, t_2)][1 - e^{-J(t', t_1)}]$$

$$(7.23)$$

It can be verified that this component represents a sequential process.
    We are now in a position to split $G_S(t, t')$ into components:

$$G_S(t, t') = G_S^{AR}(t, t') + G_S^{PF}(t, t') + G_S^{BR}(t, t')$$ (7.24)

where

$$G_S^{AR}(t, t') \equiv K \int_{-\infty}^{t} dt_1 \int_{-\infty}^{t'} dt_2 A_0(t, t_1, t', t_2)$$ (7.25)

$$G_S^{BR}(t, t') \equiv K \int_{t'}^{t} dt_1 \int_{-\infty}^{t'} dt_2 A_0(t, t_1, t', t_2)[e^{\Delta f(t, t_1, t', t_2)} - 1]$$

$$+ K \int_{-\infty}^{t'} dt_1 \int_{-\infty}^{t_1} dt_2 \{B_0(t, t', t_1, t_2)[e^{\Delta f(t, t', t_1, t_2)} - 1]$$

$$+ C_0(t, t', t_1, t_2)[e^{-\Delta f(t, t', t_1, t_2)} - 1]\}$$

$$+ K \int_{-\infty}^{t'} dt_1 \int_{-\infty}^{t_1} dt_2 \{[B_0(t, t', t_1, t_2)e^{-J(t', t_1)} - A_0(t, t_1, t', t_2)]$$

$$+ [C_0(t, t', t_1, t_2)e^{-J(t', t_1)} - A_0(t, t_2, t', t_1)]\}$$ (7.26)

Putting a strict inspection aside, we tentatively report that in the fast modulation limit G_S^{AR} and G_S^{PF} are reduced to those in the Markovian approximation [Eqs. (3.5) and (3.6)], respectively, and G_S^{BR} vanishes [set $\Delta f = 0$ and $J(t', t_1) = 2(\Delta^2/\Lambda)(t' - t_1)$]. If frequency modulation is not rapid enough, all the three correlation functions are not exempt from the non-Markovian character. The sum of the first and second integrals in G_S^{BR} is the magnitude of the deviation from the factorization approximation and the third one is (the total correlation function obtained in the factorization approximation)$-G_S^{AR} - G_S^{PF}$.

In stationary and monochromatic excitation, G_S^{AR} and G_S^{PF} are expressed as

$$G_S^{AR}(t, t') = K \exp\left[\left(i\omega_{mf} - i\Delta_I - \frac{\Gamma_f}{2}\right)(t - t')\right]|\chi(\Delta_I)|^2 \qquad (7.27)$$

$$G_S^{PF}(t, t') = 2K \exp\left[\left(i\omega_{mf} - \frac{\Gamma_m + \Gamma_f}{2}\right)(t - t') - g(t - t')\right]$$

$$\times [\text{Re}\,\chi(\Delta_I)]\eta(\Gamma_m, \Delta_I; \Delta, \Lambda) \qquad (7.28)$$

where the response function $\chi(\Delta_I)$ of which real part represents the absorption line shape is given by

$$\chi(\Delta_I) = \int_0^\infty d\tau \exp\left[\left(i\Delta_I - \frac{\Gamma_m}{2}\right)\tau - g(\tau)\right]\phi_0 \qquad (7.29)$$

and the frequency modulation factor η is

$$\eta(\Gamma_m, \Delta_I; \Delta, \Lambda) = \int_0^\infty d\tau \exp(-\Gamma_m\tau)\{1 - \exp[-J(\tau)]\} \qquad (7.30)$$

It is apparent (using the Wiener–Khintchine relation) that Eq. (7.27) gives a RL line shape and Eq. (7.28) gives a FL line shape. In the absence of spectral broadening, G_S^{AR} might be interpreted as the RL component in the factorization approximation [62].

We now lay down more general definitions of the fast and slow modulation limits by rewriting Eq. (7.29) as

$$\chi(\Delta_I) = \int_0^\infty d\tau \exp\left\{\left(i\Delta_I - \frac{\Gamma_m}{2} - \frac{\Delta^2}{\Lambda}\right)\tau - \frac{\Delta^2}{\Lambda}[\exp(-\Lambda\tau) - 1]\right\}\phi_0 \quad (7.31)$$

Since

$$\exp\left[\left(i\Delta_I - \frac{\Gamma_m}{2} - \frac{\Delta^2}{\Lambda}\right)\tau\right]$$

is the only damping factor in the above integrand, the duration time of the absorption step, τ_a, may be expressed as

$$\frac{1}{\tau_a^2} \equiv \Delta_I^2 + \left(\frac{\Gamma_m}{2} + \frac{\Delta^2}{\Lambda}\right)^2 \qquad (7.32)$$

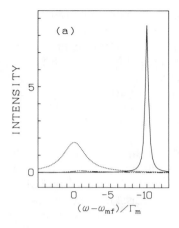

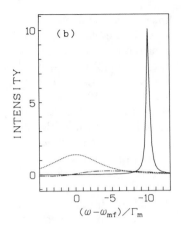

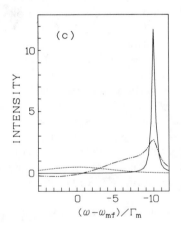

FIGURE 10. Numerical examples of the emission line shape in the stochastic model. The rates Λ are (a) 15, (b) 4, (c) 0.4 (in units of Γ_m). The common parameters are taken as follows: $\Gamma_f = 0.5$, $\Delta_I = 10$, and the modulation amplitude $\Delta = 4$. The solid, dotted, dashed-double dotted lines designate the G_S^{AR}, G_S^{PF}, and G_S^{BR} components, respectively.

We can then redefine the fast modulation limit as $1/\tau_a \ll \Lambda$ and the slow modulation limit as $1/\tau_a \gg \Lambda$. [In the slow modulation limit, $\delta\omega(t)$ hardly varies during the absorption step.] According as $1/\tau_a \ll \Lambda$ or $1/\tau_a \gg \Lambda$, $g(\tau)$ becomes Eq. (7.11) or (7.12) and the absorption line shape Re $[\chi(\Delta_I)]$ becomes Lorentzian or Gaussian. Since the off-resonance excitation limit satisfies the inequality $1/\tau_a \gg \Lambda$, as has been pointed out [30], the line shape at large detunings deviates from that in the fast modulation limit.

D. Line Shape Calculation

By putting each correlation function [Eqs. (7.26)–(7.28)] into the Wiener–Khintchine relation, one can calculate the corresponding emission line shape. The triple integrals involved can be written as the sum of products of three single integrals. The single integrals are carried out by using recurrence relations and continued fraction expressions [63] or by numerical integration. We present some numerical examples in Figs. 10a–10c to see how each component changes as the modulation rate Λ decreases. The rates Λ are (a)15, (b)4, (c)0.4 (in units of Γ_m). The common parameters are taken as follows: $\Gamma_f = 0.5$, $\Delta_I = 10$, and the modulation amplitude $\Delta = 4$. The solid, dotted, dashed-double dotted lines designate the G_S^{AR}, G_S^{PF}, and G_S^{BR} components, respectively.

It is shown in Fig. 10a that in the fast modulation limit the spectrum can be interpreted in terms of the FL and RL components alone, that is, the G_S^{BR} component is negligible [G_S^{AR} and G_S^{PF} are reduced to the corresponding correlation functions in the Markovian limit]. In this limit, the emission stage in Path $F3$ does not remember how the molecule is excited, because the fast frequency modulation, that is, the quick jump process from the Raman state to the proper excited state completely disconnects the chain of memory between the absorption and emission steps. The correlation function G_S^{PF} well describes the effect of such a quick jump process.

To be more precise, the spectral range of the effective phonon field $\mathscr{E}$ pertaining to frequency modulation, Λ, needs to cover the line width of the proper excited state. This is true in the fast modulation limit (where Λ is much larger than the detuning Δ_I and than the width of the proper excited state, $\Gamma_m/2 + \Delta^2/\Lambda$). In this case, only the proper excited state is prepared by frequency modulation. (This is deduced from the findings in Section IV that the FL component due to phase fluctuation overwhelms the RL component if the spectral broadening due to phase fluctuation covers the proper excited state sufficiently. See Fig. 3c.) As a result, via Path $F3$, only the FL component appears.

In order to cut off the chain of memory (that is, for a jump process to create the proper excited state), $\delta\omega(t)$ must vary before the emission step sets in, that is, $\Lambda > \Gamma_m$. When $\Lambda \gg \Gamma_m$, $J(t', t_1) \simeq 2(t' - t_1)\Delta^2\Lambda/(\Lambda^2 + \Delta_I^2)$ and hence G_S^{PF} has nonzero values. (This component increases until Λ exceeds Δ_I and then falls off with increasing Λ. When frequency modulation is too fast, the exchange of energy between the molecule and the bath does not proceed efficiently.) The extreme $\Lambda \gg \Gamma_m$ may be called the fast "blackout" limit. This limit condition is automatically met in the fast modulation limit. In Path R, the chain of memory between the absorption and emission steps is firm since no disturbance exists during them: only the RL component emerges. We conclude that in the fast modulation limit the emission line shape can be decomposed perfectly into the FL and RL ones (no interference-like one).

When frequency modulation is slower than longitudinal relaxation, that is, $\Lambda \ll \Gamma_m$ (the slow blackout limit), a jump to the proper excited state hardly occurs before emission: as shown in Fig. 10c, the FL component due to frequency modulation is thus small [$J \simeq 0$ and hence $G_S^{PF} \simeq 0$]. In this limit, the memory of the absorption step has not faded yet in the emission step. It is also found in Fig. 10c that G_S^{AR} gives rise to a RL component and G_S^{BR} gives a broad band around the Raman frequency. Since the slow blackout limit is a sufficient condition of the slow modulation limit, one can set $g(\tau) = \Delta^2 \tau^2/2$ (and $J = 0$) in Eq. (7.26)]: G_S^{BR} takes a RL line shape. As Λ becomes smaller and smaller, the G_S^{BR} component comes to take a "perfect" RL line shape. In the slow blackout limit, that is, in the inhomogeneous broadening limit, only the Raman state is prepared: both G_S^{AR} and G_S^{BR} exhibit RL line shapes (in this case, G_S^{BR} represents the RL component that cannot be expressed in the factorization approximation).

The remaining limit case is the slow modulation and fast blackout limit, that is, $\Gamma_m \ll \Lambda \ll 1/\tau_a$. As shown in Fig. 10b, a FL component emerges (owing to multiphonon jump processes) but it is nearly Gaussian. The G_S^{BR} component germinaates spreading between the FL and RL ones. In the slow modulation and fast blackout limit, there are phonon-induced jumps whose spectral range cannot cover the whole spectral range of the proper excited state (Λ is smaller than the detuning Δ_I or than the inhomogeneous width of the proper excited state, Δ). Consequently, such interactions with phonons shift the energy of the Raman state (those states would more or less adiabatically follow the change in the bath state, like the Raman state in RSE follows the incident light). Emission components from those states spread between the RL and FL ones. In the slow modulation regime, the G_S^{BR} component corresponds to what is called the broad Raman [64].

In the Markov approximation, as pointed out in Section VI, the ratio of the FL component to the RL component, $S^{PF}(\Delta_I)/S^R(\Delta_I)$, is a constant $2\Gamma'/\Gamma_m$ and independent of the excitation wavelength. If the case is not Markovian, however, Eq. (6.10) predicts that $S^{PF}(\Delta_I)/S^R(\Delta_I)$ would decrease with increasing Δ_I. To make it sure in the stochastic model, the RL component (solid line) and FL one (dotted line) are plotted in Fig. 11 against Δ_I. Four cases where the modulation rates Λ are (a) 200, (b) 20, (c) 2, (d) 0.2 are presented. The modulation amplitudes take in all the cases the same value $\Delta = 20$. Each component, $S^R(\Delta_I)$ or $S^{PF}(\Delta_I)$, is normalized by its value at $\Delta_I = 40$ and drawn on a scale of common logarithms. The solid and dotted lines in the case (a) are close together where Δ_I is less than 100 (which region belongs to the fast modulation limit) but under other circumstances the two lines in each case diverge as Δ_I increases. In any case, the RL component $S^R(\Delta_I)$ [$\propto |\chi(\Delta_I)/^2$] is found to decrease as $1/\Delta_I^2$. On the other hand, the FL component $S^{PF}(\Delta_I)$ decreases in the fast modulation limit as $1/\Delta_I^2$ but it comes to decrease in proportion to $1/\Delta_I^4$ or more rapidly as Λ decreases (or as Δ_I grows). The G_S^{BR} component has been eliminated from the discussion, since

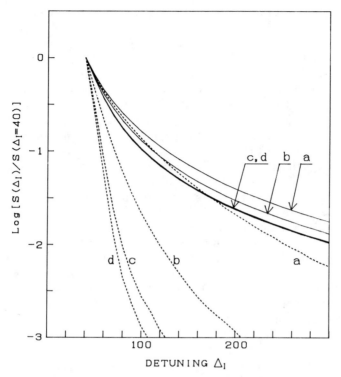

FIGURE 11. Excitation spectra of the RL component (solid line) and the FL component (dotted line). Each component, $S^R(\Delta_I)$ or $S^{PF}(\Delta_I)$, is normalized by its value at $\Delta_I = 40$ and plotted on a scale of common logarithms. The modulation rates Λ are (a) 200, (b) 20, (c) 2, (d) 0.2, respectively; in all the cases, $\Delta = 20$ is fixed (in units of Γ_m).

it is small relative to the RL component when the excitation wavelength is tuned far way from resonance (we have found that when $\Delta_I \gg \Delta, \Lambda$ the G_S^{BR} component takes a RL line shape and decreases as $1/\Delta_I^3$).

In the fast modulation limit, a jump from the Raman state to the proper excited state (see Fig. 9) is so fast that the duration of the jump, $1/\Lambda$, is shorter than the time necessary for resolving the energy difference Δ_I, namely, $1/\Delta_I$. The jump rate is thus independent of Δ_I. The growth rate of the proper excited state is then in proportion to $1/\Delta_I^2$, since it is the product of the jump rate η and the growth rate of the Raman state, Re $\chi(\Delta_I)$ [see Eq. (7.28)]. On the other hand, if $1/\Lambda > 1/\Delta_I$, η is sensitive to Δ_I. In this case, $J(\tau) \propto \tau^2/\Delta_I^2$ [see Eq. (7.18)] and η therefore decreases as $1/\Delta_I^2$ [65]: the growth rate of the proper excited state is proportional to Re $\chi(\Delta_I)/\Delta_I^2$ [66].

It has been exemplified that the classification into the three components is profitable and overall the concept of "Raman" and "fluorescence" is, in importance, not deteriorated even outside the Markovian regime. A brief summary of the three correlation functions G_S^{AR}, G_S^{PF}, G_S^{BR} is as follows: G_S^{AR}, which generates the RL component, represents the emission process where no phonon process (no frequency modulation) intervenes between the absorption and emission steps; G_S^{PF}, which generates the FL one, represents the process where a phonon-induced transition from the Raman state to the proper excited state intervenes between the two steps; G_S^{BR}, which generates the broad Raman component, represents processes where the creation and annihilation of phonons lasts a long time extending from the absorption step to the emission step.

Sue et al. [62] have separated the expression for the emission line shape into four parts I_1, I_2, I_3, and I_4. The line shape I_1 is identical with that of G_S^{AR}. The line shapes I_2 and $I_3 + I_4$ approximately correspond to G_S^{BR} and G_S^{PF}, respectively, but the meanings of I_2 and $I_3 + I_4$ are somewhat obscure. The difference originates from the fact that we have devised the jump mechanism of creating the proper excited state and crammed it into a single correlation function, namely, G_S^{PF}.

The present approach is applicable to the dynamical model as well. It would be shown that the FL component in the dynamical model can also be interpreted as a jump process like in the stochastic model.

VIII.   SUMMARY AND CONCLUSION

We have probed into the origins of resonance Raman and resonance fluorescence by primarily focusing on the emission correlation function. The way of defining "fluorescence" and "Raman" is not based on the division of the emission spectrum, but is based on the division of the emission correlation function. This way of defining is more general, since the emission correlation function is descriptive of the "bare" emission process itself (or the state of the emitted field) while the emission spectrum results from interaction between the emitted light and the detector. (The definition based on the emission spectrum would depend on the nature of the detector and on the type of measurement.) The conclusive discussions summarized below have been formed by dividing the emission correlation function into several parts so that each divided one exhibits a RL or a FL component (or an interferencelike one). Unlike previous works, our novel approach is completely free from the puzzling diagrammatic interpretation [67] and has provided much more profound understanding of RSE (for example, the approach has made it clear what kinds of molecular states produce what kinds of emission).

In the Markovian limit, the total correlation function is split into G_S^{PF} and G_S^{AR}, which separation reflects a "real" statistical distribution caused by

MPD. The function G_S^{PF} represents the FL component due to MPD. In the next step, G_S^{AR} is artificially divided into: the RL component G_S^R; two FL ones, namely, the FL one due to phase fluctuation, G_S^{EF}, and the one due to Fourier broadening, G_S^{TF}; two interferencelike ones, G_S^M and G_S^T. The definitions of these correlation functions remain unchanged whether the incident light is stationary or pulsed.

The RL component is attributed to the emission from a sort of excited state that adiabatically follows the temporal change in the incident light. This state named "Raman state" retains perfect temporal and energetic correlation with the incident light. The Raman state is modulated at the incident light frequency to give rise to the energetic correlation and its amplitude conforms to the change in the pulse amplitude (consequently, this state disappears after the pulse has decayed). In the RSE process through the Raman state, which Path R in Fig. 8 represents, no MPD collision intervenes between the absorption and emission steps: these two steps are inseparable. Path R is represented by G_S^R.

The FL components originate in the step-by-step processes via the "proper" excited state evolving at the Bohr frequency: a photon is emitted after the preparation of the proper excited state, that is, after the completion of the absorption step. It has been illustrated that a MPD collision can lift an initial state molecule to a proper excited state in cooperation with an incident light (Path $F3$). This FL component due to MPD is understood as a two step process. The molecule is initially excited into the Raman state and then converted into the proper excited state by exchanging the detuning energy $\hbar\Delta_I$ with the heat bath (the jump process from the Raman state to the proper excited state due to MPD). In this sense, the MPD collision is inelastic but in the Markovian limit it can be treated as an elastic one. What the fluctuating part of the polarizability correlation function, $\langle \delta\hat{\alpha}^\dagger \delta\hat{\alpha}(t) \rangle$, formulated by Huber [22], means turns out to be Path F3.

An incident light which is energetically broadened by phase fluctuation [68] or by Fourier broadening can prepare a proper excited state, if the energy broadening of the incident light covers the absorption width of the excited state. The proper excited state molecules prepared through the spectral broadening mechanisms may suffers a MPD collision (Path $F2$) or may not (Path $F1$). Path $F1$ is represented by the sum of G_S^{EF} and G_S^{TF}; the FL components due to phase fluctuation and due to Fourier broadening are contained in Path $F1$. After the termination of the incident light, the FL components due to these spectral broadening mechanisms thus decay at the rate of $\Gamma_m + 2\Gamma'$ because of the population leakage to Path $F2$ by MPD. On the other hand, the FL component in Path $F2$ decays slower than at Γ_m because of the population supply from Path $F1$. Needless to say, the sum of the three FL components ($F1 + F2 + F3$ or $G_S^{EF} + G_S^{TF} + G_S^{PF}$) decays at Γ_m.

There are two interference-like terms G_S^M and G_S^T that cannot be regarded as "fluorescence" or as "Raman" (these two terms represent the interference between Paths $F1$ and R). The term G_S^M, where RL and FL characters are mixed by phase fluctuation, is relatively small in any excitation condition. The term G_S^T, which originates mainly from interference between the Raman and Fourier broadening states, is exactly zero if the incident light is stationary. In stationary excitation the spectrum can therefore be interpreted in terms of only two categories, namely "fluorescence" and "Raman." In pulsed excitation, the classification into the two categories however entails a proviso. Until $\tau_p + 1/\Gamma_D$ (the time at which the finite response time $1/\Gamma_D$ of the detector has passed since the incident pulse was turned off at τ_p), G_S^T affects the observed spectrum considerably: in this time region the classification into "fluorescence" and "Raman" is of limited utility. After $\tau_p + 1/\Gamma_D$ the effects of all the interferencelike terms and of the RL one disappear: only FL emissions are detectable (which is consistent with the experimental fact).

We have also applied the above partition technique to the stochastic model and carried out line shape calculations to investigate the nature of "Raman" and "fluorescence" in non-Markovian cases. In general, the FL component due to frequency modulation can be interpreted in terms of the jump process from the initially prepared Raman state to the proper excited state. The jump process plays the key role in determining the FL/RL ratio. In non-Markovian cases, such states can be created that phonons are mounted on the Raman state. Emission components from those states (the broad Raman component) spread between the RL and FL components. It has been exemplified that the concept of "Raman" and "fluorescence" is useful even outside the Markovian regime.

In the research field of multi-photon processes including RSE, "real" or "virtual" transition has been routinely used as a scientific term, but what they mean has been only vaguely understood. The present study enables these vague terms to acquire an unequivocal meaning; the "real" transition should be read as a transition via a proper excited state and the "virtual" transition should be read as a transition via the Raman state. It deserves special emphasis that those both in the proper excited states and in the Raman state are counted in the number of excited molecules.

Acknowledgments

This work was supported in part by grants from the Ministry of Education, Science, and Culture of Japan (63606004, 63740236, 01540357) and by a grant under the Monbusho International Program (63044013). We wish to thank Mr. Y. Matsuda for assistance in preparing the original figures. We also acknowledge helpful discussions with Prof. S. H. Lin and Prof. T. Nakajima. One of the authors (HK) would like to thank Dr. S. Kinoshita for providing him with preprints prior to publication.

References

1. M. V. Klein, *Phys. Rev. B* **8**, 919 (1973).
2. R. Y. Shen, *Phys. Rev. B* **9**, 622 (1974); **14**, 1772 (1976).
3. J. R. Solin and H. Merkelo, *Phys. Rev. B* **12**, 624 (1975); **12**, 6008; **14**, 1775 (1976).
4. T. Kushida, *Solid State Commun.* **32**, 33 (1979); **32**, 209 (1979).
5. W. Holzer, W. F. Murphy, and J. H. Bernstein, *J. Chem. Phys.* **52**, 399 (1970).
6. D. G. Fouche and R. K. Chang, *Phys. Rev. Lett.* **29**, 536 (1972).
7. P. P. Shorygin, *Sov. Phys. Usp.* **16**, 99 (1974).
8. J. M. Freedman and D. L. Rousseau, *Chem. Phys. Lett.* **55**, 488 (1978).
9. R. M. Hochstrasser and C. A. Nyi, *J. Chem. Phys.* **70**, 1112 (1979).
10. P. W. Williams, D. L. Rousseau, and S. H. Dworetsky, *Phys. Rev. Lett.* **32**, 196 (1974); D. L. Rousseau, G. D. Patterson, and P. F. Williams, *Phys. Rev. Lett.* **34**, 1306 (1975); D. L. Rousseau and P. F. Williams, *J. Chem. Phys.* **64**, 3519 (1976).
11. W. Overlaet, M. Kimura, S. Kinoshita, and T. Kushida, *J. Phys. Soc. Japan* **56**, 60 (1987); T. Kushida and S. Kinoshita. *Ind. J. Phys.* **63**, S147 81989).
12. A. Omont, E. W. Smith, and J. Cooper, *Astrophys. J.* **175**, 185 (1982); **182**, 283 (1973).
13. Y. Fujimura, H. Kono, T. Nakajima, and S. H. Lin, *J. Chem. Phys.* **75**, 99 (1981).
14. S. Mukamel, *Phys. Rep.* **93**, 1 (1982).
15. S. Mukamel, A. Ben-Reuven, and J. Jortner, *Phys. Rev. A* **12**, 947 (1975). *J. Chem. Phys.* **64**, 3971 (1976).
16. T. K. Yee and T. K. Gustafson, *Phys. Rev. A* **18**, 1597 (1978); Y. R. Shen, *The Principles of Nonlinear Optics* (Wiley, New York, 1984).
17. E. Hanamura, *J. Luminescence* **12/13**, 119 (1976); E. Hanamura and T. Takagahara, *J. Phys. Soc. Japan* **47**, 410 (1979).
18. H. Kono, Y. Fujimura, and T. Nakajima, *Int. J. Quantum Chem.* **18**, 293 (1980).
19. H. Kono, Ph.D. thesis, Tohoku University, March, 1981.
20. J. S. Melinger and A. C. Albrecht, *J. Chem. Phys.* **84**, 1247 (1986).
21. J. S. Melinger and A. C. Albrecht, *J. Phys. Chem.* **91**, 2704 (1987).
22. D. L. Huber, *Phys. Rev.* **158**, 843 (1967); **170**, 418 (1968); **178**, 93 (1969); *Phys. Rev. B* **1**, 3409 (1970).
23. T. Takagahara, E. Hanamura, and R. Kubo, *J. Phys. Soc. Japan* **43**, 802 (1977); **43**, 811 (1977); **43**, 1522 (1977).
24. T. Takagahara, *Phys. Rev. A* **35**, 2493 (1987).
25. A. Hizhnyakov and I. Tehver, *Phys. Status Solidi* **21**, 755 (1967); **39**, 67 (1970).
26. (a) Y. Toyozawa, *J. Phys. Soc. Japan* **41**, 400 (1976); (b) A. Kotani and Y. Toyozawa, *ibid.* **41**, 1699 (1976); (c) Y. Toyozawa, A. Kotani, and A. Sumi, *ibid.* **42**, 1495 (1977).
27. A. Kotani, *J. Phys. Soc. Japan* **44**, 965 (1978).
28. J. H. Eberly and K. Wódkiewicz, *J. Opt. Soc. Am.* **67**, 1252 (1977).
29. G. Nienhuis, *Physica* **96C**, 391 (1979).
30. R. Kubo, in *Fluctuation, Relaxation, and Resonance in Magnetic Systems*, edited by D. Ter Haar (Oliver and Boyd, Edinburgh, 1962); *Adv. Chem. Phys.* **15**, 101 (1969).
31. (a) S. Mukamel, *J. Chem. Phys.* **71**, 2884 (1979); (b) *Chem. Phys.* **37**, 33 (1979).
32. M. Aihara and A. Kotani, *Solid State Commun.* **46**, 751 (1983).

33. A. Nakamura, S. Shimura, M. Hirai, M. Aihara, and S. Nakashima, *Phys. Rev. B* **35**, 1281 (1987).
34. R. G. Breene, Jr., *Theories of Spectral Line Shape* (Wiley, New York, 1981).
35. A. Ben-Reuven, *Adv. Chem. Phys.* **33**, 235 (1975).
36. H. Kono, Y. Fujimura, and S. H. Lin, *J. Chem. Phys.* **75**, 2569 (1981).
37. M. Baranger, *Phys. Rev.* **111**, 481 (1958); **111**, 494 (1958).
38. Y. Fujimura, in *Advances in Multi-Photon Processes and Spectroscopy*, edited by S. H. Lin (World Scientific, Singapore, 1987), Vol. 2, p. 1.
39. W. H. Louisell, *Quantum Statistical Properties of Radiation* (Wiley, New York, 1973).
40. B. R. Mollow, *Phys. Rev. A* **12**, 1919 (1975).
41. H. J. Kimble and L. Mandel, *Phys. Rev. A* **13**, 2123 (1975).
42. (a) J. L. Carlsten, A. Szöke, and M. G. Raymer, *Phys. Rev. A* **15**, 1029 (1977); (b) E. Courtens and A. Szöke, *Phys. Rev. A* **15**, 1588 (1977).
43. L. Allen and J. H. Eberly, *Optical Resonance and Two-Level Atom* (Wiley, New York, 1975).
44. D. J. Diestler and A. H. Zewail, *J. Chem. Phys.* **71**, 3103 (1979).
45. R. J. Glauber, *Phys. Rev.* **130**, 2529 (1963); **131**, 2766 (1963).
46. V. M. Titulaer and R. J. Glauber, *Phys. Rev.* **145**, 1041 (1966); A. Tramer and R. Voltz, *Excited State* **4**, 281 (1979).
47. V. S. Raman and G. J. Small, *J. Chem. Phys.* **64**, 3359 (1976).
48. J. Czub and S. Kryszewski, *J. Phys. Rev. B* **16**, 3171 (1983); M. Florjanczyk, K. Rzazewski, and S. Kryszewski, *Phys. Rev. A* **31**, 1558 (1985).
49. T. S. Ho and H. Rabitz, *Phys. Rev. A* **37**, 1576 (1988); **37**, 4184 (1988).
50. W. Heitler, *Quantum Theory of Radiation* (Oxford, New York, 1954), p. 196.
51. E. J. Heller, R. L. Sundberg, and D. J. Tannor, *J. Phys. Chem.* **86**, 1822 (1982).
52. S. O. Williams and D. G. Imre, *J. Phys. Chem.* **92**, 3363 (1988); **92**, 3374 (1988).
53. R. V. Rama Krishna and R. D. Coalson, *Chem. Phys.* **120**, 327 (1988).
54. On the other hand, RSE in femtosecond laser excitation is an example of the multilevel case. A theoretical discussion of this subject is found in H. Kono and Y. Fujimura, *J. Chem. Phys.* **91**, 5960 (1989).
55. A. Ron and A. Ron, *Chem. Phys. Lett.* **58**, 329 (1978).
56. For an elaborate description of the binary collision process in photon fields, see K. Burnett, *Phys. Rep.* **118**, 339 (1985).
57. For the condition under which the stochastic model is valid, see Ref. 26b and Y. Kayanuma, A. Kotani, and Y. Toyozawa, *Physics Monthly* (in Japanese) **8**, 126 (1987).
58. S. Mukamel, I. Oppenheim, and J. Ross, *Phys. Rev. A* **17**, 1988 (1977).
59. For the manipulation, see S. Mukamel, *Phys. Rev. A* **28**, 3480 (1983); *J. Chem. Phys.* **82**, 5398 (1985).
60. Equations (2.11) are correct irrespective of the expressions of the functions A, B, and C.
61. S. Mukamel and A. Nitzan, *J. Chem. Phys.* **66**, 2462 (1977); A. Nitzan, S. Mukamel, and A. Ben-Reuven, *Chem. Phys.* **24**, 37 (1977).
62. J. Sue, Y. J. Jing, and S. Mukamel, *J. Chem. Phys.* **85**, 462 (1986).
63. These techniques were developed by Kubo et al. See, for example, Refs. 24 and 59.
64. J. Watanabe, S. Kinoshita, and T. Kushida, *J. Chem. Phys.* **87**, 4471 (1987); T. Kushida and S. Kinoshita, *Raman Spectroscopy* **17B**, 495 (1989).

65. The form of $1/\Delta_I^2$ arises from the fact that the correlation function of the random frequency, $\langle \delta\omega(\tau)\delta\omega \rangle$, gives a Lorentzian spectral profile. If $\langle \delta\omega(\tau)\delta\omega \rangle$ decays as $\exp(-\Lambda^2\tau^2)$, the spectral profile of the effective phonon field $\mathscr{E}$ is Gaussian. In this case, the jump rate would decrease as $\exp(-\Delta_I^2/4\Lambda^2)$.

66. As Δ_I increases, the growth rate of the Raman state, $\mathrm{Re}\,\chi(\Delta_I)$ decreases at least as rapidly as $1/\Delta_I^2$. Hence, the growth rate of the proper excited state or $S^{PF}(\Delta_I)$ decreases at least as rapidly as $1/\Delta_I^4$.

67. What we have done is to disentangle the four paths $F1$, $F2$, $F3$ and R from the diagrammatic representation.

68. The emission correlation functions obtained here must be revised if the phase fluctuation function ξ is not of exponential decay as in Eq. (3.12). If $\xi(\tau) = \exp(-\Gamma_I^2\tau^2)$, the spectral profile of the incident light develops Gaussian wings that diminish faster than the Lorentzian profile associated with Eq. (3.12). In the case of Gaussian excitation, the ratio of the FL component due to phase fluctuation to the RL component, $S^{EF}(\Delta_I)/S^R(\Delta_I)$, would decreases as $\exp(-\Delta_I^2/4\Gamma_I^2)/\Delta_I^2$ (while in Lorentzian excitation it remains constant). Generalization to this problem needs to be done. However, we believe that the qualitative conclusions drawn here are not subject to substantial modification.

AUTHOR INDEX

Numbers in parentheses are reference numbers and indicate that the author's work is referred to although his name is not mentioned in the text. Numbers in *italic* show the pages on which the complete references are listed.

SUBJECT INDEX